# Switched Capacitor Filters

# Switched Capacitor Filters
## Theory, analysis and design

P.V. ANANDA MOHAN

*Deputy General Manager, Transmission Research and Development
Indian Telephone Industries Ltd
Bangalore*

V. RAMACHANDRAN

*Professor, Electrical and Computer Engineering
Concordia University
Montreal*

M.N.S. SWAMY

*Former Dean of Engineering and Computer Science
Concordia University
Montreal*

PRENTICE HALL

*London  New York  Toronto  Sydney  Tokyo  Singapore
Madrid  Mexico City  Munich*

First published 1995 by
Prentice Hall International (UK) Ltd
Campus 400, Maylands Avenue
Hemel Hempstead
Hertfordshire, HP2 7EZ
A division of
Simon & Schuster International Group

Typeset in 10/12 pt Times
by Mathematical Composition Setters Ltd, Salisbury
Designed by Claire Brodmann
Printed and bound in Great Britain
at the University Press, Cambridge

*Library of Congress cataloging-in-publication data*
Mohan, P. V. Ananda, 1949–
    Switched capacitor filters : theory, analysis and design /
  P.V. Ananda Mohan, V. Ramachandran, M. N. S. Swamy.
       p.  cm.
    Includes bibliographical references and index.
    ISBN 0-13-879818-4
    1. Switched capacitor circuits—Design. 2. Electric filters—
Design. I. Ramachandran, V., 1934– . II. Swamy, M. N. S., 1935– .
III. Title.
TK7868.S88M65 1995
    621.3815′324—dc20                                       92-42244
                                                                CIP

*British Library cataloguing in publication data*
A catalogue record for this book is available from
the British Library

ISBN 0-13-879818-4


1  2  3  4  5    99  98  97  96  95

विद्या ददाति विनयं विनयाद्याति पात्रताम् ।
पात्रत्वाद्धनमाप्नोति धनाद्धर्मं ततः सुखम् ॥

Learning gives modesty
By modesty one attains worthiness
Because of worthiness one obtains wealth
With wealth one is able to give charity
And from that one attains happiness

# Contents

# Preface

Ever since electric wave filters were introduced more than seventy years ago, filters have played a vital role in communication systems. With the advent of large scale integration (LSI) and very large scale integration (VLSI) technologies, greater emphasis has been placed on the miniaturization of the major components of communication systems, including filters. Consequently, one of the main directions of recent investigations into filters is the exploration of new technology for achieving miniaturization. The switched capacitor (SC) technique for designing filters is a recent development and this is highly suitable for the realization of complete filters on a silicon chip.

Switched capacitor filters realized using metal oxide semiconductor (MOS) technology use periodically operated switches with capacitors and operational amplifiers (OA). They are essentially based on active RC filter configurations, but do eliminate resistors through the use of switched capacitors. Further, these are sampled data in nature and, as such, require sampled data system theory for their analysis and design. Thus, SC filters are an outgrowth of active RC filter and digital filter theories.

Noticing the need for educating students in new technologies, engineering curricula have introduced courses on active RC filters and digital filters. In this book, a systematic treatment of switched capacitor filters is presented. Starting from the fundamentals of active RC filters and digital signal processing, analysis and design methods for SC filters are developed. In deriving and evaluating SC filter configurations, considerable attention has been given to theoretical aspects as well as technological requirements such as stray sensitivity, minimal chip area, capacitor required, etc.

The structure of the book is as follows: Chapter 1 provides a general introduction to the integration of filters on a silicon chip and discusses the fundamentals of sampled data signal processing. In Chapter 2, analysis of switched capacitor networks is developed. Chapter 3 discusses passive SC networks and some of their applications. Chapter 4 deals with first-order active SC networks. Chapter 5 deals with second-order SC networks using one or more operational amplifiers. The subject of component simulation type SC ladder filters is discussed in Chapter 6, while SC ladder filters based on operational simulation of LC ladders and on multiloop feedback concepts are discussed in Chapter 7. Chapter 8 deals with N-path SC filters. The concluding Chapter 9 deals with applications of SC networks and practical considerations.

There are two appendices. Appendix A discusses the use of a general-purpose circuit analysis program, SPICE, in the analysis of SC networks. Appendix B gives a computer program for the optimal design of SC filters.

We have attempted to give a wide coverage of SC filter design techniques with the aim of providing the designer with a large choice of design methods. The designer will be able to choose a particular method on technological advantages, power consumption, cost, area, etc.

The background material required to understand the subject matter of this book is basic network analysis and synthesis. The topics presented could be taught in a senior undergraduate or graduate-level course in one semester. In order to contain the size of the book, only relevant topics in active RC and digital filter design are covered in this text. By supplementing these with additional information where necessary, the book may be used to cover, in a one-year course, the design of active RC, digital and switched capacitor filters.

Several worked examples have been given to illustrate some of the ideas developed in the book. Also, problems are presented at the end of each chapter to revise the knowledge acquired by the reader as well as to provide information. An extensive list of references is provided at the end of each chapter to assist the reader.

We thank Jayne Claassen, Krishna Shewtahal and Tezeta Taye for typing the first manuscript of the book. Also, we would like to thank Silvin Gresu for helping us in drawing some of the figures on the computer.

We thank our respective wives, Radhanirmala Mohan, Kamala Ramachandran and Leela Swamy, and our children for their patience and understanding during the preparation of this book.

P.V. Ananda Mohan
V. Ramachandran
M.N.S. Swamy

# MOS technology and sampled data filters

A filter is a frequency-selective device designed to pass (or transmit) certain signal frequencies and to stop (or reject) certain other signal frequencies. Filters may be classified as low-pass (LP), high-pass (HP), band-pass (BP) or band-stop (BS) (which includes notch) types, depending on the range of frequencies passed. In these filters, the variation of amplitude with frequency is of interest. There is another type of filter called the all-pass (AP) filter for which the amplitude response is constant, whereas the phase varies with frequency.

Filters may also be classified according to the technology used in realizing them, or the application for which they are intended. We have thus a variety of filters, such as passive filters, active RC filters (which includes active **R** and active **C** filters), switched capacitor (SC) filters, crystal and ceramic filters, microwave filters, and digital filters. Since passive and active RC filters are directly connected with the design of SC filters, they will be briefly reviewed, with special emphasis on their suitability for miniaturization.

## 1.1   Passive and active filters

Passive filters use resistors, capacitors and inductors. Filters can be built using resistors and capacitors only. But the resulting RC networks can realize only simple negative real-axis poles. However, the use of inductors together with resistors and capacitors can realize network functions with complex-conjugate poles. Such RLC networks can realize filters with rapid variation of amplitude or phase response using a smaller number of elements (i.e., using a low-order filter) than an RC filter. However, it is preferable to avoid inductors in practice, because they are bulky (especially if the inductance values are large) and non-ideal, and the realization of high-quality miniaturized inductors is not found to be practical.

It is possible to realize resistances and capacitances in hybrid integrated circuits. Since RC networks by themselves cannot realize sharp-cutoff filters, highly selective amplitude or phase responses can be realized using passive RC networks along with active elements like bipolar transistors, operational amplifiers (OA), negative-impedance converters (NIC), gyrators, differential voltage-controlled current/voltage sources, or current conveyors. However, the most popular among these is the OA. Active RC filters may use one or more OAs, together with passive RC networks, to realize the desired filtering functions. The OAs used in these active RC filters are assumed to have infinite input impedance, low output impedance and infinite d.c. gain.

Active RC filters, however, have some limitations, because of the non-ideal nature of the OAs used. Their performance is dependent on the finite bandwidth of OAs. In addition, there will be variations due to temperature or power supply changes. The filter parameters are sensitive to resistor and capacitor values. These have to be considered in the design and implementation of active filters. In addition, designers are interested in considering the miniaturization of filters to the extent that they are technologically compatible with other subsystems which use digital techniques.

These active RC filters can be realized using thin-film or thick-film or silicon bipolar integrated circuit (IC) technology. Each realization has its own advantages and disadvantages. It is not the intention here to discuss the various processes in detail (these are dealt with in the literature on active filters). It is sufficient to state that metal oxide semiconductor (MOS) technology [1.1–1.3] has certain attractive features which have made it highly preferable to large scale and very large scale integration (LSI/VLSI) of electronic circuits.

## 1.2 MOS technology for analog signal processing

First, we shall briefly describe the MOS transistor structures in integrated circuits.

### 1.2.1 The basic MOS process

Figure 1.1 shows the fabrication process sequence of a p-channel MOS transistor. In the n-type substrate, two p-type islands are formed by diffusion or ion implantation to serve as source and drain. The n-region between the source and drain is called the *channel*. A thin layer of dielectric (usually silicon dioxide) is grown over the channel. On the top of the dielectric layer, aluminium is deposited to serve as the gate electrode. With a negative drain to source voltage, and zero potential difference between gate and source, there is no conduction between source and drain.

A negative gate–source voltage, $V_{GS}$, will induce a positive charge at the surface of the channel just beneath the oxide layer. When a sufficiently large negative $V_{GS}$ is applied, the field beneath the gate electrode may be sufficient to 'invert' the channel from n-type to p-type, thus causing conduction to take place. The gate voltage required to form a conducting channel is called the 'threshold voltage', denoted $V_T$, and is typically 4 V for PMOS transistors thus realized.

The above mode of operation of the field effect transistor (FET) is called the *enhancement mode*. In this mode of operation, the PMOS transistor is normally off and when a gate–source voltage is applied, a channel is formed and conduction takes place. In another mode of operation called the *depletion mode*, the device

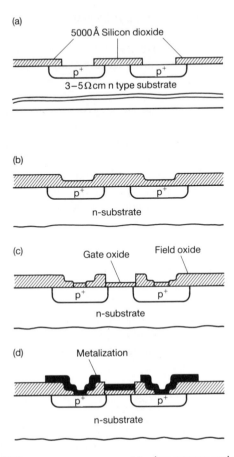

*Fig. 1.1*  Thick oxide MOS process sequence:  (a) after source-and-drain diffusion;
(b) thickening of field oxide;  (c) after growth of gate oxide;  (d) completed device
*(adapted from [1.4],* © *1977 by Addison-Wesley Publishing Company Inc. Reprinted with
permission of the publisher).*

initially will be conducting due to the presence of a channel between source and
drain (formed by diffusion). By applying a suitable gate–source voltage, the channel
can be depleted or enhanced. When $V_{GS}$ equals $V_T$, the channel depletes completely
and the device turns off.

Usually, both the depletion- and enhancement-mode MOS transistors are used in
circuit design. The above discussion has considered p-channel devices. Using a p-
type wafer as the substrate and by diffusing n-type source and drain regions, an
NMOS transistor can be realized. NMOS technology has higher-speed capabilities
for digital applications than PMOS technology, since the conduction in NMOS
transistors is through electrons which have more mobility than holes, which are the
current carriers in PMOS transistors.

There are two limitations to the above PMOS or NMOS process: first, the large threshold voltage (of about 4 V); and second, the large capacitance due to gate–drain voltage. The large threshold voltage is a disadvantage when compatibility of MOS circuits with other logic circuits like transistor-transistor logic (TTL) is desired. The threshold voltage can be decreased by using a silicon nitride layer over a thick silicon dioxide layer. Silicon nitride has a higher dielectric constant than silicon dioxide. By this process, the threshold voltage can be reduced to about 2 V. However, this still leaves the second limitation due to large overlap capacitance. This can be removed by using ion implantation.

## 1.2.2 Realization of a self-aligned gate using the ion implantation technique

In this process, a gate metal electrode is deposited which has smaller dimensions than the channel that is to be formed. By masking all the other regions except the source to drain region of the desired transistor, ion implantation can be used selectively to extend the source and drain regions up to the gate. The gate metal electrode masks the area below it in this process. Thus the source, drain and gate regions are accurately defined. The ion implantation may also be used to modify the doping of the substrate surface in the channel, thus modifying the threshold voltage of the device. In this process, the threshold voltage can be made zero or even positive. Thus depletion-mode devices also can be realized.

Another popular method used to define the gate–source and gate–drain geometries accurately is the silicon-gate process. This process is briefly described below.

## 1.2.3 The silicon-gate process

The process sequence is as follows. First, a thick oxide layer is grown over the entire wafer (1 $\mu$m thick approximately). Then, a thin (0.1 $\mu$m) oxide layer is formed over the device area by applying a mask, etching windows and regrowing thin oxide. Silicon nitride may be deposited over this layer to decrease the threshold voltage. In the next step, a layer of silicon is deposited over the entire wafer. Since the silicon does not have a reference crystal plane to settle on, it will assume an amorphous or polycrystalline form. The second mask is now applied, delineating the source and drain regions. In two etching steps, the silicon and nitride are removed in all the areas except the gate. Also the thin oxide is etched away over both the source and drain areas. Next, the wafer is diffused with p-type impurities. This step forms the source and drain regions and also heavily dopes the gate electrode to establish good conductivity.

An oxide layer is then deposited over the wafer. In a third mask, the contact windows are outlined and etched through the oxide. Contact to the gate electrode is made outside the channel region so that the width of the channel can be made very narrow. After evaporating aluminium, the fourth and last mask is then applied to delineate the interconnection pattern.

In addition to accurately defining the device geometries and decreasing the threshold voltage (because of the low work function of silicon compared to aluminium), silicon-gate technology offers a second layer of interconnection through the presence of amorphous silicon. Amorphous silicon also can be used to form cross-over interconnections.

We will next discuss the realization of both NMOS and PMOS transistors in a single chip. Such devices are useful in complementary MOS (CMOS) technology.

## 1.2.4 The CMOS process

The CMOS integrated circuit (IC) structure is shown in Figure 1.2. The p-channel device is formed as already described in an n-substrate. To realize the n-channel device, a p-well is first formed by diffusion. In this p-well, the source and drain regions of the n-channel transistor are diffused. Thus, three diffusions are necessary for CMOS technology and further MOS transistors in CMOS technology occupy considerable chip area. However, CMOS gates have the advantage of low power dissipation.

## 1.2.5 Advantages of MOS technology

In the above, various methods of realizing MOS transistors have been studied. Some of the advantages of MOS technology are as follows:

1. MOS devices are inherently self-isolating. This is achieved by virtue of the fact that for PMOS, the substrate is connected to the most positive potential, thereby isolating the channel from source and drain through reverse biased p-n junctions. As against this, in the bipolar technology, deep $p^+$ diffusion is required to isolate the adjacent transistors through reverse biased junctions. The lateral diffusion resulting from this process increases the area occupied by bipolar transistors in

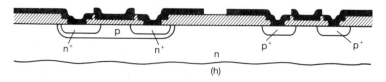

(h)

*Fig. 1.2* Complementary MOS devices *(adapted from [1.4], © 1977 by Addison-Wesley Publishing Company Inc. Reprinted with permission of the publisher).*

the standard bipolar process. The self-isolation property leads to large density of MOS components in MOS technology.

2. MOS technology requires less processing steps than bipolar technology. As an illustration, the NMOS process requires one diffusion step, whereas the bipolar process requires a minimum of three diffusions for base, emitter and collector diffusions.

3. More levels of interconnections are inherently possible in MOS technology. For example, in the silicon-gate process, two levels of interconnections naturally exist.

4. The MOS transistor has an extremely high input resistance. By virtue of this property, information can be stored on the gate–source capacitance. (It is also possible to store information on drain–source capacitance.) Further, the charge stored on gate–source capacitances can be sensed and/or transferred on to another capacitance non-destructively. These properties can be used to construct dynamic logic circuits in which the power supply is disconnected most of the time to reduce power dissipation.

5. Another interesting property of MOS transistor is its bilateral nature. Source and drain can be interchanged at will, and by suitable biasing of the gate, bidirectional transfer of information is possible.

These properties have made possible the realization of complex digital systems such as microprocessors, memories, and shift registers in MOS technology. In the processing of analog signals as well, these advantages of MOS technology can be used profitably. One of them involves the rearrangement of switches, capacitors and OAs to provide alternative configurations for design of filters with a behaviour similar to active RC or LC filters. These are known as *switched capacitor* (SC) filters. These will be briefly introduced in the next section.

## 1.3 Switched capacitor filters

These utilize a capacitor and two switches to simulate the circuit behaviour of a resistor, as shown in Figure 1.3. The operation of the switched capacitor resistor is as follows. When the switch is in the left-hand position, $C_1$ is charged to voltage $v_1$ and when the switch is thrown to the right, $C_1$ is discharged to the voltage $v_2$. The amount of charge thus flowing into (or away from) $v_2$ is $Q_e = C_1(v_2 - v_1)$. By throwing the switch back and forth every $T$ seconds, the current flow, $i$, into $v_2$ will be

$$i = \frac{C_1(v_2 - v_1)}{T} = \frac{(v_2 - v_1)}{(T/C_1)} \tag{1.1}$$

Thus, the switched capacitor simulates the behaviour of a resistor of value $(T/C_1) = R_1$ connected between voltage sources $v_1$ and $v_2$.

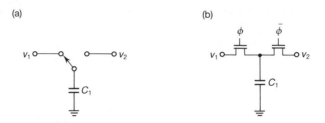

*Fig. 1.3* (a) A switched capacitor which in many cases can simulate the function of a resistor; (b) an MOS implementation of the circuit shown in (a).

The circuit of Figure 1.3 can be realized in MOS technology using two MOS switches and a capacitor. If a capacitor $C_2$ is associated with the above SC resistor $R_1$, the resulting time constant $C_2 R_1$ is

$$\tau = C_2 R_1 = \left(\frac{C_2}{C_1}\right) T \tag{1.2}$$

Thus the time constant $\tau$ is determined by the ratio of capacitors and not their absolute values, which makes it insensitive to process variations. This desirable property for integrated filter realizations, which avoids trimming, is achieved in the SC technique. It is thus seen that resistors in active RC filters can be realized by MOS switched capacitors. The availability of OAs in MOS technology can provide a solution to integrated filter realization [1.5].

Commercially available OAs use largely bipolar technology and, in some cases, a combination of MOS and bipolar technology. Fully integrated MOS OAs have, however, only recently become available, and these have boosted the possibility of SC filter realization in monolithic form. We will briefly consider the MOS OAs, capacitor and switches next.

## 1.4  MOS technology for SC filtering

### 1.4.1  MOS operational amplifiers

An OA in NMOS and CMOS technologies usually requires an area of about one-third to one-fifth of that of a corresponding bipolar OA. It can realize a voltage gain of the order of 60–80 dB and have a unity–gain bandwidth greater than 1 MHz. It typically occupies a die area less than $(1.25)10^{-7}\,\mathrm{m}^2$. One of the important requirements of such an OA is to drive a capacitive load so that it can be charged to its final value quickly. In addition, it can be designed to suit the requirements of gain, slew rate etc. It is known that the performance of an SC filter depends upon the various characteristics of the OAs used. As such, it is essential that the OA is designed to suit many requirements. The important ones are voltage gain, transient response time, noise, d.c. offset voltage, rejection of power supply noise, common mode range and output voltage range, and power dissipation. The properties of OAs and their design considerations will be discussed further in Chapter 9.

## 1.4.2  MOS capacitors [1.6]

The capacitors in SC filtering can be realized either as MOS capacitors (between a metal electrode and highly doped silicon with thin silicon dioxide as dielectric, as already described) or as polysilicon–polysilicon capacitors (between two polysilicon layers and with silicon dioxide as dielectric). These two structures are presented in Figure 1.4. In both these structures, thermally grown silicon dioxide dielectric is recommended because the deposited oxide exhibits large variation in thickness and undesirable charge-voltage hysteresis (due to dielectric relaxation effects), which affect accuracy. In the former structure, lightly doped silicon electrodes are not used because of the resulting large voltage coefficient of capacitance. In the silicon-gate process, heavy $n^+$ implanted bottom plates could be used. A library of capacitor types is presented in Table 1.1. It can be seen that the capacitor realized between poly I and poly II layers exhibits good properties.

The chip area in an MOS circuit can be reduced by realizing large capacitance per unit area. Thermally grown silicon dioxide layers on crystalline silicon are advantageous, since thin layers of about 500 Å thickness can be easily realized. The oxide

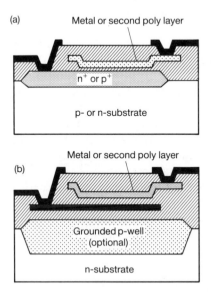

*Fig. 1.4*  MOS capacitor structures:  (a) poly-II (or metal) over a heavily doped diffused layer;  (b) poly-II (or metal) over poly-I  *(adapted from [1.6]*, © 1983 IEEE*)*.

**Table 1.1** Properties of monolithic capacitors in depletion-load NMOS (top) and CMOS (bottom) technologies  (*adapted from [1.6]*, © 1983 IEEE).

| Precision capacitor type | $t_{ox}$ (Å) | Approximate absolute accuracy $\sigma_C$ (%) | Approximate temperature coefficient (ppm $^\circ C^{-1}$) | Approximate voltage coefficient (ppm $V^{-1}$) | Relative merit |
|---|---|---|---|---|---|
| metal–n$^+$ | 500–1000 | 10 | 10–50 | 20–200 | 3 |
| poly–n$^+$ | 500–1000 | 10 | 10–50 | 20–200 | 2 |
| poly II–poly I | 1000–1500 | 20 | 10–50 | 20–200 | 1 |
| metal–poly | 1000–1500 | 20 | 10–50 | 20–200 | 4 |
| metal–n$^+$ | 500–1000 | 10 | 10–50 | 20–200 | 3 |
| metal–p$^+$ | 500–1000 | 10 | 10–50 | 20–200 | 6 |
| poly–n$^+$ | 500–1000 | 10 | 10–50 | 20–200 | 2 |
| poly–p$^+$ | 500–1000 | 10 | 10–50 | 20–200 | 5 |
| metal–poly | 1000–1500 | 20 | 10–50 | 20–200 | 4 |
| poly II–poly I | 1000–1500 | 20 | 10–50 | 20–200 | 1 |

grown on polysilicon is generally at least 1000 Å thick to achieve a minimum breakdown voltage of 40 V. In thinner layers of such polyoxides, defects tend to be present. Some technologies use silicon dioxide (thickness 1000 Å) over silicon nitride (thickness 200 Å) to form the dielectric layer. This results in larger capacitance per unit area due to the larger relative dielectric constant ($\varepsilon_{ox} = 7.5$) of silicon nitride. Further, the defects in polyoxide tend to be filled by the silicon nitride layer, thus improving the yield as well as the breakdown voltage. The plates of the capacitor are usually thick (5000 Å) so that the capacitance is constant over the range of operating signal voltages. The tolerance of monolithic MOS capacitors is usually ±10% to ±20%.

McCreary and Gray [1.7] have extensively investigated MOS capacitor array realization. The capacitor structure due to them is presented in Figure 1.5(a). Note that the capacitor area is defined by the metal edge. At every capacitor edge, a uniform etchant concentration is maintained by the floating metal strips. This ensures nearly uniform 'undercutting' effects for all the capacitors. In the calculation of the area for the capacitance, the area of interconnect over the thin oxide is included while that over the thick oxide is neglected. Also, the capacitor area is made insensitive to errors due to mask alignment by spanning the capacitor interconnect in one direction only. Smarandoiu *et al.* [1.8] have suggested alternative structures shown in Figure 1.5(b)–(c). In this method, the capacitor dimensions are defined with the thin oxide mask and thus the photolithographic definition of the capacitor is excellent.

It is important to achieve accurate capacitor ratios for MOS SC realizations because the accuracy of the time constant depends on the ratio of capacitors. Thus, the errors invariably occurring in the fabrication of capacitors can be minimized.

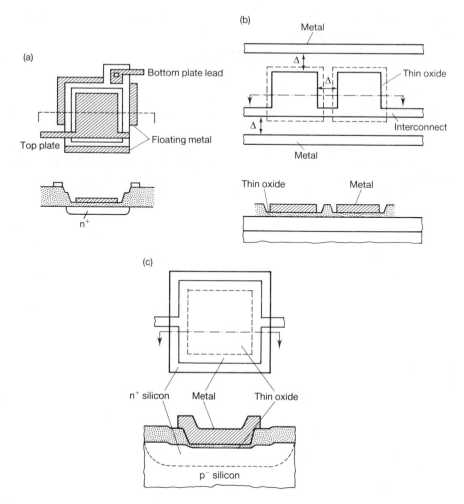

*Fig. 1.5* (a) McCreary and Gray structure *(adapted from [1.7], © 1975 IEEE)*; (b), (c) Smarandoiu *et al.* capacitor structures *(adapted from [1.8], © 1978 IEEE)*.

The capacitance of an MOS capacitor can be expressed as

$$C = \frac{\varepsilon_0 \varepsilon_{ox} WL}{t_{ox}}$$ (1.3)

where
$\varepsilon_0$ = permittivity of free space
$\varepsilon_{ox}$ = dielectric constant of silicon dioxide
$L$ = length of the electrode
$W$ = width of the electrode
$t_{ox}$ = thickness of silicon dioxide dielectric

If the variables in equation (1.3) are statistically independent, then the standard deviation of $C$ is given by

$$\sigma_C = \left[ \left( \frac{\Delta \varepsilon_{ox}}{\varepsilon_{ox}} \right)^2 + \left( \frac{\Delta L}{L} \right)^2 + \left( \frac{\Delta W}{W} \right)^2 + \left( \frac{\Delta t_{ox}}{t_{ox}} \right)^2 \right]^{1/2} \qquad (1.4)$$

where $\Delta y / y$ is the relative error of the parameter $y$. It is observed that the variance in capacitor values can occur because of any of the four terms. Those errors resulting from the variations in the dielectric constant and the thickness are termed *oxide effects*, whereas those from the other two are designated *edge effects*. For large capacitors the oxide effects are dominant, whereas for small capacitors the edge effects are dominant. The crossover point occurs for $W$ of about 20–50 $\mu$m. The magnitude of the random errors mentioned above can only be reduced by reducing the total capacitance.

The accuracy of capacitor ratios is dependent on the capacitor type. Thus capacitors made on crystalline silicon are usually superior to those with polyoxide dielectrics. Similarly, matching characteristics of polyoxide/silicon nitride capacitors are also relatively poor because of the difficulty in exactly controlling the thickness of the chemically grown silicon nitride layer.

The above-mentioned errors are due to purely random effects. In addition, systematic errors due to oxide or edge effects also exist. The oxide errors arise because of the non-uniform oxide growth conditions. If the resulting variation in oxide thickness (see Figure 1.6(a)) is approximated as a first-order gradient, then the ratio error can be computed from the following equations:

$$C_2 = 2C_1(1 + \varepsilon_C L)$$

$$C_4 = 4C_1(1 + 2\varepsilon_C L), \qquad \varepsilon_C = \frac{a}{x_0} \tag{1.5a}$$

Note that $C_2$ is normally $2C_1$. Another factor affecting the capacitor ratio is the undercutting of the mask already mentioned. This is explained in Figure 1.6(b). Thus the ratio error between $C_2$ and $C_4$ is

$$C_4 = 2C_2(1 + \varepsilon_b), \qquad \varepsilon_b = 4 \frac{\Delta x}{L_4} \tag{1.5b}$$

This latter problem can be overcome using a geometry such that the ratio of the perimeter length to the area is held constant. Thus, large capacitor values are realized by using small capacitors. These small capacitors are the smallest realizable capacitors in MOS technology, evidently obtained by the photolithographic definition and are called *unit capacitors*[*] (see Figure 1.7(b)). In cases where non-integer

---

[*] A *unit capacitance*, $C_u$, is the minimum capacitance that can be fabricated in an IC. All the other capacitances can be compared with this $C_u$, which facilitates easy evaluation of *total capacitance* and *capacitor spread*. The total capacitance is a measure of the area required on the silicon chip. The capacitor spread value gives an idea of the largest capacitor value required. In SC networks, capacitor spread of the order of 1000 is realizable. This corresponds to an accuracy of 10 bits coefficient representation in digital filters.

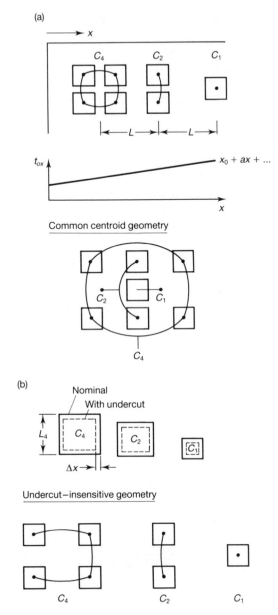

(a)

Common centroid geometry

(b)

Undercut–insensitive geometry

g. 1.6 (a) Capacitor ratio error due to oxide undercut; (b) capacitor ratio error due to hotomask undercut *(adapted from [1.7]*, © 1975 IEEE*)*.

values for the ratios are needed, a capacitor larger than the unit size can be employed (combining the fractional value of capacitor and one unit capacitor), as shown in Figure 1.7(a). The former effect due to oxide gradient can be reduced by using *centroid geometry*, in which all capacitors are symmetrically located about a common centre point. Thus capacitor ratios can be maintained accurately. In spite of the above remedies, useful for correcting ratio errors, random edge variations still remain. It is felt, however, that such accurate techniques are required only for precision analog-to-digital converters, whereas for the SC filters to be studied in this book (which exhibit low sensitivity) the common centroid layouts may not be required. Also the dummy strips around capacitors, as illustrated in Figure 1.5(a), also need not be employed. The reader is referred to [1.9–1.11] for more information on matching properties, random errors and their effects.

In both the types of capacitors presented in Figure 1.4, a sizeable *parasitic capacitance* exists between the bottom plate of the capacitor and the substrate. Typically, this bottom plate parasitic capacitance has a value of one-twentieth to one-fifth of the MOS capacitor itself. Further, since the top plate is connected to other circuitry on the silicon chip, there will be a capacitance to substrate due to interconnections. This can range from 0.1% to 1% of the desired MOS capacitance. These parasitic capacitances are unavoidable and the design of SC filters must be done in such a way that the parasitics (also called *stray capacitances* or *strays*) do not degrade the performance of the filters.

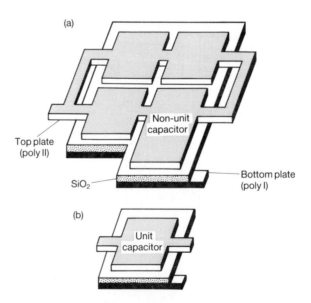

*Fig. 1.7* Typical capacitor layouts: (a) non-integer value capacitor made of unit capacitor; (b) unit capacitor *(adapted from [1.6],* © 1983 IEEE).

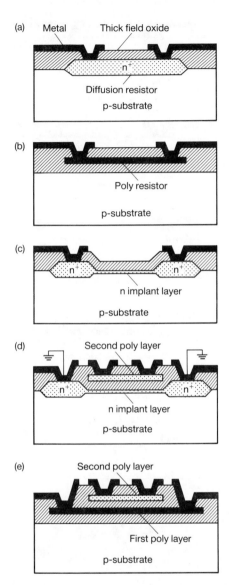

*Fig. 1.8* Resistors available in NMOS depletion technology: (a) n$^+$ diffusion resistor; (b) a polysilicon (poly-I) resistor; (c) a depletion-implant resistor; (d) poly-II resistor with substrate shield; (e) double polysilicon distributed RC structure with resistor in poly-II layer *(adapted from [1.6]*, © 1983 *IEEE)*.

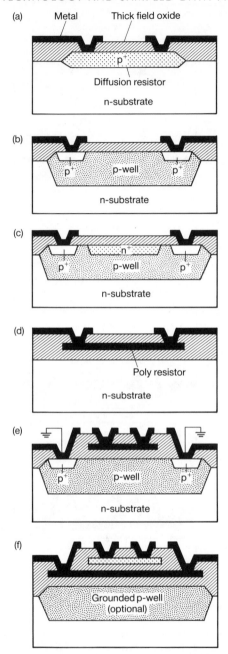

*Fig. 1.9*  Resistors available in p-well CMOS process: (a) a $p^+$ (or $n^+$) diffusion resistor;  (b) a diffused p-well resistor;  (c) a diffused p-well pinch resistor;  (d) a polysilicon resistor;  (e) a polysilicon resistor with substrate shield;  (f) a double polysilicon distributed active RC structure with resistor formed in poly-II layer *(adapted from [1.6]*, © 1983 *IEEE)*.

Another aspect of importance in monolithic MOS realization is the ratio of capacitors that can be realized. The minimal value of capacitance realizable is as low as 0.1 pF and the maximum value is decided by the area occupied on the chip. An economical size of MOS capacitor can typically occupy an area of 40 000 $\mu m^2$.

## 1.4.3 MOS resistors [1.6]

Resistors are required in MOS technology for realizing antialiasing filters for SC filters and also for smoothing filters following the SC filters. Several resistor structures are available in NMOS and CMOS technologies, which are presented in Figures 1.8 and 1.9. It has already been pointed out that precision active RC filtering needs accurate realization of resistor values. However, for applications such as smoothing and antialiasing filters, the absolute values of resistances need not be quite accurate, thereby permitting the use of any of these resistor structures. The resistors are usually realized with square or serpentine layouts, as illustrated in Figure 1.10. The properties of these resistors are summarized in Table 1.2. The choice of the particular type of resistor is dependent on the relative merits listed in Table 1.2.

## 1.4.4 MOS switches [1.12]

The MOS transistor has proved to be an excellent analog switch as compared to the bipolar switch. There are some important properties of the MOS transistor which make it suitable as a switch. They are as follows:

1. While in saturation, the MOS transistor does not have any appreciable offset voltage, while the bipolar transistor has a definite (even though small) voltage drop between collector and emitter.
2. The gate offers high input impedance.
3. In NMOS or PMOS technology, the respective transistor (Figure 1.11(a)) can handle input voltages almost equal to the supply voltage. In fact, it could be one threshold less than the input voltage.
4. When used to drive capacitive loads, the on-resistance of the MOS switch is not of much consequence. This is because the resulting time constant of the on-resistance of such a switch and the largest possible MOS capacitor is below 100 ns. This means that the capacitor can charge to within 0.1% of the final value in less than 1 $\mu s$.

However, there are some minor problems associated with the use of MOS transistors as switches:

1. There exist junction leakage currents which may result in slow voltage drifts causing reduction in the available dynamic range. As a typical example, a 10 $\mu A$

(a)                                                    (b)

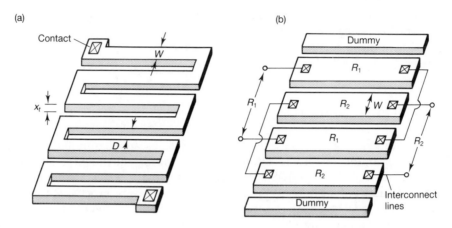

*Fig. 1.10* Topological mask layouts for resistors: (a) serpentine layout; (b) precision matched pair of resistors for gainsetting amplifier applications *(adapted from [1.6],* © 1983 IEEE*)*.

**Table 1.2** Properties of monolithic resistors in depletion-load NMOS (top) and CMOS (bottom) technologies *(adapted from [1.6],* © 1983 IEEE).

| Resistor type (see Figures 1.8 and 1.9) | Nominal sheet resistivity (ohms/□) | Absolute accuracy $\sigma_R$ (%) | Approximate temperature coefficient (ppm °C$^{-1}$) | Approximate voltage coefficient (ppm V$^{-1}$) | Relative merit |
|---|---|---|---|---|---|
| n$^+$ diffusion | 20–80 | 25–50 | 200–2000 | 50–500 | 4 |
| n$^+$ polysilicon | 50–150 | 50 | 500–1500 | 20–200 | 3 |
| n$^-$ depletion implant | 10 K | 25 | 20 K | 25 K | 5 |
| n$^+$ poly over implant | 50–150 | 50 | 500–1500 | 20–200 | 2 |
| Poly II over poly I | 50–150 | 50 | 500–1500 | 20–200 | 1 |
| p$^+$ diffusion | 50–200 | 25–50 | 200–2000 | 50–500 | 6 |
| p$^-$ well diffusion | 3–5 K | 25 | 5 K | 10 K | 1 |
| p$^-$ well pinch resistor | 5–10 K | 50 | 10 K | 20 K | 5 |
| n$^+$ polysilicon | 50–150 | 50 | 500–1500 | 20–200 | 4 |
| n$^+$ poly over p$^-$ well | 50–150 | 50 | 500–1500 | 20–200 | 3 |
| Poly II over poly I | 50–150 | 50 | 500–1500 | 20–200 | 2 |

leakage current charging a capacitor of 10 pF may cause an offset voltage of 1 mV in 1 ms.

2. Parasitic capacitances exist between gate to source and gate to drain. These are of the order of 0.01–0.1 pF. These can introduce d.c. offsets at signal nodes due to clock feedthrough effects. However, parasitic capacitances to substrate or ground can be made not to have any effect on the signals by using techniques described in detail in later chapters.

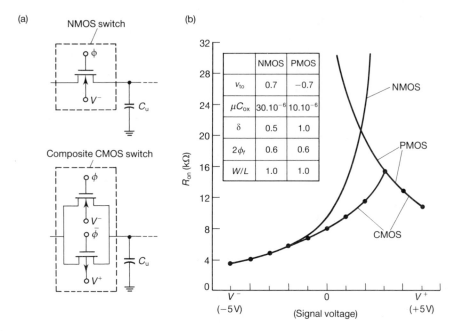

*Fig. 1.11* (a) NMOS and CMOS analog switches; (b) variation of analog MOS switch resistance with input voltage *(adapted from [1.13],* © *1983 IEEE).*

The switch size can be minimized in the process chosen. The on-resistance of a MOS switch can be derived from the voltage–current relationship of the MOS transistor operating in the non-saturation region. The variations of resistance with input voltage for the three different switches are shown in Figure 1.11(b). From the curves, it is seen that, while the use of single-channel switches may limit the dynamic range, the use of CMOS switches does not present this problem. The dimensions of these MOS switches are chosen such that the time constant associated with them while charging a unit capacitance $C_u$ via the worst case on-resistance is small. (The reader is urged to refer to [1.13] for an excellent introduction to the principles of MOS analog switches and specifically to [1.2] for a discussion on the parasitic capacitances associated with MOS transistors.)

In this book, we study in detail the theory and design of SC filters. From the foregoing discussions, it is clear that an SC filter is a discrete (or sampled data) structure in nature. As is well known, such structures are best treated using $z$-transforms. (It is expected that the reader is familiar with $z$-transforms and the relations between $s$- and $z$-domains and these are not discussed here.)

We shall now discuss the design of sampled data filters starting from the analog specifications.

## 1.5   Design of sampled data filters

The design of sampled data filters starts with a specification of the desired frequency characteristics in the analog domain. The reason for such a choice is simple: there are considerable sources from which the design information in the analog domain can be obtained. Once the analog domain specifications and transfer function specifying poles and zeros are obtained, the next step is to map these poles and zeros into the $z$-domain in such a way that the resulting sampled data filter satisfies the specifications with which we started the design procedure. In this section, the design of sampled data filters for given specifications in the analog domain will be illustrated using different approaches: the impulse invariance method; the matched-$z$ transformation method; the backward Euler integration (BEI) or $p$-transformation method; the forward Euler integration (FEI) method; the bilinear transformation (BT) method; and finally, the lossless discrete integrator (LDI) transformation technique.

### 1.5.1   The impulse invariance method

This method ensures that the impulse response, $\hat{h}(n)$, of the sampled data filter is the sampled version of the impulse response, $h(t)$, of the corresponding analog filter by defining

$$\hat{h}(n) = h(t)\,|_{t=nT} \tag{1.6}$$

where $T$ is the sample period.

The impulse response, $h(t)$, of an analog filter is defined as the inverse Laplace transform of its system function $H_a(s)$, given as

$$H_a(s) = \sum_{i=1}^{m} \frac{A_i}{s + s_i} \tag{1.7}$$

where $(-s_i)$ are the poles of the analog transfer function. Thus

$$h(t) = \mathbb{L}^{-1}\left\{ \sum_{i=1}^{m} \frac{A_i}{s + s_i} \right\} = \sum_{i=1}^{m} A_i e^{-s_i t} \tag{1.8}$$

If we desire that

$$h(nT) = h(t) \qquad \text{at} \qquad t = 0, T, 2T, \ldots \tag{1.9a}$$

then

$$h(nT) = \sum_{i=1}^{m} A_i e^{-s_i nT} \tag{1.9b}$$

Taking the $z$-transform of equation (1.9b):

$$H_d(z) = \sum_{n=0}^{\infty} h(nT)z^{-n} = \sum_{i=1}^{m} A_i \sum_{n=0}^{\infty} e^{-s_i nT} z^{-n} \qquad (1.10)$$

or

$$H_d(z) = \sum_{i=1}^{m} \frac{A_i}{1 - e^{-s_i T} z^{-1}} \qquad (1.11)$$

Thus equation (1.6), the condition that the impulse response of the sampled data filter be equal to the sampled impulse response of a given continuous filter $H_a(s)$, leads to a sampled data filter defined by equation (1.11), where the constants $A_i$ and $s_i$ are already specified by $H_a(s)$. Thus, by means of the following correspondence, $z$-transforms can be tabulated:*

$$H_a(s) = \sum_{i=1}^{m} \frac{A_i}{s + s_i} \rightarrow \sum_{i=1}^{m} \frac{A_i}{1 - e^{-s_i T} z^{-1}} = H_d(z) \qquad (1.12)$$

Although the time responses are essentially the same for analog as well as sampled data filters (due to the impulse invariance of the transform), the frequency responses will be different. This is because of the complex-plane mapping produced by the $z$-transform. The frequency response of the sampled data filter function is equal to the frequency response of the continuous function plus contributions of the response displaced by multiples of $2\pi/T$. This addition or folding-in of these terms is called *aliasing*.

The gain of an impulse invariant sampled data filter is proportional to the sampling frequency and may be considerably large. The following example illustrates this point.

*Example 1.1*
Design a sampled data filter using impulse invariance technique from a Butterworth-type analog second-order low-pass filter with a cut-off frequency of 1000 Hz and sampling frequency 10 000 Hz.

The transfer function of a second-order Butterworth low-pass filter for a cut-off frequency of 1 rad s$^{-1}$ is

$$H_a(s) = \frac{1}{s^2 + \sqrt{2}s + 1} \qquad (1.13a)$$

---

*  To be completely general, expression (1.12) must also include terms corresponding to multiple poles. A factor corresponding to a multiple pole with multiplicity $q$ and residue $A_i$, i.e., $A_i/(s + s_i)^q$ will yield an $H_d(z)$ term given by

$$\left[ A_i \frac{(-1)^{q-1}}{(q-1)!} \cdot \frac{\partial^{q-1}}{\partial a^{q-1}} \left( \frac{1}{1 - e^{-aT} z^{-1}} \right) \right] \Bigg|_{d = s_i}$$

Denormalizing equation (1.13a) for a cut-off frequency of 1000 Hz (for denormalizing, $s$ is replaced by $s/[(2\pi)(1000)]$) and expressing the same in partial fractions gives

$$H_a(s) = \frac{j2\pi \dfrac{1000}{\sqrt{2}}}{s + 2\pi \dfrac{1000}{\sqrt{2}} + j2\pi \dfrac{1000}{\sqrt{2}}} - \frac{j2\pi \dfrac{1000}{\sqrt{2}}}{s + 2\pi \dfrac{1000}{\sqrt{2}} - j2\pi \dfrac{1000}{\sqrt{2}}} \qquad (1.13b)$$

Using expression (1.12), we obtain

$$H_d(z) = \frac{j2\pi \dfrac{1000}{\sqrt{2}}}{[1 - z^{-1}e^{[-2\pi1000/\sqrt{2} + j2\pi1000/\sqrt{2}]1/10\,000}]}$$

$$- \frac{j2\pi \dfrac{1000}{\sqrt{2}}}{[1 - z^{-1}e^{[-2\pi1000/\sqrt{2} - j2\pi1000/\sqrt{2}]1/10\,000}]} \qquad (1.13c)$$

Note that $T = 1/10\,000$ s, since the sampling frequency is given as 10 000 Hz. Thus, $H_d(z)$ in equation (1.13c) becomes, after simplification:

$$H_d(z) = \frac{2449.2028z^{-1}}{1 - 1.158\,045\,9z^{-1} + 0.411\,240\,7z^{-2}} \qquad (1.13d)$$

Note that at d.c. (i.e., $z = 1$), the gain of this function is 9673.1978, which is prohibitively large. Therefore, it is customary to multiply $H_d(z)$ by $T$ (i.e., in this case by $1/10\,000$) so that the d.c. gain is unity.

Example 1.1 demonstrates the method of obtaining the $z$-domain transfer function from the $s$-domain transfer function by the impulse invariance method. In general, a high-order analog transfer function can always be written as a sum of biquadratic factors so that, by means of a table relating $H_a(s) \rightarrow H_d(z)$ (e.g., Table 1.3), the digital transfer function can be obtained, without going through all the steps outlined in the above example. An examination of Table 1.3 shows that, in the impulse invariance method, the zeros of the resulting $z$-domain transfer function will not be mapped as the zeros of the analog domain transfer function (see entry 2).

It may be noted that the above method has been also designated the *standard z-transform method*. It is applicable to the design of BP and all-pole LP filters. For other filter types, e.g., HP and BS, since a significant portion of the frequency response extends at high frequencies, the impulse invariance method maps the frequency response of the analog filter into the $z$-domain only up to $f_s/2$ and also the effect of aliasing is present. Hence, a useful alternative is the matched-$z$ transformation method considered next.

**Table 1.3** Analog and digital transfer functions related through the impulse invariance method.

| $H_a(s)$ | | $H_d(z)$ |
|---|---|---|
| 1. $\dfrac{a}{s+a}$ | $\rightarrow$ | $\dfrac{a}{1 - e^{-aT}z^{-1}}$ |
| 2. $\dfrac{s+a}{(s+a)^2 + b^2}$ | $\rightarrow$ | $\dfrac{1 - e^{-aT}(\cos bT)z^{-1}}{1 - 2e^{-aT}(\cos bT)z^{-1} + e^{-2aT}z^{-2}}$ |
| 3. $\dfrac{b}{(s+a)^2 + b^2}$ | $\rightarrow$ | $\dfrac{e^{-aT}(\sin bT)z^{-1}}{1 - 2e^{-aT}(\cos bT)z^{-1} + e^{-2aT}z^{-2}}$ |

## 1.5.2 The matched-z transformation method

In this method, the zeros and poles of the sampled data filter are matched to those of the continuous or analog filter through the relation

$$z = e^{sT} \tag{1.14a}$$

Thus a pole at $s = \alpha$ would map to a pole at $z = e^{-\alpha T}$ and a pair of complex poles given by $[(s + \alpha)^2 + \beta^2]$ corresponds to a $z$-domain quadratic factor

$$1 - 2e^{-\alpha T}z^{-1}(\cos \beta T) + e^{-2\alpha T}z^{-2} \tag{1.14b}$$

After transforming the poles and zeros, the gain, $K$, can be chosen to realize the desired gain for the sampled data transfer function:

$$H_d(z) = \frac{N_d(z)}{D_d(z)} = K \frac{\displaystyle\prod_{m=1}^{M} (1 - e^{-x_m T}z^{-1})}{\displaystyle\prod_{n=1}^{N} (1 - e^{-\alpha_n T}z^{-1})} \tag{1.15a}$$

when the analog transfer function is given as

$$H_a(s) = \frac{\displaystyle\prod_{m=1}^{M} (s + x_m)}{\displaystyle\prod_{n=1}^{N} (s + \alpha_n)} \tag{1.15b}$$

The poles of the resulting $H_d(z)$ are the same as those in the impulse invariance method, whereas the zeros are usually different. They can be either real or complex-conjugate occurring in pairs. It is usual to consider $D_d(z)$ or $N_d(z)$ to be made up of quadratic factors of the form

$$D_d(z) = z^2 - (2R \cos \theta)z + R^2 \tag{1.16}$$

where $R$ is the distance of the point $X$ representing the root from the origin in the $z$-domain and $\theta$ is the angle of the vector joining the origin to $X$ measured in the counterclockwise direction, as shown in Figure 1.12. The $D_d(z)$ in equation (1.16) can be factorized as

$$D_d(z) = z^2 - (2R \cos \theta)z + R^2$$
$$= (z - Re^{j\theta})(z - Re^{-j\theta}) \tag{1.17}$$

Since

$$z = e^{sT} = e^{\sigma T} \cdot e^{j\omega T} \tag{1.18}$$

we note that the poles on the $s$-plane ($\sigma \pm j\omega$) are obtained from

$$R = e^{\sigma T} \tag{1.19a}$$

$$\theta = \omega T \tag{1.19b}$$

From equations (1.19), we obtain

$$\sigma = f_s \ln R \tag{1.20a}$$

$$\omega = f_s \theta \tag{1.20b}$$

The quadratic factor corresponding to these roots in the $s$-domain is seen to be

$$D(s) = s^2 - 2\sigma s + (\sigma^2 + \omega^2)$$
$$= s^2 + s\left(\frac{\omega_p}{Q_p}\right) + \omega_p^2 \tag{1.21}$$

where $\omega_p$ is the pole-frequency and $Q_p$ is the pole-$Q$. Thus, we note from equations (1.20) and (1.21) that

$$\omega_p = f_s\sqrt{\theta^2 + (\ln R)^2} \tag{1.22a}$$

$$Q_p = -\frac{\sqrt{\theta^2 + (\ln R)^2}}{2 \ln R} \tag{1.22b}$$

Equations (1.22) give the $\omega_p$ and $Q_p$ of the complex-conjugate pair corresponding to the $z$-domain quadratic factor of equation (1.17). The relations (1.19)–(1.22) will

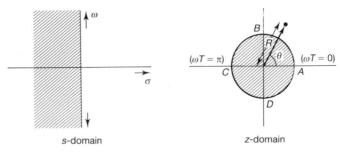

Fig. 1.12   The mapping of $z = e^{sT}$ illustrating $s$-domain and $z$-domain.

be quite useful in this book in subsequent chapters. It follows, from equations (1.19), for poles on the imaginary axis of the complex frequency plane, since $\sigma = 0$, that

$$\ln R = 0 \qquad \text{or} \qquad R = 1$$

and then equation (1.17) becomes

$$D_d(z) = z^2 - (2 \cos \theta)z + 1$$

We will now consider a design example illustrating the matched-$z$ transformation method.

*Example 1.2*
Design a notch filter for a notch at 1 kHz and $Q_p$ of 10, using a sampling frequency of 10 kHz.

The $s$-domain transfer function is obtained as

$$H_a(s) = \frac{s^2 + (2\pi)^2 1000^2}{s^2 + 2\pi \cdot 100s + (2\pi)^2 \cdot 1000^2} \tag{1.23}$$

Corresponding to equation (1.23), the real and imaginary parts of the $s$-domain roots $\sigma$ and $\omega$ are first evaluated, using equation (1.21).

$$\begin{aligned}
&\text{Zeros:} \quad \sigma = 0; \qquad \omega = 2000\pi \text{ rad s}^{-1} \\
&\text{Poles:} \quad \sigma = -100\pi \text{ rad s}^{-1}; \qquad \omega = 100\pi\sqrt{399} \text{ rad s}^{-1}
\end{aligned}$$

Thus, using equations (1.19a) and (1.19b), we obtain $R$ and $\theta$ for the numerator and denominator of the sampled data transfer function as follows:

$$\begin{aligned}
&\text{Zeros:} \quad R = 1, \qquad\qquad\quad \theta = 0.628\ 318\ 5 \\
&\text{Poles:} \quad R = 0.969\ 072\ 4, \qquad \omega = 0.627\ 532\ 6
\end{aligned}$$

Hence, the numerator and the denominator of the desired $H_d(z)$ can immediately be found:

$$H_d(z) = \frac{z^2 - 1.618\ 034z + 1}{z^2 - 1.568\ 886\ 9z + 0.939\ 101\ 3} \tag{1.24}$$

The reader can check from equations (1.24) and (1.22) whether the desired notch frequency and $Q_p$ are realized. It is now of interest to check the d.c. gain. To this end, $z = 1$ is substituted in equation (1.24), and so the d.c. gain is 1.031 742 7. Similarly, by substituting $z = -1$ in equation (1.24) to evaluate the gain at $f_s/2$ or half the sampling frequency, the corresponding gain is obtained as 1.031 370 1. Hence, if the d.c. gain is the same as that of the analog filter which we are trying to realize in sampled data form, equation (1.24) is multiplied by (1/1.031 370 1) yielding the final result as follows:

$$H_d(z) = \frac{0.969\ 584\ 1(z^2 - 1.618\ 034z + 1)}{z^2 - 1.568\ 886\ 9z + 0.939\ 101\ 3} \qquad (1.25)$$

This meets the analog specifications as desired.

The matched-$z$ transformation also, however, has certain limitations. When the zeros of an analog transfer function are at frequencies greater than half the sampling frequency, they will be aliased to a low frequency. A second case is the design of all-pole filters for which the sampled data versions obtained using the matched-$z$ transformation technique do not adequately represent the desired continuous system.

The impulse invariance and matched-$z$ transformations can be called 'exponential' transformations since they are based on the exponential relationship given in equation (1.18). We will next consider the algebraic substitution methods of sampled data filter design. The first two methods to be discussed in what follows are based on approximating the derivatives by finite differences.

### 1.5.3  Backward Euler approximation of derivatives

In this method, the derivative is approximated by the first backward difference as follows:

$$\left.\frac{dy}{dt}\right|_{t=nT} = \frac{y(nT) - y((n-1)T)}{T} \qquad (1.26)$$

Similarly, for the second derivative $d^2y/dt^2$, using the same approximation, we have:

$$\left.\frac{d^2y}{dt^2}\right|_{t=nT} = \frac{\left.\left(\frac{dy}{dt}\right)\right|_{t=nT} - \left.\left(\frac{dy}{dt}\right)\right|_{t=(n-1)T}}{T} \qquad (1.27)$$

Substituting equation (1.26) in equation (1.27), we obtain

$$\left.\frac{d^2y}{dt^2}\right|_{t=nT} = \frac{[y(nT) - y((n-1)T)] - [y((n-1)T) - y((n-2)T)]}{T^2}$$

$$= \frac{y(nT) - 2y((n-1)T) + y((n-2)T)}{T^2} \qquad (1.28)$$

Then, taking z-transforms of equations (1.26) and (1.28), we obtain

$$Z\left(\frac{dy}{dt}\bigg|_{t=nT}\right) = \left[\frac{1-z^{-1}}{T}\right]Y(z) \qquad (1.29a)$$

$$Z\left(\frac{d^2y}{dt^2}\bigg|_{t=nT}\right) = \left[\frac{1-z^{-1}}{T}\right]^2 Y(z) \qquad (1.29b)$$

In general, it follows that all the derivatives in the continuous-time transfer function can be approximated in a simple manner to yield the corresponding z-domain transfer function. We consider an example to show the application of this technique in practice.

*Example 1.3*
Design a digital filter using the backward Euler approximation of the derivative from an LP second-order Bessel filter.

The second-order Bessel filter has a transfer function given by

$$H_a(s) = \frac{Y_a(s)}{X_a(s)} = \frac{k}{s^2 + 3s + 3} \qquad (1.30a)$$

where $Y_a(s)$ is the Laplace transform of the output and $X_a(s)$ is the Laplace transform of the input. The differential equation characterizing equation (1.30a) can be immediately written as

$$\frac{d^2y(t)}{dt^2} + 3\frac{dy(t)}{dt} + 3y(t) = kx(t) \qquad (1.30b)$$

Substituting $y(nT)$ for $y(t)$ and $x(nT)$ for $x(t)$ and using the backward Euler approximation, we obtain, using equations (1.26) and (1.27),

$$\left[\frac{y(nT) - 2y((n-1)T) + y((n-2)T)}{T^2}\right]$$

$$+ 3\left[\frac{y(nT) - y((n-1)T)}{T}\right] + 3y(nT) = kx(nT) \qquad (1.30c)$$

or

$$\left[\frac{1}{T^2} + \frac{3}{T} + 3\right]y(nT) - \left[\frac{2}{T^2} + \frac{3}{T}\right]y((n-1)T) + \frac{y((n-2)T)}{T^2} = kx(nT) \qquad (1.30d)$$

Taking $z$-transforms of both sides:

$$\left[\left(\frac{1}{T^2}+\frac{3}{T}+3\right) - z^{-1}\left(\frac{2}{T^2}+\frac{3}{T}\right) + \frac{z^{-2}}{T^2}\right]Y_{\mathrm{d}}(z) = kX_{\mathrm{d}}(z) \qquad (1.30e)$$

or

$$H_{\mathrm{d}}(z) = \frac{Y_{\mathrm{d}}(z)}{X_{\mathrm{d}}(z)} = \frac{kT^2}{z^{-2} - (2+3T)z^{-1} + (3T^2+3T+1)} \qquad (1.30f)$$

Note that we can also obtain equation (1.30f) from (1.30a) directly, by substituting

$$s = \frac{1-z^{-1}}{T} \qquad (1.31)$$

Thus, we have considered in the above an algebraic transformation. We must study its mapping properties.

Two desirable properties of any algebraic transformation from $s$-domain to $z$-domain are as follows:

1. The $j\omega$-axis in the $s$-plane should be mapped to the unit circle in the $z$-domain.
2. Points in the left-half $s$-plane should be mapped inside the unit circle.

The first property preserves the frequency selective properties of the continuous system, whereas the second property ensures that stable continuous systems are mapped into stable discrete systems.

Considering now equation (1.31), we will first check for Property 1. Substituting $s = j\omega$ in equation (1.31), we obtain:

$$z = \frac{1}{1-sT} = \frac{1}{1-j\omega T} \qquad (1.32)$$

Note that the $j\omega$-axis is not mapped onto the unit circle since $z \neq 1$ for all $\omega$. Rewriting equation (1.32), we obtain, after manipulation,

$$z = \frac{1}{2}\left(1 + \frac{1+j\omega T}{1-j\omega T}\right)$$

$$= \tfrac{1}{2}(1 + e^{2j\,\tan^{-1}\,\omega T}) \qquad (1.33)$$

Taking real and imaginary parts of $z$ in equation (1.33):

$$\mathrm{Re}(z) = \frac{1}{2} + \frac{\cos(2\,\tan^{-1}\,\omega T)}{2} \qquad (1.34a)$$

$$\mathrm{Im}(z) = \frac{\sin(2\,\tan^{-1}\,\omega T)}{2} \qquad (1.34b)$$

Equations (1.34a) and (1.34b) represent a circle in the $z$-domain with radius 0.5 and centred at $z = +0.5$, as shown in Figure 1.13, since

$$(\mathrm{Re}(z) - 0.5)^2 + (\mathrm{Im}(z))^2 = (0.5)^2 \qquad (1.35)$$

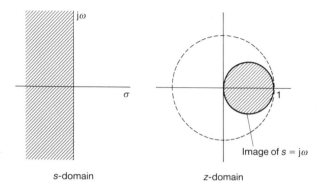

s-domain                          z-domain

*Fig. 1.13* Mapping resulting from backward Euler transformation of the derivative.

It is thus seen that the imaginary axis in the s-plane is *not* mapped into the unit circle, i.e., Property 1 is not satisfied.

To verify Property 2, we substitute $s = \sigma + j\omega$ and consider $\sigma < 0$. Then, from equation (1.31),

$$z = \frac{1}{1 - sT} = \frac{1}{(1 - \sigma T) - j\omega T}$$

Hence

$$|z| = \frac{1}{\sqrt{(1 - \sigma T)^2 + \omega^2 T^2}} < 1 \qquad (1.36)$$

Thus, the left-half of the complex frequency plane is mapped to within the unit circle in z-domain, and Property 2 is satisfied. Since the points on the inner circle and the unit circle are same approximately for $\omega T \ll 1$, the approximation is reasonable only when $\omega T \ll 1$ or when high sampling frequencies are used. This transformation has also been named the p-*transformation* in the literature [1.14].

Since the mapping provided by p-transformation yields stable sampled data filters corresponding to stable analog filters, it can be used for designing such sampled data filters. However, it should be ensured that the z-domain transfer function thus obtained must realize the same poles and zeros as the s-domain filter. This is achieved by what is known as *prewarping*, meaning that in order to obtain the s-domain poles (or zeros) at desired locations, we should choose the z-domain poles or zeros through the relationship $z = e^{sT}$. As an illustration, for a given s-domain pole at $(-\alpha_i + j\beta_i)$, the z-domain pole occurs at

$$z = e^{sT} = e^{(-\alpha_i + j\beta_i) T}$$

or

$$p = \frac{(1 - z^{-1})}{T} = \frac{1 - e^{(\alpha_i - j\beta_i) T}}{T} \qquad (1.37)$$

Thus, in the $p$-domain, the transfer function can be obtained. We will be considering application of this technique in later chapters. Another transformation using the 'forward difference' approximation for a derivative is considered next.

## 1.5.4  Forward Euler approximation of derivatives

In this case, the derivative is approximated by the forward difference, i.e.,

$$\left.\frac{dy}{dt}\right|_{t=nT} = \frac{y((n+1)T) - y(nT)}{T} \tag{1.38}$$

Following the same method as in the previous case, this corresponds to the transformation

$$s \to \frac{z-1}{T} \tag{1.39}$$

Manipulating equation (1.39), we obtain

$$z = 1 + sT = 1 + j\omega T \tag{1.40}$$

Thus the imaginary axis of the complex frequency plane is not mapped into the unit circle in the $z$-domain and Property 1 is not satisfied. Also, the left-half of the complex plane $s = -\sigma + j\omega$ is mapped such that $z = 1 - \sigma T + j\omega T$, which gives

$$|z| = \sqrt{(1-\sigma T)^2 + \omega^2 T^2} > 1, \qquad \omega^2 T^2 > 1 - (1-\sigma T)^2 \tag{1.41}$$

The mapping resulting from equation (1.41) is shown in Figure 1.14. It is evident that stable analog transfer functions do not result in stable sampled data transfer functions when forward Euler approximation is used.

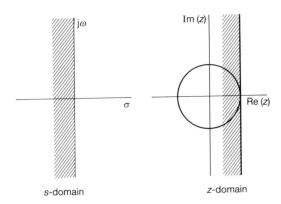

*Fig. 1.14*  Mapping resulting from forward Euler transformation of derivatives.

*Example 1.4*

Derive a sampled data filter from an analog second-order Butterworth filter using forward Euler transformation. Use a clock frequency of 10 kHz. The desired pole-frequency is 1 kHz.

The second-order Butterworth transfer function is given by

$$H_a(s) = \frac{1}{s^2 + \sqrt{2}s + 1} \tag{1.42a}$$

which has a cut-off frequency of 1 rad s$^{-1}$. Denormalizing to get a cut-off frequency of 1 kHz, we obtain

$$H_a(s) = \frac{4\pi^2 \cdot 1000^2}{s^2 + \sqrt{2} \cdot 2\pi \cdot 1000s + 4\pi^2 \cdot 1000^2} \tag{1.42b}$$

Using the forward Euler transformation (equation (1.39)) in (1.42b) gives, after simplification,

$$H_d(z) = \frac{0.394\ 784\ 2}{z^2 - 1.111\ 423\ 4z + 0.506\ 207\ 6} \tag{1.42c}$$

Using equations (1.22), the reader can check that this $H_d(z)$ corresponds to an $\omega_p$ of 1202.3671 Hz and $Q_p$ of 1.109 as against an $\omega_p$ of 1000 Hz and $Q_p$ of 0.707, which are the desired specifications.

For comparison, the sampled data transfer function obtained by using backward Euler transformation is

$$H_d(z) = \frac{\dfrac{4\pi^2}{100} z^2}{2.283\ 360\ 8z^2 - 2.888\ 576\ 6z + 1} \tag{1.42d}$$

The resulting $\omega_p$ and $Q_p$ are computed as 810.7295 Hz and 0.616 965 2, respectively. Thus there is no $Q_p$ enhancement in this case as in the case of forward Euler approximation.

It is thus seen that backward Euler approximation is preferable to forward Euler approximation, since the latter leads to $Q_p$ enhancement. We next investigate alternative procedures for obtaining sampled data transfer functions.

## 1.5.5 The bilinear transformation method

The bilinear transformation method can be explained from a numerical analysis point of view. Consider a signal $y_a(t)$ from which a sampled signal is derived at multiples of $t = nT$. It is desired to perform an approximate integration of the signal. We first assume that the variation between successive sampling instants is linear. It is noted that

$$y_a(nT) = \int_{(n-1)T}^{nT} \frac{dy_a}{dt}\,dt + y_a((n-1)T) \tag{1.43}$$

Next, the integration in equation (1.43) can be approximated using the trapezoidal rule as follows:

$$\int_{(n-1)T}^{nT} \frac{dy_a}{dt}\,dt = \frac{T}{2}\left[\frac{dy_a}{dt}\bigg|_{nT} + \frac{dy_a}{dt}\bigg|_{(n-1)T}\right] \tag{1.44}$$

Thus, from equations (1.43) and (1.44), we obtain

$$y_a(nT) = y_a((n-1)T) + \frac{T}{2}\left[\frac{dy_a}{dt}\bigg|_{nT} + \frac{dy_a}{dt}\bigg|_{(n-1)T}\right] \tag{1.45}$$

Equation (1.45) can be reinterpreted as approximating the average of derivatives by a difference. We will now examine the result of this operation on an integrator with transfer function given by

$$H_a(s) = \frac{Y_a(s)}{X_a(s)} = \frac{1}{s} \tag{1.46}$$

From equation (1.46), we obtain

$$\frac{dy_a}{dt} = x_a \tag{1.47}$$

Using equation (1.47) in equation (1.45), we obtain, after simplification,

$$H_d(z) = \frac{T}{2}\left(\frac{1+z^{-1}}{1-z^{-1}}\right) \tag{1.48}$$

Comparing equations (1.46) and (1.48), it is seen that by substituting

$$s = \frac{2}{T}\left(\frac{1-z^{-1}}{1+z^{-1}}\right) \tag{1.49}$$

we get $H_d(z)$ from $H_a(s)$. It can be shown that for an $n$th-order differential equation, the above method can be extended to yield $H_d(z)$ related to $H_a(s)$ through equation (1.49).

From equation (1.49), it follows that

$$z = \frac{1 + s\dfrac{T}{2}}{1 - s\dfrac{T}{2}} \tag{1.50}$$

We now study the mapping properties of the transformation given in equation (1.50). Thus, for $z$ on the unit circle, i.e., $z = e^{j\omega_D T}$, where we denote $\omega_D$ as the digital frequency, from equation (1.49), we have

$$s = \frac{2}{T}\left(\frac{1 - e^{-j\omega_D T}}{1 + e^{-j\omega_D T}}\right) = j\frac{2}{T}\tan\frac{\omega_D T}{2} \tag{1.51}$$

Denoting $s = \sigma + j\omega$, it is observed thus from equation (1.51) that for $z$ on the unit circle, $\sigma = 0$ and

$$\omega = \frac{2}{T}\tan\frac{\omega_D T}{2} \tag{1.52}$$

Note that $\omega$ is the complex frequency in the $s$-plane corresponding to the analog filter and $\omega_D$ corresponds to the digital filter. Equation (1.52) shows that the positive and negative imaginary axes of the $s$-plane are mapped respectively into the upper- and lower-half of the $s$-plane. Further, the left-half of the $s$-plane maps into the interior of the unit circle and the right-half of the $s$-plane maps outside the unit circle. This can be observed from equation (1.50) as follows. When the real part of $s$ is negative, from equation (1.50),

$$|z| = \left|\frac{1 - \sigma\dfrac{T}{2} + \dfrac{j\omega T}{2}}{1 + \sigma\dfrac{T}{2} - \dfrac{j\omega T}{2}}\right| < 1 \tag{1.53}$$

corresponding to the inside of the unit circle. Conversely, for points such that the real part of $s$ is positive, $|z|$ is greater than 1, corresponding to the outside of the unit circle. We thus see that the bilinear transformation avoids the problem of aliasing encountered with the use of impulse invariance because it maps the *entire* imaginary axis in the $s$-plane onto the unit circle in the $z$-domain. The price paid, however, for this is the introduction of frequency warping effect or the distortion in the frequency axis, given by equation (1.52). This can be corrected by 'prewarping' the analog filter specifications. This is illustrated by the procedure one has to follow while designing sampled data filters using bilinear transformation. The design steps are summarized below:

- **Step 1:** From the sampled data filter specifications, prewarp the critical frequencies (i.e., cut-off frequency, notch frequency, passband edge, stopband edge, etc.) using equation (1.52).
- **Step 2:** From these 'new' critical frequencies, design an analog filter using well-known methods of analog filter design, and obtain $H_a(s)$, or the analog transfer function.
- **Step 3:** Substitute in $H_a(s)$, $s = (2/T)(z - 1)/(z + 1)$ to obtain the sampled data transfer function $H_d(z)$.

The use of these three steps in the design of sampled data filters will be illustrated in the following design examples.

*Example 1.5*
Design a sampled data filter using the bilinear transformation to realize a second-order low-pass Butterworth filter with a cut-off frequency of 1000 Hz and sampling frequency of 10 kHz.

*Step 1:* For the required sampled data filter, the critical frequency is the 3 dB cut-off frequency given as 1000 Hz. Prewarping using equation (1.52) yields,

$$\omega = 2 \cdot 10\ 000\ \tan\left(\frac{2\pi \cdot 1000}{2 \cdot 10\ 000}\right) = 2\pi \cdot 1034.2515\ \text{rad s}^{-1}$$

where we have used $T = 1/f_s = 10^{-4}$ s. Thus the cut-off frequency is 1034.2515 Hz. The reader may note the warping effect.

*Step 2:* Since we intend to have a Butterworth type of response, with a cut-off frequency of 1034.2515 Hz, the transfer function in the analog domain is

$$H_a(s) = \frac{\omega_0^2}{s^2 + \sqrt{2}\omega_0 s + \omega_0^2}$$

$$= \frac{(6498.3939)^2}{s^2 + \sqrt{2} \cdot 6498.3939s + (6498.3939)^2}$$

*Step 3:* Substituting $s = (2/T)(z-1)/(z+1)$ we obtain, with $T = 10^{-4}$ s,

$$H_d(z) = \frac{0.674\ 553(z+1)^2}{z^2 - 1.142\ 980\ 6z + 0.412\ 801\ 6}$$

Thus we have obtained $H_d(z)$, which satisfies the desired specifications.

*Example 1.6*
Design a BP sampled data filter to realize a centre frequency of 1633 Hz, $Q_p = 16$ and midband gain = 10 dB. Use a sampling frequency of 8 kHz.

The analog transfer function which meets the above requirements is

$$H_a(s) = \frac{2\pi \cdot 1633 \cdot 0.197\ 642\ 4s}{s^2 + \left(\frac{2\pi \cdot 1633}{16}\right)s + (2\pi)^2 \cdot 1633^2}$$

Prewarping the centre frequency, we obtain a prewarped centre frequency of 1901.027 Hz. We next evaluate the 3 dB points of $H_a(s)$ as 1684.8285 Hz and 1582.766 Hz. Prewarping these 3 dB frequencies, we obtain new 3 dB frequencies of 1982.9967 Hz and 1823.9226 Hz. It thus follows that the new bandwidth is

159.074 17 Hz and, hence, that the prewarped $Q_p$ is 1901.027/159.074 17 or 11.950 57. The analog transfer function thus obtained is

$$H_a'(s) = \frac{2\pi \cdot 1901.027s}{s^2 + \left(\dfrac{2\pi \cdot 1901.027}{11.950\ 57}\right)s + (2\pi)^2 \cdot 1901.027^2}$$

Since the $Q_p$ has changed, we need to change the gain accordingly to meet the specification of midband gain, namely, 10 dB. After simplification, $H_a'(s)$ becomes

$$H_a'(s) = \frac{3160.6727s}{s^2 + 999.492\ 47s + 14.2671^8}$$

Substituting for $s$ the bilinear transformation, we obtain

$$H_d(z) = \frac{0.1219(1 - z^{-2})}{1 - 0.5455z^{-1} + 0.9229z^{-2}}$$

*Example 1.7*
Design a notch filter with a notch frequency at 1 kHz, $\omega_p$ at 1 kHz and $Q_p = 10$, using a sampling frequency of 10 kHz, and the bilinear transformation method.

The analog transfer function that meets the above requirements is obtained in Example 1.2 as equation (1.23). The notch (and pole) frequency becomes after prewarping,

$$\omega_{npw} = 6498.3939 \text{ rad s}^{-1}$$

The next step is to obtain the $Q_p$ specification for the analog transfer function. The 0.707 or $(-3 \text{ dB})$ responses occur at frequencies

$$\omega_{1,2} = \omega_0 \sqrt{1 + \frac{1}{2Q_p^2} \pm \sqrt{\frac{1}{4Q_p^4} + \frac{1}{Q_p^2}}} \tag{1.54}$$

where $Q_p$ is given as 10 in the specification. Hence, after denormalizing to $\omega_0 = (2\pi)(1000)$ Hz, we obtain

$$\omega_1 = 5976.8752 \text{ rad s}^{-1}$$

$$\omega_2 = 6605.1936 \text{ rad s}^{-1}$$

We wish to obtain the same bandwidth as this prototype filter in our prewarped response also. The corresponding prewarped 3 dB frequencies are

$$\omega_{1pw} = 6161.3964 \text{ rad s}^{-1}$$

$$\omega_{2pw} = 6856.3001 \text{ rad s}^{-1}$$

Thus the notch bandwidth is $694.904\ 35\ \mathrm{rad\ s^{-1}}$. The new prewarped $Q_{pw}$ is obtained as

$$Q_{pw} = \frac{\omega_{npw}}{\omega_{2pw} - \omega_{1pw}} = \frac{6498.3939}{6856.3001 - 6161.3964} = 9.351\ 494\ 2$$

Thus the prewarped analog transfer function is simply

$$H_a(s) = \frac{s^2 + (6498.3939)^2}{s^2 + \left(\dfrac{6498.3939}{9.351\ 494\ 2}\right)s + (6498.3939)^2}$$

Substituting the bilinear transformation, we obtain, after simplification,

$$H_d(z) = \frac{0.963\ 530\ 3(z^2 - 1.618\ 034z + 1)}{z^2 - 1.568\ 732\ 9z + 0.939\ 060\ 5}$$

It is of interest to compare this $H_d(z)$ with that obtained using the matched-$z$ transformation method in Example 1.2. Note that the denominators are slightly different.

It may be noted that while the bilinear transformation reproduces the amplitude response of the prototype analog filter in the amplitude response of the sampled data filter, it does not preserve the phase response. This is because of the distortion of the frequency axis, as given by equation (1.52). In the next subsection, we discuss another useful mapping transformation, the lossless discrete integrator (LDI) transformation.

## 1.5.6  The lossless discrete integrator transformation [1.15–1.21]

For the LDI transformation, the derivative is approximated differently:

$$\left.\frac{dy(t)}{dt}\right|_{t=(n+1/2)T} = \frac{y(nT + T) - y(nT)}{T} \tag{1.55}$$

The first observation that can be made from this approximation is that the sampling points of the derivative and the original time function are not the same. (Compare with the forward Euler or backward Euler approximations to observe this point.) Just as the bilinear transformation is derived from trapezoidal integration method, so it can be noted that the above approximation is derived from the midpoint integration method, as is seen by rewriting equation (1.55) as follows:

$$\int_{nT}^{(n+1)T} \frac{dy_a}{dt}\,dt = T \cdot \left.\frac{dy_a}{dt}\right|_{(n+1/2)T} \tag{1.56}$$

i.e., the area under the curve $dy_a/dt$ between sampling instants $nT$ and $(n+1)T$ is approximated by midpoint integration.

We will now consider a first-order system with

$$H_a(s) = \frac{Y_a(s)}{X_a(s)} = \frac{d_0}{c_1 s + c_0} \tag{1.57a}$$

from which the differential equation characterizing it is obtained as follows:

$$c_1 \frac{dy_a(t)}{dt} + c_0 y_a(t) = d_0 x_a(t) \tag{1.57b}$$

Using equation (1.55), we obtain at the instant $(nT + T/2)$,

$$c_1 \left[ \frac{y_a(nT + T) - y_a(nT)}{T} \right] + c_0 y_a \left( nT + \frac{T}{2} \right) = d_0 x \left( nT + \frac{T}{2} \right) \tag{1.58a}$$

Taking the $z$-transform of equation (1.58a), we obtain

$$c_1 \frac{(z - 1) Y_d(z)}{T} + c_0 Y_d(z) z^{1/2} = d_0 X_d(z) z^{1/2} \tag{1.58b}$$

or

$$H_d(z) = \frac{Y_d(z)}{X_d(z)} = \frac{d_0}{c_1 \left( \dfrac{z^{1/2} - z^{-1/2}}{T} \right) + c_0} \tag{1.58c}$$

Comparing equation (1.58c) with (1.57a), we note that

$$s \to \frac{z^{1/2} - z^{-1/2}}{T} \tag{1.59}$$

which is the LDI transformation.

The mapping properties of equation (1.59) will now be studied. Denoting once again $\omega_D$ as the frequency of the sampled data filter, the unit circle in the $z$-domain is described by $z = e^{j\omega_D T}$ and therefore substituting for $z$ in equation (1.59), we have

$$s = j \frac{2}{T} \sin \frac{\omega_D T}{2} \tag{1.60}$$

Denoting $s = \sigma + j\omega$, we observe from equation (1.60) that $\sigma = 0$ and

$$\omega = \frac{2}{T} \sin \frac{\omega_D T}{2} \tag{1.61}$$

Thus, for values of $z$ on the unit circle, we have $\sigma = 0$. Hence the imaginary axis of the $s$-plane is mapped onto the unit circle in the $z$-domain. However, note that, as the discrete time filter frequency varies in the range

$$-\pi f_s < \omega_D < \pi f_s$$

so the continuous time filter frequency ranges from

$$-2f_s < \omega < 2f_s$$

Thus only part of the imaginary axis of the $s$-plane is uniquely mapped onto the unit circle in the $z$-domain. As such, there is no aliasing of the frequency response.

The effect of transformation (1.59) is 'warping' as discussed above for the bilinear transformation, but the 'warping' for LDI and bilinear transformations is different in that bilinear transformation compresses the frequency scale whereas the LDI transformation expands it. This can be seen from the following inequalities:

$$|\sin x| < |x| < |\tan x|, \qquad \text{for} \quad -\frac{\pi}{2} < x < \frac{\pi}{2} \qquad (1.62)$$

Consequently, the frequency warping introduced by the bilinear transformation is more than that introduced by the LDI transformation. This point is illustrated in Figure 1.15.

We will next consider a design example of a second-order LP Butterworth filter using the LDI transformation.

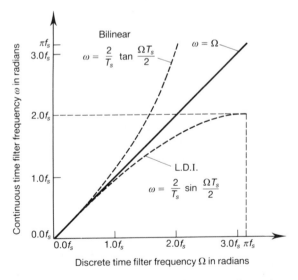

Fig. 1.15  Warping effect introduced by bilinear and LDI transformations *(adapted from [1.20], © 1980 IEEE).*

*Example 1.8*
Design a second-order LP Butterworth filter for a cut-off frequency of 1 kHz using a sampling frequency of 10 kHz. Use the LDI transformation.

Prewarping the cut-off frequency using equation (1.61), we obtain a prewarped cut-off frequency of

$$\omega_c = 6180.3399 \text{ rad s}^{-1}$$

(compare this with 6498.3939 rad s$^{-1}$ for the bilinear transformation case). The analog prewarped Butterworth transfer function can now be obtained as follows:

$$H_a(s) = \frac{(6180.3399)^2}{s^2 + (6180.3399)\sqrt{2}s + (6180.3399)^2}$$

Substituting for $s$ the LDI transformation of (1.59), we get

$$H_d(z) = \frac{(6180.3399)^2 z^{-1}}{\frac{4}{T^2}(z-1)^2 + (6180.3399)\sqrt{2}\,\frac{2}{T}(z-1)z^{-1/2} + (6180.3399)^2 z^{-1}}$$

It can be seen that the magnitude of $H_d(z)$ is a rational function in $\sin(\omega T/2)$, whereas $H_d(z)$ is not a rational function in $z$. Alternatively, $H_d(z)$ may be considered to be a high-order transfer function, of order 4 (i.e., a ratio of polynomials in $z^{-1/2}$). This results from the nature of the LDI transformation given in equation (1.61) and causes some difficulties in design.

It is thus seen that the bilinear transformation converts rational functions in the $s$-domain to rational functions in the $z$-domain, whereas the LDI transformation does not. However, it is possible to design digital filters using the LDI transformation, as will be shown next.

Considering the LDI transformation in equation (1.59) once again, we note that for every point in the $s$-domain, these are two corresponding points in the $z$-domain. We also note that if we replace $z^{1/2}$ by $(-z^{-1/2})$, the $s$-domain point will not change. In other words, the two roots of the quadratic equation in $z^{1/2}$ implied in equation (1.59),

$$z - sTz^{1/2} - 1 = 0 \tag{1.63}$$

are such that one of the roots is within the unit circle, while the other is outside the unit circle. To realize stable $z$-domain transfer functions, it follows therefore that, corresponding to each $s$-domain root, we shall choose the root in the $z$-domain within the unit circle. This procedure can be used to obtain digital transfer functions based on the LDI transformation [1.20]. Prior to using this procedure, one should prewarp the $s$-domain roots according to (1.61). This complete procedure is illustrated in the next example.

*Example 1.9*

Design an LDI-type second-order LP Butterworth filter by mapping the roots from the *s*-domain into the *z*-domain for a 3 dB frequency of 1 kHz and sampling frequency 10 kHz.

The *s*-domain roots obtained after prewarping the specifications are obtained from Example 1.8 as follows:

$$s_{1,2} = -14\ 140\ \sin\frac{\pi}{10} \pm j14\ 140\ \sin\frac{\pi}{10}$$

Then we can use equation (1.63) to obtain the *z*-domain roots within the unit circle:

$$z_{1,2} = 0.583\ 327 \pm j0.267\ 370\ 6$$

The resulting *z*-domain transfer function is

$$H_d(z) = \frac{0.245\ 098\ 7z^{-1}}{1 - 1.166\ 654z^{-1} + 0.411\ 764\ 1z^{-2}}$$

where the gain constant in the numerator is chosen so as to make the d.c. gain unity. It can be verified that the gain at 1000 Hz is $1/\sqrt{2}$ as desired; $\omega_p$ and $Q_p$ are also evaluated as 983.03 Hz and 0.6961, respectively.

Thus it has been shown that it is possible to design digital filters based on the LDI transformation but the only difference between bilinear and LDI transformations is that the latter is not 'algebraic'. It involves the computation of poles from the desired specifications. As a final remark, it is also observed that the magnitude squared response of $H_d(z)$ obtained using the LDI transformation will be of the form

$$|H_d(j\omega)|^2 = \frac{1}{1 + F\left[\dfrac{\sin\dfrac{\omega T}{2}}{\sin\dfrac{\omega_0 T}{2}}\right]^2}$$

where $\sin(\omega_0 T/2)$ corresponds to the prewarped cut-off frequency; $\omega = \omega_0$ is the 3 dB point. The term $\sin(\omega T/2)$ arises because of LDI mapping. Similarly, by replacing the sine function by the tangent function, we obtain the squared magnitude response of the bilinear transformation type design.

Even though exact mapping of the *s*-domain roots into the *z*-domain roots is possible using the method illustrated above, the LDI type mapping requires that large sampling frequencies be used to obtain satisfactory responses, as illustrated next.

It has been noted that the LDI transformation maps the continuous-time filter response only up to a radian frequency $2f_s$. Consequently, the resulting digital filters

have less stopband attenuation than those designed using the bilinear transformation [1.21]. It is also required that large sampling frequencies be used in order to
realize large stopband attenuation [1.21].

We will continue our study of the application of these transformations to design
SC filters in subsequent chapters. The next sections deal with other design aspects
of high-order sampled data filters.

## 1.6  Realization of high-order sampled data transfer functions

So far, we have studied several methods of obtaining a sampled data transfer function from the given specifications in the analog domain. Several examples have been
presented to illustrate the design procedure. From the given specifications of the
sampled data filter, the first step is to obtain an analog transfer function which meets
the requirement.

For the bilinear transformation or LDI transformation methods, the given
specifications have to be prewarped and then, using the standard filter design
tables, an $s$-domain transfer function can be found. For the impulse invariance or
matched-$z$ transformation, this prewarping is not required. Filter design tables
usually give the numerator and denominator of the $s$-domain transfer function as
high-order polynomials or in the form of quadratic factors combining pairs of complex conjugate poles. From the $H_a(s)$ thus obtained, we first obtain a high-order $z$-
domain transfer function by any of the methods considered above. The next step
is to realize this high-order sampled data filter.

### 1.6.1  Sampled data filter structures

There are several methods of realizing a given sampled data transfer function. These
use delays and multipliers in various fashions. In a *direct-form* realization, the
number of delays required equals the order of the digital transfer function. A third-
order realization is illustrated in Figure 1.16. This is *canonic* in delays and contains
feedforward and feedback paths to realize the transmission zeros and poles,
respectively. The multipliers are coefficients of the transfer function and hence the
ease of implementation. This structure can be extended to realize a high-order
sampled data filter but the resulting structure has large sensitivities to errors in
coefficient values. This is explained in the following.

The higher-order transfer function is written as

$$H_d(z) = \frac{\sum\limits_{k=0}^{M} b_k z^{-k}}{1 - \sum\limits_{k=1}^{N} a_k z^{-k}} \qquad (1.64)$$

where, as explained before, $a_k$ and $b_k$ are the multiplier values in a direct-form

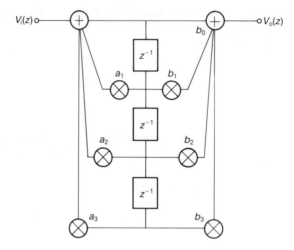

Fig. 1.16  A third-order direct-form realization of a sampled data filter.

realization. Due to the errors in realizing $a_k$ and $b_k$, $H_d(z)$ evidently changes to $\hat{H}_d(z)$, given by

$$\hat{H}_d(z) = \frac{\displaystyle\sum_{k=0}^{M} \hat{b}_k z^{-k}}{1 - \displaystyle\sum_{k=1}^{N} \hat{a}_k z^{-k}} \tag{1.65}$$

where $\hat{a}_k = a_k + \Delta a_k$ and $\hat{b}_k = b_k + \Delta b_k$. We can write the denominator polynomial of equation (1.65) in terms of its first-order factors as

$$D_d(z) = 1 - \sum_{k=1}^{N} \hat{a}_k z^{-k} = \prod_{k=1}^{N} (1 - z_k z^{-1}) \tag{1.66}$$

We next define the poles of $\hat{H}_d(z)$ as $z_i + \Delta z_i$, $i = 1, 2, ..., N$. The error $\Delta z_i$ in $z_i$ can be expressed in terms of the errors in the coefficient values as

$$\Delta z_i = \sum_{k=1}^{N} \frac{\partial z_i}{\partial a_k} \Delta a_k, \qquad i = 1, 2, ..., N \tag{1.67}$$

From equation (1.66), noting that

$$\left\{\frac{\partial D_d(z)}{\partial z_i}\right\}\Bigg|_{z=z_i} \cdot \frac{\partial z_i}{\partial a_k} = \left\{\frac{\partial D_d(z)}{\partial a_k}\right\}\Bigg|_{z=z_i} \tag{1.68}$$

we obtain

$$\frac{\partial z_i}{\partial a_k} = \frac{z_i^{N-k}}{\displaystyle\prod_{\substack{l=1 \\ l \neq i}}^{N} (z_i - z_l)} \tag{1.69}$$

Thus equation (1.69) gives the sensitivity of the $i$th pole to a change in the $k$th coefficient in the denominator polynomial of $H_d(z)$. In a similar manner the sensitivity of the zeros to the coefficient errors can also be derived.

The denominator of equation (1.69) can be interpreted as follows. The factors $(z_i - z_l)$ represent the lengths of the vectors from the pole $z_i$ to all the other poles. Thus, if the poles or zeros are tightly clustered, as in the case of high-$Q_p$ band-pass filters, the factors $(z_i - z_l)$ become small and consequently the sensitivity becomes large. Hence, the direct-form structures are not recommended for high-order sampled data filters. Instead, second-order structures can be either cascaded or paralleled to realize a high-order transfer function. Such sampled data filter structures are shown in Figure 1.17.

Note that the structure of Figure 1.17(a) is obtained by expressing a high-order $H_d(z)$ as a product of second-order (and first-order for odd-order transfer functions) whereas that of Figure 1.17(b) is obtained by a partial fraction expansion of $H_d(z)$ into second-order and/or first-order transfer functions. The second-order blocks can, in general, have the numerator and denominator of their transfer functions as quadratic expressions and hence, the general second-order transfer functions are called *biquadratic transfer functions* and the blocks are denoted as *biquads*. The cascade approach is of immense practical interest, since the sensitivities of such cascade structures are lower than that of direct-form realizations. In these, there is no interaction between the movement of complex poles of a particular biquad and other biquads. The transfer function of the overall filter is directly affected by the variation in magnitude of any particular second-order filter. As will be seen later, this may be a disadvantage. We will consider the cascade approach in some detail in the following sections.

## 1.6.2 Design of high-order filters by cascading second-order blocks

From the high-order transfer function obtained in the $s$-domain to meet the specifications, the $H_d(z)$ can be obtained by any method described in Section 1.5. The $H_a(s)$ or $H_d(s)$ thus obtained can be paired into quadratic factors, i.e., associating a second-order or first-order numerator with a second-order denominator resulting in

$$H_d(z) = \prod_{i=1}^{N} \frac{N_{id}(z)}{D_{id}(z)} \tag{1.70}$$

It is interesting to note that $N_{id}(z)$ and $D_{id}(z)$ are independent and, as such, any $N_{id}$ can be associated with any $D_{id}$ yielding numerous cascade realizations in the form of Figure 1.17(a) for a given $H_d(z)$. Hence, the designer has the option to

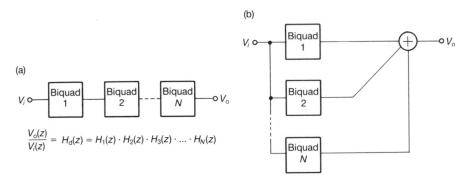

**Fig. 1.17** Realization of Nth-order transfer function using second- and first-order blocks: (a) cascade realization; (b) parallel realization.

choose any of these combinations. But, an optimum one is chosen[*] on the basis of various criteria.

A rule of thumb for achieving optimal dynamic range is to assign the pole-pair with high $Q_p$s with the nearest zeros. This guarantees that the response of each section is as flat as possible in the pass band. Further depending on the design of each biquadratic filter section, whether it contains one or more active elements, each biquad should be designed to have optimal dynamic range. This means that *both (or all) of the active elements inside the biquad block should saturate at the same level.* This requires analysis of the transfer functions of the biquads at all the possible outputs of OAs over the frequency range (zero to infinity), and then determination of the maximum amplitude of the transfer functions. With these maxima known, the components in the individual sections will be adjusted to equalize them. The dynamic range optimization technique will be illustrated in later chapters. For the present, it is intended to present formulas for the evaluation of the maximum magnitude of a second-order digital transfer function.

Before presenting these, we first examine, in the next section, the variety of biquadratic sampled data filter transfer functions that are usually needed in the cascade design of high-order sampled data transfer functions.

---

[*]  We state here, without proof, for which the reader is referred to references cited at the end of this chapter, the criterion for pairing the poles and zeros. Also, the sequence in which the various biquads thus obtained have to be cascaded is of interest. A further point to be mentioned is that the pole–zero pair selection influences the sensitivity also. The reader is referred to the work of Leuder [1.22], Halfin [1.23] and Moschytz [1.24] in this connection. A comprehensive treatment of these topics is also available in Sedra and Brackett [1.25].

## 1.7   General biquadratic sampled data transfer functions

The denominator polynomial of these is of the form

$$D_d(z) = 1 + \alpha z^{-1} + \beta z^{-2} \tag{1.71}$$

The numerator can take a variety of forms, and these are presented in Table 1.4. When we obtain low-pass and band-pass filters using bilinear transformation from analog prototype, the numerators obtained are distinct. There are several other low-pass and band-pass transfer functions as illustrated in Table 1.4. In the notation used in Table 1.4 for low-pass and band-pass transfer functions $LP_{ij}$ or $BP_{ij}$, $i$ indicates the number of $(1 + z^{-1})$ factors, and $j$ the number of $z^{-1}$ factors in the numerator. In the general case, $H_d(z)$ can be written as

$$H_d(z) = \frac{pz^2 - qz + r}{z^2 - uz + v} \tag{1.72}$$

The maximum amplitude of $H_d(z)$ can be evaluated by substituting $z = e^{j\omega T}$ in equation (1.72), and then plotting equation (1.72) against $\omega$. It will be useful to be able to evaluate just the maximum amplitude of equation (1.72), for dynamic range optimization, since often there is no necessity to have the complete plot of the frequency response (see Ananda Mohan *et al.* [1.27]).

For this purpose, we note that from $H_d(z)$ given by equation (1.72), we can obtain an equivalent $s$-domain transfer function $H_a(s)$ through the relation

$$z = \frac{1+s}{1-s} \tag{1.73}$$

Since the maxima of equation (1.72) or $H_a(s)$ will be the same due to the mapping

**Table 1.4**   Generic biquad transfer functions   (*adapted from [1.26], Copyright © 1979 AT&T. All rights reserved. Reprinted with permission*).

| Generic form | Numerator $N_d(z)$ |
|---|---|
| LP20 (bilinear transform) | $k(1 + z^{-1})^2$ |
| LP11 | $kz^{-1}(1 + z^{-1})$ |
| LP10 | $k(1 + z^{-1})$ |
| LP02 | $kz^{-2}$ |
| LP01 | $kz^{-1}$ |
| LP00 | $k$ |
| BP10 (bilinear transform) | $k(1 - z^{-1})(1 + z^{-1})$ |
| BP01 | $kz^{-1}(1 - z^{-1})$ |
| BP00 | $k(1 - z^{-1})$ |
| HP | $k(1 - z^{-1})^2$ |
| LPN | $k(1 + \varepsilon z^{-1} + z^{-2})$, $\varepsilon > \alpha/\sqrt{\beta}$, $\beta > 0$ |
| HPN | $k(1 + \varepsilon z^{-1} + z^{-2})$, $\varepsilon < \alpha/\sqrt{\beta}$, $\beta > 0$ |
| AP | $k(\beta + \alpha z^{-1} + z^{-2})$ |
| General | $k(\gamma + \varepsilon z^{-1} + \delta z^{-2})$ |

property of the bilinear $s \to z$ transformation, it suffices for us to evaluate the maximum of $H_a(s)$, in order to evaluate the maximum of $H_d(z)$. The $H_a(s)$ obtained using equation (1.73) in (1.72) is

$$H_a(s) = \frac{\dfrac{p+q+r}{1+u+v}s^2 + \dfrac{2(p-r)}{1+u+v}s + \dfrac{p-q+r}{1+u+v}}{s^2 + \dfrac{2(1-v)}{1+u+v}s + \dfrac{1-u+v}{1+u+v}} \qquad (1.74a)$$

To evaluate the maximum of $|H_a(j\omega)|$, we rewrite $H_a(s)$ in equation (1.74a) as

$$H_a(s) = \frac{as^2 + bs + c}{s^2 + ds + e} \qquad (1.74b)$$

and, assuming the maximum of $|H_a(j\omega)|$ as $k$, we obtain

$$k^2 = \frac{(c - a\omega^2)^2 + b^2\omega^2}{(e - \omega^2)^2 + d^2\omega^2} \qquad (1.75a)$$

or

$$\omega^4(a^2 - k^2) + \omega^2(b^2 - 2ac + 2k^2e - k^2d^2) + (c^2 - k^2e^2) = 0 \qquad (1.75b)$$

If $\omega$ in equation (1.75b) is to represent a peak, it must have equal roots or its discriminant must be zero. Hence, we obtain, after simplification,

$$k^4(d^4 - 4ed^2) + k^2(4b^2e - 2b^2d^2 - 8ace + 4acd^2 + 4c^2 + 4a^2e^2) + (b^4 - 4b^2ac) = 0 \qquad (1.76)$$

where the maximum amplitude $k$ of $|H_a(j\omega)|$ can be found in terms of $a$, $b$, $c$, $d$ and $e$.

For the bilinear transformation type of band-pass transfer functions ($a = c = 0$), equation (1.75b) gives the mid-band gain as ($b/d$). For bilinear transformation types of low-pass ($a = b = 0$), high-pass ($b = c = 0$), and notch ($b = 0$) transfer functions, since $b = 0$, the last term in equation (1.76) is zero, and $k$ can be immediately obtained from equation (1.76). In the case of all-pass transfer functions, the gain is obtained from equation (1.72) itself as $r$. For general biquadratic transfer functions in the $z$-domain, equation (1.76) can be solved to give $k$. The formulas thus obtained for maximal amplitude $k$ are listed in Table 1.5. The use of Table 1.5 is illustrated in the following example.

*Example 1.10*
Evaluate the maximum magnitude of the $z$-domain transfer function given by

$$H_d(z) = \frac{0.890\,93(z^2 - 1.9922z + 1)}{z^2 - 1.990\,29z + 0.997\,23}$$

and with a sampling frequency of 128 kHz.

**Table 1.5** Formulas for maximal amplitude of second-order digital transfer functions (*adapted from [1.27],* © 1983 IEEE).

1. Band-pass (bilinear)*
   $a = c = 0$
   $$\frac{p - r}{1 - v}$$

2. Low-pass (bilinear)*
   $a = b = 0$
   $$\frac{(p - q + r)(1 + u + v)}{2(1 - v)\sqrt{4v - u^2}}$$

3. High-pass
   $b = c = 0$
   $$\frac{(p + q + r)(1 - u + v)}{2(1 - v)\sqrt{4v - u^2}}$$

4. Notch
   $b = 0$
   $$\frac{2\sqrt{acd^2 + c^2 + a^2 e^2 - 2ace}}{d\sqrt{4e - d^2}}$$

5. All-pass

   $u = \dfrac{q}{r}, \; v = \dfrac{p}{r}$           $r$

6. General biquadratic function

   $$\{[-(4b^2 e - 8ace - 2b^2 d^2 + 4acd^2 + 4a^2 e^2 + 4c^2) \\ \pm \{(4b^2 e - 8ace - 2b^2 d^2 + 4acd^2 + 4c^2 + 4a^2 e^2)^2 \\ - 4(b^4 - 4b^2 ac)(d^4 - 4ed^2)\}^{1/2}] \\ \div 2(d^4 - 4ed^2)\}^{1/2}$$

*   Note that, for all other low-pass and band-pass transfer functions, formula 6 can be used.

For this example, $p = 0.890\,93$, $q = 1.774\,910\,7$, $r = 0.890\,93$, $u = 1.990\,29$, and $v = 0.997\,23$. Hence, the values of $a$, $b$, $c$, $d$ and $e$ are obtained from equation (1.74a) as $a = 0.891\,975\,6$, $b = 0$, $c = 0.001\,742\,76$, $d = 0.001\,389$, and $e = 0.001\,740\,4$.

Then, using formula 4 of Table 1.5, we get $k = 3.417\,717\,6$ or 10.674 dB. Note that the information about the sampling frequency need not be used at all in the above evaluation.

## 1.8 Sensitivity evaluation of second-order sampled data transfer functions

The next aspect of sampled data filter design to be considered is the evaluation of the sensitivity of the transfer function to errors due to inaccurate coefficients. For second-order transfer functions, $\omega_p$ and $Q_p$ are the important parameters to be considered.

## 1.8.1  The bilinear transformation method [1.28]

For a digital transfer function with a denominator given by

$$D_d(z) = z^2 - uz + v \tag{1.77}$$

by applying the bilinear $s \to z$ transformation, we obtain

$$D_a(s) = s^2 + \frac{(1-v)4f_s}{1+u+v} s + 4f_s^2 \frac{(1-u+v)}{(1+u+v)} \tag{1.78a}$$

Identifying equation (1.78a) with

$$D_a(s) = s^2 + \frac{\omega_p}{Q_p} s + \omega_p^2 \tag{1.78b}$$

we obtain the equivalent expressions,

$$\omega_p = \frac{f_s}{\delta} = \sqrt{\frac{(1-u+v)}{(1+u+v)}} \cdot 2f_s \tag{1.79a}$$

and

$$Q_p = \frac{\sqrt{(1+u+v)(1-u+v)}}{2(1-v)} = \frac{1+u+v}{4\delta(1-v)} \tag{1.79b}$$

Thus $\omega_p$ and $Q_p$ are now expressed in terms of the coefficients of $H_d(z)$.

The $\omega_p$ and $Q_p$ values may be dependent on a number of parameters, e.g., $C_1, C_2, \ldots$ It is of interest to evaluate the $\omega_p$- and $Q_p$-sensitivities, called *Schoeffler sensitivities*, to these parameters. The sensitivity of $\omega_p$ to $C_i$ is defined as

$$S_{C_i}^{\omega_p} = \left(\frac{d\omega_p}{dC_i}\right) \cdot \frac{C_i}{\omega_p} = \frac{(d\omega_p/\omega_p)}{(dC_i/C_i)} \tag{1.80a}$$

and, similarly, the sensitivity of $Q_p$ to $C_i$ as

$$S_{C_i}^{Q_p} = \left(\frac{dQ_p}{dC_i}\right) \cdot \frac{C_i}{Q_p} = \frac{(dQ_p/Q_p)}{(dC_i/C_i)} \tag{1.80b}$$

These expressions represent the ratio of percentage change in $\omega_p$ to the percentage change in $C_i$, and the ratio of percentage change in $Q_p$ to the percentage change in $C_i$. For a good design, if a parameter $C_i$ changes by 1%, the performance (as measured by an index such as $\omega_p$ or $Q_p$) must change by 1% or less. Note also that the sensitivities obtained in equations (1.80a) and (1.80b) can have positive or negative values, depending on whether $\omega_p$ or $Q_p$ increases with an increase in $C_i$ or decreases with an increase in $C_i$.

In the above treatment, the sensitivities of $\omega_p$ and $Q_p$ to the various parameters determining $u$ and $v$ have been considered. However, one is interested in the sensitivity of the magnitude of the transfer function to these parameters. Considering a second-order low-pass transfer function

$$H_a(s) = \frac{1}{s^2 + \dfrac{\omega_p}{Q_p} s + \omega_p^2} \qquad (1.81a)$$

the magnitude function is written as

$$|H_a(j\omega)|^2 = \frac{1}{\left[(\omega_p^2 - \omega^2)^2 + \dfrac{\omega^2 \omega_p^2}{Q_p^2}\right]} \qquad (1.81b)$$

Then, the sensitivities of $|H_a(j\omega)|$ to $\omega_p$ and $Q_p$ are evaluated using the definition given in equations (1.80), as

$$S_{\omega_p}^{T_a} = \frac{-\left[2(1 - \gamma^2) + \dfrac{\gamma^2}{Q_p^2}\right]}{\left[(1 - \gamma^2)^2 + \dfrac{\gamma^2}{Q_p^2}\right]} \qquad (1.82a)$$

$$S_{Q_p}^{T_a} = \frac{\dfrac{\gamma^2}{Q_p^2}}{\left[(1 - \gamma^2)^2 + \dfrac{\gamma^2}{Q_p^2}\right]} \qquad (1.82b)$$

where $T_a$ is $|H_a(j\omega)|$ and $\gamma = \omega/\omega_p$. It may be seen from equations (1.82) that both the sensitivities are functions of $\gamma$ and hence universal curves of these sensitivities as functions of $\gamma$ for various values of $Q_p$ are of practical interest. Such curves are shown in Figure 1.18. It can be noted that $S_{\omega_p}^{T_a}$ reaches a maximum of about $Q_p$ at the 3 dB frequencies $\gamma = (1 \pm 1/2Q_p)$. The maximum value of $S_{Q_p}^{T_a}$ is unity. Further, the ratio of $S_{\omega_p}^{T_a}$ to $S_{Q_p}^{T_a}$ is

$$\frac{S_{\omega_p}^{T_a}}{S_{Q_p}^{T_a}} = -\left[1 + 2Q_p^2\left(\frac{1}{\gamma^2} - 1\right)\right] \qquad (1.83)$$

For narrow band approximation, i.e., for frequencies close to resonance,

$$\frac{S_{\omega_p}^{T_a}}{S_{Q_p}^{T_a}} = \left(4 \frac{\Delta\omega}{\omega_p} Q_p^2 - 1\right) \qquad (1.84)$$

Thus, at the 3 dB frequencies, i.e., $\gamma = (1 \pm 1/2Q_p)$, the magnitude of the ratio of sensitivities is $2Q_p$. This result means that the order of magnitude of sensitivities to $\omega_p$ must be kept less than $2Q_p$ times the sensitivities to $Q_p$.

We have so far noted that the sensitivities of $\omega_p$ and $Q_p$ to various coefficients can be determined, and that the sensitivities of the transfer function to $\omega_p$ and $Q_p$

(a)

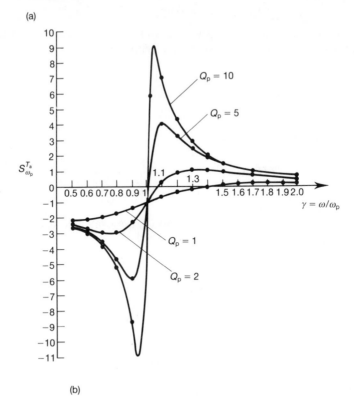

(b)

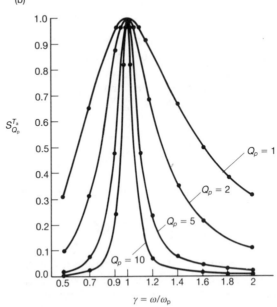

*Fig. 1.18* Universal curves for (a) $S_{\omega_p}^{T_a}$; (b) $S_{Q_p}^{T_a}$ *(adapted from [1.25]).*

can also be determined. The variation of the transfer function due to the variation in various parameters can be expressed as

$$\frac{\Delta\,|\,H_a(j\omega)\,|}{|\,H_a(j\omega)\,|} = \sum_{i=1}^{m} S_{\omega_{p_i}}^{T_a} \left[\frac{\Delta\omega_{p_i}}{\omega_{p_i}}\right] + \sum_{i=1}^{m} S_{Q_{p_i}}^{T_a} \left[\frac{\Delta Q_{p_i}}{Q_{p_i}}\right] \qquad (1.85a)$$

$$\frac{\Delta\omega_p}{\omega_p} = \sum_{i=1}^{k} S_{C_i}^{\omega_p} \frac{\Delta C_i}{C_i} \qquad (1.85b)$$

$$\frac{\Delta Q_p}{Q_p} = \sum_{i=1}^{k} S_{C_i}^{Q_p} \frac{\Delta C_i}{C_i} \qquad (1.85c)$$

Thus $\Delta\,|\,H_a(j\omega)\,|/|\,H_a(j\omega)\,|$ can be used as a criterion for comparing various realizations. Note that the expressions in equations (1.85) consider the total variation due to all the biquads and/or first-order blocks used in the realization. In general, for the evaluation of a given sampled data filter, it suffices to evaluate $\omega_p$- and $Q_p$-sensitivities and to ensure that the $\omega_p$-sensitivities are especially low. For more accurate analysis, statistical behaviour of all the elements used to realize the transfer functions must be taken into account, and be included in the sensitivity analysis. This topic can be explored further in references [1.25, 1.29–1.31].

The above characterization of the problem of sensitivity evaluation is based on bilinear $s \rightarrow z$ transformation through which an $s$-domain transfer function is obtained corresponding to the $z$-domain transfer function to be analyzed. It may be noted that the sampled data filter under consideration has a frequency response of interest up to $f_s/2$, which by virtue of the warping effect of the bilinear transformation, is mapped up to infinite frequencies.

However, the magnitude response and its variations due to any parameter are exactly preserved. Hence, the sensitivity analysis is exact, even though it is carried out in the $s$-domain.

## 1.8.2   Nishihara's method

Nishihara [1.32] has suggested for a band-pass filter realizing a bilinear transformation type transfer function of the form

$$H_d(z) = \frac{g_0(1 - z^{-2})}{1 + g_1 z^{-1} + g_2 z^{-2}} \qquad (1.86)$$

the choice of two parameters, viz., $\cos \omega_0$ and $\tan(\omega_0/2)$ (normalized so that $T = 1$ s) for sensitivity characterization. These two parameters are directly related to the measurable parameters of a band-pass filter: the resonant frequency $\omega_0$ and the bandwidth $\omega_b$. Substituting $z = e^{j\omega}$ we obtain

$$H_d(j\omega) = \frac{j2g_0 \sin \omega}{(1 + g_2)\cos \omega + g_1 + j(1 - g_2)\sin \omega} \qquad (1.87)$$

The maximum of $H_d(j\omega)$ occurs at $\omega_0$ defined by

$$\cos \omega_0 = \frac{-g_1}{1 + g_2} \tag{1.88}$$

and the bandwidth $\omega_b$ is given by

$$\tan \frac{\omega_b}{2} = \frac{1 - g_2}{1 + g_2} \tag{1.89}$$

The next step is to evaluate the sensitivity of $\cos \omega_0$ and $\tan(\omega_b/2)$ to all pertinent parameters. The transfer function sensitivity, expressed as function of $\cos \omega_0$, $\tan(\omega_b/2)$ and their sensitivities to various parameters, can next be evaluated. Thus the only difference between Nishihara's and the bilinear transformation approach studied earlier is the choice of different parameters for the sensitivity characterization. It will now be shown that the two methods are equivalent. In the bilinear transformation method, $\delta$ is the parameter expressing $\omega_p$. As mentioned before, if $\omega_0$ is the *actual* centre frequency of the band-pass filter, then

$$\delta = \frac{f_s}{\omega_p} = \frac{1}{2 \tan \dfrac{\omega_0}{2}} \tag{1.90}$$

(since $f_s = 1$ Hz for $T = 1$ s).

Next, from equations (1.79), we obtain for the given band-pass transfer function,

$$\frac{1}{4\delta^2} = \frac{1 + g_1 + g_2}{1 - g_1 + g_2}$$

$$\frac{1}{4\delta Q_p} = \frac{1 - g_2}{1 - g_1 + g_2} \tag{1.91}$$

From equations (1.90) and (1.91), after simplification, we obtain equation (1.88). Next considering the 3 dB frequencies in the actual response as $\omega_1$ and $\omega_2$, the prewarped frequencies become $2 \tan(\omega_1/2)$ and $2 \tan(\omega_2/2)$. The centre frequency is such that the band edges are geometrically symmetric, i.e., the centre frequency $\omega_p$ is $2\sqrt{\tan(\omega_1/2)\tan(\omega_2/2)}$. Thus,

$$\tan \frac{\omega_b}{2} = \tan \frac{\omega_1 - \omega_2}{2} = \frac{\tan(\omega_1/2) - \tan(\omega_2/2)}{1 + \tan(\omega_1/2)\tan(\omega_2/2)}$$

$$= \frac{\omega_p/2Q_p}{1 + \dfrac{\omega_p^2}{4}} = \frac{1/2\delta Q_p}{1 + \dfrac{1}{4\delta^2}} \tag{1.92}$$

Using equation (1.91) in (1.92), we obtain equation (1.89). It is thus seen that choosing $\delta$ and $Q_p$, or $\tan(\omega_b/2)$ and $\cos \omega_0$, as the parameters for sensitivity evaluation, the two methods will be equivalent. Further, the choice of $\delta$ and $Q_p$ allows concepts in sensitivity characterization of active RC filters to be conveniently applied.

We conclude the discussion on sensitivity evaluation by briefly mentioning the method of Gold and Rader [1.33].

## 1.8.3  The Gold and Rader method

In this method, the pole position sensitivity to the coefficients for some parameters determining the sampled data transfer function is obtained. Consider a pair of complex conjugate poles in the $(R, \theta)$-plane, $z = Re^{\pm j\theta}$, the resulting denominator of the transfer function is

$$D_d(z) = z^2 - (2R \cos \theta)z + R^2 \qquad (1.93a)$$

which is realized in practice as

$$D_d(z) = z^2 - Kz + L \qquad (1.93b)$$

It then follows that

$$R = \sqrt{L} \qquad (1.94a)$$

$$\theta = \cos^{-1} \frac{K}{2\sqrt{L}} \qquad (1.94b)$$

The shifts in $R$ and $\theta$ (assuming these shifts are small) due to inaccurate $K$ and $L$ values are obtained as

$$\Delta R = \frac{1}{2R} \Delta L \qquad (1.95a)$$

$$\Delta \theta = \frac{\Delta L}{2R^2 \tan \theta} - \frac{\Delta K}{2R \sin \theta} \qquad (1.95b)$$

We also note that since $\theta = \omega_r T$, $\Delta \omega_r = \Delta \theta f_s$. Thus, the error in resonant frequency ($\omega_r$) is proportional to the sampling rate. Secondly, from equation (1.95b), it is observed that the errors are large when $\theta$ is small, i.e., for smaller $\omega_p$ compared to the sampling frequency.

The reader is referred to literature for low-sensitivity digital filter structures.

In the above method, the shifts in pole and zero positions are obtained and these can be used to compare the various structures.

## 1.9  Conclusions

This chapter discussed MOS technology, and showed how it can be utilized to implement SC filters in practice. Since SC networks are basically sampled data structures, the different methods of obtaining the transfer functions starting from an analog function were discussed and illustrative examples worked out. Also, the methods of evaluating various sensitivities of sampled data second-order transfer functions were discussed.

## 1.10  Exercises

1.1 Compare the amplitude and phase variation with frequency introduced by a negative real pole, and a complex pole.
(*Hint*: For a pole at $s_1$, the voltage gain of any network is $H(s) = 1/(s - s_1)$. Evaluate the magnitude and phase of $H(s)$, when $s_1$ is a negative real number and a left-half-plane complex pole.)

1.2 Derive the $H_d(z)$ corresponding to an $s$-domain transfer function

$$H_a(s) = \frac{s^2}{s^2 + 0.1s + 1}$$

with $\omega_p$ *denormalized* to 1000 Hz and sampling frequency of 20 kHz. Use impulse invariance method. Discuss whether it realizes the desired specifications by plotting the frequency responses of sampled data filter as well as the specifications $H_a(s)$.

1.3 *Step-invariance and sinusoidal invariance methods of design of sampled data filters* [1.34]: In these methods, the input for the continuous-time filters is assumed to be a step or a sinusoid. The resulting time response of the continuous-time system is mapped onto the $z$-domain. The design procedure is similar to that of the impulse invariance method. Demonstrate the design procedure considering a second-order low-pass filter example.

1.4 Design a notch-filter [1.35] using the bilinear transformation for given specifications, viz., notch frequency and bandwidth. Derive a transfer function with a notch frequency of 1250 Hz and 3 dB bandwidth of 100 Hz. Choose a sampling frequency of 10 kHz.
(*Hint*: Defining $a_2 = [1 - \tan(\Omega T/2)]/[1 + \tan(\Omega T/2)]$ and $a_1 = 2\cos(\omega_0 T)/[1 + \tan(\Omega T/2)]$, where $\Omega$ is the bandwidth in radians and $\omega_0$ is the notch frequency in radians, the $z$-domain transfer function is

$$H_d(z) = \frac{(z^{-2} + 1) - 2a_1 z^{-1} + a_2(z^{-2} + 1)}{1 - a_1 z^{-1} + a_2 z^{-2}}$$

Prove this result and then use it.)

1.5 *Frequency transformations in the s-domain*: If we have a low-pass filter with a normalized cut-off frequency $\omega_c$, the high-pass, band-pass and band-reject filters can be obtained by frequency transformations. A frequency transformation $S = F(s)$ is sought such that, given a low-pass transfer function $H_a(s)$, we obtain $H\{F(s)\}$ by the above transformation. The low-pass to high-pass transformation is $S \to \omega_c/s$. Thus the low-pass response shown in Figure 1.19(a) maps to that in Figure 1.19(b). Derive similar transformations for

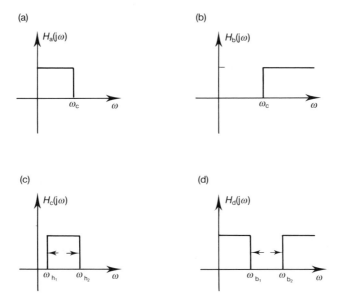

*Fig. 1.19* Frequency transformations in the s-domain.

band-pass and band-reject filters [1.37], with the requirements in Figures 1.19(c) and 1.19(d).

(*Hint*: Choose a general transformation [1.36]

$$S = F(s) = \frac{as^2 + bs + c}{ds + e}$$

and map the critical frequencies to obtain *a*, *b*, *c*, *d* and *e*.) Verify that the responses of band-pass and band-reject filters designed using frequency transformations are only geometrically symmetric.

1.6 *Digital frequency transformations*: Suppose that we have a digital low-pass filter with cut-off frequency $\theta_p = \omega_c T$. We would like to derive a high-pass filter with a cut-off frequency $\theta_c$. Then the transformation is

$$z^{-1} \rightarrow \frac{z^{-1} + \alpha}{1 + \alpha z^{-1}}$$

where

$$\alpha = \frac{-\cos\left[(\theta_p + \theta_c)/2\right]}{\cos\left[(\theta_p - \theta_c)/2\right]}$$

Derive the above result. Also, derive the transformations for band-pass and band-reject types given below.

*Band-pass*

$$z^{-1} \rightarrow -\frac{z^{-2} - \left(\dfrac{2\alpha\beta}{\beta + 1}\right)z^{-1} + \dfrac{\beta - 1}{\beta + 1}}{\left(\dfrac{\beta - 1}{\beta + 1}\right)z^{-2} - \left(\dfrac{2\alpha\beta}{\beta + 1}\right)z^{-1} + 1}$$

where

$$\alpha = \cos\theta_0 = \frac{\cos[(\theta_u + \theta_1)/2]}{\cos[(\theta_u - \theta_1)/2]}$$

$$\beta = \cos[(\theta_u - \theta_1)/2]\tan(\theta_p/2)$$

*Band-reject*

$$z^{-1} \rightarrow -\frac{z^{-2} - \left(\dfrac{2\alpha}{1 + \beta}\right)z^{-1} + \dfrac{1 - \beta}{1 + \beta}}{\left(\dfrac{1 - \beta}{1 + \beta}\right)z^{-2} - \left(\dfrac{2\alpha}{1 + \beta}\right)z^{-1} + 1}$$

where

$$\alpha = \cos\theta_0,$$

$$\beta = \tan[(\theta_u - \theta_1)/2]\tan(\theta_p/2)$$

Note that $\theta_u$, $\theta_1$ determine $\theta_0$, i.e., all three cannot be fixed independently, and $\theta_u = \omega_u T$, $\theta_1 = \omega_1 T$ are the upper and lower band edges with $\theta_0$ as the centre frequency.

1.7 *Digital ladder filters*: Cascade and direct-form realizations have been described earlier. Digital ladder filters, introduced in [1.38], are based on continued fraction expansion. In this method, if $H_d(z) = N_d(z)/D_d(z)$ is desired, then $1/D_d(z)$ is realized as a continued fraction in $z$ about $\infty$, and the realization is shown in Figure 1.20. Derive the transfer function. Note also that by a

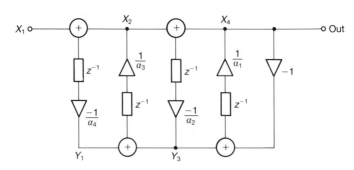

**Fig. 1.20** A digital ladder filter.

weighted sum of all the outputs $Y_1$, $X_2$, $Y_3$ and $X_4$ and the input $X_1$, $N_d(z)/D_d(z)$ can be realized.

1.8 The zero-order hold circuit is shown in Figure 1.21. Show that its transfer function is $H_h(s) = (1 - e^{-sT})/s$. Evaluate the magnitude of $H_h(s)$. Find the maximum correction required at 3.4 kHz for a sampling frequency of 8 kHz. How do you equalize the correction obtained?

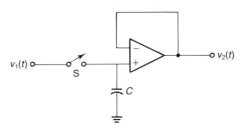

*Fig. 1.21* A zero-order hold circuit.

1.9 *Sensitivity formulas*: Sensitivity evaluation using equations (1.80) requires differentiation. Thus, by deriving a certain set of basic formulas, the labour involved in sensitivity evaluation of complex expressions can be reduced. Some useful algebraic relationships are listed below:

$$S_z^{x/y} = S_z^x - S_z^y$$
$$S_z^{x^n} = nS_z^x$$
$$S_x^{uv} = S_x^u + S_x^v$$
$$S_{1/x}^u = -S_x^u$$
$$S_x^{ku} = S_x^u, \text{ if } k \text{ is a constant.}$$

Verify the above relationships. Use them to evaluate the sensitivities of $\omega_p$ and $Q_p$ to $C_1$, $C_2$, $R_1$ and $R_2$:

$$\omega_p = \frac{1}{\sqrt{C_1 C_2 R_1 R_2}} \quad \text{and} \quad Q_p = \frac{1}{2}\sqrt{\frac{C_2 R_2}{C_1 R_1}}$$

1.10 *Low-sensitivity digital filter structure* [1.39]: For second-order digital filters with poles very near the unit circle, it is convenient to realize $b$ in the denominator of the transfer function

$$D_d(z) = z^2 - az + b$$

as $(1 - b_1)$. Similarly, for small angles $\theta$, $a$ will be approximately 2. Hence, it is convenient to realize the term $a$ as $(2 - a_1)$. Thus the resulting $D_d(z)$ is

$$D_d'(z) = z^2 - (2 - a_1)z + (2 - b_1)$$

Evaluate the $R$, $\theta$ sensitivities of the new structure using the Gold and Rader method (Section 1.8.3). Derive a structure to realize the above $D_d'(z)$.

Alternative structures could be obtained by shifting the origin of the $z$-domain to $z = 1$ by a linear shift of coordinates, e.g., $\hat{z} = z + 1$. (Note that for $\omega T \ll 1$, $\hat{z} = sT$). Suggest a structure to realize $H_d(z)$ using the above transformation.

(*Hint*: Realize $H_d(\hat{z})$ using delays in the $\hat{z}$-domain, after obtaining $H_d(\hat{z})$ by substitution for $\hat{z}$ in $H_d(z)$.)

1.11 The maximum amplitude $k$ of $H_d(z)$ in equation (1.72) can also be found directly without recourse to bilinear transformation [1.40]. Prove that $k$ is given by

$$k^4(d_0^2 - 4e_0) + k^2(4c_0 + 4a_0e_0 - 2b_0d_0) + (b_0^2 - 4a_0c_0) = 0$$

where

$$a_0 = \frac{pr}{v}$$

$$b_0 = \frac{(p + r)q}{2v}$$

$$c_0 = \frac{(p - r)^2 + q^2}{4v}$$

$$d_0 = \frac{(1 + v)u}{2v}$$

$$e_0 = \frac{(1 - v)^2 + u^2}{4v}$$

## 1.11  References

[1.1] A.G. Milnes, *Semiconductor Devices and Integrated Electronics*, Van Nostrand Reinhold Co., 1972.

[1.2] W.M. Penny and L. Lau (eds), *MOS Integrated Circuits: Theory, Fabrication, Design and Systems Application of MOS LSI*, Van Nostrand Reinhold Co., 1972.

[1.3] H.R. Camenzind, *Electronic Integrated System Design*, Van Nostrand Reinhold Co., 1972.

[1.4] A.B. Glaser and G.E. Subak-Sharpe, *Integrated Circuit Engineering: Design, Fabrication and Applications*, Addison-Wesley Publishing Co., 1977.

[1.5] D.A. Hodges, P.R. Gray and R.W. Broderson, Potential of MOS technologies for analog integrated circuits, *IEEE Journal of Solid-State Circuits*, SC-13, 3, 285–94, June 1978.

[1.6] D.J. Allstot and W.C. Black., Jr, Technological design considerations for monolithic MOS switched-capacitor filtering system, *Proc. IEEE*, 71, 8, 967–86, August 1983.

[1.7] J.L. McCreary and P.R. Gray, ALL-MOS charge redistribution analog-to-digital conversion techniques—Part I, *IEEE Journal of Solid-State Circuits*, SC-10, 6, 371–9, December 1975.

[1.8] G. Smarandoiu, D.A. Hodges, P.R. Gray and G.F. Landsburg, CMOS pulse-code-modulation voice codec, *IEEE Journal of Solid-State Circuits*, SC-13, 4, 504–10, August 1978.

[1.9] J.L. McCreary, Matching properties, and voltage and temperature dependence of MOS capacitors, *IEEE Journal of Solid-State Circuits*, SC-1, 6, 608–16, December 1981.

[1.10] J.B. Shyu, G.C. Temes and K. Yao, Random errors in MOS capacitors, *IEEE Journal of Solid-State Circuits*, SC-17, 6, 1070–5, December 1982.

[1.11] J.B. Shyu, G.C. Temes and F. Krummenacher, Random error effects in matched MOS capacitors and current sources, *IEEE Journal of Solid-State Circuits*, SC-19, 6, 948–55, December 1984.

[1.12] D.A. Hodges, Analog switches and passive elements in MOSLSI. In *Analog MOS Integrated Circuits*, IEEE Press, 1980, 14–18.

[1.13] A.S. Sedra and K.C. Smith, *Microelectronic Circuits*, Holt, Rinehart and Winston, 1987, Chapter 7.

[1.14] B.J. Hosticka and G.S. Moschytz, Practical design of switched-capacitor networks for integrated circuit implementation, *IEE Journal on Electronics, Circuits and Systems*, 3, 76–88, 1979.

[1.15] C.M. Rader and B. Gold, Digital filter design techniques in the frequency-domain, *Proc. IEEE*, 55, 2, 149–71, February 1967.

[1.16] L.T. Bruton, Low-sensitivity digital ladder filters, *IEEE Transactions on Circuits and Systems*, CAS-22, 3, 168–76, March 1975.

[1.17] M.S. Lee and C. Chang, Switched-capacitor filters using the LDI and bilinear trans-formations, *IEEE Transactions on Circuits and Systems*, CAS-28, 4, 265–70, April 1981.

[1.18] M.S. Lee and C. Chang, Low sensitivity switched-capacitor ladder filters, *IEEE Transactions on Circuits and Systems*, CAS-27, 6, 475–80, June 1980.

[1.19] R.D. Davis and T.N. Trick, Optimum design of low-pass switched-capacitor ladder filters, *IEEE Transactions on Circuits and Systems*, CAS-27, 6, 522–7, June 1980.

[1.20] T.C. Choi and R.W. Brodersen, Considerations for high-frequency switched-capacitor ladder filters, *IEEE Transactions on Circuits and Systems*, CAS-27, 6, 545–52, June 1980.

[1.21] D. Herbst, B. Hoefflinger, K. Schumacher, R. Schweer, A. Fettweis, K.A. Owenier and J. Pandel, MOS switched-capacitor filters with reduced number of operational amplifiers, *IEEE Journal of Solid-State Circuits*, SC-14, 6, 1010–19, December 1979.

[1.22] E. Leuder, A decomposition of a transfer function minimizing distortion and in-band losses, *Bell System Technical Journal*, 49, 455–69, 1970.

[1.23] S. Halfin, Simultaneous determination of ordering and amplification of cascaded subsystems, *Journal of Optimization Theory and Applications*, 6, 356–60, 1970.

[1.24] G.S. Moschytz, *Linear Integrated Networks: Design*, Van Nostrand Reinhold Co., 1975.

[1.25] A.S. Sedra and P.O. Brackett, *Filter Theory and Design—Active and Passive*, Matrix Publishers, 1978.

[1.26] P.E. Fleischer and K.R. Laker, A family of active switched-capacitor biquad building blocks, *Bell System Technical Journal*, 58, 2235–69, December 1979.

[1.27] P.V. Ananda Mohan, V. Ramachandran and M.N.S. Swamy, Formulas for dynamic range evaluation of second-order switched-capacitor filters, *IEEE Transactions on Circuits and Systems*, CAS-30, 5, 321, May 1983.

[1.28] G. Szentirmai and G.C. Temes, Switched-capacitor building blocks, *IEEE Transactions on Circuits and Systems*, CAS-27, 6, 492–501, June 1980.

[1.29] L.T. Bruton, *RC Active Circuits: Theory and Design*, Prentice Hall, 1980.

[1.30] P.E. Fleischer, Sensitivity minimization in a single amplifier biquad circuit, *IEEE Transactions on Circuits and Systems*, CAS-23, 1, 45–55, January 1976.

[1.31] A.L. Rosenblum and M.S. Ghausi, Multiparameter sensitivity in active RC networks, *IEEE Transactions on Circuits and Systems*, CAS-18, 6, 592–9, November 1971.

[1.32] A. Nishihara, Characterization of second-order discrete-time filters, *Electronics Letters*, 19, 3, 84–6, 3 February 1983.

[1.33] B. Gold and C.M. Rader, *Digital Processing of Signals*, McGraw-Hill, 1969.

[1.34] A. Antoniou, *Digital Filters: Analysis and Design*, McGraw-Hill, 1979.

[1.35] K. Hirano, S. Nishimura and S.K. Mitra, Design of digital notch filters, *IEEE Transactions on Circuits and Systems*, CAS-21, 4, 540–6, July 1974.

[1.36] D.E. Johnson, *Introduction to Filter Theory*, Prentice Hall, 1976.

[1.37] H.Y.Z. Lam, *Analog/Digital Filters: Design and Realization*, Prentice Hall, 1979.

[1.38] S.K. Mitra and R.J. Sherwood, Digital ladder networks, *IEEE Transactions on Audio and Electro-acoustics*, AU-21, 1, 30–6, February 1973.

[1.39] R.C. Agarwal and C.S. Burrus, New recursive digital filter structures having very low sensitivity and round-off noise, *IEEE Transactions on Circuits and Systems*, CAS-22, 12, 921–7, December 1975.

## 1.11.1 Further reading

[1.40] P.V. Ananda Mohan, Comments on 'Calculation of $L_\infty$ norms for scaling second-order state-space filter sections', *IEEE Transactions on Circuits and Systems*, CAS-36, 2, 310–11, February 1989.

[1.41] M.S. Ghausi, Analog active filters, *IEEE Transactions on Circuits and Systems*, CAS-31, 1, 13–31, January 1984.

[1.42] E.A. Vittoz, The design of high performance analog circuits on digital MOS chips, *IEEE Journal of Solid-State Circuits*, SC-20, 3, 657–65, June 1985.

[1.43] J.Y. Chen, CMOS—the emerging VLSI technology, *IEEE Circuits and Devices Magazine*, 2, 2, 16–31, March 1986.

[1.44] Y.P. Tsividis, Analog MOS integrated circuits—certain new ideas, trends and obstacles, *IEEE Journal of Solid-State Circuits*, SC-22, 3, 317–21, June 1987.

[1.45] C.R.W. Campbell and K.M. Reineck, A pole/zero ordering procedure in SC filter design, *IEEE Transactions on Circuits and Systems*, CAS-31, 9, 821–5, September 1984.

[1.46] J. Vlach and E. Christen, Poles, zeros and their sensitivities in switched-capacitor networks, *IEEE Transactions on Circuits and Systems*, CAS-32, 3, 279–84, March 1985.

[1.47] R.E. Bogner and A.G. Constantinides, *Introduction to Digital Filtering*, Wiley, 1975.

[1.48] H.J. Blinchikoff and A.I. Zvervev, *Filtering in the Time and Frequency Domain*, Wiley, 1976.

[1.49] S. Signell, On selectivity properties of discrete-time linear networks, *IEEE Transactions on Circuits and Systems*, CAS-31, 3, 275–80, March 1984.

[1.50] P.V. Ananda Mohan, Analysis of discrete-time filters, *IEEE Transactions on Education*, ED-32, 398–401, August 1989.

# Analysis of switched capacitor networks

In this chapter, methods for the analysis of networks consisting of switches, capacitors and OAs will be developed. These analytical methods are useful for the determination of transfer functions, input and output impedances, etc., of complex SC networks. These only cater for frequency-domain evaluation. The analytical approach to be considered can also be designated the *equivalent circuit method*, since *m*-phase SC networks can be modified via the *equivalent circuit* formulation to 2*m*-port networks in the *z*-domain from which the network parameters can be evaluated.

A network containing capacitors and only periodically operated switches will be considered first. The clock frequency used to operate the switches will have at least two phases, as shown in Figure 2.1(a). The discussion will be initially limited to the simple case of a two-phase clock. The two phases in the clock will be designated as

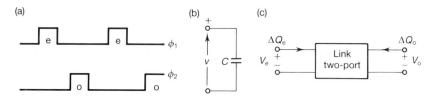

*Fig. 2.1* (a) The even and odd phases of a clock; (b) a two-terminal capacitor $C$; (c) link two-port interconnecting the 'even' and 'odd' phases.

'even' and 'odd', respectively. The switches in the SC networks are operated (i.e., closed or opened) in either the 'even' or 'odd' phases of the clock. (The even and odd phases may also be designated as the conventional $\phi_1$ and $\phi_2$ clock phases, respectively.) Consequently, in the even phase, the circuit structure is dependent on the manner in which the switches and capacitors are interconnected. Similarly, in the odd phase, the topology of the circuit is different. Thus, an SC network may be viewed as two sub-networks in each phase of the clock, depending on the inter-connection of switches and capacitors. However, these two sub-networks are not independent. This is by virtue of the energy storage property of the capacitors. Unless a capacitor is discharged, it retains the charge it has acquired in the previous phase (i.e., the previous half-cycle for a two-phase clock). As a consequence, the two circuit structures corresponding to the two phases of the clock are 'linked' by virtue of the 'memory' of the capacitors. The equivalent circuit of an SC network should represent the ideas developed thus far. In addition, the time delay between the two phases of the clock should also be taken into account. The equivalent circuit of a simple capacitor will be derived first.

## 2.1 Equivalent circuit of a capacitor in an SC network using the link two-port concept [2.1–2.3, 1.14]

The first step in deriving the equivalent circuit of a capacitor in an SC network (Figure 2.1(b)) is to write the charge conservation equation. We denote the voltages across the capacitor at the time instant $nT$ as $v(nT)$ and at an immediately previous instant as $v((n - \frac{1}{2})T)$. Then, the charge flowing into the capacitor at the instant $nT$ can be written as

$$\Delta q_{nT} = C\{v(nT) - v((n - \tfrac{1}{2})T)\} \qquad (2.1a)$$

Similarly, the charge that has flown in the previous half-cycle (i.e., between the ins-tants $(n - \frac{1}{2})T$ and $(n - 1)T$), is

$$\Delta q_{(n-1/2)T} = C\{v((n - \tfrac{1}{2})T) - v((n - 1)T)\} \qquad (2.1b)$$

(Note that the symbol $\Delta$ has been used to represent the incremental charge flow.) For convenience, all the instants $nT, (n - 1)T, (n - 2)T, \ldots$ shall be denoted as 'even', and all the instants $(n - \frac{1}{2})T, (n - \frac{3}{2})T, (n - \frac{5}{2})T, \ldots$ shall be denoted as 'odd'. Accordingly, suffixes 'e' and 'o' will be added to represent the charges and the voltages at even and odd instants, respectively. Equations (2.1) are modified to

$$\Delta q_e = C\{v_e(nT) - v_o((n - \tfrac{1}{2})T)\} \tag{2.2a}$$

$$\Delta q_o = C\{v_o((n - \tfrac{1}{2})T) - v_e((n - 1)T)\} \tag{2.2b}$$

Noting that the voltages in the 'odd' phases of the clock are delayed by a half-cycle with respect to the 'even' phases of the clock and taking the $z$-transform of both sides of equations (2.2), we obtain

$$\Delta Q_e(z) = C\{V_e(z) - z^{-1/2}V_o(z)\} \tag{2.3a}$$

$$z^{-1/2}\,\Delta Q_o(z) = C\{z^{-1/2}V_o(z) - z^{-1}V_e(z)\} \tag{2.3b}$$

Equations (2.3) can be rewritten in matrix form as

$$\begin{bmatrix} \Delta Q_e(z) \\ \Delta Q_o(z) \end{bmatrix} = \begin{bmatrix} C & -Cz^{-1/2} \\ -Cz^{-1/2} & C \end{bmatrix} \begin{bmatrix} V_e(z) \\ V_o(z) \end{bmatrix} \tag{2.4}$$

Equation (2.4) represents a two-port equivalent of a capacitor in an SC network. It is called a *link two-port* (LTP), since it 'links' the voltages across the capacitor and the charges flowing into the capacitor in the even and odd phases. A link two-port is shown in Figure 2.1(c). Equation (2.4) is a capacitance matrix representation. (Note that, strictly speaking, it is an incremental charge-voltage matrix.) A transmission matrix representation can be readily obtained as

$$\begin{bmatrix} V_e \\ \Delta Q_e \end{bmatrix} = \begin{bmatrix} z^{1/2} & \dfrac{z^{1/2}}{C} \\ C(1 - z^{-1})z^{1/2} & z^{1/2} \end{bmatrix} \begin{bmatrix} V_o \\ -\Delta Q_o \end{bmatrix} \tag{2.5}$$

Equations (2.4) and (2.5) characterize the link two-port. Using the concept of the link two-port to represent the coupling between the two time-invariant even and odd networks, any SC network can be represented by an equivalent circuit. In the following example, we consider a grounded capacitor and a series (floating) capacitor and derive their equivalent circuits.

*Example 2.1*
Derive the equivalent circuits of a grounded capacitor and a floating capacitor.

Consider the grounded capacitor of Figure 2.2(a). The behaviour of this capacitor is described by

$$V_{1e} = V_{2e} \tag{2.6a}$$

$$V_{1o} = V_{2o} \tag{2.6b}$$

These are in addition to the relationships given by equation (2.4) between the voltage across the capacitor and the charge flowing into it in each phase. Thus, equations

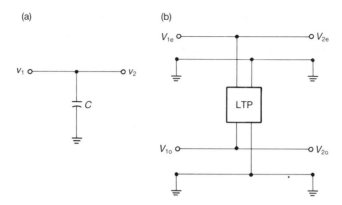

*Fig. 2.2* (a) A grounded capacitor; (b) its equivalent circuit.

(2.6a) and (2.6b) can be realized by a 'through' connection between input and output in both the phases. The resulting equivalent circuit is shown in Figure 2.2(b).

The floating capacitor (Figure 2.3(a)) will be considered next. In this case, in addition to the relationships between the charge flowing into the capacitor and the voltage across the capacitor in each phase given by equation (2.4), it is to be ensured that the charge entering the capacitor in each phase at one terminal must leave at the other terminal, i.e.,

$$\Delta Q_{1e} = \Delta Q_{2e} \tag{2.7a}$$

$$\Delta Q_{1o} = \Delta Q_{2o} \tag{2.7b}$$

This is satisfied by using 'ideal transformers', as shown in Figure 2.3(b).

The foregoing shows how to derive equivalent circuits for grounded and floating capacitors. In the next example, these will be used to analyze a simple SC network.

*Example 2.2*
Derive the equivalent circuit of the passive SC network of Figure 2.4.

It is first noted that the network of Figure 2.4(a) can be considered to be a cascade connection of four sub-networks as shown separated by dotted lines. Further, the equivalent circuits of a grounded capacitor have already been obtained. Using this together with the two time-invariant networks in both the phases, the complete equivalent circuit is obtained. These steps are summarized in Figure 2.4. In the top portion of Figure 2.4(b) the even phase interconnections between the various elements are shown, and in the bottom portion of Figure 2.4(b) those of the odd phase are shown. Next, the two time-invariant networks are interconnected by LTPs representing the behaviour of capacitors. The resulting network can be simplified and seen to be a cascade of four LTPs as shown in Figure 2.4(c). Since the

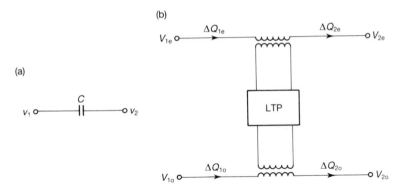

Fig. 2.3   (a) A floating capacitor;   (b) its equivalent circuit.

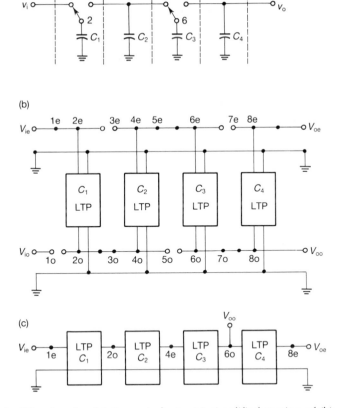

Fig. 2.4   (a) An SC network;   (b) its equivalent;   (c) simplified version of (b).

transmission matrix of the LTP is known, the circuit of Figure 2.4(c) can be analyzed. The reader is urged to derive the transfer function of the network of Figure 2.4(c) in Exercise 2.2.

The above discussion has described the LTP in terms of the capacitance matrix or the *ABCD* matrix. It is possible to represent it, however, by physical elements such as resistances, capacitances, etc. Such an equivalent circuit is shown in Figure 2.5. This is obtained by rewriting equations (2.3) as

$$\Delta Q_e(z) = C(1 - z^{-1/2})V_e(z) + Cz^{-1/2}[V_e(z) - V_o(z)] \tag{2.8a}$$

$$\Delta Q_o(z) = C(1 - z^{-1/2})V_o(z) + Cz^{-1/2}[V_o(z) - V_e(z)] \tag{2.8b}$$

The reader can verify that these equations are realized by the equivalent circuit of Figure 2.5. The network of Figure 2.5 can be substituted for the LTP in the equivalent circuits of the grounded and floating capacitors in Figures 2.2 and 2.3. In the following example, some additional results will be derived.

*Example 2.3*
Analyze the SC network of Figure 2.6(a).

Note that in Figure 2.6(a), the letters e and o indicate the phase in which the switches operate. Following the procedure illustrated in Example 2.2 and using the LTP concept, the equivalent circuits as shown in Figure 2.6(b) are obtained, which can be simplified to that in Figure 2.6(d).
    An examination of Figure 2.6(b) shows that the capacitor is discharged in every odd period, and in the even period the charge flowing is related to the voltage across the branch by[*]

$$\Delta Q_e = CV_e \tag{2.9a}$$

Thus the equivalent of Figure 2.6(d) can be obtained. It is of interest to see the effect

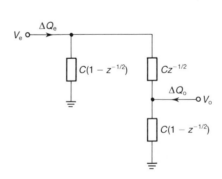

*Fig. 2.5*  Equivalent circuit of an LTP.

[*]  For simplicity hereafter, the z-transform $V_x(z)$ is represented by $V_x$.

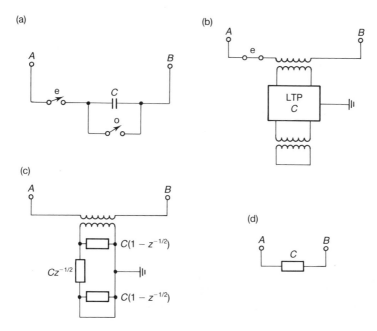

*Fig. 2.6* (a) A series-switched capacitor;  (b)–(d) its equivalent circuits.

of discharging the capacitor once in every clock period. Equation (2.9a) relates the charge flowing in one half-cycle to the voltage across the branch. In the other half-cycle, there is no charge flowing in the branch. Considering the period of a clock cycle as $T$, the net charge flowing in a cycle is the same as in equation (2.9a) and consequently the current[*] flowing in the branch is simply

$$I = \frac{\Delta Q_e + \Delta Q_o}{T} = \frac{C}{T} V_e = \frac{V_e}{R_{eq}} \qquad (2.9b)$$

In equation (2.9b), it can be identified that an equivalent resistance of value $(T/C)$ is realized. Hence, the branch in Figure 2.6 can be used to realize a series-switched capacitor type resistor of value $(T/C)$.

The next example considers the case where a capacitor is connected in series with a switch.

*Example 2.4*
Analyze the series-switched capacitor shown in Figure 2.7(a).

---

[*]  In continuous time circuits, it is known that $i = dq/dt$; whence in discrete-time circuits, $i$ can be written as $\Delta q/T$, where $\Delta q$ is the net charge flowing during a clock cycle period of $T$.

SC NETWORKS

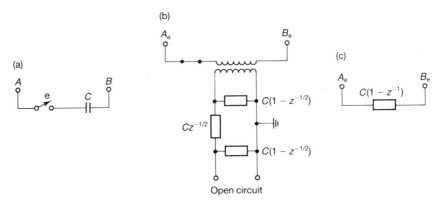

Fig. 2.7 (a) A capacitor in series with a switch; (b), (c) its z-domain equivalents.

Proceeding in a similar manner to Example 2.3, the equivalent circuit of a capacitor in series with the switch is obtained as shown in Figure 2.7(b). The branch impedance (note that in Example 2.3, the idea of current has been introduced and hence all conventional concepts can be used) exists in the even phase only, with a value $[T/C(1 - z^{-1})]$. It is noted that the impedance of the branch in Figure 2.7(a) obtained above is the impedance of a capacitor in the $p$-domain, with

$$p = \frac{(1 - z^{-1})}{T} \tag{2.10a}$$

Thus, we have

$$\frac{T}{C(1 - z^{-1})} = \frac{1}{pC} \tag{2.10b}$$

Thus, it is shown that a capacitor operating only in one phase and disconnected from the circuit in another phase represents a capacitor in the $p$-domain.

It is thus seen that the concept of LTP could be used to derive $z$-domain equivalent circuits. The resulting network has capacitive elements from which the transfer functions, etc., can be derived. As an illustration, consider the network of Figure 2.8(a) and its equivalent circuits shown in Figures 2.8(b) and 2.8(c).

The first observation that can be made is that there are two inputs, one in each phase, and two outputs, one in each phase. Consequently, there are four transfer functions for two-phase SC circuits. The four transfer functions are

$$\frac{V_{oe}}{V_{ie}}, \frac{V_{oo}}{V_{ie}}, \frac{V_{oe}}{V_{io}}, \frac{V_{oo}}{V_{io}}$$

The first suffix designates whether the various voltages pertain to the input or the output and the second designates the phase in which it is measured. For the

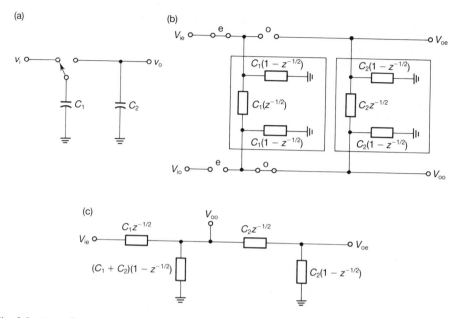

*Fig. 2.8* (a) A first-order low-pass SC network; (b), (c) its equivalent circuits.

network of Figure 2.8(c), by considering it as a all-capacitive network, the transfer functions

$$\frac{V_{oo}}{V_{io}}, \frac{V_{oe}}{V_{io}}$$

do not exist, because, in the odd input phase, the input of the circuit is disconnected from the source. The other two transfer functions are

$$\frac{V_{oo}}{V_{ie}} = \frac{C_1 z^{-1/2}}{C_1 + C_2(1 - z^{-1})} \qquad (2.11a)$$

$$\frac{V_{oe}}{V_{ie}} = \frac{C_1 z^{-1}}{C_1 + C_2(1 - z^{-1})} \qquad (2.11b)$$

It is also noted that $V_{oe} = z^{-1/2} V_{oo}$, which means that the output in the odd phase is held over a clock period. (Alternatively, the output in the even phase is just the output in the odd phase delayed by a half clock cycle.) The network of Figure 2.8(a) will be studied further in Chapter 3. (It exhibits the behaviour of first-order passive low-pass RC network.) It is thus seen that SC networks enjoy considerable flexibility and can provide a number of transfer functions depending on the manner in which input is provided and output is taken.

## 2.2 The operational amplifier in SC networks

The discussion above has considered the analysis of passive networks. The OAs in SC networks can be seen to operate in both phases of the clock, depending on the manner in which their inputs and outputs are connected. Thus the equivalent circuit of the OA in SC network will comprise two OAs, one in each phase of the clock, as shown in Figure 2.9. The analysis methods of such active SC networks using OAs shall proceed with the replacement of passive networks by their SC equivalents and each OA by two OAs operating during the two phases of the clock. These ideas are illustrated in the following.

*Example 2.5*
Derive the transfer function of active SC network of Figure 2.10(a).

The passive portions of the network of Figure 2.10(a) can be replaced by their $z$-domain equivalents first, as shown in Figure 2.10(b). The OAs in the even and odd phases are then embedded into the equivalent circuit to result in the complete equivalent circuit of Figure 2.10(b). Some simplification of the network of Figure 2.10(b) is possible. The grounded impedances at the input terminal $V_{ie}$ and the virtual ground of OA $A_o$ are of no significance; hence they can be open-circuited. It is observed next that the current through AB in the secondary of the transformer in the even path (in the equivalent circuit of $C_2$) is zero, since it drives the input terminal of an OA. Thus, the simplified equivalent circuit of Figure 2.10(c) is obtained, which can be further simplified as shown in Figure 2.10(d). Since the elements in the equivalent circuit of Figure 2.10(d) are capacitive, the transfer function can be obtained as

$$\frac{V_{oo}}{V_{ie}} = \frac{C_1}{C_2} \cdot \frac{z^{-1/2}}{(1 - z^{-1})} \tag{2.12}$$

which is the transfer function of a LDI integrator $[s \rightarrow (1/T)(1 - z^{-1})/z^{-1/2}]$. It is of interest to find the output in the even phase. Referring to Figure 2.10(c), it is

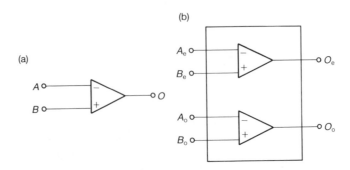

(b)

(a)

*Fig. 2.9*  (a) The OA in an SC network;  (b) its equivalent circuit in the z-domain.

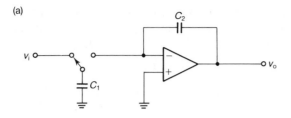

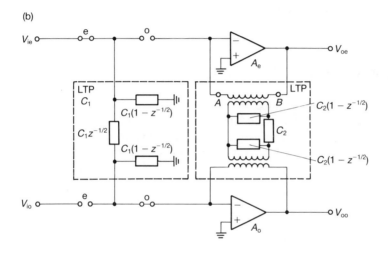

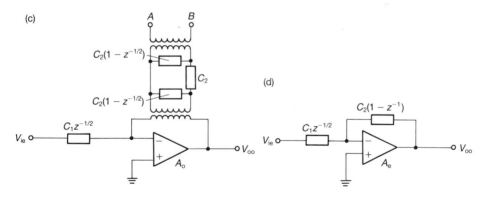

*Fig. 2.10* (a) An active SC network; (b)–(d) its equivalent circuits.

noted that since the terminal $A$ is at virtual ground, the output at $B$ is the open-circuit output voltage of the LTP of $C_2$. Since the voltage transfer function of an LTP open-circuited at the output port is $z^{-1/2}$, it follows that

$$V_{oe} = z^{-1/2}V_{oo} \qquad (2.13)$$

This gives a method of including the active elements in the equivalent circuits of SC networks. Thus any SC network using two-phase clock could be analyzed using the concept of LTP. However, the equivalent circuit of floating capacitor involves the use of transformers, which may be avoided by using a balanced lattice equivalent. This will lead to certain simplifications in the analysis of SC networks, as shown in the next section.

## 2.3 Laker's equivalent circuits [2.3]

In order to derive the balanced lattice equivalent, the starting point is once again the floating capacitor. The charge flowing into the terminals 1 and 2 of the floating capacitor in both the phases can be related to the voltages at terminals 1 and 2 measured with respect to ground. Referring to the floating capacitor shown in Figure 2.11(a), the charge conservation equations in the $z$-domain can be written as:

$$\Delta Q_{1e} = C(V_{1e} - V_{2e}) - Cz^{-1/2}(V_{1o} - V_{2o}) \qquad (2.14a)$$

$$\Delta Q_{1o} = C(V_{1o} - V_{2o}) - Cz^{-1/2}(V_{1e} - V_{2e}) \qquad (2.14b)$$

$$\Delta Q_{2e} = C(V_{2e} - V_{1e}) - Cz^{-1/2}(V_{2o} - V_{1o}) \qquad (2.14c)$$

$$\Delta Q_{2o} = C(V_{2o} - V_{1o}) - Cz^{-1/2}(V_{2e} - V_{1e}) \qquad (2.14d)$$

Equations (2.14) simply relate the charge that has flowed in the present half of the clock-cycle to the present voltage across the capacitor and the voltage across the

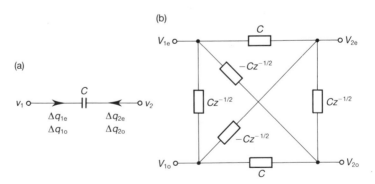

Fig. 2.11   (a) A floating capacitor in an SC circuit;   (b) its balanced lattice equivalent circuit.

capacitor a half-cycle before. A balanced lattice shown in Figure 2.11(b) precisely presents the behaviour exhibited by equations (2.14). It is thus possible to use the balanced lattice equivalent of Figure 2.11(b) in place of floating capacitor equivalent circuit of Figure 2.3(b), thereby avoiding transformers. This leads to considerable simplicity in the resulting $z$-domain equivalent circuits. It is important to note that all the elements in the balanced lattice are 'capacitive' in nature. However, through the use of certain properties of LTP discussed already, the elements in the paths between $V_{1e}$ and $V_{2e}$ as well as $V_{1o}$ and $V_{2o}$ can be represented as resistors. This is seen as follows.

For the connection shown in Figure 2.6(a), in which the capacitor is discharged and disconnected from the series path in the odd phase, it can be seen from the balanced lattice equivalent of Figure 2.11(b) that the equivalent circuit reduces to a series $C$ only. This has been interpreted as a resistor of value $T/C$ in Example 2.3. Hence, a resistor representation is used for the series elements in the odd and even paths.

In what follows, a few examples of SC networks and their equivalent circuits in the $z$-domain are considered. These are listed in Table 2.1 for convenience. For an exhaustive list, the reader is referred to [2.3]. In Table 2.1 the periodically switched capacitor is included and its significance will be discussed in the next example.

*Example 2.6*
Analyze the periodically reverse-switched capacitor [2.4, 2.5]

It is observed that there is a continuous path for the charge in both the clock phases. In each phase, the capacitor is discharged and recharged to the new branch voltage. Hence, it must be noted that the 'commutating' or 'periodically reverse-switched capacitor' transfers charge at twice the clock frequency (clock frequency is defined as the reciprocal of the sum of the times of even and odd clock periods). Thus the incremental charge flowing in any half-cycle (say, $nT$ to $(n - \frac{1}{2})T$) is:

$$\Delta q = C[V_b(nT) + V_b((n - \tfrac{1}{2})T)] \tag{2.15}$$

which after $z$-transformation yields*

$$\Delta Q(z) = \frac{C(1 + z^{-1/2})}{(1 - z^{-1/2})} V_b \tag{2.16}$$

where $V_b = V_1 - V_2$. In the $s$-domain, since the charge is the integral of the current flowing, the Laplace transformation yields, for a resistor,

$$\Delta Q(s) = \frac{1}{R} \cdot \frac{V_b(s)}{s} \tag{2.17}$$

---

* Note that $\Delta q$, the incremental charge flowing in the present half-cycle, is the difference between present charge and previous charge, i.e., $q(nT)$ and $q((n - \frac{1}{2})T)$.

**Table 2.1**  z-domain equivalent circuits of some basic SC elements.

1. Parallel-switched capacitor

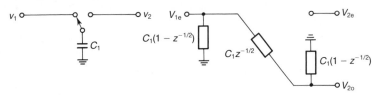

2. Series-switched capacitor

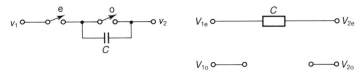

3. Parallel-switched capacitor

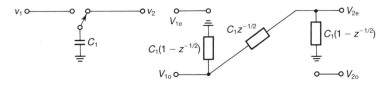

4. Series-switched capacitor

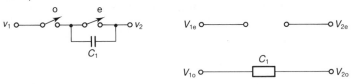

5. Series-switched capacitor
   (alternative version)

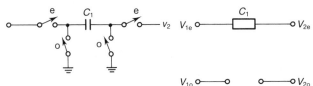

6. Series-switched capacitor

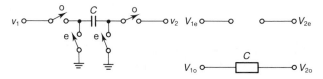

**Table 2.1**  (*continued*)

7.  Voltage-inversion type
    parallel-switched capacitor

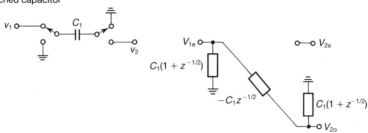

8.  Periodically reverse-switched capacitor

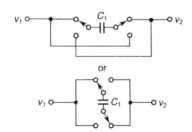

or

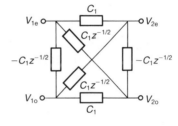

9.  Grounded capacitor

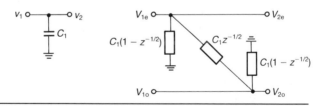

Thus, equations (2.16) and (2.17) are related by the bilinear $s \rightarrow z$ transformation with

$$R = \frac{T}{2C} \qquad (2.18)$$

It is thus shown that the periodically reverse-switched capacitor represents a resistor realizing the bilinear transformation and operating at twice the clock frequency.

It is now relevant to use the equivalent circuit given in Table 2.1 and compare the results. From the equivalent circuit, we obtain

$$\Delta Q_e(z) = C_1 [(V_{1e} - V_{2e}) + (V_{1o} - V_{2o})z^{-1/2}] \qquad (2.19a)$$

$$\Delta Q_o(z) = C_1 [(V_{1o} - V_{2o}) + (V_{1e} - V_{2e})z^{-1/2}] \qquad (2.19b)$$

It can be seen that these relationships do not show the behaviour of a bilinear transformation type resistor simulation. However, evaluating the total charge flowing in a clock cycle by summing equations (2.19a) and (2.19b), we obtain

$$\Delta Q(z) = \Delta Q_e(z) + \Delta Q_o(z)$$
$$= [C_1(1 + z^{-1/2})(V_{1e} - V_{2e})] + [C_1(1 + z^{-1/2})(V_{1o} - V_{2o})] \qquad (2.20a)$$

meaning that the circuit behaves as a trapezoidal charge integrator in both the clock phases. Alternatively, equation (2.20a) can be written as

$$\Delta Q(z) = C_1(1 + z^{-1/2})[(V_{1e} + V_{1o}) - (V_{2e} + V_{2o})] \qquad (2.20b)$$

which can be interpreted as the behaviour of a resistor in the 'bilinear transformation' sense, assuming that the terminal voltages of the 'resistor' are measured in both the even and odd clock phases or measured all the time. This means that the succeeding and preceding network elements are connected to the periodically reverse-switched capacitors all the time or in both the clock phases.

Thus the series-switched capacitor realizes a '$p$-transformation' type resistor (Example 2.4), whereas the commutating capacitor realizes a 'bilinear transformation' type resistor. Their use for realizing SC filters related to active RC filters through the respective transformations will be discussed later.

Referring once again to Laker's equivalent circuits, it may be noted that they can be used to analyze active SC networks using the equivalent odd and even OAs for each OA in the SC network. The following example studies a simple active SC network.

*Example 2.7*
Derive the transfer function of the SC network of Figure 2.12(a).

Using Table 2.1, the $z$-domain equivalent circuit can be constructed as in Figure 2.12(b). The next step is to remove some branches in the network which do not affect the transfer functions. All branches between two outputs, two voltage sources, voltage source and ground, virtual ground and ground, virtual ground and virtual ground, can be deleted. The resulting simplified equivalent circuit is shown in Figure 2.12(c). (In Figure 2.12(b), all the elements to be deleted are scored off as shown.) Also, since the branches in the $z$-domain equivalent circuits are capacitive in nature, parallel branches can be summed to simplify the circuit further. These have also been scored off in Figure 2.12(b). This reduces the circuit of Figure 2.12(b) to that of Figure 2.12(c).

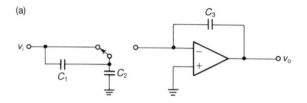

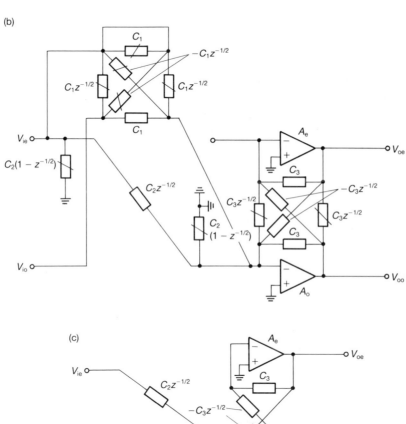

Fig. 2.12 (a) An active SC network; (b), (c) its equivalent circuits.

Performing the charge summation at the virtual ground nodes of amplifiers $A_o$ and $A_e$, we obtain:

$$C_2 V_{ie} z^{-1/2} + C_1 V_{io} + C_3 V_{oo} - C_3 V_{oe} z^{-1/2} = 0 \qquad (2.21a)$$

$$- C_3 V_{oo} z^{-1/2} + C_3 V_{oe} = 0 \qquad (2.21b)$$

Solving equations (2.21a) and (2.21b), we obtain

$$V_{oo} = \frac{C_1 V_{io} + C_2 V_{ie} z^{-1/2}}{C_3 (1 - z^{-1})} \qquad (2.22a)$$

$$V_{oe} = V_{oo} z^{-1/2} \qquad (2.22b)$$

This shows the use of Laker's equivalent circuits in the analysis of two-phase SC networks. Even though, at first glance, the method may appear to be cumbersome, through experience one can obtain the equivalent circuits in a straightforward manner. Laker's method will be used extensively in the succeeding chapters. In the above discussion, we have considered the analysis of SC circuits operating with a two-phase clock. It is of interest to develop z-domain equivalent circuits for SC networks employing a multiphase clock. This subject will be considered in the next section.

## 2.4 z-Domain equivalents for multiphase SC circuits [2.6, 2.7]

The switching signals used to operate the switches in a multiphase SC network are shown in Figure 2.13. With an $m$-phase clock, the duration for which any switch can be closed is $T/m$, where $T$ is the reciprocal of the sampling (or switching) frequency $f_s$. A capacitor in an $m$-phase SC network is now considered. In each phase, the

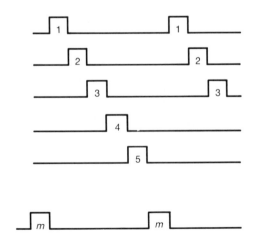

Fig. 2.13  The switching waveforms in an m-phase SC network.

charge conservation equation can be written. Thus, in phase 1, the incremental charge flowing in the capacitor at terminals 1 and 2, at the instant $nT$, is given by

$$\Delta q_{11} = C\left\{ [v_1(nT) - v_2(nT)] - \left[ v_1\left(\left(n - \frac{1}{m}\right)T\right) - v_2\left(\left(n - \frac{1}{m}\right)T\right)\right]\right\} \quad (2.23a)$$

$$\Delta q_{21} = C\left\{ [v_2(nT) - v_1(nT)] - \left[ v_2\left(\left(n - \frac{1}{m}\right)T\right) - v_1\left(\left(n - \frac{1}{m}\right)T\right)\right]\right\} \quad (2.23b)$$

Taking the $z$-transform of both these equations, we obtain

$$\Delta Q_{11}(z) = C[(V_{11} - V_{21}) - (V_{1m} - V_{2m})z^{-1/m}] \quad (2.24a)$$

$$\Delta Q_{21}(z) = C[(V_{21} - V_{11}) - (V_{2m} - V_{1m})z^{-1/m}] \quad (2.24b)$$

Similarly, the charge conservation equations can be written in the $z$-domain in other phases of the clock as follows:

$$\Delta Q_{12}(z) = C[(V_{12} - V_{22}) - (V_{11} - V_{21})z^{-1/m}]$$
$$\Delta Q_{13}(z) = C[(V_{13} - V_{23}) - (V_{12} - V_{22})z^{-1/m}]$$
$$\Delta Q_{14}(z) = C[(V_{14} - V_{24}) - (V_{13} - V_{23})z^{-1/m}] \quad (2.25a)$$

$$\cdots$$

$$\Delta Q_{1m}(z) = C[(V_{1m} - V_{2m}) - (V_{1(m-1)} - V_{2(m-1)})z^{-1/m}]$$

and

$$\Delta Q_{22}(z) = C[(V_{22} - V_{12}) - (V_{21} - V_{11})z^{-1/m}]$$
$$\Delta Q_{23}(z) = C[(V_{23} - V_{13}) - (V_{22} - V_{12})z^{-1/m}]$$
$$\Delta Q_{24}(z) = C[(V_{24} - V_{14}) - (V_{23} - V_{13})z^{-1/m}] \quad (2.25b)$$

$$\cdots$$

$$\Delta Q_{2m}(z) = C[(V_{2m} - V_{1m}) - (V_{2(m-1)} - V_{1(m-1)})z^{-1/m}]$$

The first subscript in equations (2.24) and (2.25) refers to the node in the basic SC network, whereas the second subscript indicates the phase in which the parameter under consideration is measured.

It remains to construct an equivalent circuit from equations (2.24) and (2.25). Mulawka [2.6] has suggested the equivalent circuit of Figure 2.14 for a floating capacitor in an $m$-phase SC network. This is an extension of Laker's two-phase $z$-domain equivalent to $m$-phases. The triangular blocks in Figure 2.14 are unity gain buffers to satisfy the requirement that the charge flowing in any phase is dependent on the voltages in the present phase and previous phase only. The network of Figure 2.14 is a ring structure, i.e., the charge flowing in the second phase is dependent on the voltage across the capacitor in the first phase; the charge flowing in the third phase is dependent on the voltage across the capacitor in the second phase; and similarly, the charge flowing in the first phase is dependent on the voltage across the capacitor in the $m$th phase, the last phase in the previous clock cycle.

(a)

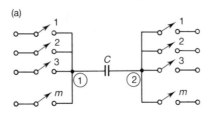

(b)

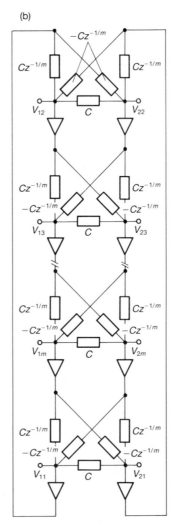

Fig. 2.14 (a) A floating capacitor in an m-phase SC network; (b) its equivalent z-domain circuit.

Some simplification can be achieved in the $z$-domain equivalent circuit of Figure 2.14, when there is no charge redistribution or charge flowing into the capacitor in one or more phases. For example, in a three-phase SC network, in one capacitor, if there is no charge flowing in the second phase, the ports corresponding to this phase can be ignored. Further, the voltage across the capacitor remains unchanged. As such, the charge flowing in the third phase is dependent on the voltages in the third and first phases only. The resulting simplified equivalent circuit is shown in Figure 2.15.

Finally, it is noted that the OA in a multiphase SC network (say, of $m$ phases) will be represented in the $z$-domain equivalent circuit as $m$ OAs appropriately connected to the $z$-domain equivalent circuit of the passive part of the original SC network.

It is possible to derive $z$-domain equivalent circuits for certain sub-networks in multiphase SC circuits which will enable easy analysis of these networks. In Table 2.2, a few examples are presented. The first example considers a grounded capacitor in a multiphase SC network, while the second, an OA with a feedback

(a)

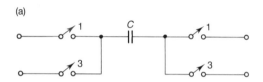

(b)

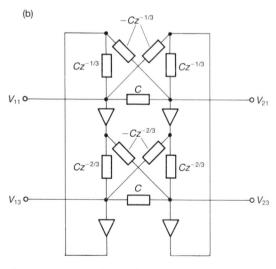

*Fig. 2.15* (a) A switched capacitor with no charge transfer in phase 2; (b) its $z$-domain equivalent circuit.

capacitor is shown with its $z$-domain equivalent circuit. In the third, a capacitor discharged in one phase is shown. The result is that the ring structure no longer exists, because the memory of the capacitor in phase 1 is destroyed. In the last two examples, we consider respectively a capacitor with one end connected to an OA output in all phases, and a capacitor with one end connected to virtual ground.

**Table 2.2**  $z$-domain equivalent circuits of SC sub-networks in multiphase SC networks.

1. Grounded capacitor

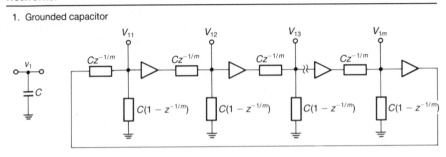

2. OA with feedback capacitor

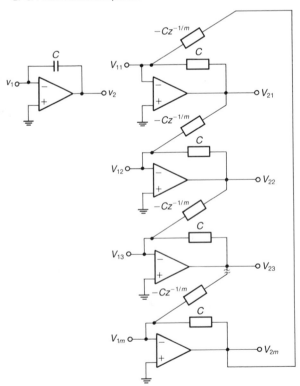

**Table 2.2**  (*continued*)  85

3.  Floating capacitor discharged in one phase (e.g., phase 1)

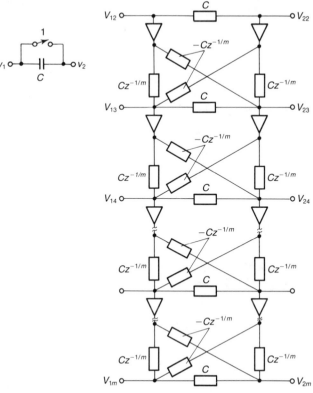

4.  Capacitor with one terminal connected to OA/voltage source

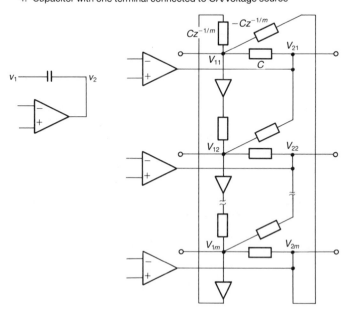

**Table 2.2** (*continued*)

5. Capacitor with one terminal to the virtual ground of an OA

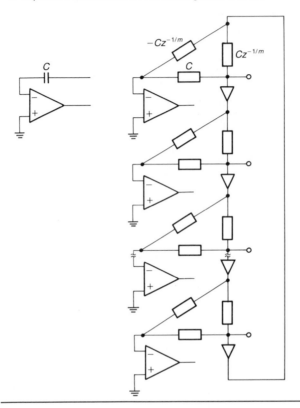

In the following example, a multiphase SC network is analyzed using Mulawka's equivalent circuits.

*Example 2.8*
Derive an expression for the charge flowing into the input port in phase 1 in the four-phase SC network of Figure 2.16(a).

The four-phase $z$-domain equivalent circuit can be drawn for the sub-networks in Figure 2.16(a) and then interconnected to result in the complete equivalent circuit shown in Figure 2.16(b). Some simplification can be achieved by deleting all branches between outputs of voltage sources, output and ground, etc. The resulting circuit in Figure 2.16(c) can be analyzed next. Using Kirchhoff's current law (KCL) at nodes $P$ and $Q$, we obtain:

$$C_0 V_{A1} z^{-1/4} - C_0 V_{B1} z^{-1/4} + C_2 V_{o2} - C_2 V_{o1} z^{-1/4} = 0 \qquad (2.26a)$$

$$C_1(V_R - V_{C_{14}})z^{-1/4} + C_0(V_{o3} - V_{C_{14}})z^{-1/4} = (C_0 + C_1)V_{C_{14}}(1 - z^{-1/4})$$

$$(2.26b)$$

It is noted that

$$V_R = V_{o1}z^{-1/2} \tag{2.26c}$$

$$V_{o1} = V_{o2}z^{-3/4} \tag{2.26d}$$

$$V_{o3} = V_{o2}z^{-1/4} \tag{2.26e}$$

After simplification, from equations (2.26) we obtain

$$V_{C_{14}} = -\frac{[C_0 z^{-1/2} + C_1 z^{-3/2}]}{(C_0 + C_1)} \cdot \frac{C_0 z^{-1/2}(V_{A1} - V_{B1})}{C_2(1 - z^{-1})} \tag{2.27}$$

Also, writing an expression for the charge entering terminal $A$ in Figure 2.16(c),

$$\Delta Q_{A1} = -C_0[V_{A1}z^{-1/4} + (V_{A1} - V_{C_{14}})z^{-1/4} + (V_{A1} - V_{B1})] \tag{2.28}$$

and substituting for $V_{C_{14}}$ from equation (2.27) in equation (2.28), the charge flowing into the input terminals is obtained as

$$\Delta Q_{A1} = C_0\left[1 + \frac{C_0(C_0 z^{-1} + C_1 z^{-2})}{(C_0 + C_1)C_2(1 - z^{-1})}\right](V_{A1} - V_{B1}) \tag{2.29}$$

The reader must have felt the tedium in deriving the result in equation (2.29). In fact, writing charge-conservation equations from Figure 2.16(a) itself may be simpler. This will be illustrated next. The charge conservation equations in various phases are as follows. In phase 1,

$$\Delta Q_{1m} = C_0[V_{11} - V_{C_{04}}z^{-1/4}] \tag{2.30a}$$

$$V_{C_{11}} = V_{o1} = V_{o2}z^{-3/4} \tag{2.30b}$$

In phase 2,

$$C_0 V_{11}z^{-1/4} + C_2 V_{o2}(1 - z^{-1}) = 0 \tag{2.30c}$$

In phase 3,

$$V_{C_{03}} = V_{o2}z^{-1/4} \tag{2.30d}$$

In phase 4,

$$(C_0 + C_1)V_{C_{04}} = C_0 V_{C_{03}}z^{-1/4} + C_1 V_{C_{11}}z^{-3/4} \tag{2.30e}$$

Solving equations (2.30) together leads to equation (2.29).

In this section, a method of analyzing multiphase SC networks using $z$-domain equivalent circuits has been demonstrated. So far, what has been considered is modelling the SC network by $z$-domain equivalent circuits, derived from charge conservation equations in various clock phases. These methods are highly suitable for routine analysis of simple SC networks by hand calculations. However, for

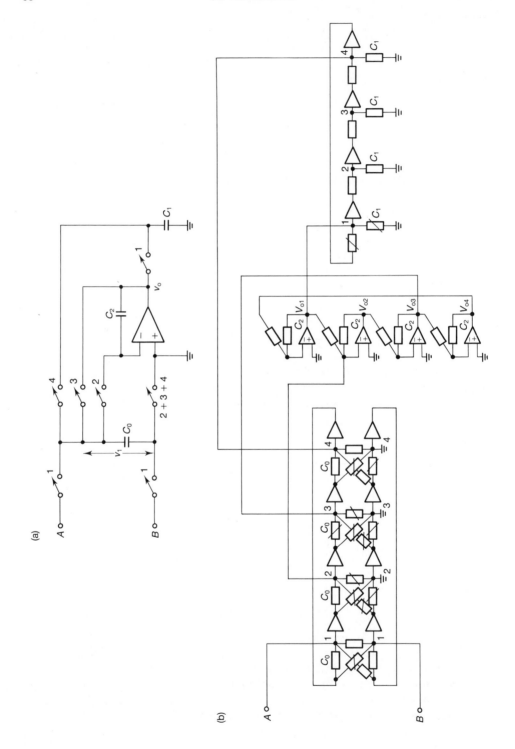

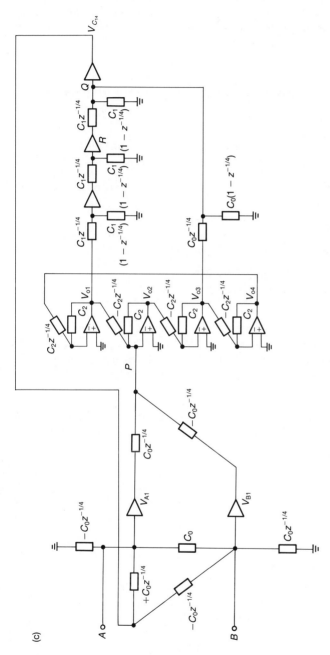

*Fig. 2.16* Four-phase SC network used in Example 2.8.

computer-aided analysis of SC networks, alternative methods may be attractive. The use of the indefinite admittance matrix in passive and active continuous-time network analysis is well known. Such a technique is also suitable for SC circuit analysis. This topic is discussed in the next section.

## 2.5  SC network analysis using indefinite admittance matrix representation [2.7, 2.8]

It is useful to recall the indefinite admittance matrix (IAM)' method of analysis of continuous-time networks. The various steps in this technique are as follows:

1. First the IAM must be written down, using only the passive part of the network. This can be done from simple knowledge of the network, i.e., the admittances connected between the various nodes of the network.
2. The active devices, usually the OAs, are next taken into account. The embedding of OAs in the passive network imposes certain constraints:

   (a) the two inputs of any infinite gain OA will be at the same potential;
   (b) the output current of the OA, which is the input current for the passive network, is no longer an independent variable.

   Hence, Nathan's method of embedding the OA could be used [2.9].
3. The complete IAM of the network obtained in the second step will now be used, after deleting the rows and columns corresponding to reference node, to evaluate the network parameters such as the transfer function, input impedance and transfer impedance.

An example of an active RC network will serve to illustrate these ideas. For a detailed treatment of the properties of the IAM, etc., the reader is referred to [2.10–2.12].

*Example 2.9*
Derive the transfer function of the Sallen and Key active filter of Figure 2.17 using the IAM method.

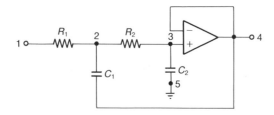

*Fig. 2.17*  Sallen–Key unity-gain OA-based active RC filter.

1. First designate all the nodes in the network of Figure 2.17. The IAM corresponding to the passive part of the network may be written, by inspection, as follows:

$$
\begin{array}{ccccc}
1 & 2 & 3 & 4 & 5
\end{array}
$$

$$
\begin{bmatrix}
\dfrac{1}{R_1} & -\dfrac{1}{R_1} & 0 & 0 & 0 \\[2ex]
-\dfrac{1}{R_1} & \dfrac{1}{R_1}+\dfrac{1}{R_2}+sC_1 & -\dfrac{1}{R_2} & -sC_1 & 0 \\[2ex]
0 & -\dfrac{1}{R_2} & \dfrac{1}{R_2}+sC_2 & 0 & -sC_2 \\[2ex]
0 & -sC_1 & 0 & sC_1 & 0 \\[2ex]
0 & 0 & -sC_2 & 0 & sC_2
\end{bmatrix}
\begin{array}{c}
1 \\[2ex] 2 \\[2ex] 3 \\[2ex] 4 \\[2ex] 5
\end{array}
$$

The sum of all the elements in any row or any column of this matrix is zero. Further, the IAM for the passive part is symmetric about the main diagonal. The diagonal elements $y_{jj}$ are the self-admittances, i.e., the sum of the admittances connected to the node $j$, whereas the off-diagonal elements $y_{ij}$ are the sign-reversed admittances between node $i$ and node $j$. The numbers of the nodes are written alongside the matrix for illustration purposes.

2. The inclusion of the OA requires that $V_3 = V_4$ so that the elements in column 3 have to be added to those in column 4 and the column corresponding to the driving node (node 3) is deleted. Next, since the current into the passive network at node 4 is no longer an independent variable, the row corresponding to node 4 is deleted. Thus the IAM of the active network of Figure 2.17 is given by:

$$
\begin{array}{cccc}
1 & 2 & 4 & 5
\end{array}
$$

$$
\mathbf{Y} =
\begin{bmatrix}
\dfrac{1}{R_1} & -\dfrac{1}{R_1} & 0 & 0 \\[2ex]
-\dfrac{1}{R_1} & \dfrac{1}{R_1}+\dfrac{1}{R_2}+sC_1 & -\left(sC_1+\dfrac{1}{R_2}\right) & 0 \\[2ex]
0 & -\dfrac{1}{R_2} & \dfrac{1}{R_2}+sC_2 & -sC_2 \\[2ex]
0 & 0 & -sC_2 & sC_2
\end{bmatrix}
\begin{array}{c}
1 \\[2ex] 2 \\[2ex] 3 \\[2ex] 5
\end{array}
\qquad (2.31)
$$

If we now ground node 5, then the row and the column corresponding to node 5 (that is, the fourth row and the fourth column in (2.31)) are deleted to obtain the definite admittance matrix (DAM) of the active RC network of Figure 2.17.

The transfer function of this network, defined as $V_4/V_1$ can be obtained to be

$$\frac{V_{45}}{V_{15}} = \frac{V_4}{V_1} = \mathrm{sgn}(4-5) \cdot \mathrm{sgn}(1-5) \cdot \frac{Y_{45}^{15}}{Y_{15}^{15}} \tag{2.32}$$

where $\mathrm{sgn}(m-n)$ means the sign of the quantity $(m-n)$ and $Y_{mn}^{ij}$ is the second cofactor of the matrix $Y$ which is $(-1)^{m+n+i+j}$ times the minor obtained by deleting columns corresponding to nodes $m$, $n$ and rows corresponding to nodes $i$, $j$ in $Y$. Thus, we have for the network of Figure 2.17,

$$\frac{V_4}{V_1} = \frac{\begin{vmatrix} -\dfrac{1}{R_1} & \dfrac{1}{R_1} + \dfrac{1}{R_2} + sC_1 \\[2mm] 0 & -\dfrac{1}{R_2} \end{vmatrix}}{\begin{vmatrix} \dfrac{1}{R_1} + \dfrac{1}{R_2} + sC_1 & -\left(sC_1 + \dfrac{1}{R_2}\right) \\[2mm] -\dfrac{1}{R_2} & \left(\dfrac{1}{R_2} + sC_2\right) \end{vmatrix}} \tag{2.33}$$

which, after simplification, becomes

$$\frac{V_4}{V_1} = \frac{\dfrac{1}{R_1 R_2}}{C_1 C_2 s^2 + C_2\left(\dfrac{1}{R_1} + \dfrac{1}{R_2}\right)s + \dfrac{1}{R_1 R_2}} \tag{2.34}$$

The reader can verify the result in equation (2.34) using other methods.

The analysis of SC networks using IAM follows a similar pattern, involving steps 1 and 2 outlined above. However, there are some differences. These will be evident when the floating capacitor in an SC network is considered. The charge-conservation equations (2.14) derived in Section 2.3 describe the floating capacitor in both phases of the clock. These can be written in matrix form as follows:

$$\begin{bmatrix} I_{1e} \\ I_{2e} \\ I_{1o} \\ I_{2o} \end{bmatrix} = \begin{bmatrix} \dfrac{C}{T} & -\dfrac{C}{T} & -\dfrac{C}{T}z^{-1/2} & \dfrac{C}{T}z^{-1/2} \\[3mm] -\dfrac{C}{T} & \dfrac{C}{T} & \dfrac{C}{T}z^{-1/2} & -\dfrac{C}{T}z^{-1/2} \\[3mm] -\dfrac{C}{T}z^{-1/2} & \dfrac{C}{T}z^{-1/2} & \dfrac{C}{T} & -\dfrac{C}{T} \\[3mm] \dfrac{C}{T}z^{-1/2} & -\dfrac{C}{T}z^{-1/2} & -\dfrac{C}{T} & \dfrac{C}{T} \end{bmatrix} \begin{bmatrix} V_{1e} \\ V_{2e} \\ V_{1o} \\ V_{2o} \end{bmatrix} \tag{2.35}$$

Note that equations (2.14) are divided on both sides by $T$ to convert the incremental charges flowing into the terminals 1 and 2 of the capacitor into currents. The matrix representation of equation (2.35) is the IAM of a floating capacitor. Thus, a capacitor in an SC network is represented by a four-port equivalent in the $z$-domain, as mentioned before. Hence, extending the same technique to a passive SC network with $n$ nodes, an IAM with $2n$ rows and $2n$ columns corresponding to $n$ even-phase ports and $n$ odd-phase ports is obtained. This is a difference between the continuous-time and SC discrete-time analysis methods using IAM. While writing the IAM for a general SC passive network, the nodes corresponding to even instants are designated 1 to $n$, whereas the nodes corresponding to odd instants are designated $(n + 1)$ to $2n$. As an illustration, in the representation of equation (2.35), the even-phase nodes are designated as 1, 2 and the odd-phase nodes are designated as 3, 4, thus obtaining the $4 \times 4$ matrix representation.

A knowledge of the network description can be used to derive the IAM. The next step is to incorporate the information regarding the switches. A switch between two nodes $i, j$ in the even phase results in the relationships

$$V_{ie} = V_{je} = V'_{ie} \qquad (2.36a)$$

by virtue of the node voltages becoming equal, and

$$I'_{ie} = I_{ie} + I_{je} \qquad (2.36b)$$

Thus, the admittances in rows $i$ and $j$ have to be added and those in columns $i$ and $j$ also have to be added to realize equations (2.36a) and (2.36b), respectively. If a switch exists in the odd phase between nodes $i$ and $j$, the above operation has to be carried out on the rows and columns $(n + i)$ and $(n + j)$, respectively. This gives the IAM of the passive SC network.

The inclusion of OAs in the SC network will impose two constraints: first, that the voltages at the non-inverting and inverting inputs are equal; and second, that the output current of the OA is a dependent variable. Hence, similar to what has been mentioned in the case of active RC networks, the elements in the columns corresponding to the input nodes of the OA are added and one column among these is deleted. Also, the row corresponding to the output node of the OA is deleted. However, it is important to note that this operation has to be done in both the phases of the clock. The next step is to remove the rows and columns corresponding to the ground node in both even and odd phases. The resulting DAM can now be used to evaluate the transfer functions, transfer impedances, etc. An example will illustrate the procedure described thus far.

*Example 2.10*
Derive the transfer functions of the SC network of Figure 2.18.

The IAM of the passive capacitor network can be written by inspection. The network in Figure 2.18 has seven nodes and hence the IAM is a $14 \times 14$ matrix. The even-phase nodes are 1 to 7, and the odd-phase nodes 8 to 14. In Figure 2.18, as

$$(2.37a)$$

| | 1 | 2 | 3 | 4 | 5 | 6 | 7 | 8 | 9 | 10 | 11 | 12 | 13 | 14 |
|---|---|---|---|---|---|---|---|---|---|---|---|---|---|---|
| 1 | 0 | 0 | 0 | 0 | 0 | 0 | 0 | 0 | 0 | 0 | 0 | 0 | 0 | 0 |
| 2 | 0 | $C_1$ | 0 | 0 | 0 | $-C_1$ | 0 | 0 | $-C_1 z^{-1/2}$ | 0 | 0 | 0 | $C_1 z^{-1/2}$ | 0 |
| 3 | 0 | 0 | $C_2$ | 0 | 0 | 0 | $-C_2$ | 0 | 0 | $-C_2 z^{-1/2}$ | 0 | 0 | 0 | $C_2 z^{-1/2}$ |
| 4 | 0 | 0 | 0 | $C_3$ | 0 | $-C_3$ | 0 | 0 | 0 | 0 | $-C_3 z^{-1/2}$ | 0 | $C_3 z^{-1/2}$ | 0 |
| 5 | 0 | 0 | 0 | 0 | $C_4$ | $-C_4$ | 0 | 0 | 0 | 0 | 0 | $-C_4 z^{-1/2}$ | $C_4 z^{-1/2}$ | 0 |
| 6 | 0 | $-C_1$ | 0 | $-C_3$ | $-C_4$ | $C_1+C_3+C_4$ | 0 | 0 | $C_1 z^{-1/2}$ | 0 | $C_3 z^{-1/2}$ | $C_4 z^{-1/2}$ | $-(C_1+C_3+C_4)z^{-1/2}$ | 0 |
| 7 | 0 | 0 | $-C_2$ | 0 | 0 | 0 | $C_2$ | 0 | 0 | $C_2 z^{-1/2}$ | 0 | 0 | 0 | $-C_2 z^{-1/2}$ |
| 8 | 0 | 0 | 0 | 0 | 0 | 0 | 0 | 0 | 0 | 0 | 0 | 0 | 0 | 0 |
| 9 | 0 | $-C_1 z^{-1/2}$ | 0 | 0 | 0 | $C_1 z^{-1/2}$ | 0 | 0 | $C_1$ | 0 | 0 | 0 | $-C_1$ | 0 |
| 10 | 0 | 0 | $-C_2 z^{-1/2}$ | 0 | 0 | 0 | $C_2 z^{-1/2}$ | 0 | 0 | $C_2$ | 0 | 0 | 0 | $-C_2$ |
| 11 | 0 | 0 | 0 | $-C_3 z^{-1/2}$ | 0 | $C_3 z^{-1/2}$ | 0 | 0 | 0 | 0 | $C_3$ | 0 | $-C_3$ | 0 |
| 12 | 0 | 0 | 0 | 0 | $-C_4 z^{-1/2}$ | $C_4 z^{-1/2}$ | 0 | 0 | 0 | 0 | 0 | $C_4$ | $-C_4$ | 0 |
| 13 | 0 | $C_1 z^{-1/2}$ | 0 | $C_3 z^{-1/2}$ | $C_4 z^{-1/2}$ | $-(C_1+C_3+C_4)z^{-1/2}$ | 0 | 0 | $-C_1$ | 0 | $-C_3$ | $-C_4$ | $C_1+C_3+C_4$ | 0 |
| 14 | 0 | 0 | $C_2 z^{-1/2}$ | 0 | 0 | 0 | $-C_2 z^{-1/2}$ | 0 | 0 | $-C_2$ | 0 | 0 | 0 | $C_2$ |

$$
\begin{array}{c|cccccccccc}
 & 1 & 3 & 5 & 6 & 7 & 8 & 9 & 12 & 13 & 14 \\
\hline
1 & C_1 & 0 & 0 & -C_1 & 0 & 0 & -C_1 z^{-1/2} & 0 & C_1 z^{-1/2} & 0 \\
3 & 0 & C_2 + C_3 & 0 & -C_3 & -C_2 & 0 & -C_2 z^{-1/2} & -C_3 z^{-1/2} & C_3 z^{-1/2} & C_2 z^{-1/2} \\
5 & 0 & 0 & C_4 & -C_4 & 0 & 0 & 0 & -C_4 z^{-1/2} & C_4 z^{-1/2} & 0 \\
6 & -C_1 & -C_3 & -C_4 & C_1 + C_3 + C_4 & 0 & 0 & C_1 z^{-1/2} & (C_3 + C_4)z^{-1/2} & -(C_1 + C_3 + C_4)z^{-1/2} & 0 \\
7 & 0 & -C_2 & 0 & 0 & C_2 & 0 & C_2 z^{-1/2} & 0 & 0 & -C_2 z^{-1/2} \\
8 & 0 & 0 & 0 & 0 & 0 & 0 & 0 & 0 & 0 & 0 \\
9 & -C_1 z^{-1/2} & -C_2 z^{-1/2} & 0 & C_1 z^{-1/2} & C_2 z^{-1/2} & 0 & (C_1 + C_2) & 0 & -C_1 & -C_2 \\
11 & 0 & -C_3 z^{-1/2} & -C_4 z^{-1/2} & (C_4 + C_3)z^{-1/2} & 0 & 0 & 0 & (C_3 + C_4) & -(C_3 + C_4) & 0 \\
13 & C_1 z^{-1/2} & C_3 z^{-1/2} & C_4 z^{-1/2} & -(C_1 + C_3 + C_4)z^{-1/2} & 0 & 0 & -C_1 & -(C_3 + C_4) & (C_1 + C_3 + C_4) & 0 \\
14 & 0 & C_2 z^{-1/2} & 0 & 0 & -C_2 z^{-1/2} & 0 & -C_2 & 0 & 0 & C_2
\end{array}
$$

(2.37b)

95

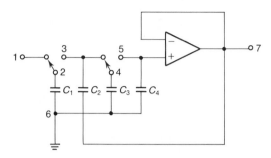

*Fig. 2.18* SC network used in IAM analysis for Example 2.10.

an example, a capacitor $C_1$ is connected between nodes 2 and 6. The sixteen elements in the matrix of equation (2.35) corresponding to $C_1$ are entered in the matrix space formed by rows and columns 2, 6, 9 and 13, respectively. This procedure could be carried out for all the capacitors in the network, resulting in the indefinite admittance matrix (2.37a).

The next step is to take into account the switches in the IAM by adding the rows and columns corresponding to the even- and odd-phase switches. As an illustration, the even-phase switch between nodes 1 and 2 will be taken into account in the IAM by adding rows 1 and 2 and columns 1 and 2. Similarly, the odd-phase switch between nodes 2 and 3 is taken into account by adding the elements in rows and columns 9 and 10. Similar operations are carried out for the even-phase switch between nodes 3 and 4, and the odd-phase switch between nodes 4 and 5.

The resulting matrix is given as expression (2.37b). So far, only the passive sub-network has been considered. Next, taking the OA into account, we need to add elements in columns corresponding to nodes 5 and 7, and 12 and 14 and retain the resulting elements in columns corresponding to nodes 7 and 14. Further, the rows corresponding to the outputs of the OA in the even and odd phases, i.e. rows corresponding to nodes 7 and 14, are deleted. Also the row as well as the column corresponding to node 8 in matrix (2.37b) can be deleted, since they have zero entries. This is because, in the odd phase, the circuit is disconnected from the source (see the even-phase switch between nodes 1 and 2 in Figure 2.18). Hence, the IAM is further simplified as in matrix (2.37c). Also, choosing node 6 as ground in Figure

$$
Y = \begin{bmatrix}
C_1 & 0 & -C_1 & 0 & -C_1 z^{-1/2} & C_1 z^{-1/2} & 0 \\
0 & C_2 + C_3 & -C_3 & -C_2 & -C_2 z^{-1/2} & C_3 z^{-1/2} & (C_2 - C_3)z^{-1/2} \\
0 & 0 & -C_4 & C_4 & 0 & C_4 z^{-1/2} & -C_4 z^{-1/2} \\
-C_1 & -C_3 & C_1 + C_3 + C_4 & -C_4 & C_1 z^{-1/2} - (C_1 + C_3 + C_4)z^{-1/2} & (C_3 + C_4)z^{-1/2} \\
-C_1 z^{-1/2} - C_2 z^{-1/2} & C_1 z^{-1/2} & C_2 z^{-1/2} & C_1 + C_2 & -C_1 & -C_2 \\
0 & -C_3 z^{-1/2} & (C_3 + C_4)z^{-1/2} & -C_4 z^{-1/2} & 0 & -(C_3 + C_4) & (C_3 + C_4) \\
C_1 z^{-1/2} & C_3 z^{-1/2} & -(C_1 + C_3 + C_4)z^{-1/2} & C_4 z^{-1/2} & -C_1 & (C_1 + C_3 + C_4) & -(C_3 + C_4)
\end{bmatrix}
\begin{matrix} 1 \\ 3 \\ 5 \\ 6 \\ 9 \\ 11 \\ 13 \end{matrix}
$$

with column headings 1, 3, 6, 7, 9, 13, 14.

$$(2.37c)$$

$$
\begin{array}{ccccc}
1 & 3 & 7 & 9 & 14 \\
\end{array}
$$

$$
Y = \begin{bmatrix}
C_1 & 0 & 0 & -C_1 z^{-1/2} & 0 \\
0 & C_2 + C_3 & -C_2 & -C_2 z^{-1/2} & (C_2 - C_3)z^{-1/2} \\
0 & 0 & C_4 & 0 & -C_4 z^{-1/2} \\
-C_1 z^{-1/2} & -C_2 z^{-1/2} & C_2 z^{-1/2} & C_1 + C_2 & -C_2 \\
0 & -C_3 z^{-1/2} & -C_4 z^{-1/2} & 0 & C_3 + C_4
\end{bmatrix}
\begin{matrix}
1 \\ 3 \\ 5 \\ 9 \\ 11
\end{matrix}
$$

$$(2.37\mathrm{d})$$

2.18, and thus deleting rows and columns 6 and 13, the DAM of the SC circuit is obtained and is given by the matrix (2.37d). It remains to obtain the necessary network functions from the IAM or DAM given by matrix (2.37c) or (2.37d), respectively. At this stage, it is useful to examine how the driving point impedances, transfer impedances and voltage transfer functions are determined for continuous-time networks. Figure 2.19 illustrates these methods for determining the driving point and transfer impedances from the IAM. The transfer impedance is defined as $Z_{tr} = V_{km}/I_{ij}$ (voltage developed at port $km$ due to a current source at port $ij$) and the driving point impedance is $V_{ij}/I_{ij}$ (voltage developed at port $ij$ due to a current source at the same port). All the other ports are open-circuited in both the above cases. It can be shown that the transfer impedance is obtained from the IAM as follows:

$$
Z_{tr} = Z_{ij}^{km} = \mathrm{sgn}(i-j) \cdot \mathrm{sgn}(k-m) \cdot \frac{Y_{km}^{ij}}{Y_j^i}
\qquad (2.38)
$$

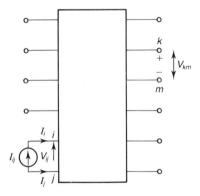

Fig. 2.19  Measurement of driving point and transfer impedances, and voltage transfer ratio

$$
Z_{dp} = \frac{V_{ij}}{I_{ij}}
$$

$$
Z_{tr} = \frac{V_{km}}{I_{ij}}
$$

$$
H_{ij}^{km} = \frac{V_{km}}{V_{ij}} = \frac{Z_{tr}}{Z_{dp}}
$$

The expression for driving point impedance follows from equation (2.38) by replacing $km$ by $ij$. The evaluation of the transfer function is straightforward once the transfer impedances and driving point impedances are known. The open-circuit voltage transfer function is $V_{km}/V_{ij}$ in Figure 2.19 and can be obtained as follows:

$$H_{ij}^{km} = \frac{V_{km}}{V_{ij}} = \text{sgn}(k-m) \cdot \text{sgn}(i-j) \cdot \frac{Y_{km}^{ij}}{Y_{ij}^{ij}} \qquad (2.39)$$

The results for the SC network are slightly different. This is because a node in the original SC network is 'split' into even-phase and odd-phase nodes in the IAM representation. Consequently, corresponding to terminals $ij$ in the continuous-time circuit, in the IAM representation, there will be four nodes $i, j$ and $n+i, n+j$ belonging to the even and odd-phase ports. Similarly, there are two pairs of terminals corresponding to terminals $k$, $m$, namely, $k, m$ and $n+k, n+m$. It follows that there are several transfer functions, transfer impedances and driving point impedances. These are illustrated in Figure 2.20. (Note that capital letters are used to designate the odd-phase terminals: $K = n+k$, $J = n+j$.)

Formulas for the driving point and transfer impedances and voltage transfer ratios can be derived in a similar manner to the derivation of equations (2.38) and (2.39) and the results are presented in Table 2.3.

Note that when evaluating the network functions from the IAM, reference nodes have to be chosen in order to make the measurements. Thus when the voltage $V_{km}$ is measured, $m$ is chosen as a reference node and, as such, $M$ will also become a reference node. Consequently, two columns will be deleted. Similarly, when a current source is connected at an input port $ij$ in the continuous-time circuit, it is required to connect two current sources at ports $ij$ and $IJ$. Thus, if we measure the voltage at port $ij$, node $j$ has to be grounded, whence node $J$ will also be grounded. In summary, in evaluating all the network functions, nodes $jJ, mM$ have to be used as reference nodes. The formulas listed in Table 2.3 show that rows $iJ$ and columns $mM$ have been deleted in all cases. (This case corresponds to grounding the network as in obtaining DAM.) The remaining row and column are deleted because the

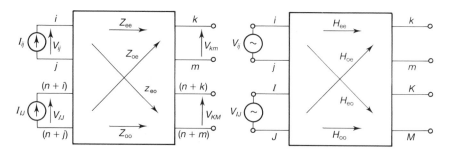

*Fig. 2.20* Four-port representation of SC networks to define transfer and driving point impedances, and transfer functions.

**Table 2.3**  Formula for network functions using the IAM *(adapted from [2.8], © 1980 IEE).*

$$Z_{\text{tr}}^{\text{ee}} = \frac{V_{km}}{I_{ij}}\bigg|_{I_{IJ}=0} = \text{sgn}(i-j) \cdot \text{sgn}(k-m) \cdot \frac{Y_{kmM}^{ijJ}}{Y_{jJ}^{jJ}}$$

$$Z_{\text{tr}}^{\text{eo}} = \frac{V_{KM}}{I_{ij}}\bigg|_{I_{IJ}=0} = \text{sgn}(i-j) \cdot \text{sgn}(K-M) \cdot \frac{Y_{KMm}^{ijJ}}{Y_{jJ}^{jJ}}$$

$$Z_{\text{tr}}^{\text{oe}} = \frac{V_{km}}{I_{IJ}}\bigg|_{I_{ij}=0} = \text{sgn}(I-J) \cdot \text{sgn}(k-m) \cdot \frac{Y_{kmM}^{IJj}}{Y_{jJ}^{jJ}}$$

$$Z_{\text{tr}}^{\text{oo}} = \frac{V_{KM}}{I_{IJ}}\bigg|_{I_{ij}=0} = \text{sgn}(I-J) \cdot \text{sgn}(K-M) \cdot \frac{Y_{KMm}^{IJj}}{Y_{jJ}^{jJ}}$$

$$Z_{\text{dp}}^{\text{ee}} = \frac{V_{ij}}{I_{ij}} = \frac{Y_{ijJ}^{ijJ}}{Y_{jJ}^{jJ}}$$

$$Z_{\text{dp}}^{\text{oo}} = \frac{V_{IJ}}{I_{IJ}} = \frac{Y_{IJj}^{IJj}}{Y_{jJ}^{jJ}}$$

$$H^{\text{ee}} = \frac{V_{km}}{V_{ij}}\bigg|_{V_{IJ}=0} = \text{sgn}(i-j) \cdot \text{sgn}(k-m) \cdot \frac{\tilde{Y}_{kmM}^{ijJ}}{\tilde{Y}_{ijJ}^{ijJ}}$$

$$H^{\text{eo}} = \frac{V_{KM}}{V_{ij}}\bigg|_{V_{IJ}=0} = \text{sgn}(i-j) \cdot \text{sgn}(K-M) \cdot \frac{\tilde{Y}_{KMm}^{ijJ}}{\tilde{Y}_{ijJ}^{ijJ}}$$

$$H^{\text{oe}} = \frac{V_{km}}{V_{IJ}}\bigg|_{V_{ij}=0} = \text{sgn}(I-J) \cdot \text{sgn}(k-m) \cdot \frac{\tilde{Y}_{kmM}^{IJj}}{\tilde{Y}_{IJj}^{IJj}}$$

$$H^{\text{oo}} = \frac{V_{KM}}{V_{IJ}}\bigg|_{V_{ij}=0} = \text{sgn}(I-J) \cdot \text{sgn}(K-M) \cdot \frac{\tilde{Y}_{KMm}^{IJj}}{\tilde{Y}_{IJj}^{IJj}}$$

voltage corresponding to the column is being evaluated in response to the current corresponding to the row.

While measuring the various driving point and transfer impedances, the current sources which are not required are open-circuited. (So, for example, $I_{IJ} = 0$ or $I_{ij} = 0$ in the formulas in Table 2.3.)

In the formulas for the transfer functions, the symbol '~' above the third cofactor of the IAM is used to imply that the IAM considered is actually obtained from the original IAM by contraction. The requirement that $V_{IJ} = 0$ when the transfer functions $H^{\text{ee}}$ and $H^{\text{eo}}$ are evaluated means that the contractions by row and column $I$ and $J$ are required. Similarly, when $V_{ij} = 0$ is required in order to evaluate $H^{\text{oe}}$ and $H^{\text{oo}}$, contraction by row and column $i$ and $j$ will be required.

It may be simpler to use the DAM for evaluating network functions. Since the reference terminal would already have been chosen and correspondingly the associated two rows and two columns are deleted, the network functions are determined by the formulas given in Table 2.4. We can use the DAM to evaluate the

**Table 2.4** Formulas for network functions using the DAM *(adapted from [2.8], © 1980 IEE).*

| | |
|---|---|
| $Z_{tr}^{ee}$ | $Y_k^i/Y$ |
| $Z_{tr}^{eo}$ | $Y_k^i/Y$ |
| $Z_{tr}^{oe}$ | $Y_k^I/Y$ |
| $Z_{tr}^{oo}$ | $Y_k^I/Y$ |
| $Z_{dp}^{ee}$ | $Y_i^i/Y$ |
| $Z_{dp}^{oo}$ | $Y_I^I/Y$ |
| $H^{ee}$ | $Y_{kI}^{iI}/Y_{iI}^{iI}$ |
| $H^{eo}$ | $Y_{KI}^{iI}/Y_{iI}^{iI}$ |
| $H^{oe}$ | $Y_{ki}^{Ii}/Y_{iI}^{iI}$ |
| $H^{oo}$ | $Y_{ki}^{Ii}/Y_{iI}^{iI}$ |

$Y$ = determinant of **Y**

transfer function of the SC network in Figure 2.18, which we have already derived in Example 2.10.

*Example 2.11*
Derive the transfer functions of the SC network of Figure 2.18 using the DAM.

The DAM already derived as expression (2.37d) is rewritten to relate the voltages and currents in even and odd phases as follows:

$$\mathbf{Y} = \begin{array}{ccccc} 1 & 3 & 5 & 9 & 14 \\ \end{array}$$

$$\mathbf{Y} = \begin{bmatrix} C_1 & 0 & 0 & -C_1 z^{-1/2} & 0 \\ 0 & C_2 + C_3 & -C_2 & -C_2 z^{-1/2} & (C_2 - C_3)z^{-1/2} \\ 0 & 0 & C_4 & 0 & -C_4 z^{-1/2} \\ -C_1 z^{-1/2} & -C_2 z^{-1/2} & C_2 z^{-1/2} & C_1 + C_2 & -C_2 \\ 0 & -C_3 z^{-1/2} & -C_4 z^{-1/2} & 0 & C_3 + C_4 \end{bmatrix} \begin{array}{c} 1 \\ 3 \\ 5 \\ 9 \\ 11 \end{array}$$

$$(2.37e)$$

We are required to find $V_5/V_1$ and $V_{14}/V_1$. Using the formulas in Table 2.4, we obtain

$$\frac{V_5}{V_1} = \frac{V_{5e}}{V_{1e}} = \frac{Y_5^1}{Y_1^1}$$

$$= \frac{z^{-2}}{\dfrac{C_2}{C_1}\left(\dfrac{C_4}{C_3}+1\right)(1-z^{-1})^2 + \left((C_3+C_4)\left(\dfrac{1}{C_1}+\dfrac{1}{C_2}\right)+\dfrac{C_4}{C_3}\right)(1-z^{-1})+1}$$

(2.40a)

and

$$\frac{V_{14}}{V_1} = \frac{V_{4o}}{V_{1e}} = \frac{Y_{14}^1}{Y_1^1} = \frac{V_{5e}}{V_{1e}}z^{1/2}$$

(2.40b)

Another example will be considered to illustrate the method further.

*Example 2.12*
Derive the transfer functions of the SC network of Figure 2.21.

The IAM can be written by inspection as follows:

|  | 1 | 2 | 3 | 4 | 5 | 6 | 7 | 8 |  |
|---|---|---|---|---|---|---|---|---|---|
| | $C_1$ | $-C_1$ | 0 | 0 | $-C_1z^{-1/2}$ | $C_1z^{-1/2}$ | 0 | 0 | 1 |
| | $-C_1$ | $C_1$ | 0 | 0 | $C_1z^{-1/2}$ | $-C_1z^{-1/2}$ | 0 | 0 | 2 |
| | 0 | 0 | $C_2$ | $-C_2$ | 0 | 0 | $-C_2z^{-1/2}$ | $C_2z^{-1/2}$ | 3 |
| | 0 | 0 | $-C_2$ | $C_2$ | 0 | 0 | $C_2z^{-1/2}$ | $-C_2z^{-1/2}$ | 4 |
| | $-C_1z^{-1/2}$ | $C_1z^{-1/2}$ | 0 | 0 | $C_1$ | $-C_1$ | 0 | 0 | 5 |
| | $C_1z^{-1/2}$ | $-C_1z^{-1/2}$ | 0 | 0 | $-C_1$ | $C_1$ | 0 | 0 | 6 |
| | 0 | 0 | $-C_2z^{-1/2}$ | $C_2z^{-1/2}$ | 0 | 0 | $C_2$ | $-C_2$ | 7 |
| | 0 | 0 | $C_2z^{-1/2}$ | $-C_2z^{-1/2}$ | 0 | 0 | $-C_2$ | $C_2$ | 8 |

(2.41a)

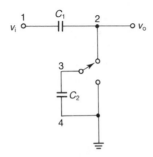

*Fig. 2.21* The first-order high-pass passive SC network used in Example 2.12.

Rows (and columns) 2 and 3 are contracted to take into account the even-phase switch (i.e., the elements in rows 2 and 3, columns 2 and 3 are added). Also, the odd-phase switch is taken into account, by contracting rows 7 and 8, columns 7 and 8, respectively. The resulting IAM is given below:

$$
\begin{array}{ccccccc}
\phantom{-}1 & \phantom{-}2 & \phantom{-}4 & \phantom{-}5 & \phantom{-}6 & \phantom{-}8 & \\
\left[\begin{array}{cccccc}
C_1 & -C_1 & 0 & -C_1 z^{-1/2} & C_1 z^{-1/2} & 0 \\
-C_2 & C_1+C_2 & -C_2 & C_1 z^{-1/2} & -C_1 z^{-1/2} & 0 \\
0 & -C_2 & C_2 & 0 & 0 & 0 \\
-C_1 z^{-1/2} & C_1 z^{-1/2} & 0 & C_1 & -C_1 & 0 \\
C_1 z^{-1/2} & -C_1 z^{-1/2} & 0 & -C_1 & C_1 & 0 \\
0 & 0 & 0 & 0 & 0 & 0
\end{array}\right] &
\begin{array}{c}
1 \\ 2 \\ 4 \\ 5 \\ 6 \\ 8
\end{array}
\end{array}
$$

$$(2.41b)$$

Next grounding the SC network at terminal 4, thus deleting rows and columns corresponding to nodes 4 and 8, we obtain the DAM:

$$
\mathbf{Y} =
\begin{array}{ccccc}
\phantom{-}1 & \phantom{-}2 & \phantom{-}5 & \phantom{-}6 & \\
\left[\begin{array}{cccc}
C_1 & -C_1 & -C_1 z^{-1/2} & C_1 z^{-1/2} \\
-C_1 & C_1+C_2 & C_1 z^{-1/2} & -C_1 z^{-1/2} \\
-C_1 z^{-1/2} & C_1 z^{-1/2} & C_1 & -C_1 \\
C_1 z^{-1/2} & -C_1 z^{-1/2} & -C_1 & C_1
\end{array}\right] &
\begin{array}{c}
1 \\ 2 \\ 5 \\ 6
\end{array}
\end{array}
$$

$$(2.41c)$$

The voltage transfer functions due to the source $V_{1e}$ are found by contraction of the matrix in equation (2.41c) to obtain $\tilde{y}$, i.e., the row and column corresponding to source $V_{1o}$ (or node 5) are deleted. Thus, the transfer functions are obtained as follows:

$$\frac{V_{2e}}{V_{1e}} = \frac{V_2}{V_1} = \frac{\tilde{y}_2^1}{\tilde{y}_1^1} = \frac{C_1(1-z^{-1})}{(C_1+C_2)-C_1 z^{-1}} \tag{2.42a}$$

$$\frac{V_{2o}}{V_{1e}} = \frac{V_6}{V_1} = \frac{\tilde{y}_6^1}{\tilde{y}_1^1} = \frac{-C_2 z^{-1/2}}{(C_1+C_2)-C_1 z^{-1}} \tag{2.42b}$$

Similarly, the transfer functions due to the source $V_{1o}$ are obtained by deleting the row and column corresponding to $V_{1e}$ and $I_{1e}$ namely row 1 and column 1:

$$\frac{V_{2e}}{V_{1o}} = \frac{V_2}{V_5} = \frac{\tilde{y}_2^5}{\tilde{y}_5^5} = 0 \tag{2.42c}$$

$$\frac{V_{2o}}{V_{1o}} = \frac{V_6}{V_5} = \frac{\tilde{y}_6^5}{\tilde{y}_5^5} = 1 \tag{2.42d}$$

While evaluating equations (2.42c) and (2.42d), the reader may have noticed that the determinant of $\tilde{y}$ is zero (in fact, the determinant of the DAM itself is zero). Thus, for the evaluation of driving point and transfer impedances, the fourth row and fourth column must be deleted to remove the dependent equation in system (2.41c). The evaluation of these is left to the reader.

We conclude this section with a few remarks. The IAM of the capacitor network (i.e., without considering the switches; see, for example, expression (2.41a)), has a certain symmetry. Thus, if the matrix in the top left-hand corner is written for capacitors similar to lumped passive networks, the remaining portions can easily be written. Next, the equicofactor property of the IAM of continuous-time networks is also true for the IAM of SC networks. For a complete and excellent treatment of this subject, the reader is referred to [2.8], where a computer program based on the above method has also been considered. It is possible to extend the IAM method for multiphase SC networks as well. The starting point is once again the charge conservation equations for each phase and the resulting IAM size will be ($mn \times mn$), corresponding to an $m$-phase $n$-node SC network. The reader is referred to [2.7] for information on this method.

So far, we have considered the analysis of SC networks in the $z$-domain by formulating equivalent circuits in the $z$-domain, and using the IAM. In the former method, the network functions in the $z$-domain can be derived by hand computation, and then the frequency response is evaluated by substituting $z = e^{j\omega T}$, where $\omega$ is the frequency of interest. In the latter method, the IAM or DAM is formed with the elements as functions of $z$, whereupon, by substituting $e^{j\omega T}$ for $z$, the network functions can be determined. The latter method has the advantage that the IAM can be formed from a description of the SC network and, as such, it satisfies the usual requirement of computer analysis programs that the network description be given in as simple a manner as possible.

Several network analysis programs are available for continuous-time networks. Hence, if an SC network is modelled by equivalent continuous-time networks, the analysis can be carried out in the frequency domain by such circuit analysis programs. In the next section, this approach of analyzing SC networks by using continuous-time equivalent circuits will be discussed.

## 2.6   SC network analysis using continuous-time two-port equivalents [2.13–2.15]

In this method, it is intended to obtain an $s$-domain equivalent circuit corresponding to an SC network. As is evident from the previous sections, SC networks operating with a two-phase clock have an important difference from continuous-time networks, which is that they are modelled by four-port equivalent circuits in the $z$-domain. However, under certain conditions they can be represented by two-port equivalents, e.g., a series-switched capacitor (considered in Example 2.3) or a periodically reverse-switched capacitor (considered in Example 2.6). In these simple cases, the continuous-time equivalent two-port can be derived, as illustrated in the following example.

*Example 2.13*
Derive a continuous-time two-port equivalent for the series-switched capacitor of Figure 2.22(a).

The charge conservation equation can be written as follows:

$$\Delta Q_e(z) = C(V_{1e} - V_{2e}) \tag{2.43}$$

where $V_{1e} - V_{2e}$ is the voltage across the capacitor. We express the incremental charge flowing through the capacitor in the $z$-domain as

$$\Delta Q_e(z) = Q_e(z)(1 - z^{-1}) \tag{2.44}$$

since it is the difference between the previous and present charges. Thus, from equations (2.43) and (2.44), we obtain

$$Q_e(z) = \frac{C(V_{1e} - V_{2e})}{(1 - z^{-1})} \tag{2.45}$$

An equivalent network in the $s$-domain can now be obtained by using inverse bilinear transformation:

$$z = \frac{1 + \dfrac{sT}{2}}{1 - \dfrac{sT}{2}}$$

to yield

$$Q(s) = C[V_1(s) - V_2(s)] \left[ \frac{1}{sT} + \frac{1}{2} \right] \tag{2.46}$$

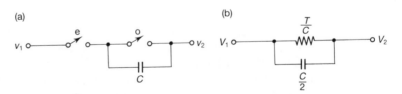

*Fig. 2.22* (a) A series-switched capacitor; (b) its equivalent analog circuit.

Noting that

$$I(s) = sQ(s) \tag{2.47}$$

we obtain, from equation (2.46),

$$I(s) = \left(\frac{C}{T} + \frac{sC}{2}\right)[V_1(s) - V_2(s)] \tag{2.48}$$

The resulting equivalent circuit is shown in Figure 2.22(b). It may be noted, however, that since the bilinear transformation has been used to obtain the equivalent analog network from the SC network, the warping of the frequency scale has to be taken into account. This is accomplished by considering the resistor value in the equivalent circuit to be $[(T/C)\tan(\omega T/2)/(\omega T/2)]$ instead of $T/C$. Alternatively, the SC network in Figure 2.22(b) can be treated as a resistor $T/C$ in parallel with a capacitor $C/2$ and analyzed first. Subsequently, the frequency scale has to be dewarped to obtain the actual behaviour.

The above example has demonstrated the method of obtaining the equivalent analog circuit (EAC). The SC circuit and EAC are equivalent by virtue of the bilinear $s \rightarrow z$ transformation used up to the Nyquist frequency (half the sampling frequency), beyond which the effects due to aliasing will be present. Secondly, the node voltages in the EAC and SC circuit at discrete instants are the same. Further, the charges flowing in each clock cycle into the branch under consideration in both analog and SC circuits are the same at the sampling instant.

Thus, for the given SC circuit, the injected charge--voltage relationships in the $z$-domain are written first. It is possible to write these in matrix notation, relating the vector of injected charges into the terminals 1 and 2 of the branch under consideration and the voltages at terminals 1 and 2 at discrete instants of time,

$$\begin{bmatrix} Q_1(z) \\ Q_2(z) \end{bmatrix} = \begin{bmatrix} \dfrac{\Delta Q_1(z)}{1 - z^{-1}} \\[2ex] \dfrac{\Delta Q_2(z)}{1 - z^{-1}} \end{bmatrix} = [C] \cdot \begin{bmatrix} V_1 \\ V_2 \end{bmatrix} \tag{2.49}$$

Note that $[C]$ is a $2 \times 2$ capacitance matrix. It is obvious that $Q_1(z)$, $Q_2(z)$, $V_1$ and $V_2$ are the $z$-transforms of sequences $q_1(nT)$, $q_2(nT)$, $v_1(nT)$ and $v_2(nT)$.

We next consider an analog circuit for which it is easier to see that the terminal currents and voltages are related through the admittance matrix:

$$[I_a(s)] = [Y_a(s)] \cdot [V_a(s)] \tag{2.50}$$

where the subscript indicates that the analog circuit is being considered. It follows that the injected charges are given by

$$[Q_a(s)] = \frac{1}{s}[I_a(s)] = \frac{1}{s}[Y_a(s)] \cdot [V_a(s)] \tag{2.51}$$

The analog circuit equations (2.51) correspond to the SC network equations (2.49). The two circuits are equivalent, if $q_a(nT)$ and $v_a(nT)$ are equal to $q(nT)$ and $v(nT)$ at the sampling instants. Note that $q_a(nT)$ and $v_a(nT)$ are obtained by taking the inverse of the Laplace transform relationships in equations (2.51). As a consequence, in the frequency domain, both the SC and analog systems have identical frequency response up to the Nyquist frequency. Further, the admittance matrix and capacitance matrix in equations (2.49) and (2.51) are related as follows:

$$\frac{[Y_a(s)]}{s} = [C] \qquad (2.52)$$

Thus the $Y$-matrix of the analog equivalent is obtained. These ideas are further illustrated in Example 2.14.

*Example 2.14*
Derive the equivalent analog circuit of a parallel-switched capacitor.

The charge-conservation equations in both the phases can be written (referring to Figure 2.23(a)) in the $z$-domain, as follows:

$$\Delta Q_{1e} = C(V_{1e} - V_{2o}z^{-1/2}) \qquad (2.53a)$$

$$\Delta Q_{2o} = C(V_{2o} - V_{1e}z^{-1/2}) \qquad (2.53b)$$

With $V_{1e} = V_{1o}z^{-1/2}$, from equations (2.53a) and (2.53b), we obtain

$$\Delta Q_{1e} = C(V_{1o} - V_{2o})z^{-1/2} \qquad (2.53c)$$

$$\Delta Q_{2o} = C(V_{2o} - V_{1o}z^{-1}) \qquad (2.53d)$$

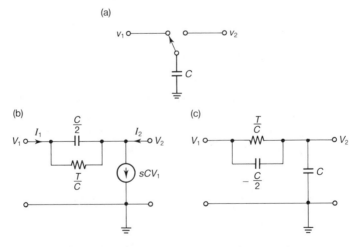

Fig. 2.23  (a) A parallel-switched capacitor;  (b), (c) continuous-time equivalents.

The charge $\Delta Q_{1e}$ could be measured either in the previous half-cycle or in the succeeding half-cycle. In the former case,

$$\Delta Q_1 = C(V_{1o} - V_{2o}) \tag{2.53e}$$

$$\Delta Q_2 = C(V_{2o} - V_{1o}z^{-1}) \tag{2.53f}$$

It may be noted that in the above two equations, the net charge that has flowed into the capacitor terminals in a clock cycle has been given. Following equations (2.49), the capacitance matrix can be obtained as follows:

$$\begin{bmatrix} I_1(s) \\ I_2(s) \end{bmatrix} = \begin{bmatrix} C \\ (1-z^{-1}) \end{bmatrix} \begin{bmatrix} 1 & -1 \\ -z^{-1} & 1 \end{bmatrix} \begin{bmatrix} V_1 \\ V_2 \end{bmatrix} \tag{2.54a}$$

Using inverse bilinear transformation, and converting charges in $s$-domain to currents, we obtain

$$\begin{bmatrix} I_1(s) \\ I_2(s) \end{bmatrix} = \begin{bmatrix} \dfrac{C}{T} + \dfrac{sC}{2} & -\left(\dfrac{C}{T} + \dfrac{sC}{2}\right) \\ -\left(\dfrac{C}{T} - \dfrac{sC}{2}\right) & \dfrac{C}{T} + \dfrac{sC}{2} \end{bmatrix} \begin{bmatrix} V_1(s) \\ V_2(s) \end{bmatrix} \tag{2.54b}$$

A simple circuit to realize the admittance matrix in equations (2.54b) is shown in Figure 2.23(b).

If the charge $\Delta Q_1$ in equation (2.53c) is measured after another half-cycle, equations (2.53c) and (2.53d) will be obtained in modified form:

$$\Delta Q_1 = C(V_{1o} - V_{2o})z^{-1} \tag{2.54c}$$

$$\Delta Q_2 = C(V_{2o} - V_{1o}z^{-1}) \tag{2.54d}$$

The resulting equivalent circuit is shown in Figure 2.23(c), with the two-port equations

$$\begin{bmatrix} I_1(s) \\ I_2(s) \end{bmatrix} = \begin{bmatrix} \dfrac{C}{T} - \dfrac{sC}{2} & -\dfrac{C}{T} + \dfrac{sC}{2} \\ -\dfrac{C}{T} + \dfrac{sC}{2} & \dfrac{C}{T} + \dfrac{sC}{2} \end{bmatrix} \begin{bmatrix} V_1(s) \\ V_2(s) \end{bmatrix} \tag{2.55}$$

The inverting switched capacitor is shown in Figure 2.24(a), together with its continuous-time two-port equivalent in Figure 2.24(b), when $V_{1e} = V_{1o}z^{-1/2}$.

In the above, continuous-time equivalents for some SC branches frequently used have been derived. Thus, for a given SC network, by first splitting the network into branches of one of the types discussed above, and then replacing each branch by its two-port $s$-domain equivalent, an equivalent analog circuit (EAC) can be obtained. The frequency response of this EAC can then be evaluated using conventional network analysis programs. It must be remembered, however, that the

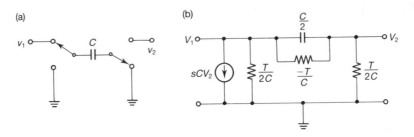

Fig. 2.24   (a) An inverting switched capacitor;   (b) its continuous-time equivalent.

restrictions that charge transfer shall occur only once in a clock cycle and that the input is sampled and held over a clock period must be satisfied. In the following, we shall analyze a particular SC network using the above method.

*Example 2.15*
Derive the transfer functions in the $s$-domain for the circuits of Figures 2.25(a) and 2.25(c).

In both the circuits of Figures 2.25(a) and 2.25(c), it is to be observed that there is charge transfer only once in a clock cycle through all the elements. Hence, they can be replaced by continuous-time two-port equivalents.
    With the condition $V_{ie} = V_{io}z^{-1/2}$, the branches $B_1$ and $B_2$ in Figure 2.25(a) can be replaced with that of Figure 2.23(c). Note that the capacitor $C$ remains as $C$ in the $s$-domain (see Exercise 2.8). Thus, the complete $s$-domain equivalent is as shown in Figure 2.25(b). The resulting transfer function is

$$\frac{V_{2o}(s)}{V_{1o}(s)} = \frac{-C_1\left(1 - \frac{sT}{2}\right)}{C_3 + (2C_2 - C_3)\frac{sT}{2}} \tag{2.56}$$

Similarly, in the case of Figure 2.25(c), the equivalent circuit of Figure 2.25(d) is obtained, and the transfer function is

$$\frac{V_{2o}(s)}{V_{1o}(s)} = \frac{-C_1\left(1 - \frac{sT}{2}\right)}{C_3 + (2C_2 + C_3)\frac{sT}{2}} \tag{2.57}$$

It is interesting to note that if the phasing of the switches in branch 2 is opposite to what has been shown, the continuous-time two-port equivalent cannot be derived. This is because, in that case, the charge transfer occurs in both the clock phases.

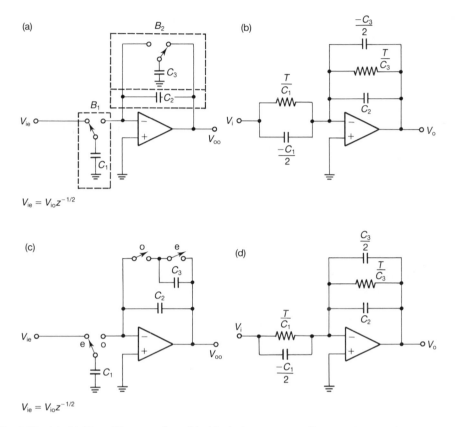

Fig. 2.25   (a), (c) Two SC networks;   (b), (d) their corresponding continuous-time two-port equivalents.

Knob and Dessoulavy [2.15] have developed continuous two-port equivalents of SC branches for various sampling conditions, using the above method. As mentioned earlier, these could be used to analyze the SC networks on available computer-aided network analysis programs, suitably modified. However, this method is applicable to only a certain class of networks. It may be of interest to the reader to note that using Laker's equivalent circuits (discussed in Section 2.2), and taking into account the sampling and holding conditions at various inputs and outputs, a z-domain two-port equivalent circuit can be constructed easily, which can be analyzed in the s-domain after bilinear transformation (see Exercise 2.5) [1.26].

The charge conservation equations in each phase of a two-phase SC network invariably contain terms of the form $z^{-1/2}$. Hence, the requirement that the inputs and outputs are sampled and held is essential for eliminating the $z^{-1/2}$ terms in the charge conservation equations, by converting them to $z^{-1}$ terms. Thereafter, bilinear transformation can be used to realize the continuous-time two-port equivalents.

While this method leads to circuits which can be analyzed by conventional circuit analysis programs, it is restricted to certain types of network with sampled and held inputs and outputs. It is possible, however, to remove these restrictions by converting four-port equivalent circuits in the $s$-domain through some transformation.

## 2.7  Analysis of SC networks using continuous-time four-port equivalents [2.16, 2.17]

It was noted earlier that, using Laker's equivalent circuits, a two-phase SC circuit can be represented by a four-port equivalent circuit in the $z$-domain. This can be converted into a four-port equivalent circuit in the $s$-domain using two methods. In the first method, the clock frequency is considered to be doubled so that each half-cycle is treated as a full cycle. Consequently, the $z^{-1/2}$ terms that occur in the charge conservation equations become $z^{-1}$ terms. Thereupon, using the bilinear transformation, analog equivalents can be obtained.

Consider, as an illustration, the balanced lattice representation of a floating capacitor in the two-phase SC circuit shown in Figure 2.26(a), where the clock frequency is considered as double that of the actual clock frequency. The corresponding continuous-time four-port equivalent is shown in Figure 2.26(b), obtained following the method described earlier used to derive the two-port equivalents. It can be seen

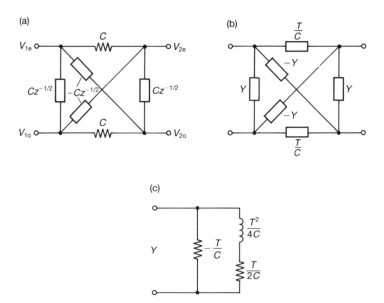

Fig. 2.26  (a) Balanced lattice equivalent of a floating capacitor;  (b) continuous-time equivalent derived from (a) using bilinear transformation;  (c) the admittance $Y$ in (b).

that the continuous-time four-port equivalent uses two types of element: a resistor, and an admittance of the form

$$Y = \frac{C}{T} \cdot \frac{[1 - sT/2]}{[1 + sT/2]} \tag{2.58}$$

The admittance $Y$ can be synthesized as shown in Figure 2.26(c). Thus, in the most complex form, the four-port continuous-time equivalent needs fourteen elements to represent a floating capacitor. Also the OAs used in the SC network will be doubled in the continuous-time equivalent circuit. An advantage in this method over the two-port equivalent circuit method is that all four transfer functions for a two-phase SC circuit can be obtained.

In the second method [2.17] for obtaining four-port continuous-time equivalent circuits, the clock frequency need not be doubled, and the LDI transformation is used instead of the bilinear transformation. Thus, the balanced lattice four-port equivalent in the $z$-domain is converted to a continuous-time equivalent. All the elements of the form $\pm Cz^{-1/2}$ in Laker's equivalent circuits are converted into resistors of value $T/C$, whereas elements of the form $C$ should be realized as impedances of the type $(T/C)z^{-1/2}$. The continuous-time equivalent for the latter is found by solving the $s \rightarrow z$ relationship implied by LDI transformation, namely, $s \rightarrow (1 - z^{-1})/Tz^{-1/2}$. Rewriting the relationship as a quadratic in $z^{-1/2}$, we obtain [1.19],

$$z^{-1} + sTz^{-1/2} - 1 = 0 \tag{2.59a}$$

or

$$z^{-1/2} = \frac{-sT \pm \sqrt{s^2 T^2 + 4}}{2} \tag{2.59b}$$

It may be recalled that LDI transformation maps the portion of the imaginary axis of the complex frequency $s$-plane, between zero and $f_s/\pi$, where $f_s$ is the sampling frequency. Hence the factor under the root sign is positive. Thus, the impedance $Cz^{-1/2}$ can be realized as a series combination of a negative inductance $T^2C/2$ and positive (or negative) frequency-dependent resistances as shown in Figure 2.27. However, only two such elements occur in the complete equivalent circuit of a floating capacitor. Thus, in this method, the number of branches required for implementation on a continuous-time circuit analysis program is eight, as against fourteen in the former method. In both these methods, it is necessary to dewarp the frequency scale to obtain the actual frequency response. However, in the latter method, the existing circuit analysis programs must be suitably modified to realize the required frequency-dependent resistors.

In the above method, the balanced lattice representation in the $z$-domain for a floating capacitor has been used to derive continuous-time two-port equivalents to facilitate analysis on existing circuit analysis programs with simple modifications. (In Appendix A, the use of the SPICE circuit analysis program for analysis of two-phase SC networks is considered in detail.) Another somewhat similar technique has

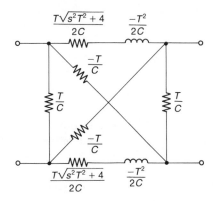

*Fig. 2.27* Continuous-time equivalent of the balanced lattice equivalent of floating capacitor obtained using the LDI transformation.

been described by Kuo *et al.* [2.18]. In this method, the Norton equivalent circuit of a floating capacitor in two-phase SC networks is used as the basis. This will be briefly considered in the following.

Referring once again to the charge conservation equations of the floating capacitor in both the phases (see equations (2.22)), the equivalent circuit using resistors and dependent current sources can be derived as shown in Figure 2.28. Thus, this new equivalent circuit is yet another representation of the link two-port. Replacing this equivalent circuit in SC networks, four-port equivalent circuits in the z-domain can be obtained. Conventional circuit analysis programs do allow dependent current sources and, as such, the equivalent circuit of Figure 2.28 is useful for SC circuit analysis.

Several methods of analyzing SC networks have been considered in the above. We next use two results derived in this chapter to show how SC circuits can be obtained directly from active RC filter structures. Three types of resistor simulation have already been described in this chapter: a parallel-switched capacitor; a series-switched capacitor; and a periodically reverse-switched capacitor. Any of these can be used to replace the resistors in active RC filters, thereby generating SC filter structures. These resulting structures can next be analyzed using the z-domain

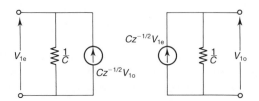

*Fig. 2.28* Norton equivalent circuit of a floating capacitor in the z-domain.

equivalent circuits developed in this chapter to obtain the various network functions. However, it will be interesting to obtain the SC network functions from the RC network functions by direct substitution of some of the $s$-$z$ domain transformations. Such methods do exist corresponding to the replacement of resistors by series-switched capacitors or periodically reverse-switched capacitors in a particular fashion. This will be considered in the next section.

## 2.8  Derivation of SC networks from RC networks

Consider an RC network for which the pertinent network equations are given by Kirchhoff's voltage law (KVL), KCL and the branch voltage–current relationships. These are written as follows:

$$A\mathbf{q} = \mathbf{0} \qquad \text{(KCL)} \qquad\qquad\qquad (2.60a)$$

$$\mathbf{v} = A^{T}\mathbf{e} \qquad \text{(KVL)} \qquad\qquad\qquad (2.60b)$$

$$Q_k(s) = \frac{G_k}{s} V_k(s) \qquad\qquad\qquad (2.60c)$$

$$Q = CV \qquad\qquad\qquad (2.60d)$$

where $A$ is the branch-to-node incidence matrix, $\mathbf{q}$ is the vector of branch charges $q_k(t)$, $\mathbf{v}$ is the vector of the branch voltages $v_k(t)$, and $\mathbf{e}$ is the vector of the node voltages $e_m(t)$. Equation (2.60c) represents the charge–voltage relationships of the resistors, whereas equation (2.60d) represents the charge–voltage relationships of the capacitors.

By replacing each resistor in the above active RC circuit by some switching arrangement and an appropriate capacitor, we obtain an SC filter with the same KCL and KVL equations, as well as charge–voltage relationships as given by equations (2.60). If the relationships of charge and voltage in the $z$-domain for the resistor are realized as:

$$Q_{kd}(z) = \frac{G_k}{F(z)} V_{kd}(z) \qquad\qquad\qquad (2.61)$$

then it follows that all the transformed voltages and charges $E_m(z)$, $V_k(z)$ and $Q_k(z)$ of the SC circuit are directly obtained from those of the RC filter simply by replacing $s$ by $F(z)$. The resulting network functions of the SC network are obtainable by simple substitution from those of the RC network. The results are valid, even if the RC network is active.

When periodically reverse-switched capacitors are used to replace the resistors, the charge conservation equation can be obtained as (see Figure 2.29(a)):

$$q_k(nT) - q_k((n-\tfrac{1}{2})T) = C_k[v_k(nT) + v_k((n-\tfrac{1}{2})T)] \qquad (2.62a)$$

which, after $z$-transform, yields

$$Q_k(z) = C_k \frac{1+z^{-1/2}}{1-z^{-1/2}} V_k \qquad (2.62b)$$

From equations (2.62) and (2.63b), we obtain, with $s \to (2/T)(1-z^{-1/2})/(1+z^{-1/2})$,

$$G_k = \frac{2C_k}{T} \qquad (2.63)$$

which gives the capacitance value to be used to realize a desired resistance.

When series-switched capacitors are used to replace resistors, the charge conservation equation can be obtained as (see Figure 2.29(b)):

$$q_k(nT) - q_k((n-1)T) = Cv(nT) \qquad (2.64)$$

The corresponding resistance values are given by

$$G_k = \frac{C_k}{T} \qquad (2.65)$$

when $p$-transformation is used, i.e., $s \to (1-z^{-1})/T$. It is instructive to examine both of the above methods [2.3, 2.5] to generate SC circuits from active RC networks.

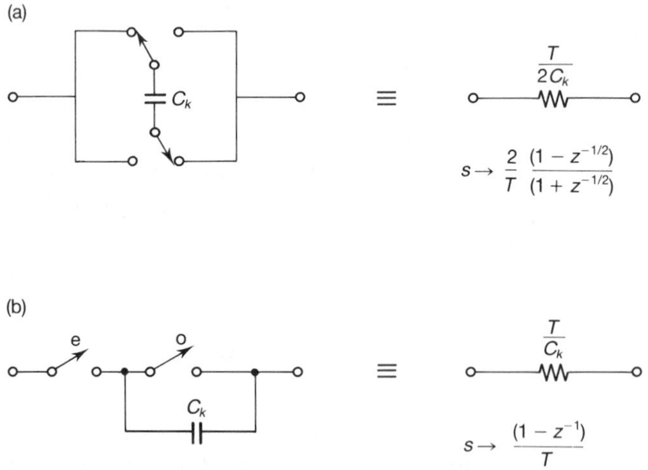

Fig. 2.29  Resistor equivalents:  (a) for bilinear transformation type SC networks; (b) $p$-transformation type of SC networks and the corresponding relationships between $s$- and $z$-domains.

*Example 2.16*

Derive SC networks corresponding to the active RC network of Figure 2.30(a).

The bilinear transformation type SC network is obtained as shown in Figure 2.30(b). Note that the capacitor values for the branches simulating resistors are given by equation (2.63). The transfer function can be readily obtained from the active RC transfer function through the bilinear transformation as follows:

$$\text{Active RC:} \quad \frac{V_o(s)}{V_i(s)} = \frac{-C_3 R_4 s}{C_2 C_3 R_1 R_4 s^2 + (C_2 + C_3) R_1 s + 1} \tag{2.66a}$$

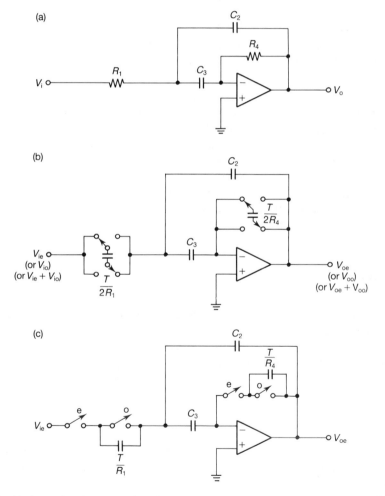

*Fig. 2.30* (a) An active RC network; (b) SC versions realizing bilinear transformation, (c) *p*-transformation.

Bilinear transformation type:

$$\frac{V_{od}}{V_{id}} = \frac{-(C_3/C_4)(1-z^{-1})}{(C_2C_3/C_1C_4)(1-z^{-1/2}) + [(C_2+C_3)/C_1](1-z^{-1}) + (1+z^{-1/2})^2} \quad (2.66b)$$

It must be noted that the bilinear transformation type of SC network derived as above works at double the actual clock frequency (see Example 2.6).

The corresponding circuit obtained through $p$-transformation is shown in Figure 2.30(c). The resulting SC transfer function is

$$\frac{V_{oe}}{V_{ie}} = \frac{-(C_3/C_4)(1-z^{-1})}{(C_2C_3/C_1C_4)(1-z^{-1})^2 + [(C_2+C_3)/C_1](1-z^{-1}) + 1} \quad (2.66c)$$

It is seen that the $p$-transformation type of circuit processes the signal in one phase. The resulting circuits can thus be designated as even and odd circuits depending on whether they use even- or odd-phase resistors. The other transfer functions in such a case are irrelevant.

The use of parallel-switched capacitors in SC networks does not lead to such a correspondence between active RC and SC transfer functions. The reason is that the necessary condition for such a correspondence, $s \to F(z)$, is not satisfied. The charge transfer takes place in this case in both odd and even phases and not in one phase alone. This is precisely the difficulty in modelling of the continuous-time two-port equivalents considered already in Section 2.5. The first two methods are amenable for such a reduction and as such are attractive. More will be said about the relative advantages of the use of the three types of SC resistors in later chapters.

## 2.9  Conclusions

In this chapter, several methods for the analysis of SC networks using two-phase or multi-phase clock have been studied. The reader is now in a position to analyze any SC circuit. In the next few chapters, these methods will be recalled, as appropriate, to explore the intricacies of SC network design.

## 2.10  Exercises

2.1 Several equivalent circuits for the LTP are possible. Balanced lattice representation and the four-port equivalent using ideal transformers have been considered. The latter can be simplified further as shown in Figures 2.31(a) and 2.31(b) for the grounded and floating cases [2.19]. Use these to analyze the circuit of Figure 2.31(c), in the even and odd phases.

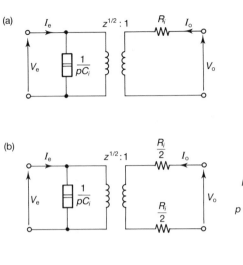

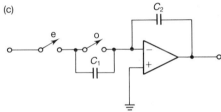

Fig. 2.31 An SC circuit for analysis using LTP equivalent circuits.

2.2 Derive the transfer function of the SC network of Figure 2.4(c), considering its LTP equivalent circuit, namely, a cascade of four LTPs. Derive the input and transfer impedances.

2.3 Derive the properties of the LTP: its input impedance, transfer impedances and transfer functions.

2.4 Derive the transfer function of the SC network of Figure 2.18 using Laker's equivalent circuit method.

2.5 The Fleischer–Laker biquad [1.26] is shown in Figure 2.32. Under the condition $V_{io} = V_{ie}z^{-1/2}$, derive the four-port equivalent circuit obtained using Laker's method. Comment on the results regarding construction of a continuous-time two-port using bilinear transformation.

2.6 SC networks using multi-phase clocks can be analyzed by the IAM method, similar to the case for two-phase SC circuits. Analyze the four-phase SC circuits of Figure 2.16 using the IAM method. Evaluate the transfer functions from the DAM obtained.

2.7 The transmission matrix representation of a capacitor in two-phase SC network is given in equation (2.5). Note, however, that it relates the voltages across the capacitor in both phases and the charges flowing through it. It is possible to

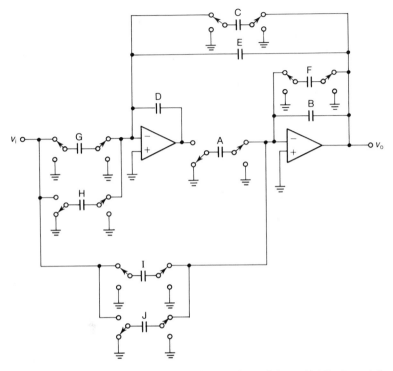

Fig. 2.32  The Fleischer–Laker biquad for analysis (*adapted from [1.26]*. Copyright © 1979 AT&T. All rights reserved. Reprinted with permission).

write an ABCD matrix for the four-port z-domain equivalent, by expressing the voltages across the capacitor in terms of terminal voltages. Then, we have

$$
\begin{bmatrix} V_{1e} \\ V_{1o} \\ I_{1e} \\ I_{1o} \end{bmatrix} = \begin{bmatrix} A & B \\ C & D \end{bmatrix} \begin{bmatrix} V_{2e} \\ V_{2o} \\ -I_{2e} \\ -I_{2o} \end{bmatrix}
$$

Each of the sub-matrices $A$, $B$, $C$ and $D$ in the above equation is of the form

$$
\begin{bmatrix} A_{ee} & A_{oe} \\ A_{eo} & A_{oo} \end{bmatrix}
$$

Discuss the procedure used to obtain each of the 16 entries in the transmission matrix of the four-port network [2.20].

2.8 Derive a continuous-time equivalent for a capacitor in an SC circuit.

2.9 An inverting amplifier using OA and resistors is shown in Figure 2.33. Derive an SC equivalent using $p$-transformation and discuss its practical feasibility.

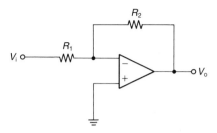

*Fig. 2.33* An inverting amplifier.


## 2.11  References

[2.1]  C.F. Kurth and G.S. Moschytz, Nodal analysis of switched-capacitor networks, *IEEE Transactions on Circuits and Systems*, CAS-26, 2, 93–105, February 1979.

[2.2]  C.F. Kurth and G.S. Moschytz, Two-port analysis of switched-capacitor networks using four-port equivalent circuits in the *z*-domain, *IEEE Transactions or Circuits and Systems*, CAS-26, 3, 166–80, March 1979.

[2.3]  K.R. Laker, Equivalent circuits for the analysis and synthesis of switched-capacitor networks, *Bell System Technical Journal*, 58, 729–69, March 1979.

[2.4]  G.C. Temes, H.J. Orchard and M. Jahanbegloo, Switched-capacitor filter design using the bilinear *z*-transform, *IEEE Transactions on Circuits and Systems*, CAS-25, 12, 1039–44, December 1978.

[2.5]  J.A. Mckinney and C.A. Halijak, The periodically reverse-switched capacitor, *IEEE Transactions on Circuit Theory*, CT-15, 3, 288–90, September 1968.

[2.6]  J.J. Mulawka, By-inspection analysis of switched-capacitor networks, *International Journal of Electronics*, 49, 359–73, November 1980.

[2.7]  E. Hokenek and G.S. Moschytz, Analysis of multi-phase switched-capacitor networks using the indefinite admittance matrix, *Proc. IEE*, Part *G*, 127, 5, 226–41, October 1980.

[2.8]  E. Hokenek and G.S. Moschytz, Analysis of general switched-capacitor networks using indefinite admittance matrix, *Proc. IEE*, Part *G*, 127, 1, 21–33, February 1980.

[2.9]  A. Nathan, Matrix analysis of networks having infinite gain operational amplifiers, *Proc. IRE*, 49, 10, 1577, October 1961.

[2.10]  L. Weinberg, *Network Analysis and Synthesis*, McGraw-Hill, 1962.

[2.11]  S.K. Mitra, *Analysis and Synthesis of Linear Active Networks*, Wiley, 1969.

[2.12]  G.S. Moschytz, *Linear Integrated Networks: Fundamentals*, Van Nostrand Reinhold, 1974.

[2.13]  A. Knob and R. Dessoulavy, Analysis of switched-capacitor networks in the frequency domain using continuous-time two-port equivalents, *IEEE Transactions on Circuits and Systems*, CAS-28, 10, 947–53, October 1981.

[2.14]  G. Muller and G.C. Temes, Simple method for analysis of a class of swiched-capacitor filters, *Electronics Letters*, 16, 22, 852–3, October 23, 1980.

[2.15]  A. Knob and R. Dessoulavy, Two-port representation of basic switched-capacitor structures by analog elements, *Proc. 4th International Symposium on Network Theory, Yugoslavia*, 419–23, September 1979.

[2.16] G.C. Temes and G. Muller, The poor man's algorithm for switched-capacitor circuit analysis, *Proc. 4th Asilomar Conference on Circuits, Systems and Computers, Pacific Grove, California*, 493–8, 1980.

[2.17] P.V. Ananda Mohan, V. Ramachandran and M.N.S. Swamy, Analysis of SC networks using continuous-time four-port equivalent circuits, *Electronics Letters*, 19, 25/26, 1094–5, 8 December 1983.

[2.18] Y.L. Kuo, M.L. Liou and J.W. Kasinkas, An equivalent circuit approach to the computer-aided analysis of switched-capacitor circuits, *IEEE Transactions on Circuits and Systems*, CAS-26, 9, 708–14, September 1979.

[2.19] G.S. Moschytz, Simplified analysis of switched-capacitor networks, *Electronics Letters*, 17, 25, 975–7, 10 December 1981.

[2.20] G.S. Moschytz and B.J. Hosticka, Transmission matrix of switched-capacitor ladder networks: application in active filter design, *Proc. IEE*, Part *G*, 127, 2, 87–98, April 1980.

## 2.11.1  Further reading

[2.21] A. Bandyopadhyay, Equivalent circuit of a switched-capacitor simulated resistor, *Proc. IEEE*, 68, 1, 178–9, January 1980.

[2.22] J.I. Sewell, Analysis of active switched-capacitor networks, *Proc. IEEE*, 68, 2, 292–3, February 1980.

[2.23] J. Vlach, K. Singhal and M. Vlach, Computer oriented formulation of equations and analysis of switched-capcitor networks, *IEEE Transactions on Circuits and Systems*, CAS-31, 9, 753–65, September 1984.

[2.24] L. Guand-Pu and S. Zhi-Guang, Analysis of switched-capacitor networks based on capacitor-switch macromodels, *IEEE Transactions on Circuits and Systems*, CAS-32, 1, 12–19, January 1985.

[2.25] A. Cichocki and R. Unbehauen, Simple technique for analysis of SC networks using general-purpose circuit simulation programs, *Electronics Letters*, 22, 18, 956–7, 22 August 1986.

[2.26] J. Vlach, Hand analysis of switched-capacitor networks, *IEEE Circuits and Devices Magazine*, 2, 6, 11–16, November 1986.

[2.27] M.F. Fahmy, M.Y. Makky and M.M. Doss, A method for frequency-domain analysis of switched-capacitor filters, *IEEE Transactions on Circuits and Systems*, CAS-34, 8, 955–60, August 1987.

[2.28] G.S. Moschytz, Elements of four-port network theory as required for SC network analysis, *International Journal of Circuit Theory and Applications*, 15, 3, 235–49, July 1987.

[2.29] L. Toth and E. Simonyi, Explicit formulas for analyzing general switched-capacitor networks, *IEEE Transactions on Circuits and Systems*, CAS-34, 12, 1564–78, December 1987.

[2.30] L.B. Wolovitz and J.I. Sewell, General analysis of large linear switched-capacitor networks, *Proc. IEE*, Part *G*, ECS, 135, 3, 119–24, June 1988.

[2.31] C.F. Lee, R.D. Davis, W.K. Jenkins and T.N. Trick, Sensitivity and nonlinear distortion analysis for switched-capacitor circuits using SCAPN, *IEEE Transactions on Circuits and Systems*, CAS-31, 2, 213–21, February 1984.

[2.32]  W.W. Poliscuk and B.L. Rojo, A note on statistical sensitivity computation in switched-capacitor networks, *IEEE Transactions on Circuits and Systems*, 35, 4, 423–5, April 1988.

[2.33]  G.S. Moschytz and U.W. Brugger, Signal-flow graph analysis of SC networks, *Proc. IEE*, Part *G*, 131, 1, 72–85, April 1984.

[2.34]  Zhou Xiao-Feng and Shen Zhi-Guang, Signal-flow graph analysis of switched-capacitor networks, *International Journal of Circuit Theory and Applications*, 13, 179–89, 1985.

[2.35]  G.S. Moschytz and J.J. Mulawka, Direct analysis of stray-insensitive switched-capacitor networks using signal-flow graphs, *Proc. IEE*, Part *G*, 133, 3, 145–53, June 1986.

[2.36]  A. Debrowski and G.S. Moschytz, Direct analysis of multiphase switched-capacitor networks using signal-flow graphs, *IEEE Transactions on Circuits and Systems*, 37, 5, 594–607, May 1990.

[2.37]  S.C. Fang, Y. Tsividis and O. Wing, State charge formulation of switched-capacitor networks containing nonlinear and time-varying elements, *IEEE Transactions on Circuits and Systems*, CAS-31, 11, 968–74, November 1984.

# Passive switched capacitor networks

Passive SC networks[*] could be derived from passive RC networks by replacing resistors with series-switched or parallel-switched or periodically reverse-switched

---

[*] Strictly speaking, the circuits described in this chapter use MOS switches and capacitors. An MOS switch, indeed, is an 'active' device requiring a clock to drive it. However, since the purpose served by the switches and the capacitor is to realize a 'resistor' which is a passive component, it is possible to visualize networks using only switches and capacitors and not active devices like OAs as 'passive' and those using OAs as 'active' networks, respectively.

capacitors. These have some properties similar to those of passive RC networks, as well as some different properties. There are also other types of passive SC network which do not have similar passive counterparts, and as such are capable of realizing additional interesting functions such as voltage amplification, realization of complex conjugate poles, etc. All these aspects of passive SC networks will be studied in this chapter in some detail.

## 3.1  Passive SC low-pass networks

### 3.1.1  First-order networks

Consider the first-order passive RC network of Figure 3.1(a), for which

$$\frac{V_o(s)}{V_i(s)} = \frac{1}{1 + sC_2 R_1} \tag{3.1}$$

The resistor $R_1$ could be replaced by a parallel-switched capacitor or by a series-switched capacitor [3.1, 3.2] as shown in Figures 3.1(b) and 3.1(c). Assuming a two-phase clock, it can be seen that, in the circuit of Figure 3.1(b), during one phase, $C_1$ is charged to the input voltage and this charge is transferred to $C_2$ during the other phase.

The charge conservation equation can be written at the instant $nT$ (i.e., when the switch in Figure 3.1(b) is thrown to the right) as follows:

$$C_2\{v_o(nT) - v_o((n-1)T)\} = C_1\{v_i((n-\tfrac{1}{2})T) - v_o(nT)\} \tag{3.2}$$

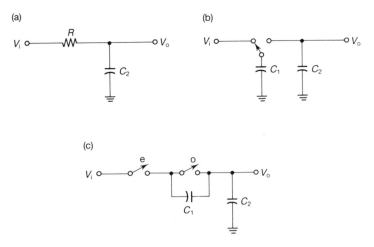

Fig. 3.1   (a) First-order passive RC LP network;   (b) SC version of (a) using parallel-switched capacitor;   (c) SC version of (a) using series-switched capacitor.

At the instant $nT$, $C_2$ was already holding a charge of $C_2 v_o((n-1)T)$, the result of a charge transfer at a previous 'such' instant. Equation (3.2) can be expressed in the $z$-domain as

$$[C_1 + C_2(1 - z^{-1})]\, V_{oo} = C_1 V_{ie} z^{-1/2} \tag{3.3}$$

or alternatively,

$$\frac{V_{oo}}{V_{ie}} = \frac{C_1 z^{-1/2}}{C_1 + C_2(1 - z^{-1})} \tag{3.4}$$

During the even phase, $C_2$ holds the charge it attained in the odd phase until the next odd phase. Hence, if the output in the even phase is considered, it is delayed by exactly half a cycle with respect to the output in the odd phase, i.e.,

$$V_{oe} = V_{oo} \cdot z^{-1/2}$$

or

$$\frac{V_{oe}}{V_{ie}} = \frac{C_1 z^{-1}}{C_1 + C_2(1 - z^{-1})} \tag{3.5}$$

The behaviour of equation (3.4) for high clock frequencies could be analyzed by noting that

$$(1 - z^{-1}) = 1 - e^{-sT} \approx sT \tag{3.6}$$

whence equation (3.4) becomes (ignoring the half-cycle delay in the numerator)

$$\frac{V_{oo}}{V_{ie}} = \frac{1}{1 + \dfrac{C_2}{C_1} sT} \tag{3.7}$$

Identifying equation (3.7) with equation (3.1), it is observed that an 'equivalent resistor' of value $T/C_1$ is realized at high clock frequencies, and hence the circuit of Figure 3.1(b) is similar to the passive network of Figure 3.1(a).

The circuit of Figure 3.1(c) using a series-switched capacitor is considered next. Noting that the charge transfer takes place during the even phase, the charge conservation equation follows as

$$C_2\{v_o(nT) - v_o((n-1)T)\} = C_1\{v_i(nT) - v_o(nT)\} \tag{3.8}$$

Therefore,

$$\frac{V_{oe}}{V_{ie}} = \frac{C_1}{C_1 + C_2(1 - z^{-1})} \tag{3.9}$$

Note that equation (3.9) is the same as equation (3.4) except for the half-cycle delay ($z^{-1/2}$ term in the numerator). Hence, at high clock frequencies, the equivalent resistor simulated is the same, i.e., $T/C_1$. In the odd phase, the output is held and hence

$$V_{oo} = V_{oe}z^{-1/2} \qquad (3.10)$$

Thus, the circuits of Figures 3.1(b) and 3.1(c) are identical as far as their amplitude-frequency characteristic is concerned, whereas their phase responses are different. This difference will have some effect on the performance of high-order active networks employing these, as will be shown in a later chapter.

## 3.1.2  Second-order networks [3.3]

The second-order SC networks using parallel-switched and series-switched capacitors, derived from the RC network of Figure 3.2(a), can now be evaluated.

For the network of Figure 3.2(b) using parallel-switched capacitors, we obtain, by means of Laker's equivalent circuits,[*]

$$\frac{V_{oe}}{V_{ie}} = \frac{z^{-2}}{\frac{C_2C_4}{C_1C_3}(1-z^{-1})^2 + \left\{\frac{C_2}{C_1} + \frac{C_4}{C_3} + (C_3 + C_4)\left(\frac{1}{C_1} + \frac{1}{C_2}\right)\right\}(1-z^{-1}) + 1} \qquad (3.11a)$$

$$V_{oo} = V_{oe}z^{-1/2} \qquad (3.11b)$$

For the circuit of Figure 3.2(c) using series-switched capacitors, we obtain

$$\frac{V_{oe}}{V_{ie}} = \frac{1}{\frac{C_2C_4}{C_1C_3}(1-z^{-1})^2 + \left(\frac{C_2 + C_4}{C_1} + \frac{C_4}{C_3}\right)(1-z^{-1}) + 1} \qquad (3.12a)$$

$$V_{oo} = V_{oe}z^{-1/2} \qquad (3.12b)$$

There are two differences between the transfer functions realized by circuits of Figures 3.2(b) and 3.2(c). The first is that there are no delay terms in the numerator of equation (3.12a). The second is that the realized denominators of the transfer functions are different, i.e., the negative real poles realized by the circuits are different. For high clock frequencies, as mentioned before, equation (3.6) (the approximation $(1 - z^{-1}) = sT$) could be used[†] to study the properties of equations

---

[*]  All the charge conservation equations and transfer functions can be obtained by using Laker's equivalent circuits. These can also be written by inspection, albeit with some difficulty.
[†]  Note that, alternatively, the substitution $(1 - z^{-1})/T \to p$ can be used to analyze the circuits in the $p$-domain. For high sampling frequencies, the $p$ and $s$ domains approximate each other. Hence, the approximation $(1 - z^{-1}) = sT$ or $pT$ means the same for high sampling frequencies.

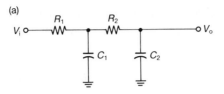

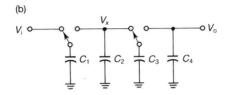

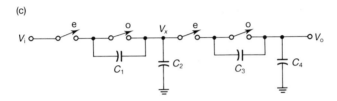

*Fig. 3.2* (a) Second-order passive RC LP network; (b) SC version of (a) using parallel-switched capacitors; (c) SC version of (b) using series-switched capacitors.

(3.11a) and (3.12a). The corresponding denominators of equations (3.11a) and (3.12a) become, respectively

$$D_1(s) = \frac{C_2C_4}{C_1C_3} s^2 T^2 + \left[\frac{C_2}{C_1} + \frac{C_4}{C_3} + (C_3 + C_4)\left(\frac{1}{C_1} + \frac{1}{C_2}\right)\right] sT + 1 \qquad (3.13a)$$

$$D_2(s) = \frac{C_2C_4}{C_1C_3} s^2 T^2 + \left[\frac{C_2 + C_4}{C_1} + \frac{C_4}{C_3}\right] sT + 1 \qquad (3.13b)$$

Identifying equations (3.13) with $(s^2 + s\omega_p/Q_p + \omega_p^2)$ to evaluate $\omega_p$ and $Q_p$, it can be seen that the $\omega_p$ for both the circuits is the same, i.e.,

$$\omega_p = f_s \sqrt{\frac{C_1C_3}{C_2C_4}} \qquad (3.14a)$$

where $f_s = 1/T$ is the sampling frequency, and also that the respective $Q_p$s are

$$Q_{p1} = \sqrt{\frac{C_2C_4}{C_1C_3}} \left[\left(\frac{C_2}{C_1} + \frac{C_4}{C_3}\right) + (C_3 + C_4)\left(\frac{1}{C_1} + \frac{1}{C_2}\right)\right]^{-1} \qquad (3.14b)$$

$$Q_{p2} = \sqrt{\frac{C_2C_4}{C_1C_3}} \left(\frac{C_2 + C_4}{C_1} + \frac{C_4}{C_3}\right)^{-1} \qquad (3.14c)$$

Thus, for the same capacitor values, it can be seen from equations (3.14b) and (3.14c) that $Q_{p2} > Q_{p1}$ or the circuit using series-switched capacitors realizes a $Q_p$ greater than that using parallel-switched capacitors.

In practice, one can use either 'tapered' or 'untapered' networks. Both these types of network use equal time constants in the two sections, i.e.,

$$\frac{C_2}{C_1} = \frac{C_4}{C_3} \tag{3.15}$$

but for untapered networks $C_2 = C_4$ and $C_1 = C_3$, whereas for tapered networks, $\rho C_2 = C_4$ and $\rho C_1 = C_3$. These two cases will now be considered separately.

*Case 1: $C_2 = C_4$; $C_1 = C_3$*
Equations (3.14b) and (3.14c) give

$$Q_{p1} = \left( 3 + \frac{2C_1}{C_2} + \frac{C_1^2}{C_2^2} \right)^{-1} \tag{3.16a}$$

$$Q_{p2} = \tfrac{1}{3} \tag{3.16b}$$

Thus for the circuit of Figure 3.2(c), $Q_p$ is independent of $C_2/C_1$ and is larger than that for the circuit of Figure 3.2(b). It will be shown in a later chapter that this difference in $Q_p$ necessitates the use of larger amplifier gain for the circuit using parallel-switched capacitors, when used to realize active SC filters.

*Case 2: $C_4 = \rho C_2$; $C_3 = \rho C_1$*
Here

$$Q_{p1} = \left( 2 + \rho + \frac{2\rho C_1}{C_2} + \frac{\rho C_1^2}{C_2^2} \right)^{-1} \tag{3.17a}$$

$$Q_{p2} = \frac{1}{2 + \rho} \tag{3.17b}$$

Thus, even with tapering, the series-switched capacitor type network realizes larger $Q_p$ than the parallel-switched capacitor type of network. But, the difference between the two $Q_p$s decreases with tapering. For example, with $C_2/C_1 = 100$ and $\rho = 1$, 2 and 10, $Q_{p1}/Q_{p2}$ values are 0.9933, 0.9949 and 0.9982, respectively, showing that $Q_{p1}$ approaches $Q_{p2}$ for large $\rho$. Tapered networks are useful in the realization of Sallen–Key second-order filters with unity amplifier gain.

### 3.1.3  Third-order networks [3.3, 3.4]

For the third-order LP SC network using parallel-switched capacitors (Figure 3.3(a)), it can be shown that

$$
\frac{V_{oe}}{V_{ie}} = z^{-3} \Bigg/ \Bigg( \left[ \frac{C_2 C_4 C_6}{C_1 C_3 C_5} \right] (1 - z^{-1})^3 + \left[ \frac{C_2}{C_1} \left\{ \frac{C_4}{C_3} + \frac{C_6}{C_5} + \left( \frac{1}{C_3} + \frac{1}{C_4} \right)(C_5 + C_6) \right\} \right.
$$
$$
\left. + \frac{C_6}{C_5} \left\{ \frac{C_4}{C_3} + (C_3 + C_4)\left( \frac{1}{C_1} + \frac{1}{C_2} \right) \right\} + \frac{C_3}{C_4} \left( \frac{1}{C_1} + \frac{1}{C_2} \right)(C_5 + C_6) \right] (1 - z^{-1})^2
$$
$$
+ \left[ \frac{C_2}{C_1} + \frac{C_4}{C_3} + \frac{C_6}{C_5} + (C_3 + C_4)\left( \frac{1}{C_1} + \frac{1}{C_2} \right) \right.
$$
$$
\left. + (C_5 + C_6)\left( \frac{1}{C_1} + \frac{1}{C_2} + \frac{1}{C_3} + \frac{1}{C_4} \right) \right] (1 - z^{-1}) + 1 \Bigg)
\tag{3.18}
$$

For identical sections, i.e.,

$$
C_2 = C_4 = C_6 = \alpha C
$$
$$
C_1 = C_3 = C_5 = C
\tag{3.19a}
$$

we obtain

$$
\frac{V_{oe}}{V_{ie}} = z^{-3} \Bigg/ \left[ \alpha^3 (1 - z^{-1})^3 + \left[ 3\alpha^2 + 2(\alpha + 1)^2 + \frac{(\alpha + 1)^2}{\alpha^2} \right] (1 - z^{-1})^2 \right.
$$
$$
\left. + \left[ 3\alpha + \frac{3(\alpha + 1)^2}{\alpha} \right] (1 - z^{-1}) + 1 \right]
\tag{3.19b}
$$

(a)

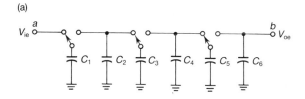

(b)

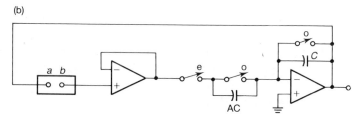

*Fig. 3.3*  (a) Third-order LP SC network;  (b) phase-shift oscillator using (a).

The reader is urged to evaluate the attenuation of this network while realizing a phase shift of $180°$, using the approximation $(1 - z^{-1}) = sT$ (i.e., higher sampling frequencies) and compare the results with an RC three-section phase-shift network (see Exercise 3.2).

A phase-shift oscillator can be realized using the SC phase-shift network of Figure 3.3(a) and this is as shown in Figure 3.3(b). Note that the OA and associated switches amplify the output signal $V_{oe}$ in the even phase and feed it back to the phase-shift network in the same phase. The amplifier realizes a gain of $-A$.

The exact behaviour of the phase-shift oscillator of Figure 3.3(b) is different from that of active RC phase-shift oscillator because of the $z^{-3}$ term in the numerator. Hence, by performing the exact analysis of the actual SC phase-shift oscillator of Figure 3.3(b), the actual amplifier gain required for oscillation and the frequency of oscillation can be obtained. We now equate the loop gain to unity to satisfy the Barkhausen criterion for oscillations. Hence, from equation (3.19a) we get:

$$\left(\alpha^3 + 5\alpha^2 + 10\alpha + 10 + \frac{5}{\alpha} + \frac{1}{\alpha^2}\right)z^3 - \left(3\alpha^3 + 10\alpha^2 + 14\alpha + 12 + \frac{7}{\alpha}\right)z^2$$

$$+ \left(3\alpha^3 + 5\alpha^2 + 4\alpha + 3 + \frac{2}{\alpha} + \frac{1}{\alpha^2}\right)z - (\alpha^3 - A)$$

$$= uz^3 - vz^2 + wz - (t - A) \qquad \text{(say)}$$

$$= 0 \qquad\qquad\qquad\qquad\qquad\qquad (3.20)$$

If equation (3.20) has to represent a 'digital oscillator', we need

$$z^3 - \frac{v}{u}z^2 + \frac{w}{u}z - \frac{(t-A)}{u} = (z^2 - rz + 1)(z - q) = z^3 - (q + r)z^2 + (1 + qr)z - q = 0$$

$$(3.21)$$

Identifying coefficients, and after suitable manipulation, the condition for oscillation can be obtained:

$$A^2 + A(v - 2t) - t(v - t) + u(w - u) = 0 \qquad (3.22a)$$

and the frequency of oscillation is

$$f_o = \frac{f_s}{2\pi}\cos^{-1}\frac{r}{2} = \frac{f_s}{2\pi}\cos^{-1}\frac{(v - t + A)}{2u} \qquad (3.22b)$$

From equations (3.22), it is noted that the amplifier gain $A$ required for oscillation is dependent only on $\alpha$, the capacitor spread. The $A$-values for various values of $\alpha$ are presented in Table 3.1. It can be seen that $A$ is minimum (8.8547) for $\alpha = \frac{3}{4}$. For $\alpha = 1$ (all capacitors equal), $A = 9$ and $f_o = 0.080\,43f_s$. The exact analysis of the phase-shift oscillator thus has shown that SC networks do have certain advantages over active RC networks in that they require lesser amplifier gain for oscillation. The reader is urged to verify that by using series-switched capacitors in phase-shift networks, a phase-shift oscillator cannot be realized.

**Table 3.1** Frequency of oscillation and gain required for oscillation for phase shift oscillator with various values of $\alpha$.

| $\alpha$ | $A$ | $f_0/f_s$ |
|---|---|---|
| 0.10 | 19.955 | 0.054 40 |
| 0.50 | 9.165 88 | 0.081 72 |
| 0.75 | 8.854 7 | 0.082 38 |
| 1.00 | 9.000 | 0.080 43 |
| 5.00 | 14.524 | 0.044 42 |
| 10.00 | 18.463 | 0.028 03 |
| 50.00 | 25.665 | 0.007 20 |
| 100.00 | 27.205 | 0.003 76 |

## 3.2 Passive SC high-pass networks [2.3, 3.3, 3.5]

The passive RC HP network of Figure 3.4(a) has transfer function

$$\frac{V_o(s)}{V_i(s)} = \frac{sCR}{1 + sCR} \tag{3.23}$$

Replacing the resistor in Figure 3.4(a) by series-switched or parallel-switched capacitor, the same SC circuit is obtained; this passive first-order HP network is shown in Figure 3.4(b). It can be shown that the circuit has transfer functions

$$\frac{V_{oe}}{V_{ie}} = \frac{C_1(1 - z^{-1})}{C_1(1 - z^{-1}) + C_2} \tag{3.24a}$$

$$V_{oo} = \frac{[C_1(1 - z^{-1}) + C_2] V_{io} - C_2 V_{ie} z^{-1/2}}{C_1(1 - z^{-1}) + C_2} \tag{3.24b}$$

It can be seen that equation (3.24a) is an HP transfer function. It is interesting to note further that when $V_{io} = 0$, $V_{oo}/V_{ie}$ is an LP transfer function [3.5]. Note also the sign change present for the output:

$$\left. \frac{V_{oo}}{V_{ie}} \right|_{V_{io}=0} = \frac{-C_2 z^{-1/2}}{C_1(1 - z^{-1}) + C_2} \tag{3.24c}$$

The circuit arrangement for realizing $V_{io} = 0$ is shown in Figure 3.4(c). Thus the addition of two switches at the input has made an LP filter to be realizable using an HP network. For other input conditions, in particular $V_{ie} = V_{io} z^{-1/2}$ (i.e., the input in the odd phase is held over a clock cycle), equation (3.24b) gives

$$\frac{V_{oo}}{V_{io}} = \frac{(C_1 + C_2)(1 - z^{-1})}{C_1(1 - z^{-1}) + C_2} \tag{3.24d}$$

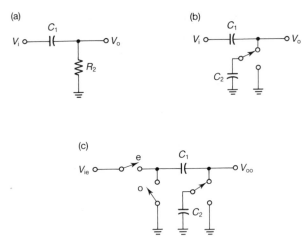

*Fig. 3.4* (a) Passive RC HP network; (b) SC HP network derived from (a); (c) SC LP network derived from (b).

Note that this transfer function has a different gain factor than that of equation (3.24a).

The second-order HP SC networks shown in Figures 3.5(b) and 3.5(c), derived from the passive RC circuit in Figure 3.5(a), are considered next. The circuits of Figures 3.5(b) and 3.5(c) are different in that the second section is switched in-phase or anti-phase with the first section. An analysis of the network of Figure 3.5(b) gives:

$$\frac{V_{oe}}{V_{ie}} = \frac{(1-z^{-1})^2}{(1-z^{-1})^2 + \left(\dfrac{C_2+C_4}{C_1} + \dfrac{C_4}{C_3}\right)(1-z^{-1}) + \dfrac{C_2C_4}{C_1C_3}} \tag{3.25a}$$

$$V_{oo} = V_{oe}z^{-1/2} + V_{io} - V_{ie}z^{-1/2} \tag{3.25b}$$

Thus a second-order HP transfer function is obtained. It may be verified from equation (3.25b) that $V_{oo}$ (output in the odd phase) does not give any useful transfer function, under any input conditions such as $V_{io} = 0$, $V_{ie} = 0$, $V_{io} = V_{ie}z^{-1/2}$, and $V_{ie} = z^{-1/2}$.

For the circuit of Figure 3.5(c), we have

$$V_{oo} = \frac{\left[\left(1 - z^{-1} + \dfrac{C_2}{C_1}\right)V_{io} - \dfrac{C_2}{C_1}V_{ie}z^{-1/2}\right](1-z^{-1})}{(1-z^{-1})^2 + \left(\dfrac{C_2+C_4}{C_1} + \dfrac{C_4}{C_3} + \dfrac{C_2C_4}{C_1^2}\right)(1-z^{-1}) + \dfrac{C_2C_4}{C_1C_3}} \tag{3.26}$$

For $V_{io} = 0$, equation (3.26) can be seen to realize a BP transfer function with a change of sign. For $V_{io} = V_{ie}z^{-1/2}$, an HP transfer function is realized. For $V_{ie} = V_{io}z^{-1/2}$ also, an HP transfer function with a gain multiplier of $1 + C_2/C_1$ is

(a)

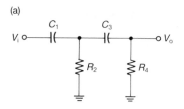

(b)

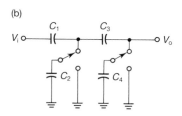

(c)

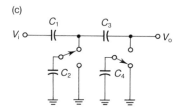

*Fig. 3.5*  Second-order HP networks:  (a) passive type;  (b) SC type with in-phase switching;  (c) SC type with anti-phase switching.

realized. It can be shown that $V_{oe}$ does not realize any useful transfer function under any input conditions. For identical sections, i.e., $C_2 = C_4$ and $C_1 = C_3$, and assuming high clock frequencies, it is observed from equations (3.25) and (3.26) that while the $\omega_p$s are the same, the respective $Q_p$s are different:

$$Q_p\,|\,\text{Figure 3.5(b)} = \tfrac{1}{3}$$

$$Q_p\,|\,\text{Figure 3.5(c)} = \frac{1}{3 + \dfrac{C_2}{C_1}} \tag{3.27}$$

showing that, as in the case of LP networks, the switch-phasing influences the $Q_p$ realizable. These HP networks can be used to realize second-order Sallen–Key active filters.

## 3.3  Passive SC band-pass networks

Passive RC BP networks are shown in Figures 3.6(a) and 3.6(b). They are different in the sense that one has an LP network followed by an HP network and the other has an HP network followed by an LP network. The SC versions of the BP network are obtained by replacing resistors with series- or parallel-switched capacitors. Eight different configurations are possible and these are shown in Figures 3.6(c)–3.6(j) to illustrate the variety of SC networks possible.

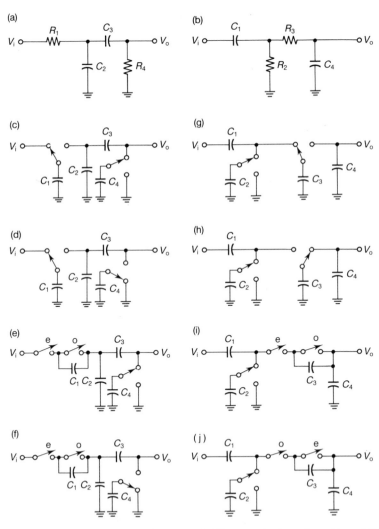

Fig. 3.6  (a), (b) Passive RC BP networks;  (c)–(j) SC networks based on (a) and (b).

The transfer functions of the circuits shown in Figures 3.6(c)–3.6(f) are as follows: for Figure 3.6(c),

$$\frac{V_{oe}}{V_{ie}} = \frac{N_1(z)}{D_1(z)} = \frac{\frac{C_1}{C_2}z^{-1}(1-z^{-1})}{(1-z^{-1})^2 + \left(\frac{C_1}{C_2} + \frac{C_1 C_4}{C_3^2} + \frac{C_4}{C_2} + \frac{C_4}{C_3}\right)(1-z^{-1}) + \frac{C_1 C_4}{C_2 C_3}} \tag{3.28a}$$

$$V_{oo}z^{-1/2} = V_{oe}\frac{[C_2(C_3+C_4) + C_3 C_4]}{C_2 C_3} \tag{3.28b}$$

for Figure 3.6(d),

$$\frac{V_{oo}}{V_{ie}} = \frac{N_2(z)}{D_2(z)} = \frac{\frac{C_1}{C_2}z^{-1/2}(1-z^{-1})}{(1-z^{-1})^2 + \left(\frac{C_1 C_3 + C_4(C_2+C_3)}{C_2 C_3}\right)(1-z^{-1}) + \frac{C_1 C_4}{C_2 C_3}} \tag{3.29a}$$

$$V_{oe} = V_{oo}z^{-1/2} \tag{3.29b}$$

for Figure 3.6(e),

$$\frac{V_{oe}}{V_{ie}} = \frac{\frac{C_1}{C_2}(1-z^{-1})}{D_2(z)} \tag{3.30a}$$

$$V_{oo} = V_{oe}z^{-1/2} \tag{3.30b}$$

and for Figure 3.6(f),

$$\frac{V_{oe}}{V_{ie}} = \frac{\frac{C_1}{C_2}z^{-1/2}(1-z^{-1})}{D_1(z)} \tag{3.31a}$$

$$V_{oe} = \frac{V_{oo}}{z^{-1/2}}\left(\frac{C_2(C_3+C_4) + C_3 C_4}{C_2 C_3}\right) \tag{3.31b}$$

It can be noted that there are only two different types of transfer function denominator. Thus, there are only two types of circuits among those of Figures 3.6(c)–3.6(f). The negative real poles realized are different accordingly. Similar analysis on the BP RC networks shown in Figures 3.6(g)–3.6(j) will show that the transfer functions are as follows: for Figure 3.6(g),

$$\frac{V_{oo}}{V_{ie}} = \frac{\frac{C_3}{C_4}z^{-1/2}(1-z^{-1})}{(1-z^{-1})^2 + \left(\frac{C_3(C_1+C_3+C_4) + C_2 C_4}{C_1 C_4}\right)(1-z^{-1}) + \frac{C_2 C_3}{C_1 C_4}} \tag{3.32a}$$

$$V_{oe} = V_{oo}z^{-1/2} \tag{3.32b}$$

for Figure 3.6(h),

$$V_{oe} = \frac{C_1 C_3 [(C_1(1-z^{-1})+C_2)V_{io} - C_2 z^{-1/2} V_{ie}] z^{-1/2}}{C_1^2 C_4 (1-z^{-1})^2 + (C_1^2 C_3 + C_4(C_1 C_2 + C_2 C_3 + C_3 C_1))(1-z^{-1})}$$

$$+ (C_3^2(C_1 + C_2) + C_1 C_2 C_3)$$

(3.33a)

$$V_{oo} = V_{oe} z^{-1/2} \tag{3.33b}$$

for Figure 3.6(i)

$$\frac{V_{oe}}{V_{ie}} = \frac{\dfrac{C_1}{C_2}(1-z^{-1})}{\dfrac{C_1 C_4}{C_2 C_3}(1-z^{-1})^2 + \left(\dfrac{C_1 + C_4}{C_2} + \dfrac{C_4}{C_3}\right)(1-z^{-1}) + 1} \tag{3.34a}$$

$$V_{oo} = V_{oe} z^{-1/2} \tag{3.34b}$$

and for Figure 3.6(j)

$$V_{oo} = \frac{C_1 C_3 [(C_1(1-z^{-1})+C_2)V_{io} + C_2 z^{-1/2} V_{ie}]}{C_1^2 C_4 (1-z^{-1})^2 + (C_1^2 C_3 + C_4(C_1 C_2 + C_2 C_3 + C_3 C_1))(1-z^{-1}) + C_1 C_2 C_3}$$

(3.35a)

$$V_{oe} = V_{oo} z^{-1/2} \tag{3.35b}$$

Thus there are three different types of transfer function denominator among the BP filters of Figures 3.6(g)–3.6(j). Note also that the output is held over a clock period. These examples have been presented to show the variety of SC circuits possible that can be derived from passive RC networks.

## 3.4   Wien-bridge SC networks and oscillators [3.3, 3.6]

Consider the passive RC Wien bridge network of Figure 3.7(a), from which SC networks can be obtained. Two networks with different transfer function denominators are shown in Figures 3.7(b) and 3.7(c). It may be noted that two other circuits are possible, but they have either of the denominators of the transfer functions of Figures 3.7(b) and 3.7(c). The transfer functions are as follows: for the SC network of Figure 3.7(b),

$$\frac{V_{oe}}{V_{ie}} = \frac{C_1 C_2 C_3 z^{-1}(1-z^{-1})}{C_2 C_3^2(1-z^{-1})^2 + [C_1 C_2(C_3 + C_4) + C_3(C_1 C_3 + C_2 C_4)](1-z^{-1}) + C_1 C_3 C_4} \tag{3.36a}$$

$$V_{oo} = \frac{C_3 + C_4}{C_3} z^{-1/2} V_{oe} \tag{3.36b}$$

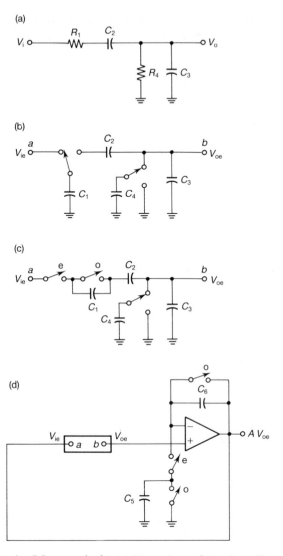

*Fig. 3.7* (a) Wien passive RC network; (b), (c) SC versions of (a); (d) oscillator using (b) and (c).

and for the SC network of Figure 3.7(c),

$$\frac{V_{oe}}{V_{ie}} = \frac{C_1 C_2 (1 - z^{-1})}{C_2 C_3 (1 - z^{-1})^2 + [C_1(C_2 + C_3) + C_2 C_4](1 - z^{-1}) + C_1 C_4} \qquad (3.37a)$$

$$V_{oo} = V_{oe} z^{-1/2} \qquad (3.37b)$$

These Wien bridge SC networks could be used to realize SC oscillators [3.3, 3.6], as shown in Figure 3.7(d). Note that the OA and associated switches amplify the output $V_{oe}$ in the even phase and feed it back to the input of the Wien network. During the odd phase, the feedback loop is open. The amplifier has gain $1 + C_5/C_6$, and the amplifier is an SC version of a non-inverting amplifier using resistors. These oscillators should be analyzed exactly to evaluate the condition for oscillation and the frequency of oscillation. Denoting $V_{oe}/V_{ie}$ in equation (3.36a) or (3.37a) as $\beta$ and using an amplifier with gain $A$, the condition for oscillation is $A\beta = 1$. Thus, the gains required for oscillation and frequencies are as follows: for the SC network of Figure 3.7(b), we have

$$A = 1 + \frac{C_4}{C_1} + \frac{C_3 + C_4}{C_2} + \frac{C_4}{C_3} \tag{3.38a}$$

$$f_o = \frac{f_s}{2\pi} \cos^{-1}\left[1 - \frac{C_1 C_4}{2C_2(C_3 + AC_1)}\right] \tag{3.38b}$$

and for the network of Figure 3.7(c), it can be shown that

$$A = 1 + \frac{C_4}{C_1} + \frac{C_3 + C_4}{C_2} \tag{3.39a}$$

$$f_o = \frac{f_s}{2\pi} \cos^{-1}\left(1 - \frac{C_1 C_4}{2C_2 C_3}\right) \tag{3.39b}$$

For $C_1 = C_4$, $C_2 = C_3$, i.e., equal time constants in both the sections, $A = 3 + 2C_1/C_2$ and $A = 3 + C_1/C_2$, respectively, for the circuits of Figures 3.7(b) and 3.7(c), thus showing that the amplifier gain is greater than 3 and is more for the circuit of Figure 3.7(b). It may also be noted that, in Figures 3.7(b) and 3.7(c), $C_2$ and the switched capacitor $C_1$ can be changed so that $C_2$ faces the input first, and then the resulting networks have different transfer functions, slightly different conditions for oscillation, and different oscillation frequencies when used in oscillators.

## 3.5  Bridged-T networks [3.3]

Consider the passive RC bridged-T network of Figure 3.8(a), which has a transfer function given by:

$$\frac{V_o(s)}{V_i(s)} = \frac{C_2 C_4 R_1 R_3 s^2 + C_4(R_1 + R_3)s + 1}{C_2 C_4 R_1 R_3 s^2 + [C_2 R_1 + C_4(R_1 + R_3)]s + 1} \tag{3.40a}$$

Equation (3.40a) realizes a minimum transmission at an angular frequency

$$\omega_N = \frac{1}{\sqrt{C_2 C_4 R_1 R_3}} \tag{3.40b}$$

of magnitude

$$\left.\frac{V_o}{V_i}\right|_{\text{Notch frequency}} = \frac{C_4(R_1 + R_3)}{C_2 R_1 + C_4(R_1 + R_3)} \tag{3.40c}$$

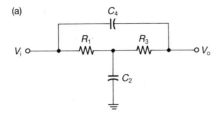

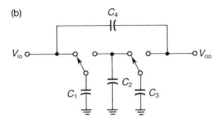

*Fig. 3.8* (a) Passive RC bridged-T network; (b) SC version of (a).

Thus a null transmission is not realizable using the bridged-T RC network of Figure 3.8(a).

Replacing resistors by parallel-switched capacitors, as shown in Figure 3.8(b), and analyzing the circuit yields

$$V_{oo} = \frac{C_1 C_2 C_3 V_{ie} z^{-3/2} + [C_4 C_2^2(1 - z^{-1}) + C_4(C_1 C_2 + C_2 C_3 + C_3 C_1)] V_{io}(1 - z^{-1})}{C_4 C_2^2(1 - z^{-1})^2 + [C_2^2 C_3 + C_3^2(C_1 + C_2) + C_4(C_1 C_2 + C_2 C_3 + C_3 C_1)](1 - z^{-1}) + C_1 C_2 C_3}$$

$$(3.41a)$$

Thus, when $V_{ie} = V_{io}z^{-1/2}$, using a sampled and hold input for the network, we obtain

$$\frac{V_{oo}}{V_{io}} = \frac{\left(1 + \dfrac{C_2 C_4}{C_1 C_3}\right)(1 - z^{-1})^2 + \left(\dfrac{C_4}{C_1} + \dfrac{C_4}{C_2} + \dfrac{C_4}{C_3} - 2\right)(1 - z^{-1}) + 1}{\left(\dfrac{C_2}{C_1}\right)\left(1 + \dfrac{C_4}{C_3}\right)(1 - z^{-1})^2 + \left(\dfrac{C_2 + C_3 + C_4}{C_1} + \dfrac{C_3 + C_4}{C_2} + \dfrac{C_4}{C_3}\right)(1 - z^{-1}) + 1}$$

$$(3.41b)$$

When

$$\frac{C_4}{C_1} + \frac{C_4}{C_2} + \frac{C_4}{C_3} = 1 \qquad\qquad (3.42)$$

the numerator of equation (3.41b) becomes

$$N(z) = \left(1 + \frac{C_2 C_4}{C_1 C_3}\right) z^{-2} - \left(1 + \frac{2C_2 C_4}{C_1 C_3}\right) z^{-1} + \left(1 + \frac{C_2 C_4}{C_1 C_3}\right)$$

showing that a perfect null is obtainable at a frequency given by

$$f_{\text{null}} = \frac{f_s}{2\pi} \cos^{-1} \left\{ \frac{\left(1 + \frac{2C_2 C_4}{C_1 C_3}\right)}{2\left(1 + \frac{C_2 C_4}{C_1 C_3}\right)} \right\} \tag{3.43}$$

It remains to be seen whether equation (3.41b) realizes an LP notch, HP notch or null transfer function. This is possible by comparing the magnitudes of equation (3.41b) at zero ($z = 1$) and $f_s/2$ ($z = -1$) frequencies. These are, respectively, unity and

$$\frac{\frac{4C_2 C_4}{C_1 C_3} + 3}{\frac{4C_2 C_4}{C_1 C_3} + 3 + \frac{6C_2 + 2C_3}{C_1} + \frac{2C_3}{C_2}} < 1 \tag{3.44}$$

Thus the circuit of Figure 3.8(b) can realize an LP notch transfer function. There are only three capacitor ratios ($C_2/C_1$, $C_3/C_1$ and $C_4/C_1$) involved in the design of the notch filter of Figure 3.8(b) to satisfy four requirements, namely, $\omega_p$, zero-frequency, $Q_p$, and condition for null. Thus the notch filter can be designed in two ways depending on the two types of specifications: given the null frequency and $\omega_p$; or given the null frequency and transmission magnitude at infinite frequency. (Note that, for passive RC networks, these specifications are the same, whereas for the SC network of Figure 3.8(b), they are different.) The design procedure is left as an exercise to the reader. Note that the bridged-T SC network can be used as the basis for developing single-amplifier SC filters.

## 3.6  All-pass and null networks [3.3]

Two simple methods of realizing first-order active all-pass networks are shown in Figures 3.9(a) and 3.9(b). The circuit of Figure 3.9(b) is popular in realizing active RC biquads [3.7]. It is of interest to see whether simpler all-pass realizations are possible using SC techniques. If the realization of Figure 3.9(a) is considered, it is observed that an inversion of the input is required to achieve an all-pass function, which suggests that an 'inverting' SC may serve the purpose. Thus, the circuit of Figure 3.9(c) is obtained which realizes a true all-pass function, when $V_{ie} = V_{io} z^{-1/2}$, given by

$$\frac{V_{oo}}{V_{io}} = \frac{C_1 - (C_1 + C_2) z^{-1}}{(C_1 + C_2) - C_1 z^{-1}} \tag{3.45}$$

($V_{oe}$ has an extra delay).

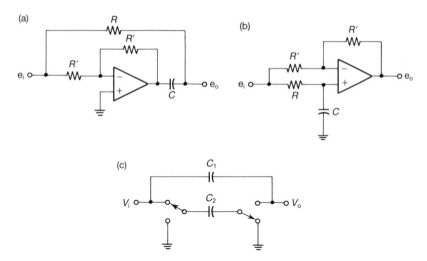

Fig. 3.9  First-order all-pass networks:  (a), (b) active RC versions; (c) SC version.

Such circuits, since they employ fewer components than those of Figures 3.9(a) and 3.9(b), can be used to replace the two all-pass networks required to realize Moschytz's modified Tarmy–Ghausi active filter [3.8], as will be discussed in Chapter 5.

It will now be shown that the voltage inversion property can be extended to realize second-order all-pass networks. The desired all-pass function is

$$T(z) = \frac{R^2z^2 - Pz + 1}{z^2 - Pz + R^2}$$  (3.46a)

which can be rewritten as

$$T(z)|_{AP} = \frac{(1 - z^{-1})^2 + (P - 2)(1 - z^{-1}) + (R^2 - P + 1)}{R^2(1 - z^{-1})^2 + (P - 2R^2)(1 - z^{-1}) + (R^2 - P + 1)}$$  (3.46b)

Using the 'p-transformation' to obtain the s-domain transfer function required to realize equation (3.46b), we obtain

$$T(p)|_{AP} = \frac{s^2T^2 + (P - 2)sT + (R^2 - P + 1)}{R^2s^2T^2 + (P - 2R^2)sT + (R^2 - P + 1)}$$  (3.46c)

Evidently, equation (3.46c) can be seen to be an HP notch which can be realized as shown in Figure 3.10(a). It may be noted that this circuit requires an inverted input which could be obtained in the SC version through an inverting SC. Further, the circuit also needs an attenuated input. It is seen that attenuation may be provided by modifying the circuit of Figure 3.10(a) as shown in Figure 3.10(b) which, after combining resistors $R_3$ and $R_4$, is simplified as in Figure 3.10(c). The SC version of Figure 3.10(c) is thus obtained, and is as shown in Figure 3.10(d).

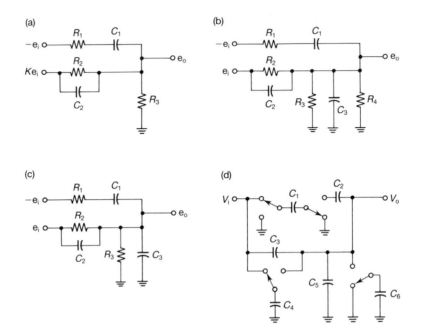

**Fig. 3.10**  (a) An HP notch in the analog domain;  (b) equivalent version of (a);  (c) network reduced from (b);  (d) SC version realizing all-pass or notch functions.

For the circuit of Figure 3.10(d), the transfer functions $V_{oe}/V_{ie}$ with $V_{io} = V_{ie}z^{-1/2}$, and $V_{oo}/V_{io}$ with $V_{ie} = V_{io}z^{-1/2}$ are delay-free, whereas the other two transfer functions, $V_{oo}/V_{io}$ and $V_{oo}/V_{ie}$ have an additional $z^{-1/2}$ term in the numerator of the transfer function. Further, $V_{oe}/V_{ie}$ has zeros and poles both dependent on all the capacitor values, whereas $V_{oo}/V_{io}$ has zeros dependent only on $C_1$, $C_2$ and $C_4$ and poles dependent on all the capacitors. Hence, only $V_{oo}/V_{io}$ is considered here in view of its simplicity for design:

$$\frac{V_{oo}}{V_{io}} = \frac{C_2(C_3 - C_1 - C_4)(1 - z^{-1})^2 + [C_1(C_3 - C_4) + C_2(C_4 - C_1)](1 - z^{-1}) + C_1C_4}{C_2(C_3 + C_5)(1 - z^{-1})^2 + [C_1(C_2 + C_3 + C_5) + C_2(C_4 + C_6)](1 - z^{-1}) + C_1(C_4 + C_6)}$$

(3.47a)

It can be shown, by comparing the magnitudes of equation (3.47a) at $z = 1$ and $z = -1$, that this can realize all types of notch and all-pass transfer functions. The design can be carried out in the 'p-domain' for given pole and zero frequencies and pole and zero $Q$s in the $s$-domain. From the $s$-domain transfer function, using matched-$z$ transformation, the transfer function is obtained as

$$\frac{V_{oo}}{V_{io}} = \frac{(1 - z^{-1})^2 + a(1 - z^{-1}) + b}{(1 - z^{-1})^2 + c(1 - z^{-1}) + d}$$

(3.47b)

Identifying $a$, $b$, $c$ and $d$ of equation (3.47b) in (3.47a), the design equations can be derived. The reader is urged to pursue this in Exercise 3.9.

For realizing notch filters, it may be noted that $C_5$ can be omitted, whereas for all-pass networks $C_5$ is essential. Hence, notch filter design with $C_5 = 0$ is possible (see Exercise 3.10).

Thus, it has been shown that notch and all-pass filters can be designed using passive SC networks. It may be mentioned that twin-T networks, which also employ only six components, are third-order networks and as such should be designed with pole-zero cancellation, whereas the circuits described herein are second-order networks and do not have passive RC counterparts.

The discussion till now has considered SC networks derived from passive RC networks. Alternative approaches are possible based on certain unique properties of SC networks. Two of these are considered in what follows.

## 3.7  Passive non-reciprocal networks

The basic gyrator realization is shown in Figure 3.11(a). It is useful to study the operation of this circuit. It requires two dependent current sources of opposite sign. It is noted that

$$V_2(s) = \frac{I_2(s)}{sC} = \frac{V_1(s)}{sCR} \tag{3.48a}$$

$$I_1(s) = \frac{V_2(s)}{R} = \frac{V_1(s)}{sCR^2} \tag{3.48b}$$

Hence

$$Z_{in} = \frac{V_1(s)}{I_1(s)} = sCR^2 \tag{3.48c}$$

showing that a lossless inductance is realized at the input terminals, when the output terminals are loaded by a capacitor.

## 3.7.1  Two-phase passive SC gyrators

An SC realization can now be derived from Figure 3.11(a), by replacing the dependent current sources with switched capacitors. We make use of the property that a capacitor can be charged to a voltage at one port and this charge can be transferred either directly or after inversion to another port. The circuit of Figure 3.11(b) uses this principle. The capacitor $C_3$ is charged to $V_1$ in the even phase and transfers this charge to $C_4$ in the odd phase. The capacitor $C_5$ is charged to $V_2$ in the even phase and, after an inversion, its charge is transferred in the odd phase to the input terminals. Some differences, however, exist between the model

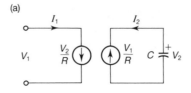

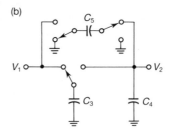

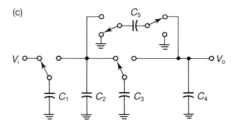

*Fig. 3.11* (a) A conceptual gyrator; (b) its SC implementation; (c) LP filter using (b).

of Figure 3.11(a) and the realization of Figure 3.11(b). First, charging of $C_3$ or $C_5$ by capacitors at the input or output terminals invariably leads to charge division or 'loading', thus causing incomplete (or lossy) transfer of charge. Second, there is a half-cycle delay associated with each charge transfer. Third, the input impedance can be evaluated in both the phases. Hence, in order to evaluate the performance of the gyrator, the circuit of Figure 3.11(c) is considered, where $C_2$ and the network of Figure 3.11(b) form a resonant circuit and $C_1$ is used as a resistor to feed the input. The transfer function of this circuit can be derived as follows:

$$\frac{V_{oo}}{V_{ie}} = C_1 \left[ C_4 + C_5 - \frac{C_4 C_5}{C_2 C_3} (C_2 + C_3) \right] z^{-3/2} \Bigg/ \Bigg\{ \left[ \frac{2C_5}{C_1} + \frac{C_3 C_5^2}{C_1 C_2 C_4} + \frac{C_2 C_4}{C_1 C_3} \right] (1 - z^{-1})^2$$

$$+ \left[ \frac{C_2}{C_1} + \frac{C_4}{C_3} + (C_3 + C_4) \left( \frac{1}{C_1} + \frac{1}{C_2} \right) + \left( \frac{C_2 + C_3 + C_4 + C_5}{C_1} \right) \right.$$

$$\times \left( \frac{1}{C_2} + \frac{1}{C_3} + \frac{1}{C_4} \right) C_5 - \frac{7 C_5}{C_1} + \frac{C_3 C_5}{C_2 C_4} \right] (1 - z^{-1}) + \left[ 1 + \left( \frac{4}{C_1} + \frac{1}{C_2} + \frac{1}{C_3} + \frac{1}{C_4} \right) C_5 \right] \Bigg\}$$

$$(3.49)$$

(Note that $V_{oe}$ does not give any useful transfer function.) It may be first noted that the circuit of Figure 3.11(c) is a modification of the second-order LP filter of Figure 3.2(b). It is interesting to find out whether the addition of $C_5$ has increased the realized $Q_p$ of this circuit. For high clock frequencies, using equation (3.6) in the denominator of equation (3.49), $\omega_p$ and $Q_p$ can be found:

$$\omega_p = f_s \frac{\sqrt{1 + \left(\dfrac{4}{C_1} + \dfrac{1}{C_2} + \dfrac{1}{C_3} + \dfrac{1}{C_4}\right) C_5}}{\sqrt{\dfrac{C_2 C_4}{C_1 C_3} + \dfrac{C_3 C_5^2}{C_1 C_2 C_4} + \dfrac{2C_5}{C_1}}} \tag{3.50a}$$

$$Q_p = \frac{\sqrt{1 + \left(\dfrac{4}{C_1} + \dfrac{1}{C_2} + \dfrac{1}{C_3} + \dfrac{1}{C_4}\right)\left(\dfrac{C_2 C_4}{C_1 C_3} + \dfrac{C_3 C_5^2}{C_1 C_2 C_4} + \dfrac{2C_5}{C_1}\right) C_5}}{\left[\dfrac{C_2}{C_1} + \dfrac{C_4}{C_3} + (C_3 + C_4)\left(\dfrac{1}{C_1} + \dfrac{1}{C_2}\right)\right.}$$

$$\left. + \left(\dfrac{C_2 + C_3 + C_4 + C_5}{C_1}\right)\left(\dfrac{1}{C_2} + \dfrac{1}{C_3} + \dfrac{1}{C_4}\right) C_5 - \dfrac{7C_5}{C_1} + \dfrac{C_3 C_5}{C_2 C_4}\right] \tag{3.50b}$$

For identical sections, i.e., $C_1 = C_3$, $C_2 = C_4$ and denoting $C_2/C_1 = \alpha$ and $C_5/C_1 = \beta$, and for large $\alpha$ (since $\alpha$ means large sampling frequencies),

$$Q_p = \frac{\sqrt{1 + 5\beta}}{3 + 2\beta} \tag{3.51a}$$

$Q_p$ is maximum ($= 0.49$) when $\beta = 1.1$. Thus it is seen that the $Q_p$ is boosted from $\frac{1}{3}$ (with $C_5 = 0$) to 0.49 (with $C_5 = 1.1 C_1$).

With tapering, i.e., $C_2/C_1 = C_4/C_3$ and $C_4 = \rho C_2$, equation (3.51a) is modified as follows:

$$Q_p = \frac{\sqrt{1 + (4 + \rho)\beta}}{2 + \dfrac{1}{\rho} + \beta + \rho\beta} \tag{3.51b}$$

which is maximum when

$$\beta = \frac{7 + 4/\rho}{\rho^2 + 5\rho + 4} \tag{3.51c}$$

As an example, with $\rho = 10$, $Q_p = 0.492$ as against $Q_p = 0.4761$ with $\beta = 0$. Thus it is seen that the $Q_p$-enhancement is more for untapered networks than for tapered networks.

It has been shown [3.9–3.13] that by increasing the number of clock phases from two to four, passive networks with complex conjugate poles can be realized. This is discussed in the next section.

## 3.7.2  Four-phase SC passive gyrator

The circuit of Figure 3.11(c) with $C_1 = 0$ can be redrawn first as in Figure 3.12(a).
In this circuit, the arrows indicate whether the capacitor terminals are inverted or
not while transferring charge from one port to another port. For convenience, $C_3$
is denoted as the 'left' capacitor and $C_5$ as the 'right' capacitor. With this notation,
in phase 1, the capacitors are as shown in Figure 3.12(a). In phase 2, the capacitors
$C_3$ and $C_5$ exchange their places: $C_3$ occupies the position of $C_5$ and $C_5$ occupies the
position of $C_3$ after inversion of terminals (Figure 3.12(b)). Suppose that the
operation is repeated in phase 3 with the result shown in Figure 3.12(c), and again
in phase 4 (Figure 3.12(d)) so that in phase 1, $C_3$ and $C_5$ are brought back to their
original positions. The resulting behaviour of the SC network can be analyzed by

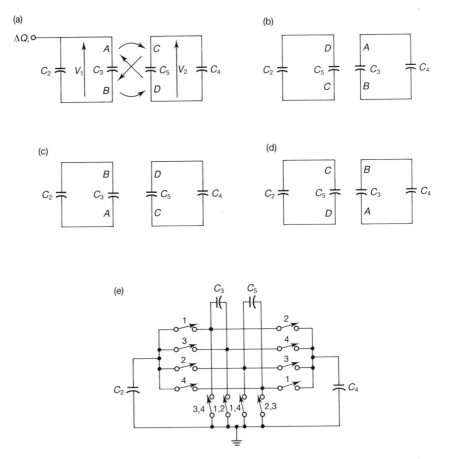

Fig. 3.12  Passive four-phase SC gyrator showing the capacitor connections in (a) phase
1;  (b) phase 2;  (c) phase 3;  (d) phase 4; and (e) the practical implementation *(adapted
from [3.12], © 1984 IEE).*

writing the charge conservation equation at any one phase, since a similar operation is taking place in all phases. Thus, in phase 1, we have

$$\Delta Q_i + C_2 V_1(1 - z^{-1/4}) + C_3 V_1 + C_3 V_2 z^{-1/4} = 0$$
$$- C_5 V_1 z^{-3/4} + C_5 V_2 + C_4 V_2(1 - z^{-1/4}) = 0 \tag{3.52}$$

where $\Delta Q_i$ is the charge input to the network. Considering the case for no input ($\Delta Q_i = 0$) the determinant of equations (3.52) yields, after simplification, the denominator of the charge transfer function,

$$D(z) = (C_2 C_4 + C_3 C_5)z^{-1/2} - [C_2(2C_4 + C_5) + C_3 C_4] z^{-1/4} + (C_2 + C_3)(C_4 + C_5) \tag{3.53a}$$

$$= cz^{-1/2} - bz^{-1/4} + a \qquad \text{(say)} \tag{3.53b}$$

It can be checked that equation (3.53a) realizes complex conjugate poles. It can be shown that the resulting $\delta$ and $Q_p$ may be evaluated from the following equations:

$$\frac{\delta}{Q_p} = \frac{a - c}{a - b + c} = \frac{1}{2}\left[\frac{C_2}{C_3} + \frac{C_4}{C_5}\right] \tag{3.54a}$$

$$\frac{1}{4}\left(4\delta^2 - \frac{2\delta}{Q_p} + 1\right) = \frac{c}{a - b + c} = \frac{C_2 C_4}{C_3 C_5} + \frac{1}{2} \tag{3.54b}$$

From equation (3.54a), it is evident that an arbitrarily large $Q_p$ can be realized by proper choice of capacitor ratios. Choosing the capacitor values to be [3.8–3.10]:

$$C_2 = C_4 = C\left(1 - \frac{1}{N}\right) \tag{3.55a}$$

$$C_3 = C_5 = \frac{C}{N}$$

we obtain

$$\delta = \frac{\sqrt{2N^2 - 2N + 1}}{2} \tag{3.55b}$$

$$\frac{\delta}{Q_p} = N - 1$$

Thus, $N$ uniquely decides the values of $\delta$ and $Q_p$. It may also be noted that the circuit works at double the clock frequency (see equation (3.53)). Denoting $C_2/C_3$ as $\alpha$ and $C_4/C_5$ as $\beta$, and using equation (3.54), the circuit can also be designed given $\delta$ and $Q_p$. The realization of this gyrator is shown in Figure 3.12(e).

In practice, the resonator can be used to realize a desired transfer function, by using additional switched/unswitched capacitors to feed the input. Consider the circuit of Figure 3.13, for which the charge conservation equations in phases 2 and 1 are, respectively, as follows:

$$-C_1 V_i z^{-1/4} + (C_1 + C_3 + C_2(1 - z^{-1/4})) V_1 + C_3 V_2 z^{-1/4} = 0 \qquad (3.56a)$$

$$-C_5 V_1 z^{-1/4} + (C_5 + C_4(1 - z^{-1/4})) V_2 = 0 \qquad (3.56b)$$

The resulting voltage transfer function is:

$$\frac{V_2(z)}{V_i(z)} = \frac{C_1 C_5 z^{-1/2}}{(C_2 C_4 + C_3 C_5) z^{-1/2} - [C_2 C_5 + C_4(C_1 + 2C_2 + C_3)] z^{-1/4} + (C_1 + C_2 + C_3)(C_4 + C_5)}$$

$$(3.57)$$

For the choice

$$C_1 = C_3 = C_5 = \frac{C}{N}$$

$$C_2 = C\left(1 - \frac{2}{N}\right) \qquad (3.58a)$$

$$C_4 = C\left(1 - \frac{1}{N}\right)$$

equation (3.57) simplifies to

$$\frac{V_2(z)}{V_i(z)} = \frac{(1/N^2) z^{-1/2}}{\left[1 + \frac{3}{N}\left(\frac{1}{N} - 1\right)\right] z^{-1/2} - \left(2 - \frac{3}{N}\right) z^{-1/4} + 1} \qquad (3.58b)$$

The values of $\delta$ and $Q_p$ obtained in this case are given by

$$\delta = \sqrt{\frac{4N^2 - 6N + 3}{12}}$$

$$(3.59)$$

$$\frac{\delta}{Q_p} = N - 1$$

The d.c. gain is $\frac{1}{3}$. The circuit, in general, can be designed for arbitrarily large $Q_p$, as desired. It may also be noted that in the circuit of Figure 3.13, the total capacitance of the two groups of capacitors is still $C$. The SC network of Figure 3.13

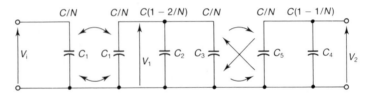

*Fig. 3.13* An SC passive low-pass filter based on a four-phase passive gyrator.

cannot, however, be designed for arbitrary values of $\delta$ and $Q_p$ in a simple manner. It is useful to note that the charges stored on the group of capacitors $(C_4, C_5)$ can be related to the input charge supplied through $C_1$ through the charge transfer function. This charge transfer function is related to the voltage transfer function (3.58b) as

$$\frac{Q_2(z)}{Q_i(z)} = \frac{CV_2}{C_1 V_i} = N \frac{V_2}{V_i} \qquad (3.60)$$

This relationship will be used in what follows.

## 3.7.3 General passive SC biquad

In the above analysis, the passive SC realization based on the gyrator model of Figure 3.11(a) has been considered. Manetti [3.12] has shown that the gyrator model can be extended to realize general biquadratic transfer functions. This structure shown in Figure 3.14(a) uses feedforward branches as well as the shunt branches at the input and output ports of the four-phase SC gyrator network. The resulting SC network can be modelled by the $z$-domain equivalent circuit of Figure 3.14(b). The reader can derive the transfer functions of the complete circuit of Figure 3.14(b) to be:

$$\frac{V_1}{V_i} = \frac{[B(H+F) - C'F]z^{-2} + [B(G+E+C') + A(H+F) - C'E]z^{-1} + A(G+E+C')}{[(B+D)(H+F) + C'^2]z^{-2}}$$
$$+ [(B+D)(E+G+C') + (H+F)(A+C+C')]z^{-1} + (A+C+C')(E+G+C')$$

$$(3.61a)$$

$$\frac{V_2}{V_i} = \frac{[F(B+D) + C'B]z^{-2} + [F(A+C+C') + E(B+D) + C'A]z^{-1} + E(A+C+C')}{D(z)}$$

$$(3.61b)$$

where $D(z)$ is the denominator of equation (3.61a), and

$$\begin{array}{ll}
A = (C_1 + C_2), & B = (C_1 - C_2) \\
C = (C_3 + C_4), & D = (C_3 - C_4) \\
E = (C_5 + C_6), & F = (C_5 - C_6) \\
G = (C_7 + C_8), & H = (C_7 - C_8)
\end{array} \qquad (3.62)$$

It can be seen that equations (3.61) realize biquadratic transfer functions. Note that eight degrees of freedom $A-H$ are available in order to realize the five coefficients of a general second-order filter transfer function. Hence, some components can be deleted. It can also be observed that the gain realizable by equations (3.61) is less than unity. As an example, Manetti [3.12] suggests the choice of $C = D = G = H = 0$,

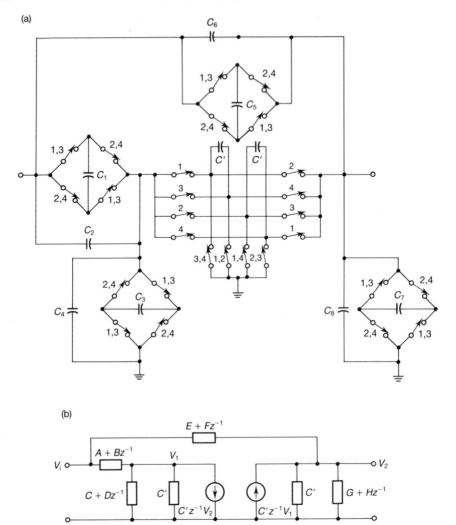

*Fig. 3.14*   (a) Manetti's passive SC biquad;   (b) its z-domain equivalent circuit.

$B = -A$, $F = -E$ in order to obtain a bilinear BP transfer function at output $V_2$. The design values for this choice are

$$A = \frac{4\delta^2 - 1}{2}$$

$$(3.63)$$

$$E = \frac{2\delta}{Q_p} + \frac{1}{2} - 2\delta^2$$

For realizability, evidently $\delta > 0.5$ and $Q_p < 4\delta/(4\delta^2 - 1)$. Further, the midband gain is not controllable. The reader is urged to verify whether other choices restore some degrees of freedom to design for all the given specifications. The generalities of the biquad of Figure 3.14(a) can also be explored in Exercise 3.12.

The biquad of Figure 3.14 can be cascaded by additional biquad sections after buffering the output, to realize high-order transfer functions. However, using yet another interesting property of SC networks, i.e., charge amplification, high-order filter design using the cascade approach is possible with isolating 'charge amplifier blocks'. This interesting approach is considered next [3.11].

## 3.7.4  Charge amplification using SC networks

The concept of charge amplification using the passive SC network of Figure 3.15 is considered first. In phase $\phi_1$, $m$ capacitors each of capacitance $C$ are connected in series, through which a charge $Q$ flows by virtue of the applied voltage. In phase $\phi_2$, all these capacitors are connected in parallel, so that the total charge available for transfer to any other network connected across them is the sum of all the $Q$s. Thus, charge amplification can be realized. It may be noted that in $\phi_1$, the capacitance $C/m$, where $m$ is the number of capacitors connected in series, 'loads' the previous circuit, which is passive, as in Figure 3.13. Similarly the 'source' capacitance of the amplified charge source $mQ$ is also finite ($mC$) which should be taken into account in the design of the succeeding stage. The cascade design of passive SC networks using these principles will be studied next. Note that the high-order LP filter design will be considered.

## 3.7.5  High-order passive SC low-pass filter design

The first stage in the high-order LP filter is shown in Figure 3.16(a), which is the same as that of Figure 3.13, except that the load capacitance is disconnected after the output voltage is held on it. It is also observed that the total capacitance in the two groups of capacitors is still maintained the same. The charge transfer function

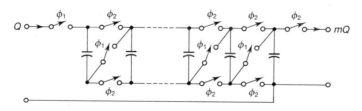

*Fig. 3.15*  Passive SC network realizing charge amplification.

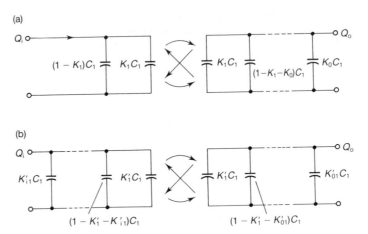

Fig. 3.16 The (a) first and (b) intermediate stages in Manetti's cascade design of passive SC LP filters.

of this structure of Figure 3.16(a) is derived from the following charge conservation equations:

$$C_1 V_1 [(1 - K_1)(1 - z^{-1/2}) + K_1)] + C_1 V_2 K_1 z^{-1/2} = -Q_i$$
$$- C_1 V_1 K_1 z^{-1/2} + C_1 V_2 [1 - (1 - K_1 - K_0)z^{-1/2}] = 0 \qquad (3.64)$$

The charge available on the output group of capacitors $Q_o$ is $C_1 V_2$, thus yielding

$$\frac{Q_o(z)}{Q_i(z)} = \frac{K_1 z^{-1/2}}{[(1 - K_1)(1 - K_1 - K_0) + K_1^2] z^{-1} - (2 - 2K_1 - K_0)z^{-1/2} + 1} \qquad (3.65)$$

It can be noted that $K_0$ and $K_1$ are the design values to realize given values of $\delta$ and $Q_p$:

$$K_1 = \frac{Q_p}{Q_p + \delta}$$

$$K_0 = \frac{4(Q_p + \delta)}{Q_p \left(4\delta^2 + \dfrac{2\delta}{Q_p} + 1\right)} - \frac{2Q_p}{(Q_p + \delta)} \qquad (3.66)$$

For realizability, $K_0 > 0$ and $K_0 + K_1 < 1$. These requirements impose the limit on the $Q_p$ values realizable for a given $\delta$, or on the $\delta$ values possible for realizing a given $Q_p$. Accordingly, the sampling frequency shall be chosen. The d.c. gain realized is

$$G_0 = \frac{\left(4\delta^2 + \dfrac{2\delta}{Q_p} + 1\right)Q_p}{4(Q_p + \delta)} \qquad (3.67)$$

The next part in the synthesis is considered in what follows, which pertains to the succeeding stage (second-order) design and the charge amplifier.

The second stage in Manetti's cascade design resembles that shown in Figure 3.16(b). In this SC network, the capacitor $K'_{i1}$ delivers charge obtained from the previous circuit after amplification to the circuit under consideration. Similarly, the capacitor $K'_{01}$ is the input capacitance of the charge amplifier (isolating section). Consequently, the SC circuit of Figure 3.16(b) has the transfer function given by

$$\frac{Q_0(z)}{Q_i(z)} = \frac{K'_1 z^{-1}}{[(1 - K'_1 - K'_{i1})(1 - K'_1 - K'_{01}) + K'^2_1]z^{-2} - (2 - 2K'_1 - K'_{i1} - K'_{01})z^{-1} + 1}$$

(3.68)

This transfer function is characterized by three unknowns $K'_1$, $K'_{i1}$ and $K'_{01}$. The design equations in terms of $\delta$ and $Q_p$ are given by

$$\frac{\delta}{Q_p} + 1 = \frac{2K'_1 + K'_{i1} + K'_{01}}{2K'^2_1 + K'_1(K'_{i1} + K'_{01}) + K'_{i1}K'_{01}}$$

(3.69a)

$$4\delta^2 + \frac{2\delta}{Q_p} + 1 = \frac{4(1 - K'_1)}{2K'^2_1 + K'_1(K'_{i1} + K'_{01}) + K'_{i1}K'_{01}}$$

(3.69b)

The reader is urged to study the synthesis of the above circuit for chosen $\delta$ and $Q_p$ values, which is quite involved. However, for the simple case when $K'_{01} = 0$, i.e., when it is intended to design a fourth-order overall filtering function, equations (3.69) yield a simple solution.

We now summarize the design procedure so far developed. From the digital transfer function needed to be realized (e.g., fourth-order), the first and second quadratic factors in the denominator can be used to obtain $K_1$, $K_0$ from (3.66) and $K_{i1}$ and $K'_1$ from (3.69). The next step is to design the charge amplifier coupling block that interfaces the first and second biquad blocks.

Assuming that $m$ capacitors are used in the charge amplifier configuration of Figure 3.15 ($m$ yet to be determined), the d.c. gain of the LP filter also can be used as an additional design parameter. The overall filter d.c. gain can be computed as follows.

The d.c. charge gain $K_{01}$ of the first stage is obtained from equation (3.65) as $[1/(2K_1 + K_{01})]$. Similarly, the d.c. charge gain of the second stage obtained from equation (3.68) is

$$K'_1 / [2K'^2_1 + K'_1(K'_{i1} + K_{02}) + K'_{i1}K_{02}]$$

The charge gains of the coupling block are, respectively, $m_1 K_{01}$, $m_2 K_{02}$, etc., where $m_1, m_2, \ldots$ are the number of capacitors used in the coupling section. For a fourth-order filter design for which the design is simple, the charge gain at d.c. evidently works out to be:

$$H_{dc} = \frac{m_1 K_{01}}{(2K_1 + K_{01})(2K'_1 + K_{02})}$$

(3.70)

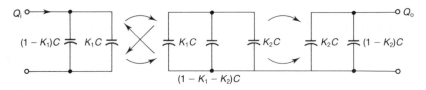

*Fig. 3.17* A third-order passive SC LP filter.

The choice of $m_1$ is examined next. Assuming the total capacitance of the preceding and succeeding groups of capacitors around the coupling section to be $C_h$ and $C_{(h+1)}$, respectively, and denoting each of the $m$ capacitors in the charge amplifier as $C$, the following relationships are obtained (see Figure 3.15):

$$K_{01}C_h = mC$$
$$K_{02}C_{h+1} = C/m \tag{3.71}$$

Thus, depending on the chosen $m$ value (obviously an integer) and the value of $C$, $C_h$ and $C_{h+1}$ can be determined, which, together with $K_1$, $K_{01}$, $K_1'$, $K_{02}$ already obtained from equations (3.66) and (3.69), yield all the design values. The approach is the same in the case of high-order filters.

In the above treatment, even-order SC passive LP filters have been considered. Odd-order SC filters can be realized by using a first-order terminating stage. A typical third-order SC LP filter circuit is shown in Figure 3.17. In this circuit, the capacitor $K_2C$ is discharged after transferring the charge into the second section and reconnected back to the first section, thus realizing the transfer function

$$\frac{Q_o(z)}{Q_i(z)} = \frac{K_1K_2z^{-2}}{[[K_1^2 + (1-K_1)(1-K_1-K_2)]\,z^{-2} - [2(1-K_1) - K_2]\,z^{-1} + 1][-(1-K_2)z^{-1} + 1]} \tag{3.72}$$

Note that equation (3.72) is the product of equation (3.65) and a first-order LP transfer function.

It appears logically possible to use the general SC biquad of Figure 3.14(a) to realize a cascade type of high-order elliptic filter design, employing the isolating charge-amplifier blocks. This is left as an exercise for the reader.

In this section, an interesting property of passive SC networks and its utility in realizing high-order SC structures have been elaborately discussed. In the next section, the voltage amplification property will be described.

## 3.8   Voltage amplification using SC networks [3.13–3.16]

This is another application where there is no analogy between SC circuits and passive RC circuits. Some passive RC circuits can provide a voltage or current gain greater than unity at a single frequency with zero phase shift, and this property has been used to realize oscillators using amplifiers with gain less than unity. The voltage amplification property to be discussed in what follows is at d.c. only.

In Figures 3.18(a) and 3.18(b), conventional and cascade voltage doublers are shown. Consider the circuit of Figure 3.18(a). During the positive half-cycle of the input voltage, $C_1$ charges to the peak value through diode $D_1$ and during the negative half-cycle $D_2$ conducts, charging $C_2$ with a polarity such that the voltages across $C_1$ and $C_2$ are twice the peak value of the input voltage. In Figure 3.18(b), during the negative half-cycle, $C_1$ charges through $D_1$. This voltage across $C_1$ aids the positive half of the input cycle to provide across $C_2$ twice the input peak voltage through $D_2$. Thus it is observed that an alternating voltage source is required. The diodes conduct in either one or another half-cycle.

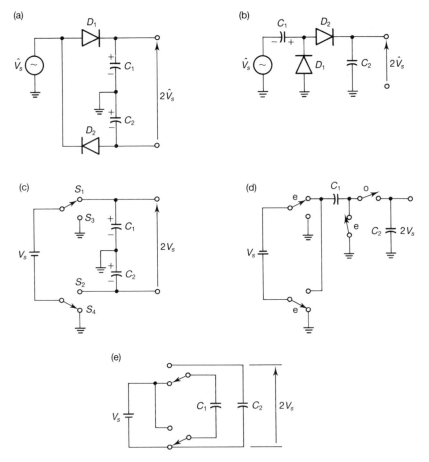

*Fig. 3.18* (a), (b) Voltage doublers using diodes and capacitors; (c)–(e) voltage doublers using switches and capacitors.

It is possible to consider the a.c. source as a d.c. source $V_s$ operating in two clock phases with direct or inverted output (i.e., $V_s$ or $-V_s$) and with diodes replaced by switches operating in different phases. Consider the realized SC circuits of Figures 3.18(c) and 3.18(d). During the connection phase shown (Figure 3.18(c)), $C_1$ charges to the positive input voltage, and in the next phase $C_2$ charges to the negative input voltage, with the resulting output voltage double that of input voltage. In Figure 3.18(d), during the even phase, $C_1$ charges to negative input voltage, and in the next phase the input voltage, aided by the voltage across $C_1$, charges $C_2$ to give twice the input voltage. Cederbaum [3.13] has suggested a simple voltage doubler (shown in Figure 3.18(e)), where during one phase $C_1$ is charged to the input voltage, and in the other phase it is connected in series with the input to provide $2V_s$ to charge the output capacitor.

The circuit of Figure 3.18(b) can be augmented by diode $D_3$ and capacitor $C_3$ to realize a voltage tripler, as shown in Figure 3.19(a). The capacitors $C_1$, $C_2$ and diodes $D_1$, $D_2$ operate as a voltage doubler, as explained above. During the negative

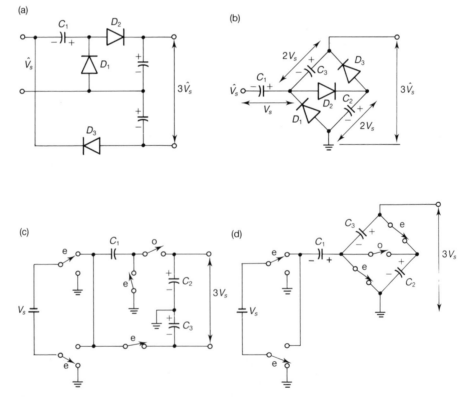

*Fig. 3.19* Voltage triplers: (a), (b) using diodes and capacitors; (c), (d) using switches and capacitors.

cycle, in addition, $C_3$ charges to negative peak voltage through $D_3$ and the voltage across $C_2$ and $C_3$ is thus $3V_s$. The reader can verify, in a similar fashion, that the circuit of Figure 3.19(b) also acts as a voltage tripler.

SC implementations follow readily from Figures 3.19(a) and 3.19(b), as shown in Figures 3.19(c) and 3.19(d). These are the doubler circuits of Figure 3.18(d) with an additional switch and capacitor. The extension of these circuits to general multiplier circuits is straightforward. Consider the multiplier [3.17] of Figure 3.20(a) and its SC equivalent [3.15, 3.16], shown in Figure 3.20(b). As explained earlier, during the odd phase, all right-hand side capacitors charge to $2V_s$. (Note that only the first capacitor has $V_s$ and all other capacitors have $2V_s$ across them.) During the odd phase, all the left-hand side capacitors (except the first) charge to $2V_s$. Thus the voltage progressively increases along the chain. Since the initial capacitors supply charge for several following capacitors, the capacitor values must be progressively increased towards the low-voltage end.

The above discussion has illustrated the possibility of voltage amplification by a factor $n$ using $n$ diodes, switches and $n$ capacitors using a two-phase clock. It is of interest to study other possible realizations. Cederbaum [3.13] has shown that using $n$ capacitors and $2n$ switches, a voltage amplification of $2^n$ can be achieved. This is based on the simple circuit of Figure 3.18(e). The principle of this circuit for two stages is illustrated in Figure 3.21(a). The clock arrangement is as shown in Figure 3.21(b). The reader can verify that the output voltage $V_0$ is $4V$ volts. This procedure could be extended to obtain $2^nV$ by using $n$-stages [3.13–3.14].

Thus voltage multiplication can be achieved using only switches and capacitors. It should be remembered that the circuits described require floating capacitors and hence parasitic capacitances will affect the performance of the circuit.

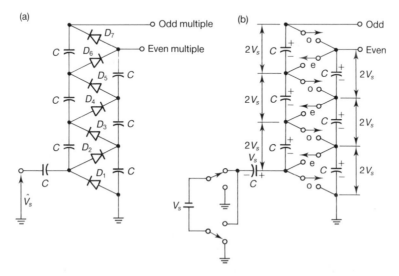

*Fig. 3.20* Voltage multipliers with: (a) diodes and capacitors; (b) switches and capacitors.

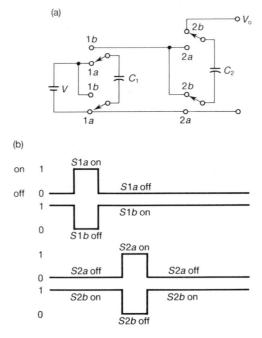

*Fig. 3.21* (a) Cederbaum's voltage amplification circuit; (b) timing waveforms.


## 3.9 Conclusions

In this chapter, passive SC networks have been studied. Some of these will be useful for realizing second-order single-amplifier filters, to be described in a later chapter. There is another class of passive SC networks called *N*-path filters and these will be studied separately, since they require different tools for their analysis.


## 3.10 Exercises

3.1 By interchanging input and ground terminals in a passive network, a 'complementary' transfer function is obtained [3.19]. As an example, in the first-order LP RC network in Figure 3.1(a), by interchanging input and ground terminals, a first-order HP network is obtained. The transfer function of the HP network will be

$$\left.\frac{V_o(s)}{V_i(s)}\right|_C = \frac{sC_2R_1}{sC_2R_1 + 1} = 1 - \left.\frac{V_o(s)}{V_i(s)}\right|_{\text{Figure 3.1(a)}}$$

The procedures can be extended to SC networks as well. Thus, from the SC network of Figure 3.1(b), we obtain the SC network of Figure 3.22 [3.18]. Analyze this SC network and compare with the HP network of Figure 3.4(b).

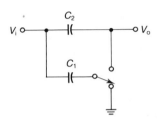

*Fig. 3.22* An SC HP network obtained from Figure 3.1(b).

3.2 Derive the attenuation of the three-section RC phase-shift network of Figure 3.23 when a phase shift of 180° is realized. Compare the results with those of the passive SC network of Figure 3.3(a).

3.3 An RC phase shift oscillator is shown in Figure 3.24. Derive the condition for oscillation and frequency of oscillation. Use the same method as in Figure 3.24 to reduce the number of OAs in the circuit of Figure 3.3(b) and compare the results.

3.4 Can series-switched capacitors be used to realize an RC phase-shift oscillator using the same circuit arrangement as in Figure 3.3(b)? Explain without deriving the transfer functions.

3.5 Using the notion of $\delta$ and $Q_p$ introduced in Chapter 1 involving bilinear transformations, derive expressions for $\delta$ and $Q_p$ in terms of capacitor ratios, for the circuits of Figures 3.2(b) and 3.2(c). Derive the bounds on the maximum $Q_p$ realizable. Compare the capacitor spread required to realize a given $\delta$ and $Q_p$.

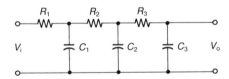

*Fig. 3.23* Three-section RC phase-shift network.

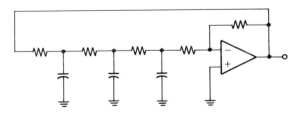

*Fig. 3.24* An RC phase-shift oscillator.

3.6 A bridged-T RC network can only realize a transmission minimum at a certain frequency. However, using 'feedforward' technique, a null network can be realized. Such a circuit is shown in Figure 3.25. Derive the transfer function and condition for obtaining null and all-pass realizations. Note that the circuit realizes only negative real poles.

3.7 Develop the design equations for the bridged-T SC notch filter of Figure 3.8(b) for two types of requirement:
(a) a given null frequency and pole frequency;
(b) a given null frequency and transmission at infinite frequency.

3.8 An alternative first-order all-pass network is shown in Figure 3.26. Analyze the difference between this all-pass network and that of Figure 3.9(b).

3.9 Develop the design equations of the second-order passive all-pass filter of Figure 3.10(d) for a given $s$-domain specification, using the matched-$z$ transformation method.

3.10 Design a notch filter using the SC network of Figure 3.10(d) and choosing $C_5 = 0$. Check whether all types of notch can be realized.

3.11 Discuss the results obtained when $C_1 = 0$, $C_2 = C_4$ and $C_3 = C_5$ in the circuit of Figure 3.11(c). Note that this choice results in a two-phase version of the SC network due to Manetti and Liberatore [3.8–3.12], shown in Figure 3.12(a), in which $C_3 = C_5 = C/N$ and $C_2 = C_4 = C(1 - 1/N)$. Also examine whether complex conjugate poles can be realized by the circuit of Figure 3.11(c), when $C_1 = 0$.

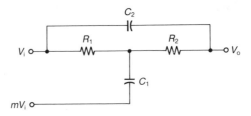

Fig. 3.25  An RC dual input network.

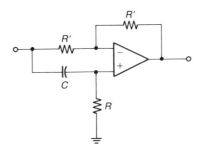

Fig. 3.26  A first-order active all-pass network.

3.12 Examine whether the passive SC biquad of Figure 3.14(a) can realize all generally required biquadratic transfer functions.

3.13 Following the design procedure outlined in Section 3.7.5, design a fourth-order Chebyshev LP filter for a 0.1 dB passband ripple. Discuss the choice of sampling frequency as well as the switching waveforms required.

## 3.11 References

[3.1] J.T. Caves, M.A. Copeland, C.F. Rahim and S.D. Rosenbaum, Sampled analog filtering using switched-capacitors as resistor equivalents, *IEEE Journal of Solid State Circuits*, SC-12, 6, 592–9, December 1977.

[3.2] R.W. Brodersen, P.R. Gray and D.A. Hodges, MOS switched-capacitor filters, *Proc. IEEE*, 67, 1, 61–75, January 1979.

[3.3] P.V. Ananda Mohan, V. Ramachandran and M.N.S. Swamy, Passive and active switched-capacitor networks, *Proc. ISCAS, Rome*, 233–6, May 1982.

[3.4] E.A. Vittoz, Micropower switched-capacitor oscillator, *IEEE Journal of Solid State Circuits*, SC-14, 3, 622–4, June 1979.

[3.5] Y.P. Tsividis, Analytical and experimental evaluation of a switched-capacitor filter correspondence, *IEEE Transactions on Circuits and Systems*, CAS-26, 2, 140–4, February 1979.

[3.6] T.R. Viswanathan, K. Singhal and G. Metzker, Application of switched-capacitor resistors in RC oscillators, *Electronics Letters*, 14, 20, 659–60, 28 September 1978.

[3.7] G.S. Moschytz, High-$Q$ factor insensitive active RC network similar to Tarmy–Ghausi circuit but using single-ended operational amplifiers, *Electronics Letters*, 8, 458–9, 7 September 1972.

[3.8] S. Manetti and A. Liberatore, Switched-capacitor low-pass filter without active components, *Electronics Letters*, 16, 23, 883–5, 6 November 1980.

[3.9] S. Manetti and A. Liberatore, Realization of non-reciprocal networks using switched-capacitors, *Proc. ISCAS, Houston*, 969–72, 1980.

[3.10] A. Liberatore, S. Manetti and A. Ricci, On the synthesis of passive switched-capacitor low-pass filters, *Proc. ISCAS, Rome*, 737–40, 1982.

[3.11] S. Manetti, New technique for the realization of passive switched-capacitor low-pass filters, *Alta Frequenza*, 51, 319–27, 1982.

[3.12] S. Manetti, Passive switched-capacitor filters: General biquad topology, *Electronics Letters*, 20, 2, 101–2, 19 January 1984.

[3.13] I. Cederbaum, Voltage-amplification in switched-capacitor networks, *Electronics Letters*, 17, 5, 194–6, 5 March 1981.

[3.14] G.H.S. Rokos, Comment on voltage amplification in switched-capacitor networks, *Electronics Letters*, 17, 14, 501, 9 July 1981.

[3.15] S. Singer, Inductance-less up dc–dc converter, *IEEE Journal of Solid State Circuits*, SC-17, 4, 778–81, August 1982.

[3.16] S. Singer, Transformer description of a family of switched systems, *Proc. IEE*, 129, Part *G*, 5, 205–10, October 1982.

[3.17] L.J. Giacoletto (ed.), *Electronic Designer's Handbook*, McGraw-Hill, 1977.

[3.18] D.L. Fried, Analog sampled-data filters, *IEEE Journal of Solid-State Circuits*, SC-7, 4, 302–4, August 1972.

[3.19] M.N.S. Swamy, On the matrix parameters of a three terminal network, *Proc. IEEE*, 54, 8, 1081–2, August 1966.

# CHAPTER FOUR

# First-order active switched capacitor networks

First-order active SC networks are useful in the realization of second-order SC sections; they can also be useful as first-order sections in the design of higher-order filters. In this chapter, we shall discuss SC integrators, differentiators, summers and delay elements. The first three of these will be extensively referred to in the subsequent chapters and as such they will be treated here in detail. While discussing each circuit, the effect of parasitic capacitance will also be considered, since ultimately these circuits should be realizable in integrated circuit (IC) form. As mentioned in Chapter 1, the design of SC filters could be according to some transformation such as forward Euler integration (FEI), backward Euler integration (BEI), bilinear transformation (BT) or LDI transformation. Consequently, there are a wide variety of first-order SC networks such as lossless and lossy differentiators, integrators, and first-order all-pass networks. All these possible realizations are studied in this chapter.

## 4.1  Lossless integrators

### 4.1.1  FEI and LDI inverting and non-inverting integrators

Consider the circuit of Figure 4.1(a) using a parallel SC integrator [3.1–3.2, 4.1–4.3]. The transfer functions of this circuit can be obtained using Laker's equivalent circuit (see Figure 4.1(b)):

$$\frac{V_{oo}}{V_{ie}} = -\frac{C_1}{C_2} \cdot \frac{z^{-1/2}}{(1 - z^{-1})} \tag{4.1a}$$

$$\frac{V_{oe}}{V_{ie}} = -\frac{C_1}{C_2} \cdot \frac{z^{-1}}{(1 - z^{-1})} \tag{4.1b}$$

Note that equation (4.1a) represents an inverting LDI integrator transfer function, whereas equation (4.1b) corresponds to an inverting FEI integrator transfer function.

The frequency response of the integrator of Figure 4.1(a) can be obtained by letting $z = e^{j\omega t}$ in equations (4.1). Thus, we have

$$\frac{V_{oo}}{V_{ie}}(\omega) = -\frac{\omega_0}{j\omega}\left[\frac{\omega T}{2\sin\left(\frac{\omega T}{2}\right)}\right] \tag{4.2a}$$

$$\frac{V_{oe}}{V_{ie}}(\omega) = -\frac{\omega_0}{j\omega}\left[\frac{\omega T}{2\sin\left(\frac{\omega T}{2}\right)}\right]e^{-j\omega T/2} \tag{4.2b}$$

where $\omega_0 = C_1/TC_2$. The first observation that can be made from equations (4.2) is that there is deviation from the continuous-time response of an integrator $(-\omega_0/j\omega)$. The error in magnitude is given by the square of the bracketed term in equations (4.2). Equation (4.2b) indicates that there is a phase of $-\omega T/2$ rad introduced in addition to the desired inverting integrator phase shift of $\pi/2$ rad. The effect of this 'excess' phase in filter design will be considered later in connection with the design of SC ladder filters.

The circuit of Figure 4.1(a), when realized in IC form, will have certain undesirable effects due to the parasitic capacitances of the top- and bottom-layer metallizations of $C_1$ to the substrate. These parasitic capacitances are shown in the circuit

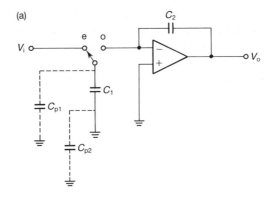

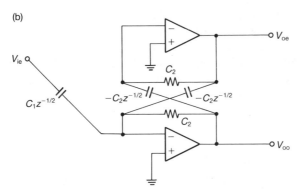

Fig. 4.1  (a) An SC inverting integrator using a parallel-switched capacitor;  (b) z-domain equivalent circuit of (a).

of Figure 4.1(a) in dotted lines. The effect of parasitic capacitance $C_{p2}$ need not be considered because it is always connected to ground. The top-plate capacitance,* $C_{p1}$, is charged to the input voltage during the even phase and transfers its charge to the output in the odd phase. In other words $C_{p1}$ is shunting $C_1$. Accordingly, equations (4.1) are modified as follows:

$$\frac{V_{oo}}{V_{ie}} = -\frac{C_1 + C_{p1}}{C_2} \cdot \frac{z^{-1/2}}{(1 - z^{-1})} \qquad (4.3a)$$

$$\frac{V_{oe}}{V_{ie}} = -\frac{C_1 + C_{p1}}{C_2} \cdot \frac{z^{-1}}{(1 - z^{-1})} \qquad (4.3b)$$

---

*   SC networks sensitive to top-plate parasitic capacitances and insensitive to bottom-plate parasitic capacitances are referred to in the literature as *bottom-plate stray-insensitive* SC networks. However, this phrase, strictly speaking, is a euphemism! Perhaps the description 'top-plate sensitive' gives a better picture of the capabilities of these SC networks. Nevertheless, the variety offered by allowing 'bottom-plate stray insensitive' designs as well makes the study of SC networks more interesting. Further, evolution of better technology with negligible parasitic capacitances may alleviate the drawbacks of top-plate parasitic sensitive designs.

For typical IC designs, the top-plate parasitic capacitance $C_{p1}$ can be 0.001 to 0.01 times the actual capacitance and hence an error in the integrator time constant of between 0.1% and 1% could result. Furthermore, since large time constants result with large $C_2/C_1$ or small $C_1$, $C_1$ needs to be larger than $C_{p1}$ to have little effect on the resulting integrator response. For typical monolithic realizations, $C_1$ ranges from about 0.5 to 2.5 pF.

By a simple modification of the circuit of Figure 4.1(a) to that of Figure 4.2(a), we obtain

$$V_{oo} = -\frac{C_1}{C_2} \cdot \frac{z^{-1/2}}{1-z^{-1}} (V_{1e} - V_{2e}) \tag{4.4a}$$

$$V_{oe} = V_{oo}z^{-1/2} \tag{4.4b}$$

Thus a differential integrator can be realized by the circuit of Figure 4.2(a). It is interesting to note that the circuit obtained with $V_1 = 0$ shown in Figure 4.2(b) realizes a non-inverting integrator [4.2].

The circuit of Figure 4.2(a) does not have any degradation in performance due to the bottom-plate parasitic capacitance, because it is always switched between either a voltage source or ground. The effect of top-plate parasitic capacitance is to change the integrator time constant of the input $V_1$ according to equations (4.3).

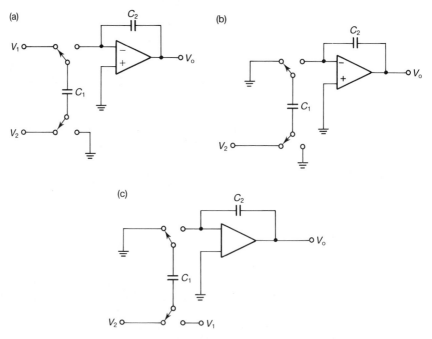

Fig. 4.2   (a) A differential SC integrator;   (b) a stray-insensitive non-inverting integrator;   (c) a stray-insensitive differential integrator.

However, interestingly enough, in the circuit of Figure 4.2(b), the top-plate parasitic capacitance also does not have any effect on the integrator time constant. The 'stray-insensitive' integrator [4.4] is widely used today in the realization of SC filters. It is thus possible to use a very small value of $C_1$ and/or $C_2$ to obtain large $C_2/C_1$, without consuming a large silicon area since the parasitic or stray capacitance has no effect on the performance of the SC integrator. It has been mentioned before that the differential integrator of Figure 4.2(a) is not stray-insensitive. A modification to achieve stray-insensitivity [4.3] is as shown in Figure 4.2(c). Under the condition $V_{io} = V_{ie}z^{-1/2}$, the transfer function realized is the same as in equation (4.4a).

## 4.1.2 BEI and LDI inverting integrators

The inverting integrator of Figure 4.3(a) uses a series-switched capacitor $C_1$ and an integrating capacitor $C_2$. The resulting transfer functions are

$$\frac{V_{oe}}{V_{ie}} = -\frac{C_1}{C_2(1 - z^{-1})}$$

$$V_{oo} = V_{oe}z^{-1/2}$$

(4.5)

showing that BEI is achieved in the even-phase output and LDI is achieved in the odd-phase output.

The parasitic capacitance $C_{p1}$ has no effect on the response of the integrator, since it is between a voltage source and ground. The parasitic $C_{p2}$ is charged to $V_{io}$ in the odd phase, and discharges into $C_2$ during the even phase, thus affecting the integrator response. The modification of the circuit to that in Figure 4.3(b) is also

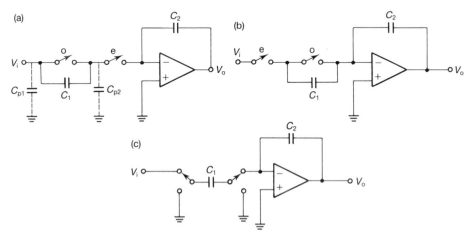

Fig. 4.3  Inverting BEI/LDI integrators using series-switched capacitors:  (a), (b) parasitic sensitive type;  (c) parasitic insensitive type (adapted from [4.4], © 1979 IEE).

sensitive to parasitic capacitance. An ingenious modification to realize a parasitic insensitive inverting integrator [4.4] is shown in Figure 4.3(c), where the capacitor plates are switched between a voltage source and ground, virtual ground and ground. This stray-insensitive inverting integrator, together with the stray-insensitive non-inverting integrator shown in Figure 4.2(b), forms a pair of complementary integrators, quite useful in high-order SC filter realizations.

*Example 4.1*
Design a non-inverting LDI type SC lossless integrator for a time constant of 1 ms and using a sampling frequency of 8 kHz.

The prewarped $\omega_p$ corresponding to the desired time constant 1 ms is

$$\omega_o = \frac{2}{T} \sin \frac{\omega T}{2} = 6122.9349 \text{ rad s}^{-1}$$

Thus, the integrator transfer function in the $s$-domain is

$$H_a(s) = \frac{6122.9349}{s}$$

Using the LDI transformation, the digital transfer function is obtained as

$$H_d(z) = \frac{6122.9349 T z^{-1/2}}{(1 - z^{-1})} = 0.765\,366\,9\,\frac{z^{-1/2}}{1 - z^{-1}}$$

Matching $H_d(z)$ with the LDI integrator transfer function, we obtain

$$\frac{C_1}{C_2} = 0.765\,366\,9$$

The desired circuit is as shown in Figure 4.2(b). The output shall be taken in the odd phase and the realization is stray-insensitive.

## Integrators for large clock frequency to pole–frequency ratio

In some instances, it is desirable to have an SC integrator with reasonable capacitor spread while using high clock frequencies. The circuit in Figure 4.4(a) is suitable for such a requirement. The transfer function of this integrator is given by

$$\frac{V_{oo}}{V_{ie}} = \frac{C_3 - C_1}{C_2} \cdot \frac{z^{-1/2}}{(1 - z^{-1})} \tag{4.6}$$

Thus a large capacitor ratio can be achieved without the need for a large capacitor spread [4.5]. Note, however, that the realized time constant is highly sensitive to $C_1$

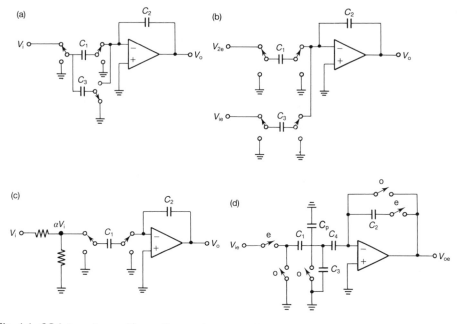

*Fig. 4.4* SC integrators with small capacitor spread while realizing large time constant:
(a) inverting integrator (von Grunigen *et al.*, *adapted from [4.5]*, © 1980 IEE);
(b) differential integrator (von Grunigen *et al.*, *adapted from [4.5]*, © 1980 IEE);
(c) inverting integrator (Kuraishi *et al.*, *adapted from [4.6]*, © 1982 IEEE);  (d) T-cell
integrator (von Peteghem, Sansen, adapted from [4.8], © 1983 IEE).

and $C_3$. A differential LDI integrator is realized by modifying the circuit of
Figure 4.4(a) as in Figure 4.4(b), and with $C_3 = C_1$:

$$V_{oo} = \left[ \frac{C_3 V_{1e} - C_1 V_{2e}}{C_2} \right] \cdot \frac{z^{-1/2}}{(1 - z^{-1})} \tag{4.7}$$

In another technique for realizing large integrator time constants, a resistive
potential divider is employed [4.6] as shown in Figure 4.4(c). This circuit is not
affected by the stray capacitances and the transfer function is given by

$$\frac{V_{oe}}{V_{ie}} = - \frac{\alpha C_1}{C_2} \cdot \frac{1}{(1 - z^{-1})} \tag{4.8a}$$

It is thus seen that $\alpha$ can be used to reduce the capacitor spread. As an illustration,
when $\alpha = 0.1$ and $C_2 = 10C_1$, a capacitor ratio of 100 can be achieved. A capacitive
potential divider can as well be used in place of resistors in Figure 4.4(c) [4.7, 4.8].
The resulting circuit, shown in Figure 4.4(d), termed a 'T-cell integrator', realizes
the transfer function

$$\frac{V_{oe}}{V_{ie}} = - \left( \frac{C_1}{C_1 + C_3 + C_4} \right) \cdot \frac{C_4}{C_2} \frac{1}{1 - z^{-1}} \tag{4.8b}$$

A possible choice is $C_1 = C_4 = \alpha C$ and $C_2 = C_3 = C$, yielding

$$\frac{V_{oe}}{V_{ie}} = -\frac{\alpha^2}{1 + 2\alpha} \cdot \frac{1}{1 - z^{-1}} \qquad (4.8c)$$

It can be seen that an effective capacitor ratio of 100 can be realized with an $\alpha$ of 0.1104, i.e., a capacitor spread of 9. Note that in the circuit of Figure 4.4(d), in the odd phase, $C_1$, $C_3$, $C_4$ and the parasitic $C_p$ are discharged so as to prevent leakage charge build-up at the junction of the three capacitors.

## Parasitic compensated integrators

A method of compensating for the effect of parasitic capacitance in the integrator of Figure 4.1(a) has been described in [4.9, 4.10]. For the circuit in Figure 4.5(a), taking into account the parasitic capacitances, we obtain

$$\frac{V_{oo}}{V_{ie}} = \frac{-2C_1(2C_1 + C_{p2})}{C_2(4C_1 + C_{p1} + C_{p2})} \cdot \frac{z^{-1/2}}{(1 - z^{-1})} \qquad (4.9a)$$

$$V_{oe} = V_{oo}z^{-1/2} \qquad (4.9b)$$

By arranging the geometries of the switches connected to nodes 1 and 2 and by proper routeing associated with nodes 1 and 2, $C_{p1} = C_{p2}$ can be achieved, whence the transfer functions become

$$\frac{V_{oo}}{V_{ie}} = -\frac{C_1}{C_2} \cdot \frac{z^{-1/2}}{(1 - z^{-1})} \qquad (4.9c)$$

$$V_{oe} = V_{oo}z^{-1/2} \qquad (4.9d)$$

and are hence insensitive to parasitic capacitances. Note that equations (4.9c) and (4.9d) are the same as equations (4.1).

It can be shown that a parasitic compensated non-inverting integrator can be realized by the circuit of Figure 4.5(b), under the condition $C_{p1} = C_{p2}$ [4.9, 4.10]. The transfer functions of this circuit are

$$\frac{V_{oe}}{V_{ie}} = \frac{C_1}{C_2} \cdot \frac{z^{-1}}{(1 - z^{-1})} \qquad (4.10a)$$

$$\frac{V_{oo}}{V_{ie}} = \frac{C_1}{C_2} \cdot \frac{z^{-3/2}}{(1 - z^{-1})} \qquad (4.10b)$$

Note the extra half-cycle delay produced by the circuit of Figure 4.5(b) in the odd-phase output; this will be shown, in a later chapter, to be useful in the design of multiplexed SC biquadratic filtering using one OA.

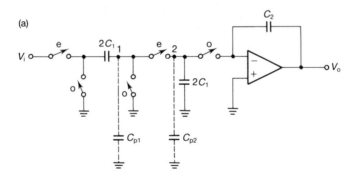

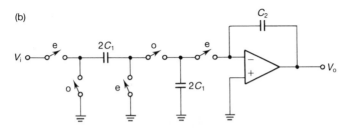

*Fig. 4.5* Parasitic compensated SC integrators (Laker, Fleischer, Ganesan): (a) inverting integrator; (b) non-inverting integrator. (Copyright © 1982 AT&T. All rights reserved. Reprinted with permission.)

## 4.1.3 Bilinear integrators

It may be recalled that the bilinear transformation maps the $s$-domain to the $z$-domain through the relationship

$$s \rightarrow \frac{2}{T} \cdot \frac{z-1}{z+1}$$

or

$$\frac{1}{s} \rightarrow \frac{T}{2} \cdot \frac{1+z^{-1}}{1-z^{-1}}$$

It was seen in Sections 4.1.1 and 4.1.2 that the use of parallel-switched or series-switched capacitors can realize FEI and BEI type integrators respectively, i.e.,

$$\frac{1}{s} \rightarrow \frac{z^{-1}}{1-z^{-1}}$$

$$\frac{1}{s} \rightarrow \frac{1}{1-z^{-1}}$$

and hence a combination of these should realize the bilinear transformation. This method [4.11] is illustrated in Figure 4.6.

The transfer functions of this integrator are given by

$$\frac{V_{oo}}{V_{io}} = -\frac{C_1}{C_2} \cdot \frac{1+z^{-1}}{1-z^{-1}}$$

$$V_{oe} = V_{oo}z^{-1/2}$$

(4.11)

for $C_1 = C_3$, and for $V_{ie} = V_{io}z^{-1/2}$. Note that a sampled and held input is required together with a matching condition on the capacitors. Further, the circuit is sensitive to parasitic capacitances.

Another inverting bilinear integrator is shown in Figure 4.7. This circuit proposed by Simonyi [1.28] has the transfer function given by

$$\left.\frac{V_{oo}}{V_{io}}\right|_{V_{ie}=V_{io}z^{-1/2}} = -\frac{C_1}{C_3} \cdot \frac{1+z^{-1}}{1-z^{-1}}$$

(4.12a)

$$V_{oe} = V_{oo}z^{-1/2}$$

(4.12b)

(Note that $V_{oe}/V_{ie}$ is independent of the input in the odd phase and thus a sampled and held input is not required.) This bilinear integrator is also affected by the top-plate parasitic capacitance $C_p$ of capacitor $2C_1$.

Knob [4.12] has described a parasitic insensitive inverting bilinear integrator, shown in Figure 4.8(a). For $V_{io} = V_{ie}z^{-1/2}$, we obtain

$$\frac{V_{oe}}{V_{ie}} = -\frac{C_1}{C_2} \cdot \frac{1+z^{-1}}{1-z^{-1}}$$

(4.13a)

Note that this circuit uses the stray-insensitive inverting integrator of Figure 4.3(c) with two series-switched capacitors, one working in the even phase and the other in the odd phase. Note also that the output is not held over a clock period. A non-inverting bilinear lossless integrator is realized by changing the switch phasings in

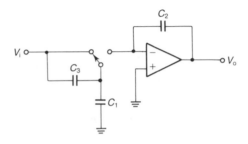

Fig. 4.6  Bilinear inverting integrator (Rahim, Copeland and Chan, *adapted from [4.11]*, © 1978, IEEE).

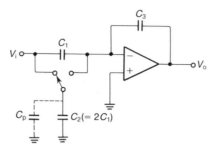

*Fig. 4.7* Simonyi's bilinear inverting integrator *(adapted from [1.28], © 1980, IEEE).*

Figure 4.8(a) to that in Figure 4.8(b). For this SC network, with $V_{ie} = V_{io}z^{-1/2}$, we obtain

$$\frac{V_{oe}}{V_{io}} = \frac{C_1}{C_2}\left(\frac{1+z^{-1}}{1-z^{-1}}\right)z^{-1/2} \tag{4.13b}$$

Note the additional half-cycle delay term in this equation.

A stray-insensitive bilinear lossless integrator with output held over a clock period and using a sampled and held input ($V_{io} = V_{ie}z^{-1/2}$) is shown in Figure 4.9 [1.17]. It can be shown that

$$\frac{V_{oe}}{V_{ie}} = -\frac{C_1}{C_3}\cdot\frac{1+z^{-1}}{1-z^{-1}} \tag{4.14a}$$

$$V_{oo} = V_{oe}z^{-1/2} \tag{4.14b}$$

*Example 4.2*
Design an inverting bilinear SC lossless integrator for the same requirements as in Example 4.1.

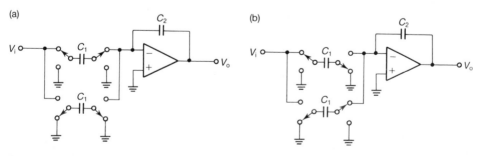

*Fig. 4.8* Knob's bilinear integrators: (a) inverting; (b) non-inverting *(adapted from [4.12], © 1980, IEE).*

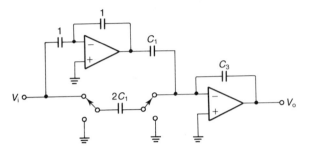

*Fig. 4.9* Lee–Chang bilinear integrator *(adapted from [1.17], © 1981, IEEE).*

The prewarped pole frequency in this case is

$$\Omega_c = \frac{2}{T} \tan \frac{\omega_o T}{2} = 6627.417 \text{ rad s}^{-1}$$

Thus the desired $H_d(z)$ is

$$H_d(z) = \frac{6627.417 T(z+1)}{2(z-1)} = 0.414\,213\,6 \frac{(z+1)}{(z-1)}$$

Several realizations can be obtained using the various integrators discussed before. It is instructive to compare them. For the integrator of Figure 4.6, the design values are $C_1 = C_3$; $C_1/C_2 = 0.414\,213\,6$. Evidently, the smaller capacitances among these are $C_1$ and $C_3$. Denoting them as $C_u$ (i.e., unit capacitance), we have

$$C_1 = C_3 = C_u$$

$$C_2 = 2.414\,21 C_u$$

Thus the total capacitance is $4.414 C_u$. Consider now Simonyi's integrator for which the design equation is $C_1/C_3 = 0.414\,213\,6$. Thus the total capacitance is once again evaluated by taking the minimum capacitance value as unity and scaling other capacitors relative to this. The result shows that

$$C_3 = C_u$$

$$C_1 = 0.414\,213\,6 C_u$$

$$C_2 = 0.828\,427\,2 C_u$$

The total capacitance is thus $5.414 C_u$, larger than in the integrator of Figure 4.6. Knob's integrator, stray-insensitive but with output not held over a clock period, also has the same total capacitance as the integrator of Figure 4.6. The reader is urged to evaluate the other integrators in a similar manner.

## 4.2  Lossy integrators

Lossy integrators or first-order LP networks are useful in the design of high-order filters or, alternatively, in the cascade design of SC filters.

As in the case of SC lossless integrators discussed in the previous section, some of these could realize bilinear first-order transfer functions. However, none of them can realize LDI transformed first-order transfer functions and this limitation has been the reason for considerable research effort in the design of SC high-order ladder filters using LDI transformation (see Chapters 6 and 7). These instead realize a first-order BEI or FEI type lossy integrator transfer function.

### 4.2.1  BEI or FEI lossy integrators  [4.3, 4.13–4.16]

Consider the basic active RC lossy integrator of Figure 4.10, in which the resistors could be replaced by parallel-switched and/or series-switched capacitors to generate several SC lossy integrator configurations. Some of these, which are only bottom-plate stray-insensitive, together with their transfer functions in both phases, are presented in Table 4.1. It may be noted that for some of these networks, the output is held over one clock period.

Completely stray-insensitive realizations are obtained from the active RC network of Figure 4.10 by replacing resistors by series-switched capacitors switched between voltage source and ground. These are also shown in Table 4.1, together with their transfer functions. Once again, it can be seen that, for some of the circuits, the output is held over a clock period.

A study of the various transfer functions in Table 4.1 shows that none of these can realize an exact LDI transformed lossy integrator transfer function, as already mentioned. It may be recalled that a LDI lossy integrator should have a transfer function of the form

$$H(z) = \frac{1}{(z^{1/2} - z^{-1/2}) + k} \tag{4.15a}$$

corresponding to an active RC transfer function of $H_a(s) = 1/(1 + sCR)$, related

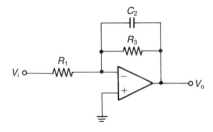

*Fig. 4.10*  An active RC lossy integrator.

**Table 4.1**  SC lossy integrators.

1.

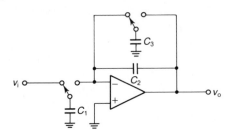

$$\frac{V_{oe}}{V_{ie}} = \frac{(C_3 - C_2)C_1 z^{-1}}{C_2(C_2 + C_3 z^{-1} - C_2 z^{-1})}$$

$$\frac{V_{oo}}{V_{ie}} = \frac{-C_1 z^{-1/2}}{(C_2 + C_3 z^{-1} - C_2 z^{-1})}$$

2.

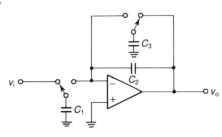

$$\frac{V_{oo}}{V_{ie}} = \frac{-C_1 z^{-1/2}}{C_2 + (C_3 - C_2)z^{-1}}$$

$$V_{oe} = V_{oo} z^{-1/2}$$

3.

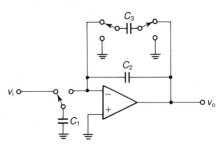

$$\frac{V_{oe}}{V_{ie}} = \frac{-C_1 z^{-1}}{C_2 + C_3 - C_2 z^{-1}}$$

$$\frac{V_{oo}}{V_{ie}} = \frac{-C_1(C_2 + C_3)z^{-1/2}}{C_2(C_2 + C_3 - C_2 z^{-1})}$$

4.

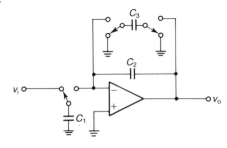

$$\frac{V_{oo}}{V_{ie}} = \frac{-C_1 z^{-1/2}}{C_3 + C_2 - C_2 z^{-1}}$$

$$V_{oe} = V_{oo} z^{-1/2}$$

**Table 4.1** (*continued*)

5.

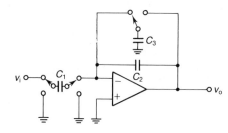

$$\frac{V_{oe}}{V_{ie}} = \frac{-C_1}{C_2 + (C_3 - C_2)z^{-1}}$$

$$V_{oo} = V_{oe}z^{-1/2}$$

6.

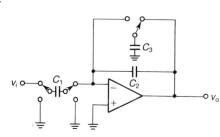

$$\frac{V_{oe}}{V_{ie}} = \frac{-C_1}{C_2 + (C_3 - C_2)z^{-1}}$$

$$\frac{V_{oo}}{V_{ie}} = \frac{(C_3 - C_2)C_1z^{-1/2}}{C_2[C_2 + (C_3 - C_2)z^{-1}]}$$

7.

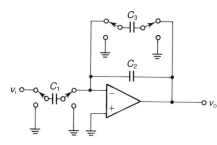

$$\frac{V_{oe}}{V_{ie}} = \frac{-C_1}{C_3 + C_2(1 - z^{-1})}$$

$$V_{oo} = V_{oe}z^{-1/2}$$

8.

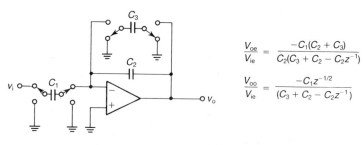

$$\frac{V_{oe}}{V_{ie}} = \frac{-C_1(C_2 + C_3)}{C_2(C_3 + C_2 - C_2z^{-1})}$$

$$\frac{V_{oo}}{V_{ie}} = \frac{-C_1z^{-1/2}}{(C_3 + C_2 - C_2z^{-1})}$$

**Table 4.1** (continued)

9.

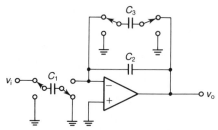

$$\frac{V_{oe}}{V_{ie}} = \frac{C_1 z^{-1}}{C_3 + C_2 - C_2 z^{-1}}$$

$$\frac{V_{oo}}{V_{ie}} = \frac{C_1 (C_2 + C_3) z^{-1/2}}{C_2 (C_3 + C_2 - C_2 z^{-1})}$$

10.

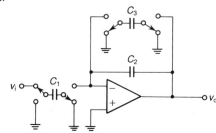

$$\frac{V_{oo}}{V_{ie}} = \frac{C_1 z^{-1/2}}{C_3 + C_2 - C_2 z^{-1}}$$

$$V_{oe} = V_{oo} z^{-1/2}$$

11.

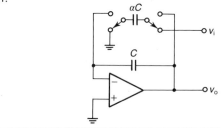

$$\frac{V_{oo}}{V_{ie}} = \frac{\alpha z^{-1/2}}{(\alpha + 1) - z^{-1}}$$

$$V_{oe} = V_{oo} z^{-1/2}$$

through LDI transformation, that is, $s \to (z^{1/2} - z^{-1/2})/T$. The lossy integrators of Table 4.1 realize in contrast 'non-ideal' transfer functions of the form

$$H_1(z) = \frac{1}{(z^{1/2} - z^{-1/2}) + kz^{1/2}}$$

or

$$H_2(z) = \frac{1}{(z^{1/2} - z^{-1/2}) + kz^{-1/2}} \tag{4.15b}$$

Consequently, these 'non-ideal' lossy integrators can be considered to realize an equivalent active RC circuit as shown in Figure 4.10, with $R_1 = T/C_1$, $R_3 = (T/C_3) z^{\pm 1/2}$.

Thus, it may be noted that the realized leak resistances of the integrators $R_3$ are *complex*, since

$$R_3 = \frac{T}{C_3} z^{\pm 1/2} = \frac{T}{C_3} \left( \cos \frac{\omega T}{2} \pm j \sin \frac{\omega T}{2} \right) \tag{4.16}$$

Hence, the design of SC filters using the lossy non-ideal LDI integrators of Table 4.1 should take into account these non-ideal effects, as discussed in Chapters 6 and 7.

All the SC lossy integrators considered above are based on the active RC circuit of Figure 4.10. Hence, they need a minimum of three capacitors. A two-capacitor stray-insensitive SC integrator, first described by Bosshart [4.17] (pointed out by Bingham [4.18]), is shown as the last entry in Table 4.1. The transfer function of this circuit is

$$\frac{V_{oe}}{V_{ie}} = \frac{\alpha}{(\alpha + 1)z - 1} \tag{4.17}$$

It is interesting to note that the d.c. gain is unity, independent of the value of $\alpha$, the ratio of the capacitances. Note, however, this means that the d.c. gain is not independently controllable!

We next consider bilinear lossy integrator realizations.

## 4.2.2  Bilinear lossy integrators

Bilinear lossy integrators should be based on bilinear lossless integrators, since the former reduce to the latter as a limiting case.

The circuit of Figure 4.11(a) [4.19], based on Knob's bilinear lossless integrator, has the transfer function

$$\frac{V_{oe}}{V_{ie}} = \frac{-C_1(1 + z^{-1})}{C_3 + C_2(1 - z^{-1})} \tag{4.18}$$

(with $V_{io} = V_{ie}z^{-1/2}$). Note that the output is not held over a clock period.

Another lossy bilinear integrator insensitive to stray capacitances, derived from the Lee and Chang [1.17] bilinear lossless integrator, is shown in Figure 4.11(b). With $V_{io} = V_{ie}z^{-1/2}$, we obtain

$$\frac{V_{oe}}{V_{ie}} = \frac{-C_1(1 + z^{-1})/2}{C_3 + C_2(1 - z^{-1})} \tag{4.19}$$

As an example, we shall next consider the design of an SC lossy integrator.

*Example 4.3*
Design inverting SC lossy integrators using (a) $p$-transformation, (b) LDI transformation and (c) bilinear transformation, for a cutoff frequency 1 kHz, sampling frequency 8 kHz and d.c. gain of unity.

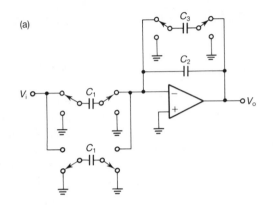

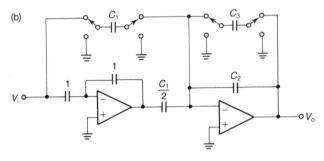

Fig. 4.11 Bilinear lossy integrator: (a) Ananda Mohan, Ramachandran and Swamy *(adapted from [4.19]*, © 1982, IEE); (b) Lee–Chang type *(adapted from [1.17]*, © 1981, IEEE).

(a) The $H_a(s)$ required is of the form $2\pi \cdot 1000/(s + 2\pi \cdot 1000)$. The $p$-transformation type $H_d(s)$ is obtained by using the relation $z_i = e^{s_i t}$ to obtain the $z$-domain pole. Thus the required $H_d(z)$ with a d.c. gain of unity is

$$H_d(z) = \frac{0.544\ 061\ 9}{1 - 0.455\ 938\ 1z^{-1}}$$

Choosing the stray-insensitive realization of Table 4.1, entry 7, we obtain $C_1 = C_3 = 1.193\ 280\ 2C_2$.

(b) Using the design method illustrated in Chapter 1, we obtain

$$H_d(z) = \frac{-0.535\ 361\ 8z^{-1/2}}{1 - 0.464\ 638\ 2z^{-1}}$$

Choosing the design of Table 4.1, entry 7 once again, we obtain $C_1 = C_3 = 1.152\ 212C_2$.

(c) The desired $H_d(z)$ using the bilinear transformation method is

$$H_d(z) = \frac{0.292\ 893\ 2(1 - z^{-1})}{1 - 0.414\ 213\ 6z^{-1}}$$

Using the lossy bilinear integrator of Figure 4.11, we obtain $C_1 = 0.292\ 893\ 2C_3$ and $C_3 = 1.414\ 213\ 3C_2$.

## 4.3 Circuits realizing finite zeros

In this section, we deal with first-order SC networks realizing finite zeros, i.e., ideal differentiators, first-order HP and AP networks. The starting point for development of the various circuits is the first-order active RC network of Figure 4.12(a). A simpler version of this circuit, a differentiator (i.e., $C_3 = 0$ as in Figure 4.12(b)), will be considered first.

An SC differentiator is shown in Figure 4.13(a). (Note that a parallel-switched capacitor cannot be used in the feedback path, because in either of the two phases the OA feedback loop is open, thus causing the OA to latch up). The transfer function of this differentiator [1.14] is

$$\frac{V_{oe}}{V_{ie}} = -\frac{C_1}{C_2}(1 - z^{-1}) \tag{4.20}$$

No other variation of this circuit is possible. Next, consider the stray insensitive SC network [4.20] shown in Figure 4.13(b), whose transfer functions are

$$\frac{V_{oe}}{V_{ie}} = \frac{-C_1(1 - z^{-1})}{C_3 + C_2(1 - z^{-1})} \tag{4.21a}$$

$$V_{oo} = V_{oe}z^{-1/2} \tag{4.21b}$$

with

$$V_{io} = V_{ie}z^{-1/2} \tag{4.21c}$$

Note that the HP networks discussed thus far realize zeros at the origin. However, it is often required to realize finite zeros. A stray-insensitive circuit [4.21] for achieving this purpose is shown in Figure 4.14(a). The transfer function of this circuit is

$$\frac{V_{oe}}{V_{ie}} = \frac{-(C_4 - C_1 z^{-1})}{C_3 + C_2(1 - z^{-1})} \tag{4.22}$$

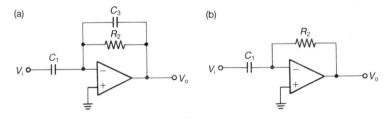

*Fig. 4.12* (a) Active RC HP network; (b) active RC differentiator.

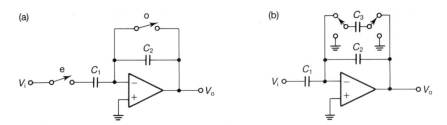

Fig. 4.13  (a) SC differentiator;  (b) SC HP network.

Evidently, with $C_1 = C_4$, an HP function is realizable, in which case the circuit can be simplified as that in Figure 4.14(b). The circuits of Figures 4.14(a) and 4.14(b) do not have the output held over a clock period.

A network with output held over a clock period and capable of realizing finite zeros [4.19] is shown in Figure 4.14(c), for which, with $V_{io} = V_{ie}z^{-1/2}$, we have

$$\frac{V_{oe}}{V_{ie}} = \frac{-(C_4 - C_1 z^{-1})}{C_3 + C_2(1 - z^{-1})} \tag{4.23a}$$

$$V_{oo} = V_{oe}z^{-1/2} \tag{4.23b}$$

An interesting application of these networks is to realize first-order all-pass transfer functions. Equations (4.22) and (4.23) realize an all-pass transfer function when

$$\frac{C_2}{C_2 + C_3} = \frac{C_4}{C_1} \tag{4.24}$$

The utility of these all-pass networks to realize second-order SC filters will be discussed in later chapters. We will next consider a design example.

*Example 4.4*
Design a first-order SC all-pass circuit for 1 kHz. Use an 8 kHz sampling frequency and choose unity gain.

The desired all-pass transfer function is obtained by matched-$z$ transformation:

$$H_d(z) = -\left[\frac{0.455\ 938}{1 - 0.455\ 938\ 1z^{-1}}\right]$$

Using the stray-insensitive realization of Figure 4.14(c), we obtain the design equations from equation (4.23a):

$$\frac{C_3}{C_2} = 1.193\ 280\ 2$$

$$\frac{C_1}{C_2} = 2.193\ 280\ 2$$

$$C_4 = C_2$$

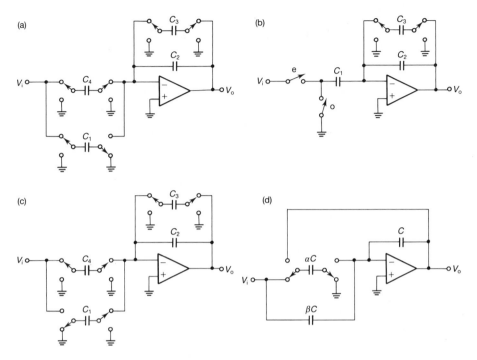

*Fig. 4.14* Stray-insensitive circuits realizing finite zeros: (a), (b) *adapted from Brugger ([4.21], © 1981, IEEE);* (c) *adapted from Ananda Mohan, Ramachandran and Swamy ([4.19], © 1982 IEE);* (d) *adapted from Bingham ([4.18], © 1984, IEEE).*

A three-capacitor stray-insensitive first-order SC network capable of realizing finite zeros [4.18] can be derived from the lossy integrator of entry 11 of Table 4.1. This circuit, shown in Figure 4.14(d), realizes the transfer function

$$\frac{V_{oo}}{V_{io}} = \frac{(\alpha + \beta)z^{-1} - \beta}{(1 + \alpha) - z^{-1}}$$

with $V_{ie} = V_{io}z^{-1/2}$. Also, $V_{oe} = V_{oo}z^{-1/2}$. Evidently, when $\beta = 1$, an all-pass transfer function is realized. Note also that the gain is independent of $\alpha$ and $\beta$ values, and is unity. Observe that an arbitrary gain cannot be achieved.

## 4.4 General first-order SC network [4.19]

It should be mentioned that a variety of circuits capable of realizing finite zeros are possible using one OA, the simplest of which is shown in Figure 4.14. To demonstrate that this is possible, a general SC network (Figure 4.15) is considered. This uses two general SC branches in the input and feedback paths of an OA and consists of seven elements in each branch: series-switched capacitors $C_1$ and $C_2$, parallel-switched capacitors $C_6$ and $C_7$, inverting switched capacitors $C_3$ and $C_4$, and an

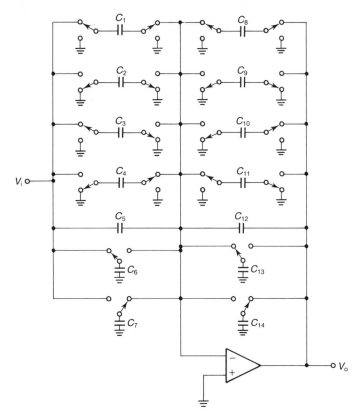

Fig. 4.15 A general first-order SC network *(adapted from [4.19],* © *1982, IEE).*

unswitched capacitor $C_5$ in the input branch and capacitors $C_8$ to $C_{14}$ in the feedback branch. The transfer functions of this circuit are given by

$$V_{oe} = \frac{-\left[\left[(C_1+C_5)(C_9+C_{12}) - (C_6-C_3-C_5)(C_{14}-C_{11}-C_{12})z^{-1}\right]V_{ie} + \left[(C_9+C_{12})(C_7-C_4-C_5) - (C_2+C_5)(C_{14}-C_{11}-C_{12})\right]\right]V_{io}z^{-1/2}}{\left[(C_8+C_{12})(C_9+C_{12}) - (C_{11}+C_{12}-C_{14})(C_{10}+C_{12}-C_{13})z^{-1}\right]}$$

(4.25a)

$$V_{oo} = \frac{-\left[\left[(C_8+C_{12})(C_6-C_3-C_5) - (C_1+C_5)(C_{13}-C_{10}-C_{12})V_{ie}z^{-1/2}\right] + \left[(C_2+C_5)(C_8+C_{12}) - (C_{13}-C_{10}-C_{14})(C_{10}+C_{12}-C_{13})z^{-1}\right]V_{io}\right]}{\left[(C_8+C_{12})(C_9+C_{12}) - (C_{11}+C_{12}-C_{14})(C_{10}+C_{12}-C_{13})z^{-1}\right]}$$

(4.25b)

We will first consider $V_{oe}$ only. For delay-free transfer functions (i.e., no $z^{-1/2}$ terms in the numerator), $V_{io}$ has to be $V_{ie}z^{-1/2}$. Then equation (4.25a) reduces to

$$\frac{V_{oe}}{V_{ie}} = \frac{\begin{aligned}-[(C_1 + C_5)(C_9 + C_{12}) + [(C_7 - C_4 - C_5)(C_9 + C_{12}) \\ + (C_3 - C_2 - C_6)(C_{14} - C_{11} - C_{12})]\,z^{-1}]\end{aligned}}{[(C_8 + C_{12})(C_9 + C_{12}) - (C_{11} + C_{12} - C_{14})(C_{10} + C_{12} - C_{13})z^{-1}]}$$

$$(4.25c)$$

It is immediately noted that non-inverting bilinear transfer functions cannot be realized because of the negative $z^0$ term in the numerator. All other inverting bilinear transfer functions can be realized by the circuit shown in Figure 4.15, because of several degrees of freedom available in the numerator to choose the coefficient of $z^{-1}$ as positive or negative. Note that $C_{12}$ cannot be zero, otherwise the OA may latch (because of the d.c. instability), since the feedback loop is open.*

It is also desirable to have stray-insensitive designs, wherever possible and in only cases where some advantage is gained. Then $C_6$, $C_7$, $C_{13}$, $C_{14}$ may be made non-zero. Consider first, $C_6 = C_7 = C_{13} = C_{14} = 0$ [4.19], when equation (4.25c) becomes

$$\frac{V_{oe}}{V_{ie}} = \frac{[(C_1 + C_5)(C_9 + C_{12}) + [(C_2 - C_3)(C_{11} + C_{12}) - (C_4 + C_5)(C_9 + C_{12})]\,z^{-1}]}{[(C_8 + C_{12})(C_9 + C_{12}) - (C_{11} + C_{12})(C_{10} + C_{12})z^{-1}]}$$

$$(4.25d)$$

Thus $C_2$ is essential only for realizing bilinear lossy integrator transfer functions. The remaining degrees of freedom can be chosen to simplify the SC structures required to realize the desired transfer function. Choosing $C_9 = 0$ simplifies the numerator of the transfer function. For stability, $(C_{11} + C_{12})(C_{10} + C_{12}) < (C_8 + C_{12})(C_9 + C_{12})$ and with $C_9$ already set to zero, $C_{10}$ and $C_{11}$ can be also zero. Then, equation (4.25d) becomes

$$\frac{V_{oe}}{V_{ie}} = \frac{(C_1 + C_5) + (C_2 - C_3 - C_4 - C_5)z^{-1}}{C_8 + C_{12} - C_{12}z^{-1}} \qquad (4.25e)$$

Even with $C_3 = C_4 = 0$, equation (4.25e) can generate all the usually desired first-order transfer functions.

The use of $C_{13}$ (or $C_{14}$) allows realization of a denominator of the form $C_8 + C_{12} - (C_{12} - C_{13})z^{-1}$, which with $C_8 = C_{13}$ (or $C_{14}$) can realize an 'exact' bilinear first-order denominator in the sense that the time constant is decided by only one ratio of capacitances $C_x/C_y$. This is of some practical utility in the design of 'exactly' bilinear switched capacitor ladder filters, as has been mentioned earlier.

---

* It is possible to remove $C_{12}$ altogether, and yet keep the OA d.c. stable by an arrangement called 'XY feedback', introduced in [4.12]. This requires two capacitors and two switches additionally, in place of $C_{12}$. However, for minimal capacitance design, $C_{12}$ is present, obviating the need for 'XY feedback' and saving switches and one capacitor. Note also that whether $C_{12}$ is present or not, its function can be accomplished by the switched capacitors $C_8$, $C_9$, $C_{10}$, $C_{11}$, which for instance realize $C_{12}$ when they are equal. We discuss the use of 'XY feedback' in Chapter 5.

We will next consider the requirement that the SC circuits have their output held over one clock period, i.e., $V_{oe} = V_{oo}z^{-1/2}$ or $V_{oo} = V_{oe}z^{-1/2}$. Considering the case for $V_{oe} = V_{oo}z^{-1/2}$ and writing the charge conservation equations for Figure 4.15, we obtain

$$(C_8 + C_{12})V_{oe} + (C_{14} - C_{11} - C_{12})V_{oo}z^{-1/2} = -(C_1 + C_5)V_{ie} - (C_7 - C_4 - C_5)V_{io}z^{-1/2}$$
$$(4.26)$$
$$(C_{13} - C_{10} - C_{12})V_{oe}z^{-1/2} + (C_9 + C_{12})V_{oo} = -(C_6 - C_3 - C_5)V_{ie}z^{-1/2} - (C_2 + C_5)V_{io}$$

For $V_{oe} = V_{oo}z^{-1/2}$, we need

$$C_1 = C_4 = C_7 = C_8 = C_{11} = C_{14} = 0$$

and $V_{ie} = V_{io}z^{-1/2}$. The first condition can be easily obtained by inspection of Figure 4.15 as well, meaning that no charge transfer shall take place in the even phase. (Even $C_5$ can be zero, without any loss of generality.) Then,

$$\frac{V_{oo}}{V_{io}} = \frac{-[C_2 + (C_6 - C_3)z^{-1}]}{C_9 + C_{12} + (C_{13} - C_{10} - C_{12})z^{-1}} \qquad (4.27)$$

For stray-insensitive designs, since $C_6$ and $C_{13}$ are zero, it can be seen that circuits with finite zeros with the output held over a clock period are realizable, excluding inverting bilinear integrators.

The results are similar for the odd-phase output, too. This completes the discussion on first-order SC networks.

A conventional method of realizing digital filters using delays, multipliers and adders is also of interest in SC filter design and the basic element in these circuits is a delay. Hence, in the next section, we will study several circuit arrangements to realize inverting or non-inverting half-cycle or full-cycle delay elements.

## 4.5  SC delay circuits

One circuit for realizing a half-cycle delay $(z^{-1/2})$ is shown in Figure 4.16(a), which is stray-insensitive and non-inverting. Note that it is the SC integrator of Figure 4.1(b) with the memory of the feedback capacitors destroyed during one phase. It can be shown that $V_{oo} = \mp V_{ie}z^{-1/2}$ for the circuit of Figure 4.16(a). If $V_{ie} = V_{io}z^{-1/2}$, then the transfer function $\mp z^{-1}$ could be realizable, which requires a sample and hold to precede the circuit of Figure 4.16(a).

The circuit of Figure 4.16(a) can be simplified as in Figure 4.16(b), reducing the number of switches from four to two. These circuits require two OAs to realize a delay of $z^{-1}$, as shown in Figure 4.16(c) [4.22, 4.23].

We next consider a single OA non-inverting full-cycle delay circuit (Figure 4.16(d)) for which $V_{oe}/V_{ie}$ is $z^{-1}$ [4.24]. This circuit, however, is only bottom-plate stray-insensitive.

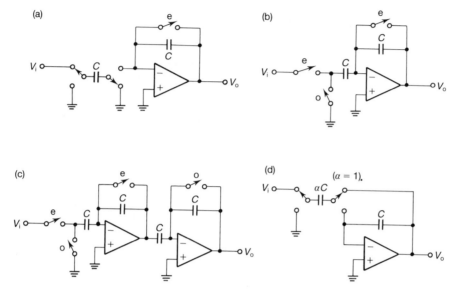

Fig. 4.16  SC delay circuits;  (c) *adapted from [4.23],* © 1978 IEEE;  (d) *adapted from [4.24],* © 1980 IEEE.

Next, we consider a stray-insensitive inverting half-cycle delay circuit. This circuit [4.22] uses a four-phase clock, and is shown in Figure 4.17(a). The switches are closed during the phases shown by the side of the switches. For example, (2 + 4) indicates that the switch is closed in the second and fourth phases of the clock. The operation of the circuit of Figure 4.17(a) is as follows. During phase 4, $C_1$ and $C_2$ are discharged. During phase 1, $C_1$ transfers a charge of $C_1 V_i$ to $C_2$. During phase 2, $C_1$ is discharged while $C_2$ retains its charge, and during phase 3 $C_2$ retains its charge. Finally, during phase 4, $C_2$ is discharged. It must now be clear that during phases 1 and 3 the output exists, and during phases 2 and 4 the output is zero. It is possible to secure a half-cycle delay, when we sample the output only during phase 3. Thus a switch is required at the output as shown, which is closed only during phase 3. This circuit is also stray-insensitive.

We next consider an elegant stray-insensitive delay realization [4.25], shown in Figure 4.17(b), using a three-phase clock. During Phase 1, the input is sampled and held on $C_{sh}$. During phase 2 the feedback capacitor $C_{fb}$ is discharged, and during phase 3 the charge on $C_{sh}$ is transferred to $C_{fb}$. Hence, if the output is sampled during phase 1, a full-cycle delay is achieved. The gain of the delay circuit is controlled by $C_{sh}/C_{fb}$.

A stray-insensitive SC delay using a two-phase clock is shown in Figure 4.17(c) [4.26]. In this circuit, in the even phase, the charge $C_{1i} V_{io} z^{-1/2}$ injected by $C_{1i}$ is stored on $C_{1a}$, which is transferred to $C_{2i}$ in the odd phase. In the next even phase, the charge on $C_{2i}$ is fed to the virtual ground of the next OA, used to realize another delay element, in a chain of delay elements. It can be derived that a non-inverting

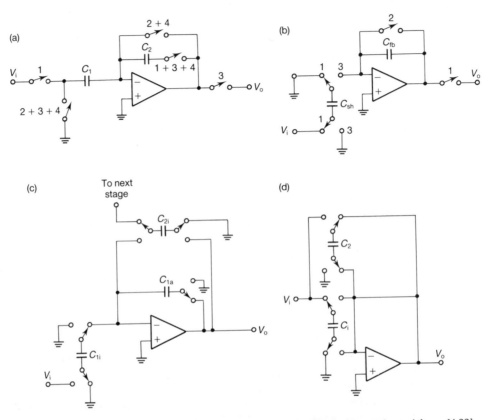

Fig. 4.17  Stray-insensitive SC circuits due to:  (a) Young and Hodges *(adapted from [4.22]*, © 1979, IEEE);  (b) Enomoto et al. *(adapted from [4.25]*, © 1982, IEE);  (c) Gillingham *(adapted from [4.26]*, © 1984, IEEE);  (d) Nagaraj *(adapted from [4.27]*, © 1984, IEE).

delay with unity gain is realized when $C_{1i} = C_{2i}$ (independent of the value of $C_{1a}$). Note that when a two-phase non-overlapping clock is used, in between the even and odd clock phases, the 'XY feedback' technique must be used.

The stray-insensitive circuit of Figure 4.17(d) [4.27] employs a two-phase clock. Its operation is as follows. In phase 1, $C_1$ is charged to the input voltage and connected across the OA in phase 2. In phase 2, the input is sampled by $C_2$ which is connected across the OA in phase 1 to provide a half-cycle delay $z^{-1/2}$. The circuit thus works at double the clock frequency and needs 'XY feedback' to stabilize the OA during the interval between phases 1 and 2. The advantage of the circuit is that the capacitance ratio does not affect its performance, unlike the other circuits. The offset compensation of this circuit has been considered in [4.34].

Some attempts have been made to realize SC delay functions using unity gain buffer-connected OAs as well. The reader is referred to [4.28–4.31] for details of these circuits.

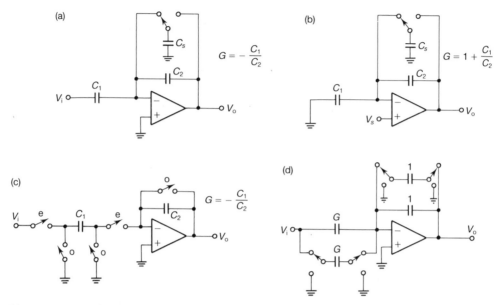

Fig. 4.18  SC amplifiers;  (d) *(adapted from [4.32],* © 1983, IEEE).

## 4.6  SC amplifiers

It may be noted that SC amplifiers could be used in certain applications. Some SC amplifiers are shown in Figure 4.18. That in Figure 4.18(a) is similar to conventional inverting amplifiers, and it needs an additional switched capacitor to bias the OA in the stable region. Usually, even without the stabilizing capacitor $C_s$, the Off-resistance of the MOS switches and the stray capacitor are able to stabilize the OA. This circuit has a gain of $-C_1/C_2$. Non-inverting amplifiers are realized by modifying the circuit to that of Figure 4.18(b). Once again, $C_s$ serves to stabilize the OA. The next amplifier to be considered is shown in Figure 4.18(c). This is a lossless integrator with the memory of the feedback capacitor $C_2$ destroyed by an additional switch. The operation is simple. In the odd phase, $C_1$ and $C_2$ are discharged. In the even phase, input is amplified by $-C_1/C_2$. The circuit is stray-insensitive and is unconditionally stable. Note that we have considered the use of such amplifiers in Chapter 3 in connection with Wien-bridge and phase-shift oscillators. We finally consider the SC amplifier [4.32] of Figure 4.18(d), which realizes a gain of $G$. This circuit can provide output held over a clock period.

## 4.7  Direct design of high-order SC filters from *z*-domain transfer functions [4.33–4.34]

So far, several active SC networks useful for realizing first-order digital transfer

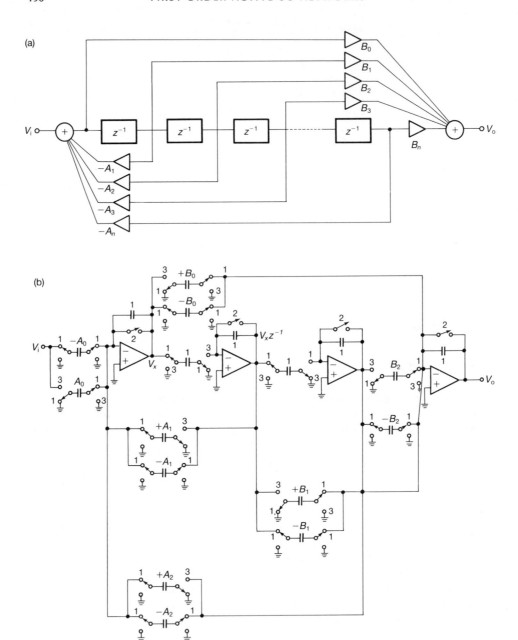

*Fig. 4.19* (a) Recursive sampled data filter structure; (b) SC implementation using Enomoto *et al.* delay circuit.

functions have been described. The design of high-order sampled data filters can be done in several ways. In one method to be considered in this section, the high-order filter is 'directly' realized by an SC structure. The $z$-domain transfer function could be obtained by any one of the methods described in Chapter 1. An $n$th-order transfer function thus obtained is of the form

$$\frac{V_o(z)}{V_i(z)} = \frac{B_0 + B_1 z^{-1} + B_2 z^{-2} + \cdots + B_n z^{-n}}{A_0 + A_1 z^{-1} + A_2 z^{-2} + \cdots + A_n z^{-n}} \tag{4.28}$$

A recursive structure that can realize equation (4.28) is shown in Figure 4.19(a). It is immediately evident that the realization requires delay elements and summers. The summers must be capable of inversion and weighting of the inputs, if necessary. Stray-insensitive realizations of the transfer function (4.28) are desirable. Using a three-phase clock, stray-insensitive non-inverting delays have been shown to be realizable, as in Figure 4.17(b). The use of these to implement equation (4.28) will be considered next. As an illustration, a second-order recursive filter structure is considered. Since the coefficients can be positive or negative in general, series branches to realize either sign are shown in Figure 4.19(b). The outputs of the various delay elements can be sampled in phase 1 if a positive coefficient is desired, and in phase 3 if a negative coefficient is desired. Accordingly, a series-switched capacitor or an inverting switched capacitor is used. Note that the several switches can be combined in Figure 4.19(b). The method can be extended to high-order filters.

High-order $z$-domain transfer functions can also be synthesized by using the same topology as Figure 4.19(a), but with the delay blocks replaced by stray-insensitive first-order blocks. One such configuration has been arrived at from different considerations by Karagoz and Acar [4.33]. For this structure, shown in Figure 4.20(d), the capacitor ratios will not directly correspond to the coefficients of the transfer function, unlike the direct realization using delays. The signal-flow graph presented in Figure 4.20(c) uses first-order networks, each realizing the transfer function, $z^{-1}/(1 - z^{-1})$ (corresponding to forward Euler integrators). The formulas* for computing the capacitor ratios, to realize a given $z$-domain transfer function described by equation (4.28), as in Figure 4.20(d), are as follows:

$$\hat{B}_{n-p} = \sum_{i=0}^{p} B_{n-i}\binom{n-i}{p-i}, \qquad p = 0, 1, \ldots, n$$

$$\tag{4.29}$$

$$\hat{A}_{n-p} = \sum_{i=0}^{p} A_{n-i}\binom{n-i}{p-i} + \binom{n}{p}, \qquad p = 0, 1, 2, \ldots, n$$

where $n$ is the order of the voltage transfer function, and $\binom{n-i}{p-i}$ and $\binom{n}{p}$ are binomial coefficients.

* A procedure for obtaining various $\hat{B}$ and $\hat{A}$ values based on synthetic division has been given in [4.54], together with SC implementation using FEI and BEI blocks.

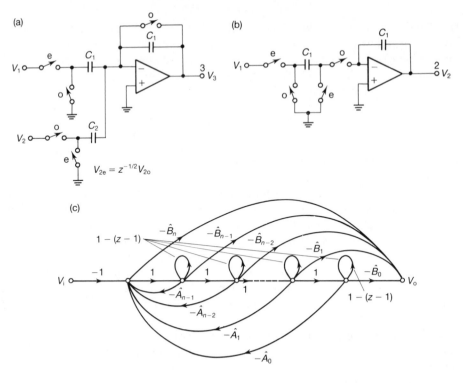

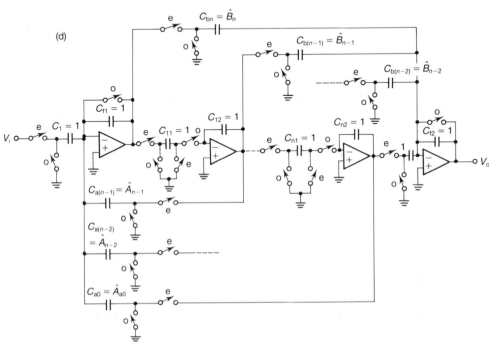

The first-order SC networks realizing the input and output summers and forward Euler integrators are shown in Figures 4.20(a) and 4.20(b). Note that for the summer, positive or negative weighting of the inputs should be possible. The two branches shown in Figure 4.20(a) perform these operations as desired. To realize an $n$th-order transfer function, $n + 2$ OAs are required, as shown in Figure 4.20(d). The method can be extended to the use of backward Euler integrators as well. However, since no stray-insensitive non-inverting bilinear lossless integrator using one OA is possible, direct-form bilinear integrators, using one OA per integrator, are not possible. A design example is considered next.

*Example 4.5*
Design a stray-insensitive direct-form SC circuit to realize the second-order digital transfer function

$$H(z) = \frac{0.9777(z^2 - 1.9978z + 1)}{(z^2 - 1.9533z + 0.9554)}$$

Choosing the realization of Figure 4.20, we first need to obtain a transfer function in the new domain,

$$u = \frac{z^{-1}}{1 - z^{-1}}$$

since we intend to use integrators of the type realizing $H(z) \equiv u$. Thus, substituting $(1 + u)/u$ for $z$ or by using equation (4.29), the transfer function in $u$ is obtained:

$$H(u) \equiv \frac{1.9553u^2 + 1.9553u + 888.81}{0.0011u^2 + 0.0467u + 1}$$

Thus the final SC circuit can be obtained from Figure 4.20(d). Note that the circuit requires a rather large capacitor spread. The reader is urged to scale the capacitors whenever possible for minimum total capacitance (i.e., to make the smallest capacitance value unity) and to evaluate the sensitivities of the above circuit to various capacitor ratios.

## 4.8  First-order SC networks using unity-gain amplifiers

In some applications, passive first-order SC networks studied in Chapter 3 with 'buffered' output are useful. A general first-order network is shown in Figure 4.21(a). Note that an HP and an LP network are realized when $V_1 = 0$ and $V_2 = 0$, respectively. The SC implementation of these follows in a straightforward manner by replacing the resistors with series-switched or parallel-switched capacitors. The resulting circuits are sensitive to top-plate parasitic capacitances (see, for

---

Fig. 4.20  (a) A stray-insensitive summer;  (b) an FEI integrator;  (c) signal flow graph of a high-order direct form filter;  (d) its implementation (c) and (d) *(adapted from [4.33], ©* 1982, IEEE).

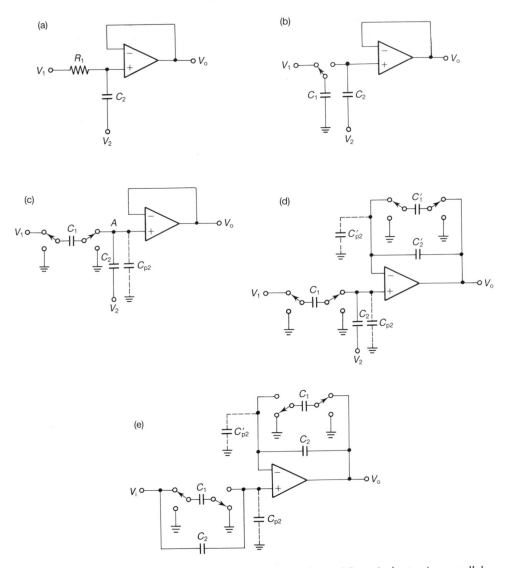

Fig. 4.21   (a) A general first-order active RC network;   (b) its SC equivalent using parallel-switched capacitors;   (c) its SC equivalent using series-switched capacitors;   (d) parasitic compensation of (c);   (e) parasitic compensation of first-order all-pass network.

example, Figure 4.21(b)). It is possible to compensate [4.35] the effect of the parasitic capacitance in certain SC structures obtained from Figure 4.21(a). The important property of these structures is that the parasitic capacitance is 'lumped' at one node, the non-inverting input of the OA. Consider the p-transformation type of circuits used to realize Figure 4.21(a), as in Figure 4.21(c). It is evident that the only parasitic capacitances that affect the performance are those of $C_1$, $C_2$ and any other stray capacitance due to wiring, etc., lumped at node $A$. It is possible to compensate this by modifying the circuit of Figure 4.21(c) as in Figure 4.21(d). Under the condition $C_{p2} = C'_{p2}$, i.e., assuming that identical parasitic capacitances exist at both the inverting and non-inverting inputs of the OA, the transfer function realized is the ideal desired one. It is thus required to route the wiring and arrange layouts such that $C_1 = C'_1$, $C_2 = C'_2$ and $C_{p2} = C'_{p2}$ are achieved.

The above method can also be used to compensate other SC networks satisfying the above-mentioned property. As another illustration, consider the first-order all-pass network shown in Figure 4.21(e). The application of these first-order networks to realize parasitic-compensated second-order filters will be considered later.

## 4.9  Conclusions

In this chapter, several first-order networks useful in the realization of biquadratic and high-order filters are studied in detail. By now, the reader should be familiar with the various alternatives available depending on stray-insensitivity, minimal capacitor count or circuit type (LDI, bilinear, BEI, FEI, etc.,) and thus should be able to appreciate the variety of high-order circuits possible. These will be considered later.

## 4.10  Exercises

4.1 *Self-equalizing sample and hold*:
   The sample and hold circuit can be shown to introduce a response of the shape $(\sin x)/x$ where $x = \omega T/2$. Hence, it has to be corrected to eliminate the undesirable droop in the passband response. One method is to simulate an $x/\sin x$ response by a circuit called an *equalizer*. It is possible to realize the function of a sample and hold circuit as well as an equalizer in a single block. Such an SC first-order lossy integrator is shown in Figure 4.22. Derive the transfer function. Suppose that a low-pass equiripple approximation is desired in the passband. Find the design values [4.36].

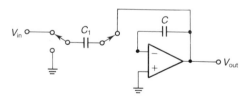

*Fig. 4.22*  An SC first-order lossy integrator.

4.2 The OA used in practice has finite d.c. gain $A_0$. Find the transfer function of the SC integrator of Figure 4.1(a), using such an amplifier and discuss the effects.

4.3 The delay circuit of Figure 4.16(d) has several applications [4.24]. Analyze its behaviour for $1 < \alpha < 2$; plot the magnitude response and develop the design equations for realizing a BP response of given $Q$ and centre frequency. Discuss the effect of parasitic capacitances. Note that this circuit with $\alpha = 1.115$ realizes a self-equalizing sample and hold [4.36] similar to that in Exercise 4.1.

4.4 It is required to design a first-order AP SC network. Derive the necessary active RC networks, which can be converted through

(a) $p$-transformation
(b) bilinear transformation

to SC networks. Compare the flexibility of the resulting networks with the stray-insensitive circuits of Figures 4.14(a) and 4.14(c).

4.5 (a) Derive an SC stray-insensitive circuit [4.37] to realize a first-order transfer function of the form

$$H(z) = \frac{-(a_0 - a_1 z^{-1})}{(1 - bz^{-1})}$$

Its signal-flow graph is as given in Figure 4.23(a).

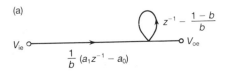

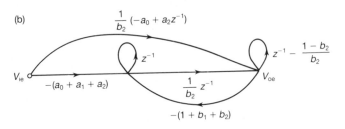

Fig. 4.23 (a) Signal flow graph of a transfer function of the form $H(z) = -(a_0 - a_1 z^{-1})/(1 - bz^{-1})$; (b) signal flow graph of a general biquadratic digital transfer function.

(b) A second-order general biquadratic digital transfer function of the form

$$H(z) = \frac{-(a_2 z^{-2} + a_1 z^{-1} + a_0)}{(b_2 z^{-2} + b_1 z^{-1} + 1)}$$

can be realized by the signal-flow graph of Figure 4.23(b) [4.43]. Realize the signal flow graph (SFG) using the stray-insensitive circuit blocks derived in (a).

4.6 Analyze the SC networks of Figure 4.24 and show that they realize bilinear inverting integrators [4.38]. Compare these with other bilinear inverting integrators discussed in this chapter.

4.7 A differential integrator is shown in Figure 4.25 in active RC form. Derive SC networks using bilinear transformation [4.39] and $p$-transformation and compare these with the differential integrator of Figure 4.2(a).

4.8 The bilinear integrator is shown in Figure 4.26(a). A method of compensating the effect of bottom-plate parasitic capacitance is shown in Figure 4.26(b). Analyze the operation of the circuit and derive the conditions for compensation [4.40] . Note that even with this compensation scheme, the top-plate parasitic capacitances still exist.

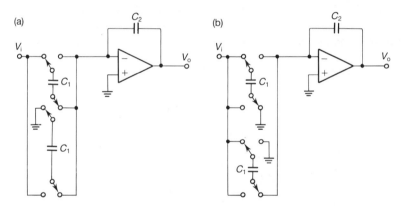

Fig. 4.24   SC networks realizing bilinear inverting integrators *(adapted from [4.38],* © 1978, IEE).

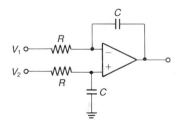

Fig. 4.25   An active RC differential integrator.

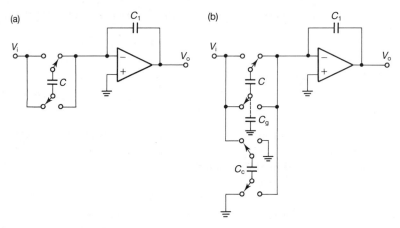

*Fig. 4.26* (a) An SC bilinear integrator; (b) the bilinear integrator of (a) compensated for bottom-plate parasitic capacitance; *(adapted from [4.40], © 1979, IEE).*

4.9 The non-inverting bilinear integrator of Figure 4.8(b) requires a sample and held input requiring an additional OA switch and capacitor. It is possible to reduce the OA count, by the modified circuits shown in Figure 4.27. In these circuits, the parasitic capacitances affect the performance. Examine whether parasitic compensation is possible by proper choice of capacitor values [4.41].

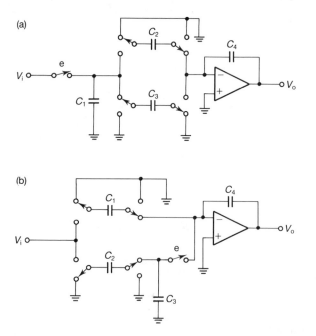

*Fig. 4.27* Non-inverting bilinear integrators; *(adapted from [4.41], © 1983, IEE).*

4.10 In SC networks, it is desirable to reduce the number of switches and capacitors. An illustration is the lossy integrator of Table 4.1, entry 11. This circuit, however, cannot be derived from the general SC network of Figure 4.15. The new SC network of Figure 4.28 is general and uses a reduced number of capacitors. Examine its generality by deriving the transfer function.

4.11 Derive the capacitor values required for realizing a $z^{-1}$ element, from the general stray-insensitive SC network of Figure 4.15 (ignore $C_6$, $C_7$, $C_{13}$ and $C_{14}$). Note also that $C_{12}$ need not be zero if 'XY feedback' is intended to be used. Compare the resulting delay circuit (if any) with the other stray-insensitive delay elements.

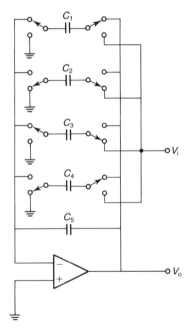

*Fig. 4.28* A general first-order SC network.

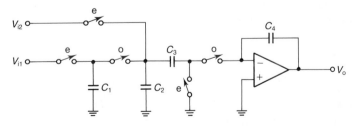

*Fig. 4.29* A parasitic compensated SC differential integrator; *(adapted from [4.42],* © 1983, IEE).

4.12 A differential integrator which can be parasitic compensated [4.42] is shown in Figure 4.29. Derive the conditions of parasitic compensation.

## 4.11   References

[4.1] B.J. Hosticka, R.W. Brodersen and P.R. Gray, MOS sampled-data recursive filters using switched-capacitor integrators, *IEEE Journal of Solid-State Circuits*, SC-12, 6, 600–8, December 1977.

[4.2] G.M. Jacobs, D.J. Allstot, R.W. Brodersen and P.R. Gray, Design techniques for MOS switched-capacitor ladder filters, *IEEE Transactions on Circuits and Systems*, CAS-25, 12, 1014–20, December 1978.

[4.3] D.J. Allstot, R.W. Brodersen and P.R. Gray, MOS switched capacitor ladder filters, *IEEE Journal of Solid-State Circuits*, SC-13, 6, 806–14, December 1978.

[4.4] K. Martin and A.S. Sedra, Stray-insensitive switched-capacitor filters based on the bilinear z-transform, *Electronics Letters*, 15, 13, 365–6, 21 June 1979.

[4.5] D.C. von Grunigen, U.W. Brugger and G.S. Moschytz, Novel stray-insensitive switched-capacitor integrator, *Electronics Letters*, 16, 10, 395–7, 9 May 1980.

[4.6] Y. Kuraishi, T. Makabe and K. Nakayama, A single chip NMOS analog front-end LSI for modems, *IEEE Journal of Solid-State Circuits*, SC-17, 6, 1039–44, December 1982.

[4.7] R.T. Kaneshiro, T.C. Choi, R.W. Brodersen and P.R. Gray, High frequency switched-capacitor filtering techniques, *Proc. ISCAS*, 797–802, 1983.

[4.8] P.M. Van Peteghem and W. Sansen, T cell SC integrator synthesizes very large capacitance ratios, *Electronics Letters*, 19, 14, 541–3, 7 July 1983.

[4.9] P.E. Fleischer, A. Ganesan and K.R. Laker, Parasitic compensated switched-capacitor circuits, *Electronics Letters*, 17, 24, 929–31, 26 November 1981.

[4.10] K.R. Laker, P.E. Fleischer and A. Ganesan, Parasitic insensitive biphase switched-capacitor filters realized with one operational amplifier per pole pair, *Bell System Technical Journal*, 61, 685–707, May–June 1982.

[4.11] C.F. Rahim, M.A. Copeland and C.H. Chan, A functional MOS circuit for achieving the bilinear transformation in switched-capacitor filters, *IEEE Journal of Solid-State Circuits*, SC-13, 6, 906–9, December 1978.

[4.12] A. Knob, Novel stray-insensitive switched-capacitor integrator realizing the bilinear z-transform, *Electronics Letters*, 16, 5, 173–4, 28 February 1980.

[4.13] K. Martin and A.S. Sedra, Transfer function deviations due to resistor–SC equivalence assumption in switched-capacitor simulation of LC ladders, *Electronics Letters*, 16, 10, 387–9, 9 May 1980.

[4.14] S.O. Scanlan, Analysis and synthesis of switched-capacitor state-variable filters, *IEEE Transactions on Circuits and Systems*, CAS-28, 2, 85–93, February 1981.

[4.15] M.S. Tawfik, C. Terrier, C. Caillon and J. Borel, Exact design of switched-capacitor bandpass ladder filters, *Electronics Letters*, 18, 25, 1101–3, 9 December 1982.

[4.16] K. Martin, Improved circuits for the realization of switched-capacitor filters, *IEEE Transactions on Circuits and Systems*, CAS-27, 4, 237–44, April 1980.

[4.17] P.W. Bosshart, A multiplexed switched-capacitor filter bank, *IEEE Journal of Solid-State Circuits*, SC-15, 6, 939–45, December 1980.

[4.18] J. Bingham, Applications of a 'direct transfer' SC integrator, *IEEE Transactions on Circuits and Systems*, CAS-31, 4, 419–20, April 1984.

[4.19] P.V. Ananda Mohan, V. Ramachandran and M.N.S. Swamy, A general stray-insensitive first-order active SC network, *Electronics Letters*, 18, 1, 1–2, 7 January 1982.

[4.20] R. Gregorian, Switched-capacitor filter design using cascaded sections, *IEEE Transactions on Circuits and Systems*, CAS-27, 6, 515–21, June 1980.

[4.21] U.W. Brugger, D.C. von Grunigen and G.S. Moschytz, A comprehensive procedure for the design of cascade switched-capacitor filters, *IEEE Transactions on Circuits and Systems*, CAS-28, 8, 803–10, August 1981.

[4.22] I.A Young and D.A. Hodges, MOS switched-capacitor analog sampled-data direct form recursive filters, *IEEE Journal of Solid-State Circuits*, SC-14, 6, 1020–33, December 1979.

[4.23] R.H. McCharles and D.A. Hodges, Charge circuits for analog LSI, *IEEE Transactions on Circuits and Systems*, CAS-25, 7, 490–7, July 1978.

[4.24] R. Dessoulavy, A. Knob, F. Krummenacher and E.A. Vittoz, A synchronous switched-capacitor filter, *IEEE Journal of Solid-State Circuits*, SC-15, 3, 301–5, June 1980.

[4.25] T. Enomoto, T. Ishihara and Masi-Aki-Yasumoto, Integrated tapped MOS analogue delay line using switched-capacitor technique, *Electronics Letters*, 18, 5, 193–5, 4 March 1982.

[4.26] P. Gillingham, Stray-free switched-capacitor unit delay circuit, *Electronics Letters*, 20, 7, 308–9, 29 March 1984.

[4.27] K. Nagaraj, A switched-capacitor delay circuit that is insensitive to capacitor mismatch and stray capacitance, *Electronics Letters*, 20, 16, 663, 2 August 1984.

[4.28] J.J. Mulawka, Switched-capacitor analogue delays comprising unity gain buffer, *Electronics Letters*, 17, 7, 276–7, 2 April 1981.

[4.29] J.J. Mulawka, Switched-capacitor implementation of analogue sampled-data recursive filters, *International Journal of Electronics*, 51, 2, 173–80, August 1981.

[4.30] B.B. Bhattacharyya and R. Raut, Low sensitivity realizations of sampled-data filters using unity gain amplifiers and switched capacitors, *Proc. European Conference on Circuit Theory and Design*, 808–13, 1981.

[4.31] B.G. Pain, Alternative approach to the design of switched-capacitor filters, *Electronics Letters*, 15, 14, 438–9, 5 July 1979. (See also comment by J.J. Hill, *Electronics Letters*, 15, 20, 650, 27 September 1979.)

[4.32] R. Gregorian, K.W. Martin and G.C. Temes, Swiched-capacitor circuit design, *Proc. IEEE*, 71, 8, 941–66, August 1983.

[4.33] C. Karagoz and A. Acar, Stray-insensitive switched-capacitor network realization for $n$th-order voltage transfer functions, *Electronics Letters*, 18, 22, 966–7, 28 October 1982.

[4.34] C. Karagoz and C. Acar, Active switched-capacitor network realization for $n$th order voltage transfer functions: A signal flow-graph approach, *International Journal of Circuit Theory and Applications*, 10, 377–403, 1982.

[4.35] P.V. Ananda Mohan, V. Ramachandran and M.N.S. Swamy, Parasitic-compensated second-order switched-capacitor filters, *Electronics Letters*, 18, 12, 531–3, 10 June 1982.

[4.36] R. Gregorian and G.C. Temes, Self-equalizing sample and hold circuits, *Electronics Letters*, 15, 13, 367–8, 21 June 1979.

[4.37]  G. Kabuli and F. Anday, Stray-insensitive switched-capacitor network realization for voltage transfer functions, *Electronics Letters*, 19, 2, 60–1, 20 January 1983.

[4.38]  G.C. Temes and I.A. Young, An improved switched-capacitor integrator, *Electronics Letters*, 14, 9, 287–8, 27 April 1978.

[4.39]  G.C. Temes, The derivation of switched-capacitor filters from active RC prototypes, *Electronics Letters*, 14, 12, 361–2, 8 June 1978.

[4.40]  G.C. Temes and R. Gregorian, Compensation of parasitic capacitances in switched-capacitor filters, *Electronics Letters*, 15, 13, 377–9, 21 June 1979.

[4.41]  S. Erikson and H. Akhlagh, Noninverting parasitic compensated bilinear SC integrator with only one amplifier, *Electronics Letters*, 19, 12, 450–2, 9 June 1983.

[4.42]  T. Inoue and F. Ueno, Parasitic compensated building blocks for switched-capacitor filters, *Electronic Letters*, 19, 23, 970–1, 10 November 1983.

## 4.11.1  Further reading

[4.43]  H. Jamal and F.E. Holmes, Novel SC integrator realizing bilinear $z$-transform, *Electronics Letters*, 18, 9, 390–1, 29 April 1982.

[4.44]  H. Jamal and F.E. Holmes, MOS switched-capacitor integrator eliminating operational amplifiers, *Electronics Letters*, 17, 24, 925–6, 26 November 1981.

[4.45]  J.I. Arreola, Y.P. Tsividis, E. Sanchez-Sinencio and P.E. Allen, Simple implementation of sampled-data filters using current multipliers, switches and capacitors, *Electronics Letters*, 15, 24, 780–2, 22 November 1979.

[4.46]  W.M.C. Sansen and P.M. von Peteghem, An area-efficient approach to the design of very large time constants in switched-capacitor integrators, *IEEE Journal of Solid-State Circuits*, SC-19, 5, 772–80, October 1984.

[4.47]  J.C.M. Bermudez and B.B. Bhattacharyya, Parasitic insensitive toggle-switched-capacitor and its application to switched-capacitor networks, *Electronics Letters*, 18, 17, 734–5, 19 August 1982.

[4.48]  F. Maloberti and F. Montecchi, Comment on 'Parasitic insensitive toggle-switched capacitor and its application to switched-capacitor networks', *Electronics Letters*, 18, 24, 1061, 25 November 1982.

[4.49]  A.E. Said, Stray-free switched-capacitor building block that realizes delay, constant multiplier or summer circuit, *Electronics Letters*, 21, 4, 167–8, 14 February 1985.

[4.50]  D.C. von Grunigen, U.W. Brugger and B.J. Hosticka, Bottom-plate stray-insensitive bilinear switched-capacitor integrators, *Electronics Letters*, 16, 1, 25–6, 3 January 1980.

[4.51]  Y. Horio and S. Mori, Switched-capacitor lossless discrete differentiator with modified sample and hold sequence, *Electronics Letters*, 21, 22, 1036–7, 24 October 1985.

[4.52]  V.F. Dias and J.E. Franca, Parasitic-compensated switched-capacitor delay lines, *Electronics Letters*, 24, 7, 377–9, 31 March 1988.

[4.53]  A. Debrowski, U. Menzi and G.S. Moschytz, Offset compensated switched-capacitor delay circuit that is insensitive to stray capacitance and to capacitor mismatch, *Electronics Letters*, 25, 10, 623–5, 11 May 1989.

[4.54] A.M. Davis and R.R. Smith, Design of state-variable SC filters by means of polynomial transformation, *IEEE Transactions on Circuits and Systems*, CAS-33, 12, 1248–51, December 1986.

[4.55] G. Espinoza Flores-Verdad and F. Montecchi, SC circuit for very large and accurate time constant integrators with low capacitance ratios, *Electronics Letters*, 24, 16, 1025–7, 4 August 1988.

[4.56] H. Qinting, A novel technique for the reduction of capacitor spread in high-$Q$ SC circuits, *IEEE Transactions on Circuits and Systems*, 36, 1, 121–6, January 1989.

[4.57] K. Nagaraj, A parasitic-insensitive area-efficient approach to realizing very large time-constants in switched-capacitor circuits, *IEEE Transactions on Circuits and Systems*, 36, 9, 1210–16, September 1989.

[4.58] G. Pschalinos and I. Haritantis, SC bilinear integrators, *IEEE Transactions on Circuits and Systems*, 36, 11, 1493–4, November 1989.

[4.59] C.Y. Wu, T.C. Yu and S.S. Chang, New monolithic switched-capacitor differentiators with good noise rejection, *IEEE Journal of Solid-State Circuits*, SC-24, 1, 177–80, February 1989.

[4.60] T.C. Yu, C.Y. Wu and S.S. Chang, Realizations of IIR/FIR and $N$-path filters using a novel switched-capacitor technique, *IEEE Transactions on Circuits and Systems*, 37, 1, 91–106, January 1990.

[4.61] A. de la Plaza, High-frequency switched-capacitor filter using unity gain buffers, *IEEE Journal of Solid-State Circuits*, SC-21, 3, 470–7, June 1986.

[4.62] F. Maloberti, F. Montecchi, G. Torelli and E. Halasz, Bilinear design of fully differential switched-capacitor ladder filters, *Proc. IEE*, Part $G$, 132, 6, 266–72, December 1985.

[4.63] H.M. Yassine, Four first-order switched-capacitor filters, *Electronics Letters*, 20, 24, 1023–4, 22 November 1984.

CHAPTER FIVE

# Second-order active SC filters

Second-order SC filters can be realized using one or more OAs. Single-amplifier SC filters have the advantage of low power consumption and possible savings in the chip area required for integration. For given specifications, the filters can be designed using the several possible SC structures. The objectives of the design could be to reduce the spread in capacitor values, improve dynamic range and to realize as large a $Q_p$ as possible for a given sampling frequency, with low $Q_p$ and $\omega_p$ sensitivities. Single-amplifier filters can be classified into the following types:

(a) based on Sallen–Key (unity-gain amplifier/amplifier with gain greater than unity) active RC filters [5.1];
(b) based on multiple feedback active RC filters [5.2];
(c) impedance simulation type;
(d) multiplexed second-order filters.

Type (d) circuits realize arbitrary biquadratic transfer functions, while the others realize specific types of transfer functions such as LP, HP, BP etc. Such SC filters can be either bottom-plate stray-insensitive or parasitic-compensated but cannot be completely stray-insensitive [1.14, 3.3, 4.1, 5.3–5.7]. Hence, stray-insensitive designs must use a minimum of two OAs. Further, single-amplifier bottom-plate stray-insensitive designs usually require a large spread in capacitor values. Hence, using two OAs in bottom-plate stray-insensitive designs can considerably reduce the capacitor spread, and total capacitance, while of course, requiring an additional OA [1.28, 5.4, 5.6, 5.8–5.12]. In this chapter, we consider the totally stray-insensitive designs. Some three or four OA SC filter designs will also be considered, in view of some of their useful properties such as the ability to provide simultaneously more than one transfer function, requirement of low capacitor spread, etc.

## 5.1  Stray-insensitive two-OA biquads

Several authors have independently described two-OA SC biquads which are stray-insensitive; these employ the Tow–Thomas active RC structure (Figure 5.1) [1.24] as the means by which one OA is eliminated (by using a voltage-inverting switched-capacitor). However, stray-insensitivity demands that only certain types of switched capacitor can be used for such voltage inversion or even for resistance simulation.

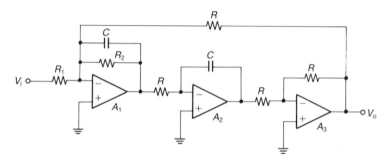

*Fig. 5.1*  Tow–Thomas active RC biquad.

During the discussion on stray-insensitive integrators in Chapter 4, we studied a pair of complementary integrators (see Figure 4.2(b) and Figure 4.3(c)) which are the basis for realizing stray-insensitive SC biquads. Considering that a lossless integrator and a lossy integrator are required, one of them capable of inversion, it can be immediately seen that these can be paired in several ways to realize biquads.

Fleischer and Laker [1.26], Martin [4.16], Martin and Sedra [4.4, 5.13] and Gregorian [4.20] have independently described biquad realizations using the above concept. The interesting feature of these biquads is that they use the same basic two-integrator loop; the differences lie in the feedforward signal used in the various designs to realize the desired transfer functions.

## 5.1.1  The Fleischer–Laker biquad

It is convenient to study Fleischer–Laker biquad topology, since it is general and can realize any biquadratic SC transfer function. This circuit is shown in Figure 5.2. It may be noted that only two types of switched capacitor are used. The capacitors $G$, $H$, $I$ and $J$ provide the feedforward input, whereas $E$ and the switched capacitor $F$ provide damping for the two-integrator loop. Note that, in principle, this type of capacitive feedback can be used in the active RC filter due to Tow and Thomas. But this circuit may not be preferable, since it requires three capacitors in all. In the SC version, since only capacitors, switches and OAs are used to realize transfer functions, the requirement of an additional capacitor is reasonable. This is one important difference between the capabilities of active SC networks and active RC networks. The SC circuit has two transfer functions when $V_{io} = V_{ie}z^{-1/2}$, with both the OA outputs held over a clock period. This is because of the manner in which the switched capacitors are connected. Note that $A$, $C$, $F$, $G$, $H$ and $I$ perform the charge transfer in the even phase only. In the odd phase, the integrators hold their charge and hence the outputs are held over a clock period. Thus, the two outputs $V_0$ and $V_0'$ at two different nodes are considered and the corresponding transfer functions designated as $T$ and $T'$, respectively. Analysis yields

$$T = \frac{V_0}{V_i} = \frac{DI + (AG - DI - DJ)z^{-1} + (DJ - AH)z^{-2}}{D(F + B) + [A(C + E) - DF - 2DB]\, z^{-1} + (DB - AE)z^{-2}} \quad (5.1)$$

$$T' = \frac{V_0'}{V_i} = \frac{[I(C + E) - G(F + B)] + [H(F + B) + BG - JC - E(I + J)]\, z^{-1} + (EJ - BH)z^{-2}}{D(F + B) + [A(C + E) - DF - 2DB]\, z^{-1} + (DB - AE)z^{-2}}$$

$$(5.2)$$

It is thus seen that $T$ and $T'$ realize biquadratic transfer functions. The desired transfer function can be now matched with $T$ or $T'$ to yield the capacitor ratios. The advantage of the Fleischer–Laker biquad is that the transfer functions contain product terms of order 2 (see equations (5.1) and (5.2)). Since the outputs of the biquad of Figure 5.2 are held over a clock period, it is easy to cascade subsequent stages.

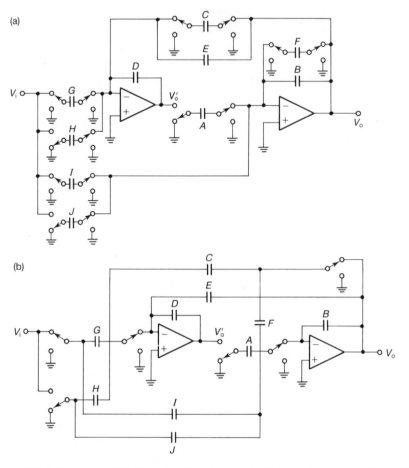

Fig. 5.2  (a) Fleischer–Laker SC biquad;  (b) minimum switch version of (a)  *(from [1.26])*
Copyright © 1979 A.T. & T. All rights reserved. Reprinted with permission.

The design of the biquad shown in Figure 5.2 can start with the choice $A = B = D = 1$ to simplify equations (5.1) and (5.2) as follows:

$$T = \frac{I + (G - I - J)z^{-1} + (J - H)z^{-2}}{(F + 1) + (C + E - F - 2)z^{-1} + (1 - E)z^{-2}} \tag{5.3a}$$

$$T' = \frac{[I(C + E) - G(F + 1)] + [H(F + 1) + G - JC - E(I + J)]z^{-1} + (EJ - H)z^{-2}}{(F + 1) + (C + E - F - 2)z^{-1} + (1 - E)z^{-2}} \tag{5.3b}$$

It is evident that the denominator of equations (5.3a) and (5.3b) is considerably simplified. It is next observed that either $E$ or $F$ will be adequate to provide the desired complex poles and, for either choice, there are enough degrees of freedom in the numerator to realize all general SC filter transfer functions. Thus either $E$ or $F$ can be set to zero and the coefficients of the desired SC transfer function can be identified with equations (5.3a) and (5.3b).

Let us briefly digress at this point to consider a second-order denominator of the form

$$D(z) = a - bz^{-1} + cz^{-2} \qquad (5.4a)$$

It can be shown that

$$\frac{\delta}{Q_p} = \frac{a-c}{a-b+c} \qquad (5.4b)$$

and

$$\frac{4\delta^2 - (2\delta/Q_p) + 1}{4} = \frac{c}{a-b+c} \qquad (5.4c)$$

## Evaluation of capacitor spread of E and F circuits [5.14]

We can obtain $\delta$ and $Q_p$ from equations (5.2) and (5.4):

$$\frac{\delta}{Q_p} = \frac{DF + AE}{AC}$$

$$\frac{4\delta^2 - \dfrac{2\delta}{Q_p} + 1}{4} = \frac{DB - AE}{AC}$$

It is thus seen that $Q_p > \delta$ can be realized. For the $E$ circuit, $F = 0$, resulting in the following design equations:

$$\frac{E}{C} = \frac{\delta}{Q_p}$$

$$\frac{DB - AE}{AC} = \frac{4\delta^2 - \dfrac{2\delta}{Q_p} + 1}{4} \qquad (5.5)$$

For the $F$ circuit, $E = 0$, leading to

$$\frac{DF}{AC} = \frac{\delta}{Q_p}$$

$$\frac{DB}{AC} = \frac{4\delta^2 - \dfrac{2\delta}{Q_p} + 1}{4} \qquad (5.6)$$

For approximately optimal dynamic range design, it is known in the case of the active RC Tow–Thomas biquad that the time constants of both the integrators are equal. This result can be extended to the SC networks as well, assuming that high $Q_p$s are under consideration [4.4, 4.16]. Thus, $(A/B) = (C/D)$ can be chosen. Further, since each group of capacitors around each OA can be scaled together, the integrating capacitors can be considered as unity to facilitate evaluation of capacitor spread. Thus, $B = D = 1$, $A = C$. Using these relationships in equations (5.5) and (5.6), the design equations for the $E$ and $F$ circuits are as follows:

$$E = \frac{2\delta/Q_p}{\sqrt{4\delta^2 + \dfrac{2\delta}{Q_p} + 1}}, \qquad A = \frac{2}{\sqrt{4\delta^2 + \dfrac{2\delta}{Q_p} + 1}} \qquad \text{for the E circuit} \qquad (5.7a)$$

$$F = \frac{4\delta/Q_p}{4\delta^2 - \dfrac{2\delta}{Q_p} + 1}, \qquad A = \frac{2}{\sqrt{4\delta^2 - \dfrac{2\delta}{Q_p} + 1}} \qquad \text{for the F circuit} \qquad (5.7b)$$

It is important to note the absence of a square root in the denominator of the formula for $F$ in equation (5.7b). Thus the approximate formulas for large $Q_p$s for the $E$ and $F$ circuits are as follows:

$$A = \frac{2}{\sqrt{4\delta^2 + 1}}, \qquad E = \frac{2\delta/Q_p}{\sqrt{4\delta^2 + 1}} \qquad \text{for the E circuit} \qquad (5.7c)$$

$$A = \frac{2}{\sqrt{4\delta^2 + 1}}, \qquad F = \frac{4\delta/Q_p}{4\delta^2 + 1} \qquad \text{for the F circuit} \qquad (5.7d)$$

It is obvious that for $\delta < \sqrt{3}/2$, the $F$ circuit has less capacitor spread, and for $\delta > \sqrt{3}/2$, the $E$ circuit has less capacitor spread. The above derivation has considered only the denominator. An examination of equation (5.1) shows that the zeros are not dependent on $E$ and $F$ for the transfer function at output $T$. Hence, the capacitor values presented in equations (5.7c) and (5.7d) are meaningful. (Note, however, that the gain of the $F$ circuit is $1/(1 + F)$ times that of the $E$ circuit. For large $Q_p$s, since $F$ is very small ($F \ll 1$), equations (5.7c) and (5.7d) can be used to compare the $E$ and $F$ circuits with output at $T$.)

However, it must be noted that for low-$Q_p$ designs, the choice of equal time constants for both the integrators in order to have an optimal dynamic range is not valid, and hence an accurate estimate of the maxima of $T$ and $T'$ must be made. The circuit must then be scaled for optimal dynamic range and minimum total capacitance and then compared. Similarly for the $T'$ designs, since the zeros are dependent on $E$ and $F$, the above results cannot be applied, as mentioned before. Consequently, a realistic comparison is made only after the design of the circuit is *complete*, as will be shown in what follows.

## Design procedure

First, the values of $F$ (or $E$), $G$, $H$, $I$, $J$, $C$ are determined by matching the desired SC transfer functions with equations (5.3a) and (5.3b).

The next step is to compute the maxima of $T$ and $T'$. While the desired transfer function has a known maximum value, the transfer function corresponding to the other output of the OA can have an altogether different maximum value. In a 'good' design, when the input frequency is swept throughout the frequency range of interest, the active devices in the filter will saturate for the same input level. This ensures that the SC filter has optimal dynamic range. The formulas for determining the maximum gain of a second-order digital transfer function listed in Table 1.5 can be used for evaluating the maxima of $T$ and $T'$ first.

To achieve the optimal dynamic range, the maxima of $T$ and $T'$ are now equalized as already mentioned. Noting that if $T$ is the desired transfer function, and that $T'_{max}$ has to be scaled to $\mu T'_{max}$, the scaling is such that

$$T'_{max} \rightarrow \mu T'_{max}$$

$$(A, D) \rightarrow \left( \frac{A}{\mu}, \frac{D}{\mu} \right)$$

Then, $\mu T'_{max} = T_{max}$ is achieved. This scaling ensures that $T$ remains unchanged. Alternatively, if $T'$ is the desired transfer function with a maximum value of $T'_{max}$, $T_{max}$ is scaled to $T'_{max}$ such that

$$\mu T_{max} = T'_{max}$$

This requires that the following capacitors be scaled as follows:

$$(B, C, E, F) \rightarrow \left( \frac{B}{\mu}, \frac{C}{\mu}, \frac{E}{\mu}, \frac{F}{\mu} \right)$$

The reader can easily verify this observation.

The next step is to consider the groups of capacitors connected to the virtual ground inputs of OAs together so that the smallest capacitance is unity.

Next, the total capacitance and capacitor value spread are evaluated.

Finally, it is noted that, in the Fleischer–Laker biquad, certain switches can be combined in the topology of Figure 5.2(a) so that the switch count can be reduced. Such a circuit with reduced switch count is shown in Figure 5.2(b). Also certain SC branches can be replaced by simpler SC branches. As an illustration, for designs requiring $I = J$, the use of a single unswitched capacitor $I$ will suffice, as illustrated in Figure 5.3. Other similar equivalences are also shown in Figure 5.3. The total capacitance evaluation is made only after these simplifications. The reader must appreciate the fact that this biquad is best recommended for monolithic integration in view of its desirable properties: first, that only two OAs are used; second, that outputs are held over a clock period; third, that the capacitor ratios can easily be evaluated; fourth, that the optimal dynamic range can be achieved; and fifth, that it is completely general. We will consider some examples to illustrate the design procedure.

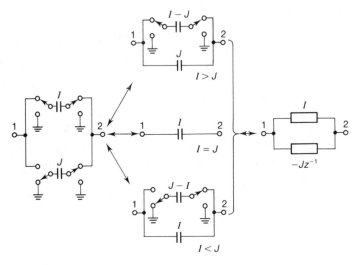

*Fig. 5.3* SC element transformations (nodes 1 and 2 are voltage sources or virtual grounds) *(from [1.26])* Copyright © 1979 A.T. & T. All rights reserved. Reprinted with permission.

*Example 5.1*
Design a bilinear BP SC filter for a centre frequency of 1633 Hz, sampling frequency 8000 Hz, a $Q_p$ of 16 and midband gain of 10 dB, using a Fleischer–Laker biquad *F* circuit with output at *T*.

Using the design methods described in Chapter 1, the required digital transfer function is

$$T(z) = \frac{0.1219(1 - z^{-1})(1 + z^{-1})}{1 - 0.5455z^{-1} + 0.9229z^{-2}}$$

From equation (5.3a) the capacitor values are identified as (with $E = 0$),

$$F = 0.083\ 541 \qquad A = B = D = 1$$

$$C = 1.492\ 469\ 4 \qquad I = G = H$$

$$I = 0.132\ 083\ 6 \qquad J = 0$$

Using these values in equation (5.3b) (with $E = 0$), $T'$ is obtained as

$$T' = \frac{0.049\ 848\ 4z^2 - 0.253\ 983\ 6z - 0.1219}{z^2 - 0.5455z + 0.9229}$$

Thus $T$ and $T'$ are obtained which can be used to evaluate the respective maxima in the frequency response. Using formulas from Chapter 1, the maxima for $T$ and $T'$ are

$$T_{max} = 3.162\ 277\ 7$$

$$T'_{max} = 3.865\ 506\ 3$$

(Note that for $T$, $T_{max}$ need not be computed, because the specifications give it as 10 dB).

Then, to reduce $T'_{max}$ to $T_{max}$, it is required to scale

$$(T'_{max}) = 3.865\ 506\ 3 \rightarrow (\mu T'_{max}) = 3.162\ 277\ 7$$

giving $\mu = 0.818\ 075\ 9$. Therefore, $A$ and $D$ are scaled to $A/\mu$, $D/\mu$ to give

$$A = D = 1.222\ 380\ 4$$

The present capacitor values are

$$A = D = 1.222\ 380\ 4 \quad F = 0.083\ 541$$

$$B = 1 \qquad\qquad G = H = 0.132\ 083\ 6$$

$$C = 1.492\ 469\ 4 \quad I = 0.132\ 083\ 6$$

The next step is to rescale the capacitors in the above two groups $(A, F, B, I)$, $(C, D, G, H)$ so that the minimum capacitance is unity. This calculation yields

$$A = 14.632\ 102 C_u \qquad F = 1.0 C_u$$

$$B = 11.970\ 17 C_u \qquad G = 1.0 C_u$$

$$C = 11.299\ 43 C_u \qquad H = 1.0 C_u$$

$$D = 9.254\ 596\ 3 C_u \qquad I = 1.581\ 063\ 2 C_u$$

The total capacitance is $51.737\ 362 C_u$, where $C_u$ is the unit capacitance and the capacitor spread is $14.632\ 102$. This completes the design.

Further it is noted that $T$ has realized an inverting BP function. If a non-inverting BP function is needed, $T'$ is used. Normally, a sign is not specified for the BP transfer function in cascade designs, but in designing leap-frog ladder type BP filter simulation, a non-inverting or inverting BP function is needed. This is illustrated next with the design at $T'$.

*Example 5.2*
Repeat the design of Fleischer–Laker biquad to realize the desired transfer function at $T'$. Use the $F$ circuit.

Identifying equation (5.3b) (with $E = 0$) with the desired digital transfer function, the $F$ and $C$ values will be the same, whereas the other capacitor values are different. For simplicity, let $G = 0$. The resulting capacitor values are

$$C = 1.492\ 469\ 4$$

$$F = 0.083\ 541$$

$$H = 0.132\ 083\ 6$$

$$I = 0.088\ 500\ 1$$

$$J = 0.095\ 893\ 4$$

Using these values in equation (5.3a), $T$ is obtained as

$$T = -\frac{0.081\ 676\ 7 - 0.170\ 176\ 8z^{-1} - 0.033\ 399\ 99z^{-2}}{1 - 0.5455z^{-1} + 0.9229z^{-2}}$$

The maximum value of $T$ is evaluated as 2.5898. Thus, $T_{max}$ has to be increased to 3.16 so that both the OAs saturate at the same level. Hence,

$$T_{max} \rightarrow \mu T_{max}$$

$$(B, C, E, F) \rightarrow \left(\frac{B}{\mu}, \frac{C}{\mu}, \frac{E}{\mu}, \frac{F}{\mu}\right)$$

In this case, $\mu$ is obviously 1.220 131 9. Thus, the capacitor values become

$$B = 0.819\ 583\ 5 \qquad E = 0$$

$$C = 1.2232 \qquad F = 0.068\ 469$$

The capacitor values in both the groups are known. Thus, scaling the above capacitor values to make the minimum capacitance in each group of capacitors unity, the final set of design values will be

$$A = 14.605\ 150C_u \qquad F = H = C_u$$

$$B = 11.970\ 140C_u \qquad I = 1.2925C_u$$

$$C = 9.2608C_u \qquad J = 1.4005C_u$$

$$D = 7.5709C_u$$

The capacitor spread is 14.605 and the total capacitance is $48.10C_u$.

*Example 5.3*
Repeat the design of the Fleischer–Laker biquad to realize a BP output at $T$ and $T'$ using the E circuit.

Proceeding in the same manner as in Examples 5.1 and 5.2, the transfer function at $T'$ is obtained as

$$T' = \frac{0.055\ 403\ 6 + 0.234\ 401\ 5z^{-1} - 0.1219z^{-2}}{1 - 0.5455z^{-1} + 0.9229z^{-2}}$$

This realizes the desired BP transfer function at $T$. The maximum value of $T'$ is 4.125 99. After scaling for dynamic range optimization and making the smallest capacitance unity, finally we obtain

$$A = 10.711\ 189 C_\mathrm{u} \qquad E = I = C_\mathrm{u}$$

$$B = 8.203\ 445\ 4 C_\mathrm{u} \qquad G = H = 1.581\ 063\ 6 C_\mathrm{u}$$

$$C = 17.865\ 11 C_\mathrm{u}$$

$$D = 16.935\ 07 C_\mathrm{u}$$

Thus the capacitance spread is 17.865 11 and the total capacitance is $58.876 C_\mathrm{u}$.
    Similarly, if a BP function at $T'$ is desired, $T$ is obtained as

$$T = \frac{1.581\ 063\ 6 - 0.740\ 570\ 2z^{-1} + 1.581\ 063\ 6z^{-2}}{1 - 0.5455z^{-1} + 0.9229z^{-2}}$$

and the peak value of $T$ is 2.696 678 6. After scaling for dynamic range optimization and making the minimum capacitance unity in both groups of capacitors, we obtain

$$A = 1.1718 C_\mathrm{u} \qquad D = 15.198 C_\mathrm{u}$$

$$B = E = C_\mathrm{u} \qquad G = 36.803 C_\mathrm{u}$$

$$C = 17.864 C_\mathrm{u} \qquad I = J = 1.8526 C_\mathrm{u}$$

Therefore, the capacitance spread is 36.8 and the total capacitance is $74.89 C_\mathrm{u}$. (It is interesting to note, however, that since $I$ and $J$ are equal, they can be combined into a single unswitched capacitor, as already explained.)

    The above examples of the design of a bilinear BP filter using $T$ and $T'$ in $E$ and $F$ circuits have shown the variety of choices available. In practical situations, the choice of a particular circuit is decided by the total area required on the chip, taking into account the number of switches and capacitors. Interestingly, the reader may also note that $T'$ design of the $E$ circuit has realized a notch transfer function and a bilinear BP transfer function simultaneously! Note also that for $T'$ design, other choices of capacitor values are possible, since the numerator contains four capacitor values $G$, $H$, $I$ and $J$, whereas only three terms need to be realized. Hence, the cases $G = 0$ or $H = 0$ or $I = 0$ or $J = 0$ could be used. All the resulting $T$ and $T'$ designs have to be compared for the final selection of a particular circuit.

**Table 5.1** Zero-placement formulas for $T_E$ and $T_F$ (adapted from [1.26])

| Filter type | Design equation | Simple solution |
|---|---|---|
| LP20 | $I = \|K\|$ <br> $G - I - J = 2\|K\|$ <br> $J - H = \|K\|$ | $I = J = \|K\|$ <br> $G = 4\|K\| \quad H = 0$ |
| LP11 | $I = 0$ <br> $G - I - J = \pm\|K\|$ <br> $J - H = \pm\|K\|$ | $H = I = 0 \quad J = \|K\|$ <br> $G = 2\|K\|$ |
| LP10 | $I = \|K\|$ <br> $G - I - J = \|K\|$ <br> $J - H = 0$ | $I = \|K\| \quad H = J = 0$ <br> $G = 2\|K\|$ |
| LP02 | $I = 0$ <br> $G - I - J = 0$ <br> $J - H = \pm\|K\|$ | $G = I = J = 0$ <br> $H = \|K\|$ |
| LP01 | $I = 0$ <br> $G - I - J = \pm\|K\|$ <br> $J - H = 0$ | $H = I = J = 0$ <br> $G = \|K\|$ |
| LP00 | $I = \|K\|$ <br> $G - I - J = 0$ <br> $J - H = 0$ | $H = J = 0$ <br> $I = G = \|K\|$ |
| BP10 | $I = \|K\|$ <br> $G - I - J = 0$ <br> $J - H = -\|K\|$ | $G = H = I = \|K\| \quad J = 0$ |
| BP01 | $I = 0$ <br> $G - I - J = \pm\|K\|$ <br> $J - H = \mp\|K\|$ | $G = H = I = 0 \quad J = \|K\|$ |
| BP00 | $I = \|K\|$ <br> $G - I - J = -\|K\|$ <br> $J - H = 0$ | $I = \|K\| \quad G = H = J = 0$ |
| HP | $I = \|K\|$ <br> $G - I - J = -2\|K\|$ <br> $J - H = \|K\|$ | $I = J = \|K\|$ <br> $G = H = 0$ |
| HPN and LPN | $I = \|K\|$ <br> $G - I - J = \|K\|\varepsilon$ <br> $J - H = \|K\|$ | $I = J = \|K\|$ <br> $G = \|K\|(2 + \varepsilon) \quad H = 0$ |
| AP <br> ($\beta > 0$) | $I = \|K\|\beta$ <br> $G - I - J = \|K\|\alpha$ <br> $J - H = \|K\|$ | $I = \|K\|\beta \quad J = \|K\|$ <br> $G = \|K\|(1 + \beta + \alpha) = \|K\|C$ <br> $H = 0$ |
| General <br> ($\gamma > 0$) | $I = \gamma$ <br> $G - I - J = \varepsilon$ <br> $J - H = \delta$ | $I = \gamma$ <br> $J = \delta + x$ <br> $G = \gamma + \delta + \varepsilon + x$ <br> $H = x \geqslant 0$ |

**Table 5.2** Zero-placement formulas for $T_E$ [1.26]. Copyright © 1979, A.T. & T. All rights reserved. Reprinted with permission.

| Filter type | Design equation | Simple solution |
|---|---|---|
| LP20 | $IC + IE - G = \pm \lvert K \rvert$ <br> $H + G - JC - JE - IE = \pm 2 \lvert K \rvert$ <br> $EJ - H = \pm \lvert K \rvert$ | $I = \dfrac{\lvert K \rvert (4E + C)}{EC} \qquad J = \dfrac{\lvert K \rvert}{E}$ <br> $G = \dfrac{\lvert K \rvert (2E + C)^2}{EC} \qquad H = 0$ |
| LP11 | $IC + IE - G = 0$ <br> $H + G - JC - JE - IE = \pm \lvert K \rvert$ <br> $EJ - H = \pm \lvert K \rvert$ | $I = \dfrac{\lvert K \rvert (2E + C)}{EC} \qquad J = \dfrac{\lvert K \rvert}{E}$ <br> $G = \dfrac{\lvert K \rvert (E + C)(2E + C)}{EC} \qquad H = 0$ |
| LP10 | $IC + IE - G = \pm \lvert K \rvert$ <br> $H + G - JC - JE - IE = \pm \lvert K \rvert$ <br> $EJ - H = 0$ | $I = \dfrac{2 \lvert K \rvert}{C} \qquad H = J = 0$ <br> $G = \dfrac{\lvert K \rvert (E + C)^2}{EC}$ |
| LP02 | $IC + IE - G = 0$ <br> $H + G - JC - JE - IE = 0$ <br> $EJ - H = \pm \lvert K \rvert$ | $I = \dfrac{\lvert K \rvert (E + C)}{EC} \qquad J = \dfrac{\lvert K \rvert}{E}$ <br> $G = \dfrac{\lvert K \rvert (E + C)^2}{EC} \qquad H = 0$ |
| LP01 | $IC + IE - G = 0$ <br> $H + G - JC - JE - IE = \pm \lvert K \rvert$ <br> $EJ - H = 0$ | $I = \dfrac{\lvert K \rvert}{C} \qquad H = J = 0$ <br> $G = \dfrac{\lvert K \rvert (E + C)}{C}$ |
| LP00 | $IC + IE - G = \pm \lvert K \rvert$ <br> $H + G - JC - JE - IE = 0$ <br> $EJ - H = 0$ | $I = \dfrac{\lvert K \rvert}{C} \qquad H = J = 0$ <br> $G = \dfrac{\lvert K \rvert E}{C}$ |
| BP10 | $IC + IE - G = \pm \lvert K \rvert$ <br> $H + G - JC - JE - IE = 0$ <br> $EJ - H = \mp \lvert K \rvert$ | $I = J = \dfrac{\lvert K \rvert}{E}$ <br> $G = \dfrac{\lvert K \rvert (2E + C)}{E} \qquad H = 0$ |
| BP01 | $IC + IE - G = 0$ <br> $H + G - JC - JE - IE = \pm \lvert K \rvert$ <br> $EJ - H = \mp \lvert K \rvert$ | $G = I = J = 0$ <br> $H = \lvert K \rvert$ |

*continued*

**Table 5.2** (*continued*)

| Filter type | Design equation | Simple solution |
|---|---|---|
| BP00 | $IC + IE - G = \pm\,\lvert K\rvert$<br>$H + G - JC - JE - IE = \mp\,\lvert K\rvert$<br>$EJ - H = 0$ | $H = I = J = 0$<br>$G = \lvert K\rvert$ |
| HP | $IC + IE - G = \pm\,\lvert K\rvert$<br>$H + G - JC - JE - IE = \mp 2\lvert K\rvert$ | $I = J = 0$<br>$G = H = \lvert K\rvert$ |
| HPN and LPN | $IC + IE - G = \pm\,\lvert K\rvert$<br>$H + G - JC - JE - IE = \pm\lvert K\rvert\,\varepsilon$<br>$EJ - H = \pm\,\lvert K\rvert$ | See general solution below |
| AP | $IC + IE - G = \pm\,\lvert K\rvert\,\beta$<br>$H + G - JC - JE - IE = \pm\lvert K\rvert\,\alpha$<br>$EJ - H = \pm\,\lvert K\rvert$ | See general solution below |
| General<br>$\delta > 0$ | $IC + IE - G = \gamma$<br>$H + G - JC - JE - IE = \varepsilon$<br>$EJ - H = \delta$ | $I = \dfrac{\gamma + \delta + \varepsilon}{C} + \dfrac{\delta}{E}\qquad J = \dfrac{\delta}{E}$<br>$G = I(C + E) - \gamma\qquad H = 0$ |

**Table 5.3** Zero-placement formulas for $T_{\hat F}$ [1.26].

| Filter type | Design equation | Simple solution | |
|---|---|---|---|
| LP20 | $\hat{G}\hat{F} + \hat{G} - \hat{I}\hat{C} = \lvert K\rvert(1 + \hat{F})$<br>$\hat{J}\hat{C} - \hat{F}\hat{H} - \hat{H} - \hat{G} = 2\lvert K\rvert(1 + \hat{F})$<br>$\hat{H} = \lvert K\rvert(1 + \hat{F})$ | $\hat{I} = 0$<br><br>$\hat{G} = \lvert K\rvert$ | $\hat{J} = \dfrac{\lvert K\rvert\,(2 + \hat{F})^2}{\hat{C}}$<br>$\hat{H} = \lvert K\rvert(1 + \hat{F})$ |
| LP11 | $\hat{G}\hat{F} + \hat{G} - \hat{I}\hat{C} = 0$<br>$\hat{J}\hat{C} - \hat{F}\hat{H} - \hat{H} - \hat{G} = \lvert K\rvert(1 + \hat{F})$<br>$\hat{H} = \lvert K\rvert(1 + \hat{F})$ | $\hat{I} = 0$<br><br>$\hat{G} = 0$ | $\hat{J} = \dfrac{\lvert K\rvert\,(1 + \hat{F})(2 + \hat{F})}{\hat{C}}$<br>$\hat{H} = \lvert K\rvert(1 + \hat{F})$ |
| LP10 | $\hat{G}\hat{F} + \hat{G} - \hat{I}\hat{C} = \pm\,\lvert K\rvert(1 + \hat{F})$<br>$\hat{J}\hat{C} - \hat{F}\hat{H} - \hat{H} - \hat{G} = \pm\,\lvert K\rvert(1 + \hat{F})$<br>$\hat{H} = 0$ | $\hat{I} = 0$<br><br>$\hat{G} = \lvert K\rvert$ | $\hat{J} = \dfrac{\lvert K\rvert\,(2 + \hat{F})}{\hat{C}}$<br>$\hat{H} = 0$ |
| LP02 | $\hat{G}\hat{F} + \hat{G} - \hat{I}\hat{C} = 0$<br>$\hat{J}\hat{C} - \hat{F}\hat{H} - \hat{H} - \hat{G} = 0$<br>$\hat{H} = \lvert K\rvert(1 + \hat{F})$ | $\hat{I} = 0$<br><br>$\hat{G} = 0$ | $\hat{J} = \dfrac{\lvert K\rvert\,(1 + \hat{F})^2}{\hat{C}}$<br>$\hat{H} = \lvert K\rvert(1 + \hat{F})$ |
| LP01 | $\hat{G}\hat{F} + \hat{G} - \hat{I}\hat{C} = 0$<br>$\hat{J}\hat{C} - \hat{F}\hat{H} - \hat{H} - \hat{G} = \pm\,\lvert K\rvert(1 + \hat{F})$<br>$\hat{H} = 0$ | $\hat{I} = 0$<br><br>$\hat{G} = 0$ | $\hat{J} = \dfrac{\lvert K\rvert\,(1 + \hat{F})}{\hat{C}}$<br>$\hat{H} = 0$ |

**Table 5.3** *(continued)*

| Filter type | Design equation | Simple solution |
|---|---|---|
| LP00 | $\hat{G}\hat{F} + \hat{G} - \hat{I}\hat{C} = \pm \lvert K \rvert (1 + \hat{F})$ <br> $\hat{J}\hat{C} - \hat{F}\hat{H} - \hat{H} - \hat{G} = 0$ <br> $\hat{H} = 0$ | $\hat{I} = \dfrac{\lvert K \rvert (1 + \hat{F})}{\hat{C}}$ $\qquad \hat{J} = 0$ <br><br> $\hat{G} = 0 \qquad \hat{H} = 0$ |
| BP10 | $\hat{G}\hat{F} + \hat{G} - \hat{I}\hat{C} = - \lvert K \rvert (1 + \hat{F})$ <br> $\hat{J}\hat{C} - \hat{F}\hat{H} - \hat{H} - \hat{G} = 0$ <br> $\hat{H} = \lvert K \rvert (1 + \hat{F})$ | $\hat{I} = \dfrac{\lvert K \rvert (1 + \hat{F})}{\hat{C}} \qquad \hat{J} = \dfrac{\lvert K \rvert (1 + \hat{F})^2}{\hat{C}}$ <br><br> $\hat{G} = 0 \qquad \hat{H} = \lvert K \rvert (1 + \hat{F})$ |
| BP01 | $\hat{G}\hat{F} + \hat{G} - \hat{I}\hat{C} = 0$ <br> $\hat{J}\hat{C} - \hat{F}\hat{H} - \hat{H} - \hat{G} = - \lvert K \rvert (1 + \hat{F})$ <br> $\hat{H} = \lvert K \rvert (1 + \hat{F})$ | $\hat{I} = 0 \qquad \hat{J} = \dfrac{\lvert K \rvert \hat{F}(1 + \hat{F})}{\hat{C}}$ <br><br> $\hat{G} = 0 \qquad \hat{H} = \lvert K \rvert (1 + \hat{F})$ |
| BP00 | $\hat{G}\hat{F} + \hat{G} - \hat{I}\hat{C} = \pm \lvert K \rvert (1 + \hat{F})$ <br> $\hat{J}\hat{C} - \hat{F}\hat{H} - \hat{H} - \hat{G} = \mp \lvert K \rvert (1 + \hat{F})$ <br> $\hat{H} = 0$ | $\hat{I} = \hat{J} = \dfrac{\lvert K \rvert (1 + \hat{F})}{\hat{C}}$ <br><br> $\hat{G} = \hat{H} = 0$ |
| HP | $\hat{G}\hat{F} + \hat{G} - \hat{I}\hat{C} = \lvert K \rvert (1 + \hat{F})$ <br> $\hat{J}\hat{C} - \hat{F}\hat{H} - \hat{H} - \hat{G} = - 2 \lvert K \rvert (1 + \hat{F})$ <br> $\hat{H} = \lvert K \rvert (1 + \hat{F})$ | $\hat{I} = 0 \qquad \hat{J} = \dfrac{\lvert K \rvert \hat{F}^2}{\hat{C}}$ <br><br> $\hat{G} = \lvert K \rvert \qquad \hat{H} = \lvert K \rvert (1 + \hat{F})$ |
| HPN and LPN | $\hat{G}\hat{F} + \hat{G} - \hat{I}\hat{C} = \lvert K \rvert (1 + \hat{F})$ <br> $\hat{J}\hat{C} - \hat{F}\hat{H} - \hat{H} - \hat{G} = - \lvert K \rvert \varepsilon(1 + \hat{F})$ <br> $\hat{H} = \lvert K \rvert (1 + \hat{F})$ | See general solution below |
| AP | $\hat{G}\hat{F} + \hat{G} - \hat{I}\hat{C} = \lvert K \rvert \beta(1 + \hat{F})$ <br> $\hat{J}\hat{C} - \hat{F}\hat{H} - \hat{H} - \hat{G} = \lvert K \rvert \alpha(1 + \hat{F})$ <br> $\hat{H} = \lvert K \rvert (1 + \hat{F})$ | See general solution below |
| General $\delta > 0$ | $\hat{G}\hat{F} + \hat{F} - \hat{I}\hat{C} = \gamma(1 + \hat{F})$ <br> $\hat{J}\hat{C} - \hat{F}\hat{H} - \hat{H} - \hat{G} = \varepsilon(1 + \hat{F})$ <br> $\hat{H} = \delta(1 + \hat{F})$ | $I = x \geqslant 0$ <br><br> $\hat{J} = \dfrac{\delta(1 + \hat{F})^2 + \varepsilon(1 + \hat{F}) + \gamma}{\hat{C}}$ <br><br> $\quad + \dfrac{x}{1 + \hat{F}}$ <br><br> $\hat{G} = \gamma + \dfrac{\hat{C}}{1 + \hat{F}} x$ <br><br> $\hat{H} = \delta(1 + \hat{F})$ |

*Note:* $\hat{G} = G(1 + \hat{F})$, $\hat{H} = H(1 + \hat{F})$, $\hat{I} = I(1 + \hat{F})$, and $\hat{J} = J(1 + \hat{F})$

## Formulas for zero placement

Because of the versatility of the Fleischer–Laker biquad topology, it is beneficial to have the formulas for placement of zeros in terms of the various capacitor values. These formulas for $T$ and $T'$ designs for $E$ and $F$ circuits are presented in Tables 5.1, 5.2 and 5.3. Note that for all the digital transfer functions listed in Table 1.4, the corresponding zero placement formulas are available in these tables.

## Pole sensitivities

It is of interest to evaluate the pole sensitivities of the Fleischer–Laker biquad to various capacitor ratios. The sensitivities evaluated using the bilinear transformation method are presented in Table 5.4. These results point out that the $E$ circuit is more sensitive than the $F$ circuit.

**Table 5.4**  Fleischer–Laker biquad sensitivities

| $E$ circuit | $F$ circuit |
|---|---|
| $\omega_p$ sensitivities | |
| $-S_B^{\omega_p} = -S_D^{\omega_p} = S_A^{\omega_p}$ | $-S_D^{\omega_p} = S_A^{\omega_p} = S_C^{\omega_p} = \dfrac{1}{2} + \dfrac{1}{8\delta^2}$ |
| $= \left(\dfrac{1}{2} + \dfrac{1}{8\delta^2} + \dfrac{1}{4\delta Q_p}\right)$ | $S_B^{\omega_p} = \left(\dfrac{1}{2} - \dfrac{1}{4\delta Q_p} + \dfrac{1}{8\delta^2}\right)$ |
| $S_C^{\omega_p} = \dfrac{1}{2} + \dfrac{1}{8\delta^2}$ | $S_F^{\omega_p} = -\dfrac{1}{4\delta Q_p}$ |
| $S_E^{\omega_p} = \dfrac{1}{4\delta Q_p}$ | |
| $Q_p$ sensitivities | |
| $-S_A^{Q_p} = S_B^{Q_p} = S_D^{Q_p}$ | $S_A^{Q_p} = S_C^{Q_p} = -S_D^{Q_p} = \left(\dfrac{1}{2} - \dfrac{1}{8\delta^2}\right)$ |
| $= \dfrac{1}{2} + \dfrac{1}{4\delta Q_p} + \dfrac{1}{8\delta^2}$ | $S_B^{Q_p} = \left(\dfrac{1}{2} - \dfrac{1}{4\delta Q_p} + \dfrac{1}{8\delta^2}\right)$ |
| $S_E^{Q_p} = -\left(1 + \dfrac{1}{4\delta Q_p}\right)$ | $S_F^{Q_p} = \left(-1 + \dfrac{1}{4\delta Q_p}\right)$ |
| $S_C^{Q_p} = \left(\dfrac{1}{2} - \dfrac{1}{8\delta^2}\right)$ | |

## 5.1.2  Other two-OA stray-insensitive filters based on two-integrator loop

Several authors have independently proposed two-amplifier biquads similar to the Fleischer–Laker biquad, wherein realization of specific transfer functions such as notch, inverting BP, etc., has been considered. These will be briefly presented here, for the sake of completeness.

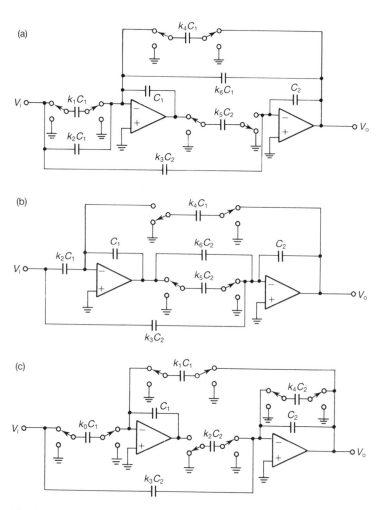

*Fig. 5.4*   (a) Martin–Sedra biquad (except for positive BP transfer function) *(adapted from [4.16],* © *1980 IEEE);*  (b) Martin–Sedra bilinear positive BP filter *(adapted from [4.16],* © *1980 IEEE);*  (c) Martin's notch filter *(adapted from [4.4],* © *1979 IEE).*

Consider the Martin–Sedra biquad [4.16] of Figure 5.4(a) realizing notch/BP functions, for which

$$\frac{V_o}{V_i} = -\frac{k_3 z^2 + (-2k_3 + k_1 k_5 + k_2 k_5)z - (-k_2 k_5 + k_3)}{z^2 + (-2 + k_4 k_5 + k_5 k_6)z + (1 - k_5 k_6)} \tag{5.8}$$

A look at Figures 5.2 and 5.4(a) shows that both employ two integrators, of which one is inverting and the other non-inverting. The circuit of Figure 5.4(a) uses $E$-type damping. The only differences are in the feedforward elements. To obtain a positive BP function, Martin and Sedra [4.16] have recommended the circuit of Figure 5.4(b). The design details for the circuits of Figures 5.4(a) and (b) will be considered in Section 7.5.5.

Martin [4.4] has considered the design of an SC notch filter using the realization of Figure 5.4(c), with a transfer function given by

$$\frac{V_o}{V_i} = \frac{-[k_3 z^2 + (k_0 k_2 - 2k_3)z + k_3]}{(1 + k_4)z^2 + (k_1 k_2 - k_4 - 2)z + 1} \tag{5.9}$$

The reader can see that the circuit is similar to the $F$-type Fleischer–Laker biquad, with only the feedforward elements different ($I = J$; hence, both have been combined into a single unswitched capacitor). Gregorian [4.20] has also described a notch filter which is similar to Martin's notch filter of Figure 5.4(c).

## 5.1.3  General stray-insensitive two-OA SC networks

It has been pointed out in Chapter 4 that a general first-order stray-insensitive network can be realized using one OA and ten (switched/unswitched) capacitors. By the use of the combination of two such first-order networks [5.15] (Figure 5.5(a)), with each branch replaced by a general stray-insensitive SC branch (Figure 5.5(b)), a general stray-insensitive SC network is obtained. This circuit is quite cumbersome to analyze. However, it can be simplified by choosing one or more of the following criteria: outputs are held over a clock period with sampled and held inputs; inputs and outputs pertain to one clock phase only, say the even phase; $Q_p$ sensitivity is as low as possible; and the capacitor spread required to realize given specifications is as small as possible, at the expense of sensitivity.

The first two requirements lead to certain simplifications resulting in the SC networks of Figures 5.5(c) and 5.5(d). Consider first the SC network of Figure 5.5(c) due to El-Masry [5.16], for which the denominator of the transfer function can be obtained as

$$\begin{aligned}
D(z) = {}& z^2 [(C_1 + C_2)(C_7 + C_8) - (C_4 + C_5)(C_{10} + C_{11})] \\
& - z [(C_1 + C_3)(C_7 + C_8) + (C_1 + C_2)(C_7 + C_9) \\
& - (C_4 + C_6)(C_{10} + C_{11}) - (C_4 + C_5)(C_{10} + C_{12})] \\
& + [(C_1 + C_3)(C_7 + C_9) - (C_4 + C_6)(C_{10} + C_{12})]
\end{aligned} \tag{5.10}$$

One method of simplifying this circuit is to ensure low $Q_p$ sensitivity. It is known that for a $D(z)$ given by

$$D(z) = a - bz^{-1} + cz^{-2}$$

$$\frac{\delta}{Q_p} = \frac{a-c}{a-b+c} \tag{5.11}$$

Hence the term $a - c$ shall not have difference terms to ensure low $Q_p$ sensitivity. Thus we examine the value of $a - c$ in equation (5.10):

$$a - c = (C_1 + C_2)(C_7 + C_8) - (C_4 + C_5)(C_{10} + C_{11}) - (C_1 + C_3)(C_7 + C_9)$$
$$+ (C_4 + C_6)(C_{10} + C_{12}) \tag{5.12}$$

The difference terms can be avoided by making the sign of $a - c$ either positive or negative. In the former case, the negative terms are unconditionally made zero (i.e., while requiring no condition on matching of capacitor values) and in the latter case,

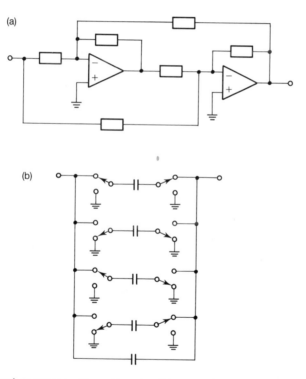

*Fig. 5.5* (a) General stray-insensitive network; (b) general stray-insensitive branch; (c) (overleaf) stray-insensitive biquad with outputs held over a clock period and requiring a sampled and held input *(adapted from [5.16], © 1983 IEEE)*; (d) (overleaf) stray-insensitive biquad with input and output sampled in the even phase *(adapted from [5.16], © 1983 IEEE)*.

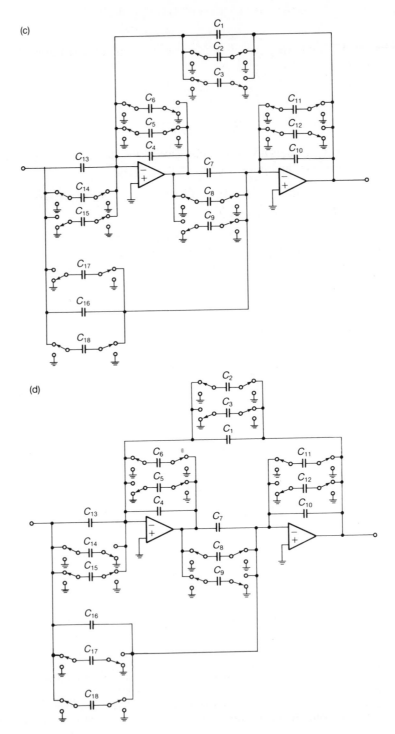

(c)

(d)

Fig. 5.5 (continued)

the positive terms are unconditionally made zero. We have thus the following possible realizations (after cancelling the $C_1C_7$ and $C_4C_{10}$ terms in equation (5.12)):

(i) $C_1$, $C_2$, $C_6$, $C_{12}$ zero;
(ii) $C_2$, $C_6$, $C_8$, $C_{12}$ zero;
(iii) $C_3$, $C_5$, $C_9$, $C_{11}$ zero;
(iv) $C_5$, $C_7$, $C_9$, $C_{11}$ zero.

However, it can be shown that only the first design can realize a complex-conjugate pole pair with arbitrary specifications. The resulting $\delta/Q_p$ value of choice (i) is

$$\frac{\delta}{Q_p} = \frac{C_3(C_7 + C_9) + C_4C_{11} + C_5(C_{10} + C_{11})}{C_5C_{11} + C_3(C_8 - C_9)} \qquad (5.13)$$

Note that the capacitors $C_5$, $C_7$, $C_9$ and $C_{11}$ damp the basic resonator formed by the other capacitors. Any one of these could be used to realize the desired $\delta$ and $Q_p$ values. Thus, the $E$ and $F$ circuits are obtained when we choose $C_5$, $C_7$ or $C_{11}$; however, when $C_9$ is chosen as the damping capacitor, a new structure is obtained as shown in Figure 5.6.

Simple modifications of the El-Masry biquad result when the input is considered to be taken in the even phase (i.e., not sampled and held) and similarly, the output is sampled in the even phase. This requirement leads to another general SC biquad, shown in Figure 5.5(d), which has the same transfer function as that of Figure 5.5(c) [5.16]. However, this circuit, in simplified form, requires fourteen switches (after reducing the switch count by removing the redundant switches). Note that Gillingham's biquad [5.17] belongs to the circuit class considered by El-Masry in Figure 5.5(d).

In the above development, low $Q_p$ sensitivity was the criterion. When this requirement is relaxed (as it can be by using any of the deleted capacitors $C_1$, $C_2$, $C_6$ or $C_{12}$ in addition to one of the damping capacitors $C_5$, $C_7$, $C_9$ or $C_{11}$), the capacitor spread can be reduced to a large extent, at the expense of $Q_p$ sensitivity. The Fischer–Moschytz biquad is thus obtained [5.18], which uses $C_{12}$, as shown in

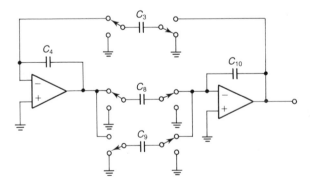

*Fig. 5.6*  A stray-insensitive SC structure with low $Q_p$ sensitivity.

Figure 5.7. This circuit has one degree of freedom more than necessary (i.e., one more capacitor) to realize the given $\delta$ and $Q_p$ values. The design procedure is simplified by assuming that $F$ and $A$ functions are combined by 'multiplexing' the switched capacitor $A$. (This technique was first suggested by Bosshart [4.17] and this has been studied in Chapter 4 as Bingham's 'direct transfer integrator'.) The denominator of the resulting transfer function is

$$D(z) = D(A + B)z^2 - (AD + 2BD + PD - AC)z + B(D + P) \qquad (5.14)$$

Assuming $B = D = 1$, the design equations are obtained as follows by using equations (5.4):

$$\frac{\delta}{Q_p} = \frac{A - P}{AC} \qquad (5.15a)$$

$$\frac{4\delta^2 - \dfrac{2\delta}{Q_p} + 1}{4} = \frac{1 + P}{AC} \qquad (5.15b)$$

By choosing equal time constants for the integrators (i.e., $A = C$), the design equations can be obtained. Note, however, that this choice may not optimize the dynamic range. It is noted that large $Q_p$s are obtained in the circuit of Figure 5.7 through small values of difference in capacitor values $A$ and $P$. In view of the resulting large $Q_p$ sensitivity, this design may not be attractive.

We next consider further modifications of Fleischer–Laker biquad topology that facilitate multiplexing.

## 5.1.4 Multiplexed SC filters

In some applications requiring a bank of filters, say, BP filters with different $\omega_p$s and $Q_p$s, it is possible to employ Fleischer–Laker topology using two OAs and multiplexing $D$ and $B$. This is possible because these capacitors are actually updated in one clock phase only. During the remaining portion of the clock cycle, these

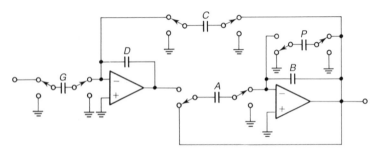

*Fig. 5.7*  Fischer–Moschytz SC biquad.

capacitors can be disconnected from the respective OAs, and OAs can be used with other switched capacitors to perform the filtering operation for another channel. Such a multiplexed filter bank [4.17] is illustrated in Figure 5.8(a). Strictly speaking, only one switch need be used to disconnect the integrating capacitor, but the use of two switches helps to remove certain undesirable effects, as explained next. The input switch eliminates the charge division that may take place between the integrating capacitor and the stray capacitances at the virtual ground of the OA and $C_{sub}$ (see Figure 5.8(a)). Similarly, without the output switch, under some signal conditions, the voltage on the input side of the integrating capacitor can go below the negative supply, turning on the isolation switch. When these two switches in series with the integrating capacitors $C$ are open, charge division takes place between $C_{sub}$ and $C$. However, both the charges on these capacitors determine the voltage on $C$ during the next phase, thus rendering the circuit stray-insensitive.

A completely multiplexed SC filter is shown in Figure 5.8(b). Note that the multiplexing of the capacitors can be arranged to realize a desired transfer function in each time slot or channel. For multiplexing $n$ biquads, $n$ time slots are required with each slot consisting of phase 1 and phase 2 as usual.

Due to the presence of stray capacitances across the input–output path of OAs, there is 'crosstalk' or interference between adjacent channels being multiplexed. When an integrating capacitor is connected to the OA, the stray capacitance shares some charge with it. The stray capacitance holding this charge contributes the crosstalk by interfering with the integrating capacitor in the next phase. The effect of such crosstalk is modelled by the equivalent SC circuit of Figure 5.8(c). Crosstalk can be eliminated by having a clamp across the OA output and input so that the stray capacitances are discharged as shown in Figure 5.8(a). The reader is referred to [4.17] for more details.

The multiplexing technique studied above can be extended to lead to the single-amplifier biquad due to Laker, Fleischer and Ganesan [4.10]. Consider once again the Fleischer–Laker biquad of Figure 5.2. It has already been pointed out that the integrating capacitors are updated once per clock cycle. Thus each OA is used to perform the integration function once per clock cycle, in the even phase. Hence, it is not directly possible to use a multiplexing scheme to reduce the OA count. Instead, if matters are arranged such that each OA works in a different phase, it is possible to multiplex the OA, as will be shown next.

Consider the $z$-domain equivalent circuit of the Fleischer–Laker biquad drawn in Figure 5.9(a) (considering only the pole-forming loop). Note that the delay associated with capacitor $A$ can be distributed between itself and the capacitors $C$ and $E$, as shown in Figure 5.9(b). The next task is to obtain an SC realization of the $z$-domain equivalent circuit of Figure 5.9(b). The circuit to be realized can only be parasitic-compensated, since the realization of $Cz^{-1/2}$ requires a parallel-switched capacitor.

The circuit in the even phase can be the same as in the original Fleischer–Laker biquad consisting of capacitors $A$, $B$ and $F$. The output in the even phase is used

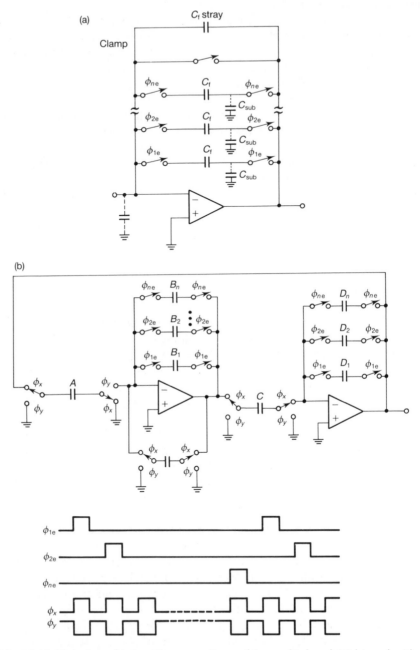

*Fig. 5.8* (a) Multiplexing of integrating capacitors; (b) a multiplexed SC biquad with clock waveforms; (c) (overleaf) modelling of crosstalk between adjacent channels *(adapted from [4.17],* © 1980 IEEE).

(c)

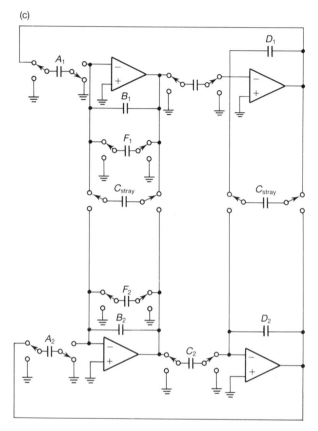

*Fig. 5.8  (continued)*

to realize an odd-phase transfer function, viz. (see Figure 5.9(a)),

$$\frac{V_{oo}}{V_{oe}} = -\left[\frac{Cz^{-1/2} + E(1 - z^{-1})z^{-1/2}}{D(1 - z^{-1})}\right] \qquad (5.16a)$$

Only the first term remains for the F-circuit ($E = 0$), which results in a parasitic-compensated SC integrator due to Fleischer, Laker and Ganesan described in Chapter 4 and redrawn in Figure 5.10(a). It may be noted that, for the E circuit, the numerator can be rearranged in equation (5.16a) as follows:

$$\frac{V_{oo}}{V_{oe}} = -\left[\frac{(C + E)z^{-1/2} - Ez^{-3/2}}{D(1 - z^{-1})}\right] \qquad (5.16b)$$

The term $-(C + E)z^{-1/2}/D(1 - z^{-1})$ can be realized by the circuit of Figure 5.10(a) with the time constant decided by $(C + E)/D$. The second term is realized by the parasitic-compensated integrator due to Ananda Mohan, Ramachandran and

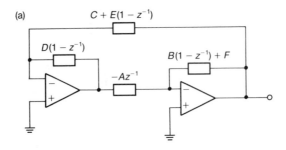

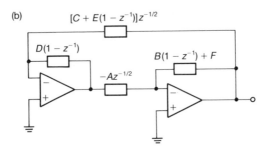

*Fig. 5.9* (a) z-domain equivalent circuit of the Fleischer–Laker biquad (pole-forming loop only); (b) equivalent of (a) to facilitate multiplexing *(from [1.26]* Copyright © 1979 AT&T. All rights reserved. Reprinted with permission).

Swamy [5.19] shown in Figure 5.10(b), which is an extension of the circuit of Figure 5.10(a). It may be noted that the additional clock phase $\phi_3$ (see Figure 5.10(d)) is required to discharge the middle capacitor in between the even and odd phases. It can be verified that by arranging the layouts of capacitors, switches and routeing of the connections such that the stray capacitance associated with capacitors $4E$ are matched, the desired transfer functions can be obtained. Note that the total capacitance required to realize this SC branch is $12E$ instead of $E$ in the original Fleischer–Laker biquad.

We next consider the realization of the feedforward inputs. One restriction on the inputs exists for multiplexed biquads, namely, that input can be sampled in only one phase, but can be fed to the virtual ground of the OA in both the phases. Thus a general network of the type shown in Figure 5.10(c) is possible. Note that the branch $Q(z)/V(z)$ relationships in the even and odd phases are respectively $(I - Jz^{-1})$ and $(Gz^{-1/2} - Hz^{-3/2})$. The resulting transfer functions of this circuit are the same as those of Fleischer–Laker biquad except that $T'$ output has an additional $z^{-1/2}$ term in the numerator.

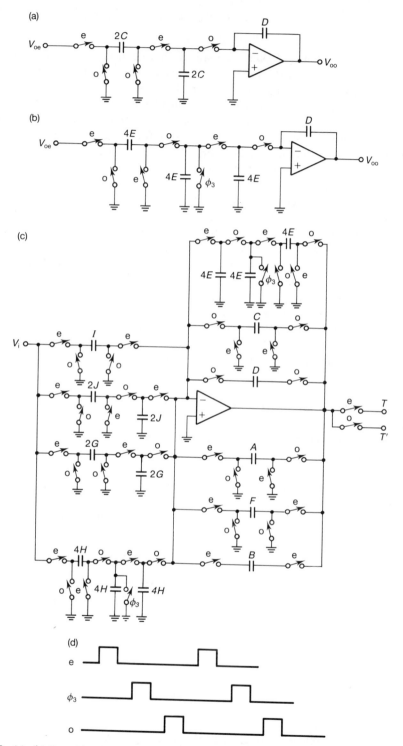

*Fig. 5.10* (a), (b) Parasitic compensated integrators *(adapted from [5.19], © 1986 IEEE)*; (c) general multiplexed biquad equivalent to the Fleischer–Laker–Ganesan biquad *(adapted from [5.19], © 1986 IEEE)*; (d) timing waveforms.

Note that Fleischer, Laker and Ganesan have considered the multiplexed biquad of Figure 5.10(c) with the feedforward capacitors $G$, $I$ and $J$ only. Consequently, certain transfer functions may not be realized, e.g., a bilinear BP one. It has been mentioned before that $E$- and $F$-type circuits have their own advantages, depending on the value of $\delta$ required. Consequently, the general structure of Figure 5.10(c), which includes $E$-type damping as well, is useful. Note also that for the evaluation of total capacitance, all the capacitors $4E$, etc., shall be considered and then scaled for minimum total capacitance.

It may also be noted that for cascading of single-amplifier biquads no additional sample and hold circuits are necessary.

It must be noted that the feedback loop of the OA has to be closed at all times to prevent instability. During even or odd phase, the feedback loop is closed by capacitors $B$ and $D$, respectively, but during the period between non-overlapping switching even and odd clock pulses, the feedback loop is open, resulting in instability. An elegant solution is to use 'XY feedback' around the OA, as shown in Figure 5.11. It may be noted that $X$ and $Y$ do not disturb the functioning of the circuit during even and odd phases. In between even and odd clock phases, however, $XY$ series connection will provide a closed feedback path, thereby stabilizing the OA. Another method is to use an unswitched capacitor $M$ across the feedback path, but this modifies the transfer functions greatly.

## 5.1.5  Electrically programmable SC filters [5.20, 5.21]

The two-amplifier biquad is attractive for electrically programmable SC BP filter realizations. For an electrically programmable second-order BP filter, the three parameters of interest are the centre frequency, the gain at centre frequency and $Q_p$. Thus, capacitor arrays which can be switched by logic circuits can be used to vary independently the $Q_p$, gain and $\omega_p$ values of the BP filters. For this purpose, it is first necessary to select a topology that has such flexibility. Choosing the

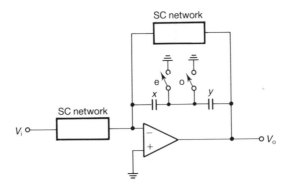

*Fig. 5.11*  XY feedback technique used to stabilize the OA.

Fleischer–Laker biquad, we note that for approximate optimal range design, i.e., the design with the two time constants of integrators being equal, the gain is decided by the input switched capacitors ($G$, $H$, $I$ and $J$). The resonant frequency is decided by both the integrating capacitors ($B$ and $D$) and the $Q_p$ value is decided by the damping capacitors ($E$ or $F$). Thus, the three parameters can be controlled by these various degrees of freedom available, while maintaining optimal dynamic range approximately. This is precisely the strategy used to design programmable SC BP filters.

One problem, however, arises for high-$Q_p$ BP filter realization. For high-$Q_p$ structures, the capacitor simulating the damping ($E$ or $F$) becomes extremely small, as shown next. For the Tow–Thomas biquad shown in Figure 5.1, the $Q_p$ value is $R_2/R_1$ and hence the resistor spread is $Q_p$. On the other hand, for the $F$-type Fleischer–Laker biquad, the minimum spread required in capacitor values for an optimal dynamic range design is obtained from equation (5.7b) as

$$F \approx \frac{1}{Q_p \delta} \tag{5.17}$$

It is thus desirable to reduce the capacitor spread considerably. An elegant solution is to realize the damping resistance separate from the two-integrator loop. An active RC circuit realizing this function is shown in Figure 5.12[1.24]. The resulting gain and $Q_p$ values are given by

$$G = \frac{R_5}{R_4} \quad \text{and} \quad Q_p = \frac{R_5}{R_3} \tag{5.18}$$

It is thus seen that $R_3$ can be used to control the $Q_p$ value, while $R_4$ can be used to control the gain. It is also noted that a band-reject or notch transfer function is realized at the output of OA $A_1$. In the SC version, the corresponding circuit can be simplified by making use of the inverting property of switched capacitors, thus resulting in three-OA configurations. Two such three-OA programmable SC structures are shown in Figures 5.13(a) and 5.13(b).

The circuit of Figure 5.13(a) uses a two-integrator loop which is sensitive to top-plate parasitic capacitances. The capacitive summer around OA $A_1$ needs to

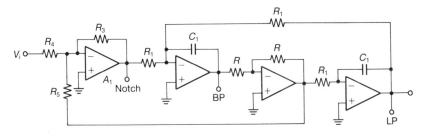

Fig. 5.12 A modified Tow–Thomas biquad realizing the damping resistance separate from the two-integrator loop.

be stabilized by a parasitic SC resistor. Even stray capacitance will usually be sufficient to bias the circuit. Binary weighted capacitor arrays such as those shown in
Figure 5.13(e) can be used to program the various capacitors controlling $Q_p$, $\omega_p$ and
$G$ values. Note that the circuit of Figure 5.13(a) actually realizes a third-order digital
transfer function, because of the parasitic capacitance used to stabilize the OA. The
reader is urged to verify this result. Note that LP, BP and notch transfer functions
are realized. The LP transfer function obtained is given by

$$\frac{V_3}{V_1} = \frac{-\dfrac{C_G}{C_Q} \cdot \dfrac{C_A^2}{C_{fo}^2} \cdot z^{-1}}{\left(1 - \dfrac{C_A^3}{C_Q C_{fo}^2}\right) z^{-2} - \left(2 - \dfrac{C_A^3}{C_Q C_{fo}^2} - \dfrac{C_A^2}{C_{fo}^2}\right) z^{-1} + 1} \tag{5.19}$$

(a)

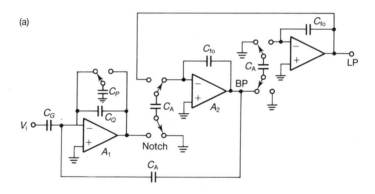

(b)

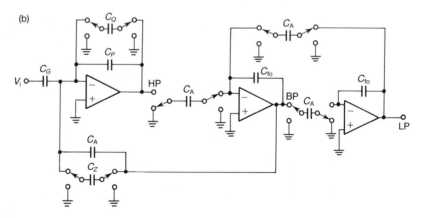

Fig. 5.13  Electrically programmable filters:  (a) Allstot *et al.* structure *(adapted from
[5.20], © 1979 IEEE)*;  (b) Cox *et al.* structure *(adapted from [5.21], © 1980 IEEE)*;  (c) Cox
structure *(adapted from [5.22], © 1983 IEEE)*;  (d) Ananda Mohan *et al.* structure *(adapted
from [5.11], © 1984 IEE)*;  (e) binary-weighted capacitor array realization.

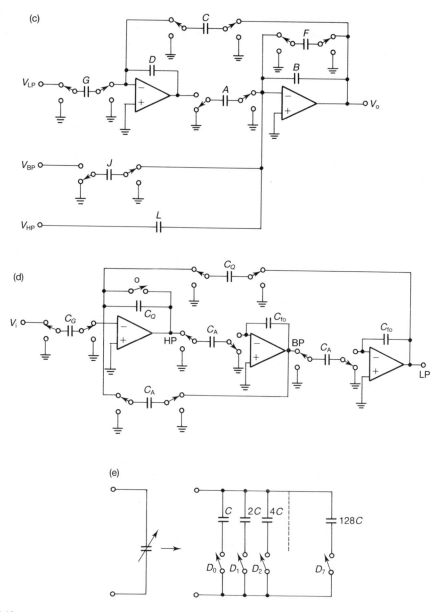

Fig. 5.13 (continued)

It can be seen that the denominator is of $E$ type as in the Fleischer–Laker biquad, even though we started with a biquad using $F$-type damping. The $\omega_p$ can be controlled by $C_{fo}$, $Q_p$ by $C_Q$ and the gain of the filter by $C_G$. The capacity spread required will be the same as that for the $E$-type Fleischer–Laker biquad.

The top-plate parasitic capacitance associated with the $C_A$ at the input of OA $A_2$ can be eliminated by using a modified Fleischer–Laker biquad as shown in Figure 5.13(b). Note that in place of $E$- or $F$-type damping, the damping is realized in a stray-insensitive manner using an additional OA as in the SC network of Figure 5.13(a). Note, however, that the capacitors $C_Z$ and $C_P$ are required for this purpose. For tuning $Q_p$, thus for exact design, $C_P$ and $C_Q$ shall be programmable arrays. Thus, this SC network requires six programmable arrays (two for $\omega_p$, two for $Q_p$, and two for gain controls). This SC network also realizes an $E$-type transfer function.

Another electrically programmable filter, shown in Figure 5.13(c), has been described by Cox [5.22], which is an $F$-type Fleischer–Laker biquad with three separate input terminals to realize the three transfer functions, viz., LP, HP and BP types. Thus, by appropriately using one or more of the input terminals, all the generally required biquadratic transfer functions can be realized. Note, however, that the $F$-type biquad is useful only for $\delta < \sqrt{3}/2$, and also that the capacitor spread required is large ($=Q_p\delta$), which was the reason for realizing the damping using another OA in the circuits of Figures 5.13(a) and 5.13(b). Hence, it is advisable to use an $E$-type Fleischer–Laker biquad for reduced capacitor spread for most of the usual $\delta$ values required (greater than $\sqrt{3}/2$). It is also recommended to have both $E$- and $F$-type damping available together with four-capacitor arrays for $G$, $H$, $I$ and $J$ in the Fleischer–Laker biquad for realizing completely general programmable SC filters.

A modification of the three-OA circuit of Figure 5.13(b) as in Figure 5.13(d) is possible. This circuit is equivalent to a Kerwin, Heulsman and Newcomb (KHN) active RC biquad and is completely stray-insensitive. It also realizes an $E$-type transfer function and it requires two capacitor arrays for $Q_p$ programming, one for gain control and two for $\omega_p$ programming, thus requiring five capacitor arrays only. Further, it is stray-insensitive and realizes LP, BP and HP transfer functions simultaneously. The design equations are the same as those of the circuits of Figures 5.13(a) and 5.13(b). The LP transfer function realized is given by

$$\frac{V_0}{V_1} = \frac{-\dfrac{C_G}{C_Q} \cdot \dfrac{C_A^2}{C_{fo}^2} \cdot z^{-1}}{\left(1 - \dfrac{C_A}{C_{fo}}\right)z^{-2} - \left(2 - \dfrac{C_A^3}{C_Q C_{fo}^2} - \dfrac{C_A}{C_{fo}}\right)z^{-1} + 1} \qquad (5.20)$$

One possible realization of a variable capacitor is shown in Figure 5.13(e). This completes the discussion on electrically programmable second-order SC networks.

## 5.1.6 Stray-insensitive all-pole low-pass and high-pass filters based on p-transformation [5.23]

SC filters derived based on Bach's RC active filters of Figures 5.14(a) and 5.14(b) will be considered. Stray-insensitive SC equivalent realizations can be obtained from the modified Bach active RC filters shown in Figures 5.14(c) and 5.14(d). These employ the $p$-transformation method to realize SC filters from active RC filters. In a manner similar to the corresponding active RC filters, these SC filters can be scaled for the optimal dynamic range.

The stray-insensitive SC LP filter under consideration is shown in Figure 5.15(a). The transfer function of this SC LP filter can be obtained as

$$\frac{V_{2e}}{V_{ie}} = \cfrac{1}{\frac{C_4C_6}{C_1C_3}(1-z^{-1})^2 + \left(\frac{C_4C_5 + C_6C_7}{C_1C_3} - \frac{C_2}{C_1}\right)(1-z^{-1}) + \frac{C_5C_7}{C_1C_3}} \qquad (5.21a)$$

$$V_{1e} = -\left(\frac{C_7}{C_3} + \frac{C_4}{C_3}(1-z^{-1})\right)V_{2e} \qquad (5.21b)$$

Note that $V_{2e}$ and $V_{1e}$ are held over a clock period. Furthermore, in order to simplify the design, $C_1 = C_5$, $C_2 = C_6$, $C_3 = C_7$ may be used. Also, the gain can be set as desired by not choosing $C_1 = C_5$. The SC network of Figure 5.15(a) can be scaled for optimal dynamic range in a manner similar to that discussed in the case of the

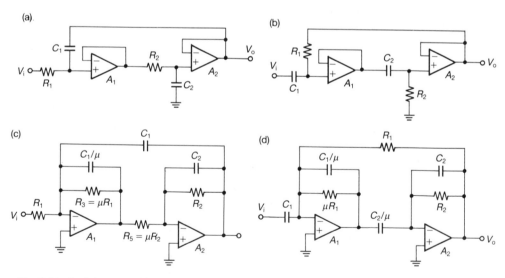

Fig. 5.14  Bach's (a) LP filter; (b) HP filter; (c), (d) active filters equivalent to (a) and (b), respectively, but using OAs with grounded non-inverting input. (Note that $\mu = 1$ in the unscaled case.)

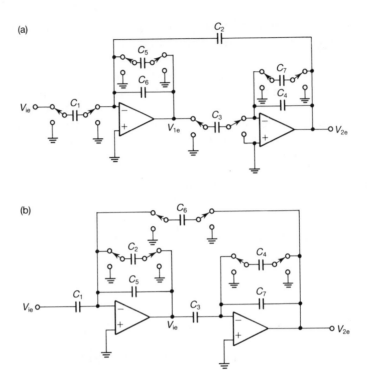

*Fig. 5.15* Stray-insensitive structures derived from Bach's active RC filter: (a) LP SC filter; (b) HP SC filter. *(Adapted from [5.23], © 1983 IEEE.)*

Fleischer–Laker biquad. The design of this SC filter can be carried out in the *p*-domain directly.

From the modified Bach HP active RC filter of Figure 5.14(d), a stray-insensitive SC HP configuration, as shown in Figure 5.15(b), can be obtained. The transfer functions of this SC HP filter, assuming $C_1 = C_5$, $C_2 = C_6$ and $C_3 = C_7$, are

$$\frac{V_{2e}}{V_{ie}} = \frac{(1 - z^{-1})^2}{(1 - z^{-1})^2 + \dfrac{C_4}{C_3}(1 - z^{-1}) + \dfrac{C_2C_4}{C_1C_3}} \tag{5.22a}$$

$$\frac{V_{1e}}{V_{ie}} = \frac{(1 - z^{-1})\left(1 - z^{-1} + \dfrac{C_4}{C_3}\right)}{(1 - z^{-1})^2 + \dfrac{C_4}{C_3}(1 - z^{-1}) + \dfrac{C_2C_4}{C_1C_3}} \tag{5.22b}$$

The design of this SC filter can be carried out in a similar manner to the LP filter. However, bilinear transformation type SC HP filters can be obtained since the numerator is of the form $(1 - z^{-1})^2$.

*Example 5.4*
Design an SC HP filter with $\omega_p = 2\pi \times 10^3$ rad s$^{-1}$ and $Q_p = 5$, and using a sampling frequency of 64 kHz.

The bilinearly pre-warped $\omega_p$ is 6288.2368 rad s$^{-1}$ and $Q_p$ is 5 as given. Hence $\delta$ ($=f_s/\omega_p$) is found to be 10.177 734. Using the design equations (5.4), we obtain

$$\frac{C_1}{C_2} = 1.035\ 546\ 8 \qquad \frac{C_3}{C_4} = 99.289\ 087$$

Then the transfer functions $T$ and $T'$ are obtained as

$$T = \frac{V_{2e}}{V_{ie}} = \frac{102.818\ 49(1 - z^{-1})^2}{102.818\ 49(1 - z^{-1})^2 + 1.035\ 546\ 8(1 - z^{-1}) + 1}$$

$$T' = \frac{V_{1e}}{V_{ie}} = \frac{1.035\ 546\ 8(1 - z^{-1})\ [99.289\ 087(1 - z^{-1}) + 1]}{102.818\ 49(1 - z^{-1})^2 + 1.035\ 546\ 8(1 - z^{-1}) + 1}$$

The maxima are found to be 5.015 109 3 and 5.075 126, respectively. The capacitor value scaling is thus not required because both the maxima are approximately equal. Thus the design values are

$$C_1 = C_5 = 1.035\ 546\ 8C_u \qquad C_4 = C_u$$

$$C_2 = C_6 = C_u$$

$$C_3 = C_7 = 99.289\ 087C_u$$

where $C_u$ is the unit capacitance. The total capacitance is 203.645 28$C_u$ and the capacitor spread is 99.289 087.

It is thus seen that stray-insensitive LP and HP SC filters can be realized in a modified Bach configuration using inverting integrators. The reader is urged to prove that dynamic range optimization is usually not necessary, since the maxima of the transfer functions are almost same. Note, however, that these stray-insensitive filters have large $Q_p$ sensitivities.

It may be remarked that the large $Q_p$ sensitivity is typical of two-OA based filters using two inverting sections. (Note that we have proved this result during the discussion on El-Masry's general SC biquad in Section 5.1.3.) However, since large $Q_p$ sensitivity can be tolerated to a larger extent than large $\omega_p$ sensitivity, the large $Q_p$ sensitivity of the SC networks of Figures 5.15(a) and 5.15(b) may not be a disadvantage. In the next subsection, we study three-OA SC networks based on Moschytz's modified Tarmy–Ghausi active RC realizations.

## 5.2  Three-OA SC band-pass filter based on Moschytz's modified Tarmy–Ghausi configuration

In the active RC version, the BP filter under consideration uses an inverting summer and two first-order all-pass networks, as shown in Figure 5.16(a). Bottom-plate stray-insensitive SC realizations can be obtained from this active RC configuration, as shown in Figure 5.16(b). Note that the OA $A_1$ is a capacitive inverting summer, stabilized using a parasitic capacitor. The SC all-pass first-order networks are used

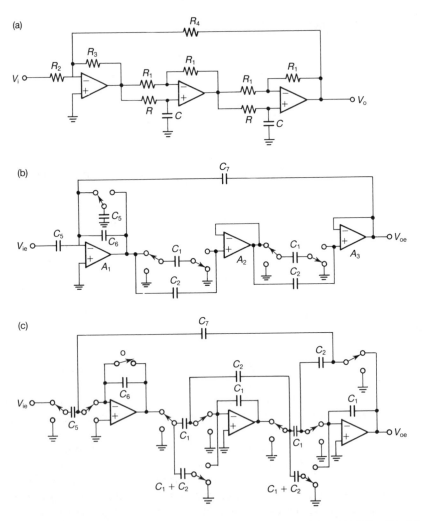

Fig. 5.16  (a) Moschytz's modified Tarmy–Ghausi filter *(adapted from [1.24]);*  (b), (c) SC filters derived from (a) Ananda Mohan et al. *(adapted from [4.35, 5.24]*, © IEE 1982, 1983).

together with isolating unity gain buffers to realize the two all-pass networks. The transfer function of this three-OA BP filter can be derived as

$$\frac{V_{oe}}{V_{ie}} = \frac{-C_5(C_1 - (C_1 + C_2)z^{-1})^2}{[C_7C_1^2 + C_6(C_1 + C_2)^2] - 2C_1(C_1 + C_2)(C_6 + C_7)z^{-1} + [C_6C_1^2 + C_7(C_1 + C_2)^2]z^{-2}}$$

(5.23)

Note that all the outputs are held over a clock period. Further, since the circuit uses two all-pass networks of gain unity, all the OAs saturate at the same output level. This is a unique advantage since no scaling for optimal dynamic range is necessary. The circuit realizes finite real zeros, however. The design equations can be derived as

$$\frac{C_1}{C_2} = \delta - \frac{1}{2}$$

(5.24a)

$$\frac{C_6}{C_7} = \frac{2Q_p + 1}{2Q_p - 1}$$

(5.24b)

$$\frac{C_5}{C_6} = \frac{2G_0}{2Q_p + 1}$$

(5.24c)

where $G_0$ is the midband gain.

It is thus evident that $\omega_p$ is dependent on $C_2/C_1$, while $Q_p$ is decided by $C_6/C_7$. The capacitor spread required to realize the desired $\delta$ and $Q_p$ is either $\delta - 0.5$, or 3, whichever is greater (considering $Q_p > 1$ is desired). The circuit of Figure 5.16(b) thus enjoys extremely low capacitor spread. The $Q_p$ sensitivity, however, is large:

$$S_{C_6/C_7}^{Q_p} \approx \frac{-(2Q_p + 1)^2}{4Q_p} \approx -Q_p$$

(5.25)

However, since $\omega_p$ sensitivities are small, the circuit may be suitable for practical applications.

It is preferable to have a stray-insensitive realization using the same principle as in the BP filter of Figure 5.16(a). Such a configuration is obtained by using two first-order stray-insensitive all-pass circuits due to Brugger et al. [4.21] (studied in Chapter 4), in place of the two all-pass networks in Figure 5.16(a). The resulting stray-insensitive SC BP filter is shown in Figure 5.16(c). The design equations for this circuit are the same as those of Figure 5.16(b) given in equations (5.24). Further, we need the capacitor values for the first-order all-pass networks as shown in Figure 5.16(c). Note also that the circuit uses an 'even' summer. Hence, the output available at the output of $A_1$ goes to zero in every odd phase, whereas the outputs of the other two OAs are not held over a clock period. Thus, an additional sample and hold may or may not be necessary, depending on the stages that follow the circuit under consideration. We next consider a design example.

*Example 5.5*
Design a BP filter with $Q_p = 16$, $\omega_p = 1633$ Hz, sampling frequency 8 kHz and 10 dB midband gain.

The required $\delta$ and $Q_p$ values are evaluated from the transfer function obtained in Example 5.1 as

$$\delta = 0.6693$$

$$Q_p = 11.9586$$

Using the design equations (5.24), we obtain the various capacitor ratios:

$$\frac{C_1}{C_2} = 0.169\ 344\ 8$$

$$\frac{C_7}{C_6} = 0.919\ 766\ 4$$

$$\frac{C_5}{C_6} = 0.2537$$

The capacitor values, after scaling for minimum total capacitance for the circuit of Figure 5.16(c) (note that the first-order all-pass networks have the capacitor values shown in Figure 5.16(c)), are

$$C_5 = C_8 = C_{10} = C_u$$

$$C_6 = 3.941\ 348C_u$$

$$C_7 = 3.625C_u$$

$$C_9 = 6.905\ 111\ 9C_u$$

$$C_{11} = 5.905\ 111\ 9C_u$$

Thus the total capacitance is $38.18C_u$ and the capacitor spread is 6.905. The best Fleischer–Laker biquad of Figure 5.2 requires a total capacitance of $51.737C_u$ and a capacitor spread of 14.632, as derived in Example 5.1. The circuit of Figure 5.16(c) uses 19 switches, 11 capacitors and three OAs, whereas the Fleischer–Laker biquad requires 12 switches, eight capacitors and only two OAs.

In the above two sections, the design of bottom-plate stray-insensitive as well as completely stray-insensitive designs has been considered. However, it is also possible to compensate the top-plate parasitic capacitances present in certain circumstances.

## 5.3  Parasitic-compensated second-order filters [4.35]

We have already studied such methods with reference to the Laker, Fleischer and

Ganesan biquad using a two-phase clocking scheme. In this section, we deal with another circuit design technique applicable to a class of SC networks. This method has been studied with reference to first-order networks in Chapter 4. It is easy to extend this method to second-order SC networks. As mentioned in Chapter 4, the important requirement for the applicability of such a compensation scheme is

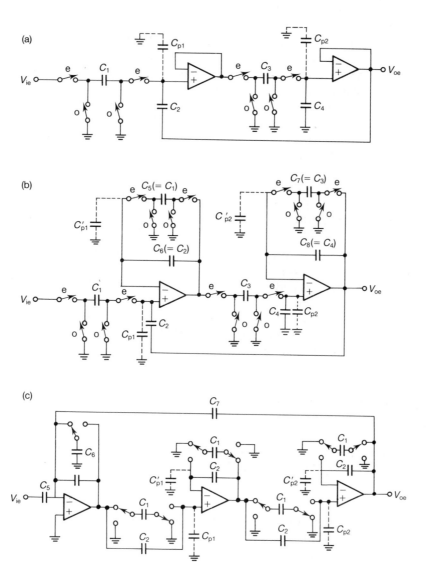

*Fig. 5.17* (a) Bach's LP SC filter; (b) parasitic compensated version of (a); (c) parasitic compensated version of Figure 5.16(b).

that the parasitic capacitance shall be lumped at one node, in only one phase of the clock.

Consider, for example, Bach's $p$-transformation type LP SC network, shown in Figure 5.17(a) for convenience. Note that the switching arrangement is chosen so that the parasitic capacitance is lumped at the non-inverting inputs of the OAs. By modifying the circuit as shown in Figure 5.17(b), under the conditions $C_1 = C_5$, $C_2 = C_6$, $C_3 = C_7$, $C_4 = C_8$ and $C_{p1} = C'_{p1}$, $C_{p2} = C'_{p2}$, the desired transfer function can be obtained. The matching conditions are easily satisfied because of the excellent matching characteristics of the capacitors $C_1-C_8$ in MOS technology. The parasitic capacitances are matched by routeing the wiring associated with the inverting and non-inverting nodes in a 'similar' manner. (Note, however, the order of the filter is doubled, assuming non-ideal capacitors.)

This technique can be applied to other SC realizations, too. We consider the BP filter of Figure 5.16(b) as another example. The modified realization is shown in Figure 5.17(c). Under the matching condition on the component values, shown in Figure 5.17(c), exact compensation can be achieved.

## 5.4  Derivation of stray-insensitive SC networks from stray-sensitive SC networks

Several single-amplifier SC networks have been described in the literature which are sensitive to top-plate parasitic capacitance, as well as affected by parasitic capacitance at an internal node. It is possible to simulate the voltages at this internal node at the outputs of extra OAs, using appropriate stray-insensitive SC branches (thus realizing the nodal charge conservation equation). This is shown by demonstrating the Hasler and Saghafi technique [5.25] using two examples. For a rigorous theoretical development, the reader is referred to their paper.

*Example 5.6*
Derive a stray-insensitive SC network from the single-amplifier SC network of Figure 5.18(a) using the nodal voltage simulation technique.

First the charge conservation equation at node $x$ in the odd phase using Laker's equivalent circuit method is obtained:

$$C_1 V_{ie} z^{-1/2} + C_5 V_{oe} z^{-1/2} - [C_1 + C_3 + C_5 + C_2(1 - z^{-1})] V_{xo} = 0 \qquad (5.26)$$

This charge conservation equation can be realized by a stray-insensitive sub-circuit shown in Figure 5.18(b), where it is required to generate the input $V_{oe}$ next. In the circuit of Figure 5.18(a), we also have

$$\frac{C_3(-V_{xo})z^{-1/2}}{C_4(1 - z^{-1})} = V'_{oe} \qquad (5.27)$$

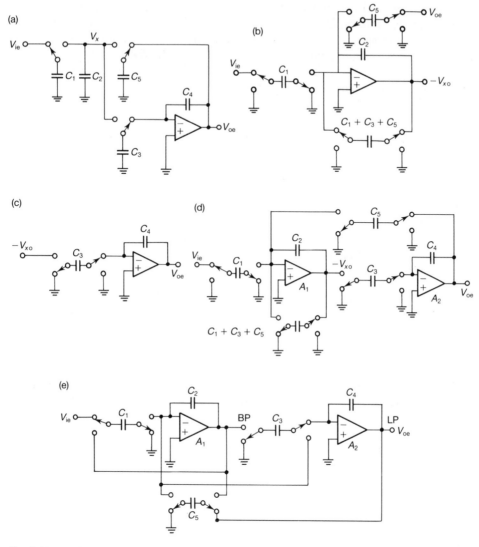

*Fig. 5.18* (a) Stray-insensitive single-amplifier SC network; (b)–(e) derivation of stray-insensitive SC network from (a).

Since the voltage $-V_{xo}$ is available at the output of OA $A_1$ in Figure 5.18(b), a stray-insensitive circuit (Figure 5.18(c)) can be used to realize equation (5.27). The resulting complete SC network is as shown in Figure 5.18(d), which is evidently stray-insensitive.

Unfortunately, the SC network of Figure 5.18(d) is different from that of Figure 5.18(a) in two respects. It uses six capacitors, as against five in the prototype

SC network. Secondly, it requires matching of some capacitors, which leads to large $Q_p$ sensitivities. Interestingly, however, by manipulation of the switch terminals, which are 'grounded', in the circuit of Figure 5.18(d) (i.e., by multiplexing), we can save one capacitor. Such a multiplexed circuit is shown in Figure 5.18(e), which has the same design equations as the single-amplifier circuit and hence has the same spread in capacitor value. Note, however, that two transfer functions are available at the outputs of OAs $A_1$ and $A_2$ (BP and LP transfer functions, respectively).

The nodal voltage simulation technique is further illustrated using the example below.

*Example 5.7*
Derive a stray-insensitive SC network equivalent to the SC network of Figure 5.19(a).

Using Laker's equivalent circuit method, we obtain the following charge conservation equations:

$$\begin{bmatrix} -(C_1+C_3+C_4+C_2(1-z^{-1})) & C_3+C_4 & C_3 z^{-1/2} \\ C_3+C_4 & -(C_3+C_4+C_5(1-z^{-1})) & -C_3 z^{-1/2} \\ C_4 z^{-1/2} & C_4 z^{-1/2} & C_6(1-z^{-1}) \end{bmatrix} \begin{bmatrix} V_{xe} \\ V_{ye} \\ V_{oo} \end{bmatrix} = \begin{bmatrix} -V_{ie}C_1 \\ 0 \\ 0 \end{bmatrix}$$

(5.28)

A stray-insensitive SC network that satisfies these equations can be realized as in Figure 5.19(b). This circuit is obtained by examining the first equation in system (5.28) and attempting to realize it in a stray-insensitive manner and using one OA having $-V_{xe}$ at the output. Thus the element in the first row and first column of the matrix in system (5.28) needs a sign change, which leads us to change the signs of the $V_{xe}$ coefficients in the second and third rows. Thus a new set of equations is obtained:

$$\begin{bmatrix} C_1+C_3+C_4+C_2(1-z^{-1}) & C_3+C_4 & C_3 z^{-1/2} \\ -(C_3+C_4) & -(C_3+C_4+C_5(1-z^{-1})) & -C_3 z^{-1/2} \\ -C_4 z^{-1/2} & -C_4 z^{-1/2} & C_6(1-z^{-1}) \end{bmatrix} \begin{bmatrix} -V_{xe} \\ V_{ye} \\ V_{oo} \end{bmatrix} = \begin{bmatrix} -V_{ie}C_1 \\ 0 \\ 0 \end{bmatrix}$$

(5.29)

These equations can be realized with $-V_{xe}$, $V_{ye}$, $V_{oo}$ as the three outputs of three OAs and by interconnecting the resulting sub-networks. This results in the SC network of Figure 5.19(b).

Note that this circuit uses more capacitors than the original circuit and needs three OAs. Some of the capacitors can be deleted by means of the multiplexing technique. (Also, capacitors of value, e.g., $C_1+C_3+C_4$ can be realized by three separate capacitors or one capacitor.)

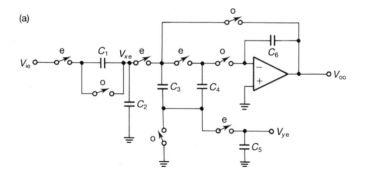

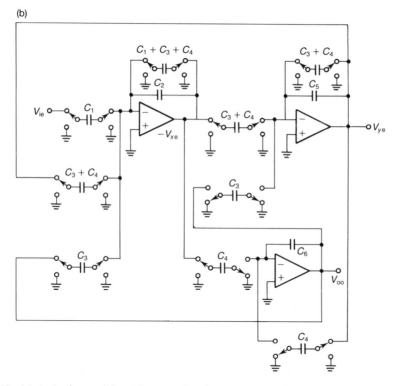

Fig. 5.19 (a) A single-amplifier SC network; (b) stray-insensitive SC network derived from (a).

The $p$-transformation type of circuits can be rendered stray-insensitive using the above approach. As an illustration, consider the SC LP filter of Figure 5.20(a), from which the SC network of Figure 5.20(b) is derived. Note that an inversion of voltage is required, needing an extra OA, whereas in the case of Figure 5.18(e), the inversion associated with a half-cycle delay could be conveniently used to reduce the OA count.

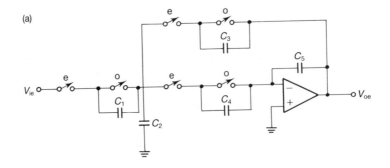

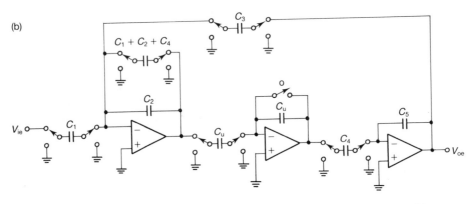

*Fig. 5.20* (a) A *p*-transformation type single-amplifier SC filter; (b) stray-insensitive equivalent of (a).

It may be remarked that usually the number of OAs needed depends on the number of different nodal voltages in the *z*-domain equivalent circuits (not sampled and held) which need to be simulated. In some cases additional OAs may be needed, as in the circuit of Figure 5.20(b). It is noted, finally, that even though the method considered above is attractive for generating stray-insensitive circuits, the matching required in some cases, as well as the large capacitor spread of single-amplifier circuits, may discourage its use.

## 5.5  Synthesis of multiphase single-amplifier SC networks

### 5.5.1  Mulawka's synthesis procedure

It may be noted that, in the previous sections, specific configurations were chosen and transfer functions were derived and matched with desired transfer functions to identify capacitor ratios. In Mulawka's method [5.26], the transfer function is given, and the topology of the circuit is assumed only to a small extent, in that, to preserve stability, a shunt capacitor in the feedback path of OA is assumed. The remaining SC networks in the input and feedback paths have to be synthesized. This means that SC networks 1 and 2 have to be determined in Figure 5.21(a) to realize a given SC transfer function. It will now be shown that Figure 5.21(a) can be represented by the *ring model* of Figure 5.21(b), considering that an *n*-phase clock is used. In this ring model, solid branches represent the OA and the capacitor $C_0$, whereas the dashed lines represent the SC network blocks. There are two types of node in the ring model. Nodes $A_1, A_2, ..., A_n$ represent the OA inverting terminals in each of the *n* phases and are *simple* nodes, whereas $B_1, B_2, ..., B_n$ represent the outputs of the OA in each phase and are called *supernodes*. A supernode is a node with at least one *self-loop*. The use of supernodes here allows the virtual ground concept to be represented in the OA model. To understand this, consider the OA model shown in Figure 5.22(a). For infinite-gain OAs, it is known that $V_1 = V_2$. The graph shown in Figure 5.22(b) represents the equality

$$- V_1 + V_2 + V_3 = V_3 \qquad (5.30)$$

or $V_1 = V_2$, as desired.

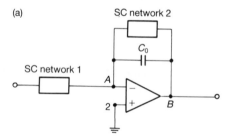

(a)

SC network 2

SC network 1

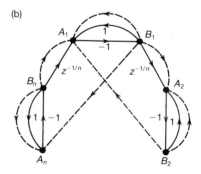

(b)

Fig. 5.21 (a) Single-amplifier SC network; (b) its ring model.

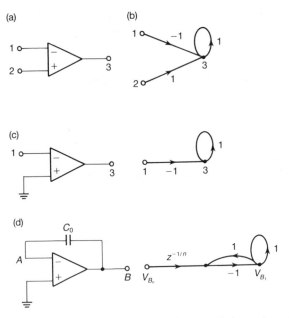

Fig. 5.22  (a) An OA;  (b) model of (a);  (c) OA with grounded non-inverting input and its model;  (d) OA with feedback capacitor $C_o$ and its model.

Since, in the circuit of Figure 5.21(a) under consideration, terminal 2 (non-inverting input) is grounded, the model of OA simplifies to that shown in Figure 5.22(c). The *memory* of $C_o$ shall be represented in the ring model. Since terminal $A$ is 'virtual ground', it is noted that $C_o$ stores the charge present in any phase till the next phase, e.g., output $B_n$ is felt at $B_1$ after $z^{-1/n}$ delay. This capacitor and OA combination is represented precisely by the signal flow graph shown in Figure 5.22(d), where it is observed that

$$[V_{B_n}(z^{-1/n})(-1) + V_{B_1}(1)](-1) + V_{B_1}(1) = V_{B_1} \qquad (5.31a)$$

or that

$$V_{B_1} = V_{B_n}z^{-1/n} \qquad (5.31b)$$

which is true. Thus the capacitor and OA structure can be represented by the ring model. The other SC branches can be presented by dotted lines, as shown, and similarly the input can be fed to any input node $A$ through suitable transmittance.

The synthesis procedure is as follows. The numerator of the desired transfer function can be realized through a feedforward branch separately because of the 'virtual ground' effect of the OA. From the denominator of the transfer function, a simple graph is formed, which is then transformed by writing signal-flow graphs, till a ring-model form of Figure 5.21(b) is reached. Subsequently, the SC networks can be drawn easily. This is illustrated in the following example.

*Example 5.8*
Realize the SC transfer function

$$H_1(z) = \frac{\alpha_1(1 - z^{-1})}{1 - (1 + \alpha_2)z^{-1} + \alpha_3 z^{-2}} \qquad (5.32)$$

Assume that a three-phase clock is used.

The denominator may be represented by the signal flow graph shown in Figure 5.23(a), where the nodes $B_1$ and $B_3$ are distinguished. This graph can be drawn by noting that the denominator of equation (5.32) represents the loop gain of the SC network from any node to the same node. For example, consider that the output needs to be taken in phase 3 at node $B_3$. The denominator $D(z)$ of equation (5.32) is

$$V_{B_3} = V_{B_3}[z^{-1}(1 + \alpha_2) - \alpha_3 z^{-2}] \qquad (5.33)$$

Next the bracketed term on the right can be manipulated in several ways. One way is to have

$$V_{B_3} = V_{B_3}z^{-1/3}[\alpha_2 z^{-2/3} + z^{-2/3} - \alpha_3 z^{-5/3}] \qquad (5.34)$$

resulting in the graph of Figure 5.23(a). Next, node $B_2$ is introduced corresponding to second phase, as shown in Figure 5.23(b). Then, the OA is introduced between all $B$ terminals, as shown in the ring model of Figure 5.23(b). The resulting structure is shown in Figure 5.23(c). The next step is to introduce the input according to numerator of equation (5.32) to node $A_1$, as shown in Figure 5.23(d). The last step is to realize the required SC circuit. This is simply done as follows. It is noted that, from $B_1$ to $A_3$, we need an inverting branch transmittance delayed by two clock phases. The simple method is to charge a capacitor in one phase and discharge it after inversion in third phase. This arrangement is shown in Figure 5.23(d). Similarly, $\alpha_3 z^{-4/3}$ is required from $B_1$ to $A_2$. Thus the signal is sampled in phase 1, stored in phase 3 and transferred in phase 2. The final circuit is shown in Figure 5.23(e). The resulting circuit is only bottom-plate stray-insensitive.

Two other single-amplifier SC filter synthesis techniques have been proposed in the literature. These are considered next.

## 5.5.2  The Vaidyanathan–Mitra method

This method [5.27] is quite general and can be used to realize any $n$th-order transfer function, using a somewhat complex clocking scheme. In this method, it is assumed that the feedback capacitor is always present, though it is not always necessary, when XY feedback can be used. The circuit configuration of a second-order SC structure is shown in Figure 5.24(a), together with the clocking scheme in Figure 5.24(b).

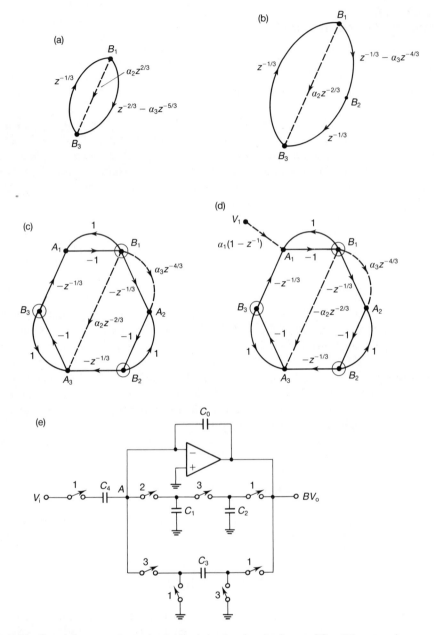

*Fig. 5.23* Signal-flow graph method of synthesis of a single amplifier SC network.

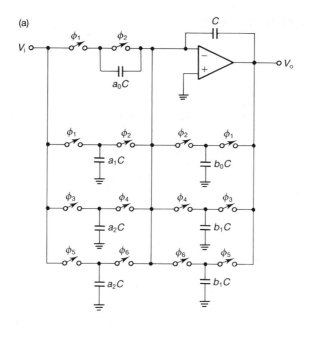

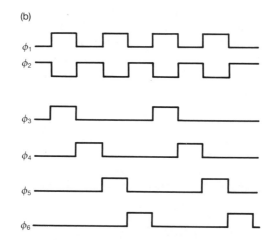

Fig. 5.24  (a) Vaidyanathan–Mitra single-amplifier network; (b) clocking waveforms *(adapted from [5.27], © 1983 IEEE).*

The output is available in $\phi_1$ and the input is sampled in phase $\phi_1$. In order to realize a desired transfer function, it is sufficient to realize the numerator and denominator as the $\Delta Q$–$V$ relationships of the feedforward and feedback branches, respectively. The realization of the numerator is considered next. The three terms in the numerator of $H(z)$ given by

$$H(z) = \frac{a_0 + a_1 z^{-1} + a_2 z^{-2}}{1 + (b_0 - 1)z^{-1} + b_1 z^{-2}} \tag{5.35}$$

are realized as three feedforward branches. The branch containing capacitor $a_0 C$ is a simple series-switched capacitor and thus the charge input in $\phi_1$ is $V_i a_0 C$. The branch containing $a_1 C$ is a toggle-switched capacitor transferring its charge on to the feedback capacitor $C$ in $\phi_2$, which, when sampled in $\phi_1$ at the output of the OA, realizes a clock-cycle delay. The realization of the $z^{-2}$ term, however, requires two branches working on alternate input samples to provide the two-clock cycle delay for all the input samples. This is easily understood with reference to the switching waveforms seen in Figure 5.24(b).

The realization is similar for the denominator, except for the fact that the presence of $C$ in the feedback path results in the minimal form, a denominator of the transfer function $C(1 - z^{-1})$. This is not a disadvantage since the $z^{-1}$ term can be modified by the appropriate branch $b_1 C$. The complete circuit realization is as in Figure 5.24(a).

In instances where the signs of $z^{-1}$ and $z^{-2}$ terms in the numerator and denominator need additional inversion, one can use the inverting switched capacitor, which leads to parasitic-insensitive designs. Note also that the branch $a_0$ can be realized as stray-insensitive.

The extension of the above technique to high-order filters is straightforward. Nevertheless, since the SC structure of Figure 5.24(a) is a direct-form structure, the coefficient sensitivity problem has to be considered. As an illustration, for large $Q_p$ designs, the second-order direct form recursive structure is very sensitive and hence, one can use the low-sensitivity structure due to Aggarwal and Burrus [1.39]. This requires $D(z)$ to be realized as

$$D(z) = 1 - (2 - a)z^{-1} + (1 - b)z^{-2} \tag{5.36}$$

showing that the $z^{-1}$ term needs two branches, and the $z^{-2}$ term four branches with appropriate inversions being used.

The reader must have noted the ease of design using the SC block of Figure 5.24(a). We next consider Saad's technique [5.28].

### 5.5.3  Saad's single-amplifier SC structure

This method [5.28] needs one clock phase more than the Vaidyanathan–Mitra method. The second-order SC structure obtained using this method and its clocking scheme are shown in Figure 5.25. It requires a sampled and held input. Further, the output is held over a clock period except for one clock phase $\phi_0$ during which the feedback capacitor is discharged. The remaining switched capacitors realize the various terms in the numerator and denominator as in the case of Figure 5.24(a). Since the output is held except in clock phase $\phi_0$, it is easy to realize a cascade design.

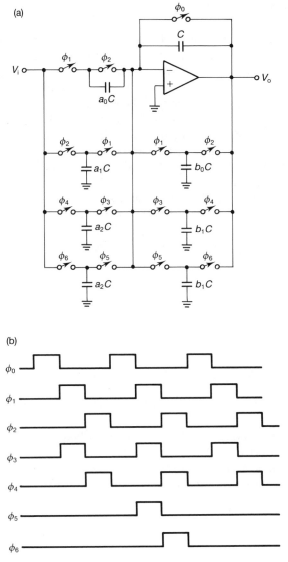

*Fig. 5.25*  (a) Saad's single-amplifier SC structure;  (b) clocking waveforms.

High-order direct-form structures could also be directly realized as in the case of the Vaidyanathan–Mitra method using additional capacitors to realize the $z^{-3}$, $z^{-4}$, etc., terms in the numerator and denominator. The presence of one phase $\phi_0$ in which the OA is connected as a buffer may be useful for offset compensation, as will be explained in a later chapter.

## 5.6  Conclusions

We conclude the discussion of second-order SC filters with a few remarks. A large variety of second-order realizations with complete stray-insensitivity or insensitivity only to bottom-plate stray capacitances are possible. Only a few representative ones are discussed here. The required filter structure can be chosen on the basis of total area on the IC chip or capacitor spread. The sensitivity of the transfer functions to various capacitances shall be determined. To this end, the statistics of capacitors also should be taken into account, together with the sensitivity formulas for the transfer function. Finally, the circuits can be simulated and the tolerance of the performance can be evaluated using Monte Carlo techniques to predict the expected deviation in performance.

Biquad designs can be used for cascade design of high-order filters. Sensitivity analysis of cascade designs shows that they are inferior to designs based on RLC networks by either component simulation or operational simulation techniques. Such high-order design techniques will be studied in detail next.

## 5.7  Exercises

5.1 A state-variable SC filter is shown in Figure 5.26. Derive the transfer functions. Compare its performance with the Fleischer–Laker biquad.

5.2 Analyze the second-order SC filter of Figure 5.27. Compare its performance with the SC filter of Figure 5.26.

5.3 Derive the transfer functions of the general SC biquad of Figure 5.5. Establish whether any advantages result over the Fleischer–Laker biquad. Synthesize for low $Q_p$ sensitivity.

5.4 A general biquadratic structure, due to Gillingham [5.17], is shown in Figure 5.28. Derive the transfer functions. An important feature of this circuit is that it processes signals in each phase, i.e., the OAs are used in both the phases. The outputs are no longer sampled and held. Note that the input is

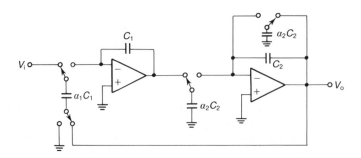

*Fig. 5.26*  A state-variable SC filter.

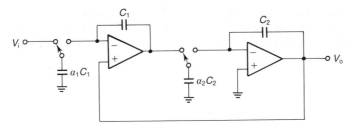

*Fig. 5.27* A second-order SC filter.

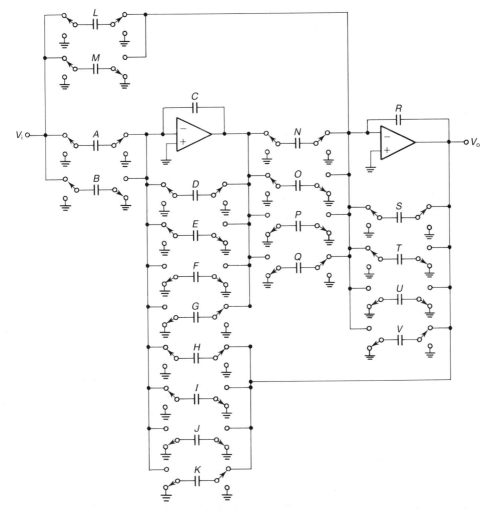

*Fig. 5.28* A general biquadratic structure, due to Gillingham (*from [5.17]*, © 1981 IEE).

sampled in one phase only. Discuss the requirements for stability. What are the capacitors that must be present for any realization?

5.5 Study the SC biquad obtained from the $F$-type Fleischer–Laker biquad, by also damping the first integrator, i.e., shunting a switched capacitor similar to $F$ across $D$. Derive the design equations, and evaluate its sensitivities.

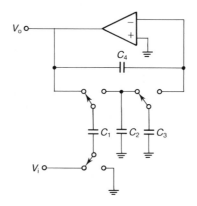

Fig. 5.29  An SC bridged-T LP network.

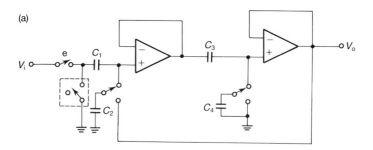

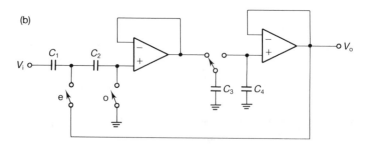

Fig. 5.30  (a) An SC network realizing both HP and BP second-order transfer functions  (adapted from [5.4], © 1983 IEE);  (b) Mulawka–Ghausi SC BP filter  (adapted from [5.5], © 1980 IEE).

5.6 Study the Fleischer–Laker biquad with both $E$- and $F$-type damping used simultaneously. Compare with $E$- and $F$-type biquads regarding capacitor spread and total capacitance.

5.7 The low-pass filter of Figure 5.29 is an application of the SC bridged-T network (see Chapter 3). Derive the design equations for realizing given values of $\delta$ and $Q_p$.

5.8 Show that the SC network of Figure 5.30(a) realizes both HP and BP second-order transfer functions by deleting or including the odd switch shown in the dotted lines [5.4], by deriving the transfer functions. The Mulawka–Ghausi [5.5] BP filter is shown in Figure 5.30(b). Derive its design equations and compare with those of Figure 5.30(a) regarding capacitor spread and total capacitance.

5.9 A multiple feedback biquad using capacitance magnification is shown in Figure 5.31(a). An SC circuit derived from this, which can realize LP, HP and BP transfer functions, is shown in Figure 5.31(b) [5.6]. Derive the transfer functions of this SC network and develop the design equations.

5.10 Derive the design equations of the Szentirmai–Temes two-OA biquad [1.28] of Figure 5.32 and verify whether it realizes all desired digital transfer functions.

5.11 The second-order $z$-domain poles are expressed by $D(z) = 1 + \alpha z^{-1} + \beta z^{-2}$, where

$$\alpha = -2 + \frac{\omega_p}{Q_p f_s} + \frac{\omega_p^2}{f_s^2}$$

$$\beta = 1 - \frac{\omega_p}{Q_p f_s}$$

These approximate expressions are valid for high $Q_p$s and high sampling frequencies. Derive the above expressions, and use them to develop the design equations for the Fleischer–Laker biquad.

5.12 The magnitude of a second-order discrete transfer function

$$H_d(z) = \frac{a_0 + a_1 z^{-1} + a_2 z^{-2}}{1 + b_1 z^{-1} + b_2 z^{-2}}$$

can be written in the form

$$H(\omega) = G \frac{\left(1 - p_0 \sin^2\left(\frac{\omega T}{2}\right)\right) + j p_1 \sin(\omega T)}{\left(1 - q_0 \sin^2\left(\frac{\omega T}{2}\right)\right) + j q_1 \sin(\omega T)}$$

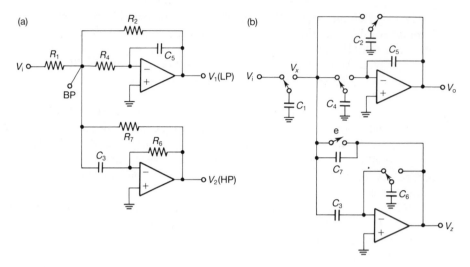

*Fig. 5.31* (a) A multiple feedback biquad *(adapted from Ananda Mohan* et al.) using capacitance magnification;  (b) an SC circuit derived from (a) realizing LP, HP and BP transfer functions  *(from [5.6])*.

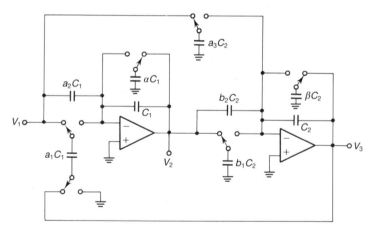

*Fig. 5.32* The Szentirmai–Temes two-OA biquad  *(adapted from [1.28],* © 1980 IEEE*)*.

This form will enable insight into the nature of the magnitude response by comparing with the second-order analog transfer function, viz.,

$$H_a(\omega) = G \frac{1 - \dfrac{\omega^2}{\omega_z^2} + j\,\dfrac{\omega}{\omega_z Q_z}}{1 - \dfrac{\omega^2}{\omega_p^2} + j\,\dfrac{\omega}{\omega_p Q_p}}$$

Observe that (a) the real parts of the denominator and the numerator are zero at $\omega_p$ and $\omega_z$, respectively; (b) the imaginary parts of the denominators of $H_d(\omega)$ and $H_a(\omega)$ are equal at $\omega_p$; (c) d.c. gains are equal for $H_d(\omega)$ and $H_a(\omega)$; and (d) the absolute values of the numerators are equal. Derive these results so as to yield the design equations for the Fleischer–Laker biquad.

5.13 Derive the condition for achieving optimal dynamic range while realizing an all-pass transfer function at $T$ output in the Fleischer–Laker biquad. Observe whether it is the same as that while realizing other transfer functions at $T$ output. Derive these results assuming high sampling frequencies and high $Q_p$s.

## 5.8 References

[5.1] R.P. Sallen and E.L. Key, A practical method of designing RC active filters, *IRE Transactions on Circuit Theory*, CT-2, 74–85, 1955.

[5.2] G.C. Temes and S.K. Mitra (eds), *Modern Filter Theory and Design*, Wiley, 1973.

[5.3] B.J. Hosticka and G.S. Moschytz, Switched-capacitor filters using FDNR-like super-capacitance, *IEEE Transactions on Circuits and Systems*, CAS-27, 6, 569–73, June 1980.

[5.4] P.V. Ananda Mohan, V. Ramachandran and M.N.S. Swamy, High-pass and band-pass second-order switched-capacitor filters, *Proc. IEE*, Part *G*, *Electronic Circuits and Systems*, 130, 1, 1–6, February 1983.

[5.5] J.J. Mulawka and M.S. Ghausi, Second-order function realization with switched-capacitor network and unity gain amplifier, *Proc. IEE*, Part *G*, *Electronic Circuits and Systems*, 127, 4, 187–90, August 1980.

[5.6] P.V. Ananda Mohan, V. Ramachandran and M.N.S. Swamy, Multiple-feedback biquadratic switched-capacitor filters, *Proc. 25th Midwest Symposium on Circuits and Systems, Houghton*, 435–9, 1982.

[5.7] W.J. Jenkins, T.N. Trick and E.I. El-Masry, New realization for switched-capacitor filters, *Proc. 12th Asilomar Conference on Circuits, Systems and Computers, California*, 697–8, 1978.

[5.8] T.R. Viswanathan, S.M. Faruque and J. Vlach, Switched-capacitor biquads based on switched-capacitor transconductance, *Electronics Letters*, 16, 2, 63–4, 17 January 1980.

[5.9] P.V. Ananda Mohan, V. Ramachandran and M.N.S. Swamy, A novel two-amplifier universal active SC filter, *Proc. IEEE*, 73, 8, 1330–1, August 1985.

[5.10] P.V. Ananda Mohan, V. Ramachandran and M.N.S. Swamy, New general biquadratic active RC and switched-capacitor filters, *Proc. European Conference on Circuit Theory and Design, Stuttgart*, 102–4, September 1983.

[5.11] P.V. Ananda Mohan, V. Ramachandran and M.N.S. Swamy, New general biquadratic active RC and switched-capacitor filters, *Proc. IEE*, Part *G*, *Electronic Circuits and Systems*, 131, 2, 51–5, April 1984.

[5.12] Y.K. Co, M. Ismail, G.C. Temes and G. Szentirmai, Correction to 'Switched-capacitor building blocks', *IEEE Transactions on Circuits and Systems*, CAS-30, 4, 253, April 1983.

[5.13] K. Martin and A.S. Sedra, Exact design of switched-capacitor band-pass filters using coupled-biquad structures, *IEEE Transactions on Circuits and Systems*, CAS-27, 6, 469–75, June 1980.

[5.14] P.V. Ananda Mohan, V. Ramachandran and M.N.S. Swamy, Capacitor spread evaluation of stray-insensitive SC biquads, *IEEE Transactions on Circuits and Systems*, CAS-30, 11, 847, November 1983.

[5.15] W.F. Lovering, Analog computer simulation of transfer functions, *Proc. IEEE*, 53, 3, 306–7, March 1965.

[5.16] E.I. El-Masry, Stray-insensitive state-space switched-capacitor filters, *IEEE Transactions on Circuits and Systems*, CAS-30, 7, 474–88, July 1983.

[5.17] P. Gillingham, Stray-insensitive switched-capacitor biquads with reduced number of capacitors, *Electronics Letters*, 17, 4, 171–3, 19 February 1981.

[5.18] G. Fischer and G.S. Moschytz, High-*Q* SC biquads with a minimum capacity spread, *Electronics Letters*, 18, 25, 1087–9, 9 December 1981.

[5.19] P.V. Ananda Mohan, V. Ramachandran and M.N.S. Swamy, Parasitic-compensated single amplifier SC biquad equivalent to Fleischer–Laker SC biquad, *IEEE Transactions on Circuits and Systems*, CAS-33, 4, 458–60, April 1986.

[5.20] D.J. Allstot, R.W. Brodersen and P.R. Gray, An electrically programmable switched-capacitor filter, *IEEE Journal of Solid-State Circuits*, SC-14, 6, 1034–41, December 1979.

[5.21] D.B. Cox, L.T. Lin, R.S. Florek and H.F. Tseng, A real-time programmable switched-capacitor filter, *IEEE Journal of Solid-State Circuits*, SC-15, 6, 972–7, December 1980.

[5.22] D.B. Cox, A digital programmable switched-capacitor universal active filter/oscillator, *IEEE Journal of Solid-State Circuits*, SC-18, 4, 383–9, August 1983.

[5.23] P.V. Ananda Mohan, V. Ramachandran and M.N.S. Swamy, Stray-insensitive high-order switched-capacitor filters, *Proc. ISCAS*, 807–10, May 1983.

[5.24] P.V. Ananda Mohan, V. Ramachandran and M.N.S. Swamy, Novel stray-insensitive SC band-pass filter, *Electronics Letters*, 19, 16, 615–16, 4 August 1983.

[5.25] M. Hasler and M. Saghafi, Stray capacitance eliminating transformations for switched-capacitor circuits, *International Journal of Circuit Theory and Applications*, 11, 3, 321–38, July 1983.

[5.26] J. Mulawka, Synthesis of single-amplifier switched-capacitor networks, *Electronics Letters*, 17, 14, 510–12, 9 July 1981.

[5.27] P.P. Vaidyanathan and S.K. Mitra, Design and analysis of switched-capacitor filters using a single operational amplifier, *Proc. 26th Midwest Symposium on Circuits and Systems*, 435–9, 1983.

[5.28] E.M. Saad, Switched-capacitor filter circuits, design and improvements, *Proc. Summer School on Circuit Theory, Prague*, 416–20, 1982.

## 5.8.1 Further reading

[5.29] C.S. Gargour and V. Ramachandran, Design of some new switched-capacitor biquad filters, *Summer Symposium on Circuit Theory, Prague*, 2254–8, July 1982.

[5.30] C.S. Gargour and V. Ramachandran, Design of stray-insensitive switched-capacitor filter using a bilinear integrator as a generating circuit, *26th Midwest Symposium on Circuits and Systems*, August 1983, 454–7.

[5.31] C.S. Gargour, B. Nowrouzian and V. Ramachandran, Design of biquadratic stray-insensitive switched-capacitor filters using unit delays and finite gain amplifiers as basic building blocks, *27th Midwest Symposium on Circuits and Systems*, June 1984, 284–7.

[5.32] Z.X. Zhou, A simplified design-oriented analysis of switched-capacitor active filters, *International Journal of Circuit Theory and Applications*, 12, 3, 179–89, July 1984.

[5.33] E. Sanchez-Sinencio, R.L. Geiger and J. Silva-Martinez, Tradeoffs between passive sensitivity, output voltage swing and total capacitance in biquadratic SC filters, *IEEE Transactions on Circuits and Systems*, CAS-31, 11, 984–7, November 1984.

[5.34] G.M. Jacobs, All-pass biquadratic switched-capacitor filters, *IEEE Transactions on Circuits and Systems*, CAS-32, 1, 1–12, January 1985.

[5.35] J.C.M. Bermudez and B.B. Bhattacharyya, A systematic procedure for generation and design of stray-insensitive SC biquads, *IEEE Transactions on Circuits and Systems*, CAS-32, 8, 767–83, August 1985.

[5.36] C.S. Gargour, V. Ramachandran and M. Ahmadi, Design of stray-insensitive switched-capacitor biquads using a new first order generating circuit, *28th Midwest Symposium on Circuits and Systems*, August 1985.

[5.37] H. Qiuting and W. Sansen, A low-sensitivity, low-capacitance ratio realization of high-$Q$ biquads, *IEEE Transactions on Circuits and Systems*, CAS-33, 10, 1039–42, October 1986.

[5.38] A.M. Davis and R.R. Smith, Design of state-variable SC filters by means of polynomial transformation, *IEEE Transactions on Circuits and Systems*, CAS-33, 12, 1248–51, December 1986.

[5.39] F. Anday, Synthesis of switched-capacitor active-filters: $z$-domain equivalent admittance approach, *International Journal of Circuit Theory and Applications*, 15, 1, 85–7, January 1987.

[5.40] B.B. Bhattacharyya and T.S. Rathore, An economically digitally programmable switched-capacitor biquad, *IEEE Journal of Solid-State Circuits*, SC-22, 4, 627–9, August 1987.

[5.41] Q. Huang and W. Sansen, Design techniques for improved capacitor area efficiency of switched-capacitor biquads, *IEEE Transactions on Circuits and Systems*, CAS-34, 12, 1590–9, December 1987.

[5.42] P.V. Ananda Mohan, V. Ramachandran and M.N.S. Swamy, Nodal voltage simulation of active RC networks, *IEEE Transactions on Circuits and Systems*, CAS-32, 10, 1085–8, October 1985.

[5.43] P.V. Ananda Mohan, V. Ramachandran and M.N.S. Swamy, New programmable switched-capacitor filter based on Tow–Thomas active RC biquad, *Electronics Letters*, 22, 5, 280–1, 27 February 1986.

[5.44] G.W. Roberts, W.M. Snelgrove and A.S. Sedra, Switched-capacitor realization of $n$th order transfer function using a single multiplexed Opamp, *IEEE Transactions on Circuits and Systems*, CAS-34, 2, 140–8, February 1987.

[5.45] C. Xuexiang, E. Sanchez-Sinencio and R.L. Geiger, Pole–zero pairing strategies for cascaded switched-capacitor filters, *Proc. IEE*, Part *G*, 134, 199–204, August 1987.

[5.46] J.F. Duque-Carrillo, J. Silva-Martinez and E. Sanchez-Sinencio, Programmable switched-capacitor bump equalizer architecture, *IEEE Journal of Solid-State Circuits*, SC-25, 1035–9, August 1990.

[5.47] U. Weder and A. Moescheitzer, Comments on the 'Design techniques for improved capacitor area efficiency in switched-capacitor biquads', *IEEE Transactions on Circuits and Systems*, CAS-37, 5, 666–8, May 1990.

# CHAPTER SIX

# Switched capacitor ladder filters based on impedance simulation

High-order SC filters could be realized as cascade connections of second-order and first-order SC networks. Experience with active RC filter designs has shown that cascade designs are inferior to designs based on *component simulation* or *operational simulation* of RLC ladder filters. The low-sensitivity property of doubly-terminated RLC networks lends itself to realization of low-sensitivity SC networks through component simulation or operational simulation. In the component simulation method, the $L$ and $R$ values are simulated by active or passive SC networks, and the design basically consists of replacing the $L$ and $R$ components by these simulated networks. In the operational simulation method, there is no direct correspondence between components of the RLC filter and SC filter; only the nodal equations as well as the branch voltage–current relationships of the RLC filter are simulated. In other words, the 'internal working' of the RLC filter is modelled. This has come to be known as the 'leap-frog ladder method'. In the component simulation method (which is the subject matter of this chapter), since the *nodes* which exist in the basic RLC structure also exist in the SC network and since these nodes are floating (i.e., separated from the ground), it is inevitable that, in the SC structure, they have parasitic capacitances to ground. Thus, stray-insensitive designs can only exist inside the simulation networks realizing $L$ and $R$, but not at the 'terminals' used to connect these to the ladder network. Hence, it must be emphasized that no stray-insensitive design is possible with the component simulation method. Nevertheless, we will study here the various design techniques and the possible advantages they offer to the designer.

The capacitances, inductances and resistances simulated by switched capacitors should be related to the analog (or $s$-domain) capacitances, inductances and resistances through some transformation, for example, the bilinear or LDI transformation, so that the simulated SC ladder filter will have corresponding transfer functions related to $s$-domain transfer functions through bilinear or LDI transformations. Hence, the inductance and resistance designs corresponding to these transformations will be discussed. It will be shown that simulation of the resistor poses some difficulties, and methods of overcoming these will be studied. Some of the ideas developed in this chapter will be useful for Chapter 7 on the operational simulation method, a design technique which is also widely used.

A closely related approach to SC ladder filter design is based on the use of special devices known as *voltage inverter switches* (VIS). The use of passive SC branches to realize $R$, $L$ and $C$ elements together with VIS elements leads to low-sensitivity SC ladder filters realizing bilinear transformation type transfer functions. These also will be studied in detail in this chapter.

Use of suitable impedance transformation in an RLC network to convert all resistances into capacitances, inductances into resistances and capacitances into frequency-dependent negative resistance (FDNR) elements is well known [1.29]. The realization of FDNR elements and their application to SC filter design are studied in Section 6.7.

## 6.1 Lossy grounded and floating inductor simulation

### 6.1.1 Hosticka–Moschytz lossy inductance [6.1]

The Hosticka–Moschytz grounded inductance circuit is shown in Figure 6.1(a). For an applied voltage $V_{ie}$, the incremental charge flowing in the $n$th cycle ($\Delta Q$) is given by

$$\Delta Q(z) = \left[ C_2 + C_3 + \frac{C_2 C_3 z^{-1}}{C_4 (1 - z^{-1})} \right] V_{ie} \tag{6.1}$$

Equation (6.1) can be rewritten as

$$\Delta Q(z) = \left[ \left( C_2 + C_3 - \frac{C_2 C_3}{C_4} \right) + \frac{C_2 C_3}{C_4 (1 - z^{-1})} \right] V_{ie} \tag{6.2}$$

Thus, the 'charge-domain admittance' ($\Delta Q(z)/V_{ie}$) is

$$\frac{\Delta Q(z)}{V_{ie}} = C_2 + C_3 - \frac{C_2 C_3}{C_4} + \frac{C_2 C_3}{C_4 (1 - z^{-1})} \tag{6.3}$$

It remains to be seen whether equation (6.3) represents a lossy or lossless inductor.

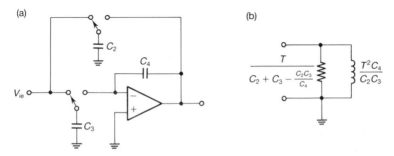

*Fig. 6.1*  (a) Hosticka–Moschytz grounded inductance;  (b) *p*-domain equivalent circuit. *(Adapted from [6.1],* © *1978 IEE.)*

Shunting a capacitor $C_1$ at the input port whose charge-domain admittance is $C_1(1 - z^{-1})$, the total admittance is

$$Y_q(z) = C_1(1 - z^{-1}) + C_2 + C_3 - \frac{C_2 C_3}{C_4} + \frac{C_2 C_3}{C_4(1 - z^{-1})}$$

$$= \frac{C_1 C_4 (1 - z^{-1})^2 + \left(C_2 + C_3 - \dfrac{C_2 C_3}{C_4}\right) C_4 (1 - z^{-1}) + C_2 C_3}{C_4(1 - z^{-1})} \tag{6.4}$$

This must be 'reactive', i.e., zeros of the realized admittance must be on the imaginary axis of the complex-frequency plane, for realizing an oscillator. Rewriting the numerator of equation (6.4) as

$$C_1 C_4 z^{-2} - [2C_1 C_4 - C_2(C_3 - C_4)] z^{-1} + C_3 C_4 + C_4(C_1 + C_2 + C_3)$$

it is observed that an oscillator can be realized only when $C_4(C_2 + C_3) = 0$, in the limiting case. Equation (6.4) can be analyzed in the $p$-domain as well, using the $p$-transformation $(1 - z^{-1}) \rightarrow pT$. Then the numerator of equation (6.4) becomes

$$p^2 T^2 (C_1 C_4) + pT[C_4(C_2 + C_3) - C_2 C_3] + C_2 C_3$$

In the $p$-domain, a lossless resonator is obtainable when $C_4(C_2 + C_3) = C_2 C_3$. However, $p$-domain losslessness means that a maximum $Q_p$ of about δ is realizable for the inductor. By making $C_4(C_2 + C_3) < C_2 C_3$, i.e., the coefficient of the $p$-term negative, as large a $Q_p$ as possible can be achieved, but at the expense of capacitor spread (since $C_4$ becomes too small).

In the $p$-domain, the equivalent circuit is obtained by substituting in equation (6.3) $pT = (1 - z^{-1})$ and noting that equation (6.3) represents a parallel connection of a resistor and an inductor, as shown in Figure 6.1(b). (Note that equation (6.3) represents incremental charge flow and that $\Delta Q$ can be considered as $TI$, where $T$ is the sampling period). For ideal inductor realization, $C_2 + C_3 = 0$, or the shunting resistance in the $p$-domain is given by $R = -L/T$.

It is of interest to realize floating lossy inductors as well. In the active RC case, grounded inductances can be converted to floating inductances, by lifting the terminal connected to ground off the ground and by adding suitable additional circuitry [6.2] (see Exercise 6.2). However, in SC versions, using the unique facilities available with switched capacitors, a floating inductance can be realized in a simple manner.

Consider the circuit of Figure 6.2(a) [6.3]. The input differential voltage, $V_A - V_B$, is sampled by $C_3$ in the even phase and integrated by $C_4$ in the odd phase. The output of OA charges $C_2$ in the odd phase, and in the even phase $C_2$ discharges into the input terminals. Thus, the circuit works 'internally' as a 'grounded' circuit, whereas at the input end, just by the addition of four switches, it can be used to sample a floating voltage. The equivalent circuit can be derived which shows that a floating lossy inductor is realized as given in Figure 6.2(b). The inductance and shunting resistance values are the same as those of the grounded lossy inductor of

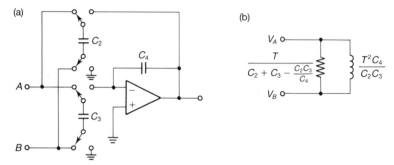

*Fig. 6.2* (a) Brugger–Hosticka floating inductance;  (b) *p*-domain equivalent of (a). *(Adapted from [6.3] © 1979 IEE.)*

Figure 6.1(a). Thus, the 'bilateral' property of a floating inductance is simply achieved with a few additional switches.

## 6.1.2  Lee's lossless inductance realization

An interesting floating lossless inductance circuit [6.4] is shown in Figure 6.3. During the even phase, the input differential voltage, $V_A - V_B$, is sampled by $C_1$ and integrated by $C_2$ in the odd phase. The output of OA $A_3$ in the odd phase charges capacitors $C_3$ which are discharged into the input terminals in the even phase by the buffers $A_1$ and $A_2$.

It can be shown that the charges flowing into terminal $A$ and going out of terminal $B$ are given by

$$\Delta Q_1(z) = \Delta Q_2(z) = \frac{C_1 C_3}{C_2}\left(\frac{z^{-1}}{1-z^{-1}}\right) V_L \qquad (6.5)$$

The 'discrete time impedance' realized by the circuit of Figure 6.3 is an LDI-transformed *s*-domain lossless inductor, as will be shown in the next example.

*Example 6.1*
Derive the discrete time impedances of inductors realizing LDI and bilinear transformations.

For an inductor in the *s*-domain,

$$I(s) = \frac{V(s)}{sL} \qquad (6.6)$$

where $I$ is the current through the inductor and $V$ is the voltage across it. Since in the time-domain $i = dq/dt$, we have:

$$I(s) = sQ(s) \qquad (6.7)$$

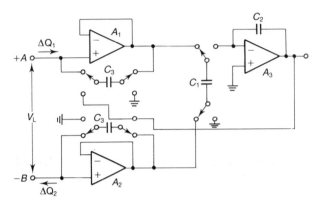

*Fig. 6.3* Lee's lossless floating inductance *(adapted from [6.4], © 1979 IEE).*

Hence, from equations (6.6) and (6.7), we get

$$Q(s) = \frac{I(s)}{s} = \frac{V(s)}{s^2 L} \tag{6.8}$$

Depending on the transformation to be employed, i.e., for the LDI transformation and the bilinear transformation (BT),

$$s \to \frac{z^{1/2} - z^{-1/2}}{T} \quad \text{or} \quad s \to \frac{2}{T} \frac{(1 - z^{-1})}{(1 + z^{-1})}$$

equation (6.8) is transformed as

$$Q(z)_{\text{LDIT}} = \frac{T^2 z^{-1}}{L \cdot (1 - z^{-1})^2} \cdot V_d \tag{6.9a}$$

$$Q(z)_{\text{BT}} = \frac{T^2 (1 + z^{-1})^2}{L \cdot 4(1 - z^{-1})^2} \cdot V_d \tag{6.9b}$$

Denoting the incremental charge flowing during one clock cycle as $\Delta Q(z)$, we have in the $z$-domain,

$$\Delta Q(z) = (1 - z^{-1})Q(z) \tag{6.10}$$

so that

$$\Delta Q(z)_{\text{LDIT}} = \frac{T^2}{L} \cdot \frac{z^{-1}}{1 - z^{-1}} \cdot V_d \tag{6.11a}$$

$$\Delta Q(z)_{\text{BT}} = \frac{T^2}{4L} \cdot \frac{(1 + z^{-1})^2}{(1 - z^{-1})} \cdot V_d \tag{6.11b}$$

Thus equations (6.11) give the charge–voltage relationships for discrete time lossless inductances.

Equation (6.5) can be compared with equation (6.11a) to show that the circuit of Figure 6.3 realizes an LDI-type lossless inductance. The inductance value can be identified in terms of capacitor ratios as

$$L = \frac{C_2 T^2}{C_1 C_3} \tag{6.12}$$

### 6.1.3 Viswanathan, Faruque, Singhal and Vlach realizations

It is interesting to note that a grounded inductance is obtained by grounding the terminal $B$ in Figure 6.3. All the other irrelevant components can be removed, leading to the circuit of Figure 6.4(a). In the circuit of Figure 6.4(a), it is noted that during the odd phase OA $A_1$ is idle, while in the even phase OA $A_2$ is idle. The circuit can now be simplified by the use of multiplexing; the multiplexed version is shown in Figure 6.4(b). This is the circuit independently proposed by Viswanathan *et al.* [6.5]. They derived it by a different approach, considering the concept of a 'transconductance' element or voltage-to-current converter [6.6–6.7]. The block diagram of a simulated inductance using this technique is shown in Figure 6.5(a). The voltage $V_{in}$ is integrated to provide $V_1$ which is converted into a current by means of a voltage-to-current converter and fed back to the input with proper polarity. The second block in Figure 6.5(a) is a transconductance and in the SC form is realized as shown in Figure 6.5(b).

The reader can verify that in Figures 6.3 and 6.4(a), $C_3$ and $A_1$ perform precisely the same function (the load evidently is the external circuit). By replacing the inverting integrator in Figure 6.4(a) with a non-inverting integrator based on SC transconductance, the grounded inductance of Figure 6.4(c) is obtained, after multiplexing [6.6]. This circuit requires fewer switches than that of Figure 6.4(b).

The concept of grounded inductance realization using transconductances and integrators can be extended to realize economical floating inductances as well. Faruque, Vlach and Viswanathan's realization [6.7] for this purpose is shown in Figure 6.4(d). In the even phase, the input voltage is sampled and held on the series connected capacitors $C_{m1}$ and $C_{m1}^*$. The charge on $C_{m1}$ and $C_{m1}^*$ thus obtained is discharged separately by the buffers $A_1$ and $A_2$ in the odd phase into the integrating capacitor $C_3$. The voltage developed in the odd phase on $C_3$ is held on the series connection of $C_{m2}$ and $C_{m2}^*$, which are discharged separately in the even phase. The charge going out of the terminal $B$ ($\Delta Q_1$) or entering the terminal $A$ is

$$\Delta Q_1(z) = \frac{C_{m2} C_{m2}^* C_{m1} C_{m1}^* (V_{Be} - V_{Ae}) z^{-1}}{(C_{m1} + C_{m1}^*)(C_{m2} + C_{m2}^*) C_3 (1 - z^{-1})} \tag{6.13}$$

which is the charge–voltage relationship to be satisfied by an LDI inductor. The interesting point in this circuit is that the series connection of capacitors $C_{m2}$ and $C_{m2}^*$ ensures equal charge to flow through $C_{m2}$ and $C_{m2}^*$, and similarly through $C_{m1}$ and $C_{m1}^*$, so that the charges entering terminal $A$ and leaving terminal $B$ are identical.

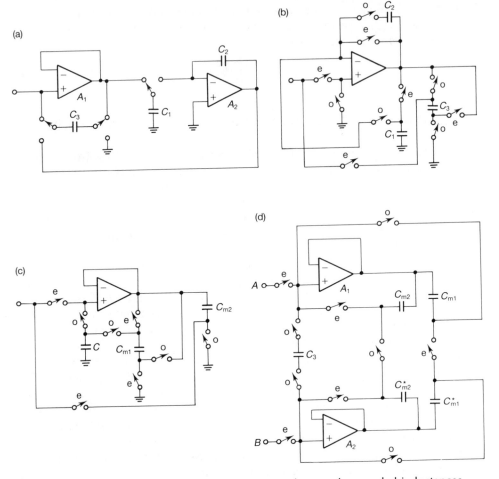

Fig. 6.4 (a) Lee's grounded inductance; (b) Viswanathan et al. grounded inductances (adapted from [6.5], © 1980 IEEE); (c) Viswanathan et al. grounded inductances (adapted from [6.6], © 1979 IEE); (d) Faruque et al. lossless floating inductance (adapted from [6.7], © 1981 IEE).

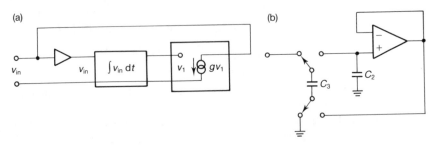

Fig. 6.5 (a) Inductance simulation method using integrator and transconductance; (b) SC transconductance.

## SC gyrators based on 'transconductance' elements

Grounded inductances can be realized by using the concept of gyrators as well [6.5--6.7]. These also use the transconductance described earlier. Conceptually, a gyrator is realizable with two voltage-controlled current sources, as shown in Figure 6.6(a). The SC implementation is as shown in Figure 6.6(b). In the even phase, $C_1$ is charged to $V_1$ and is discharged by the buffer $A_2$ in the odd phase into the output terminal 2. In the odd phase, $C_2$ is charged to $V_2$ which is discharged in the even phase by the buffer $A_1$ into the input terminal. The simplified circuit using the technique of multiplexing is shown in Figure 6.6(c). It can be shown that the charges $\Delta Q_1$ and $\Delta Q_2$ taken in at terminals 1 and 2 are, respectively,

$$\Delta Q_1(z) = C_2 V_2 z^{-1/2}$$
$$\Delta Q_2(z) = -C_1 V_1 z^{-1/2}$$

(6.14)

The reader is urged to verify that terminating the port 2 by a capacitor $C$ yields an inductive impedance at the input terminals of LDI type. Since the circuit does not 'appear' to have continuity of charge flow, it immediately follows that the input and output have to be held over a clock period (see Figure 6.6(c); in each slot the circuit is connected to either input or output only, and no through connection exists).

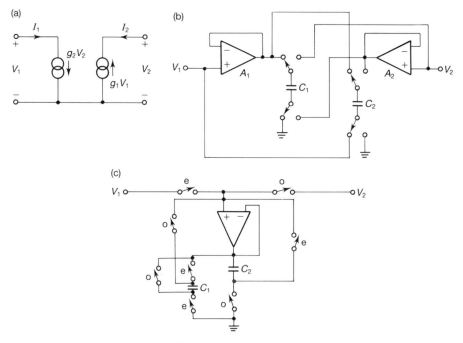

Fig. 6.6   (a) A grounded gyrator;   (b) SC gyrator using two OAs;   (c) SC gyrator using one OA;   (b) and (c) *(both adapted from [6.6], © 1979 IEE).*

### 6.1.4 Lee's SC parallel resonator simulation circuit [6.8]

Next, the floating SC parallel resonator of Figure 6.7 is considered. Assuming that the input in the odd phase is a sampled and held version of the input in the even phase, the incremental charge flow is given by

$$\Delta Q(z) = \left[ C_3(1 - z^{-1}) + \frac{C_1 C_3}{C_2} \frac{z^{-1}}{1 - z^{-1}} \right] V \tag{6.15}$$

Thus, the circuit realizes an LDI inductance shunted by a capacitance, i.e., an LC parallel resonator.

The above discussion has considered the realization of lossless LDI inductances and resonators. These are useful for realizing LDI type ladder SC filters, as will be shown later. It is also possible to synthesize bilinear lossless inductances and resonators.

### 6.1.5 Spahlinger's bilinear lossless inductance

A bilinear lossless inductance using a two-phase clock [6.9] is shown in Figure 6.8. The total incremental charge flowing through this circuit in a clock cycle can be related to the sum of the even and odd-phase input voltages in a clock cycle:

$$\frac{\Delta Q(z)}{V} = \frac{\Delta Q_e(z) + \Delta Q_o(z)}{V_{io} + V_{ie}} = \frac{C_1(1 + z^{-1/2})^2}{(1 - z^{-1/2})} \tag{6.16}$$

By comparing equations (6.16) and (6.11b), a bilinear lossless inductor is realized, the only difference being that the clock frequency is doubled.

The circuit of Figure 6.8 is another example demonstrating the usefulness of the 'inverting' SC resistor. The circuit of Figure 6.8 can be modified to realize a bilinear

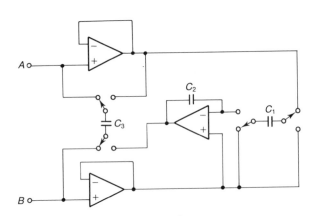

*Fig. 6.7* Lee's LC parallel resonator *(adapted from [6.8],* © 1980 IEE).

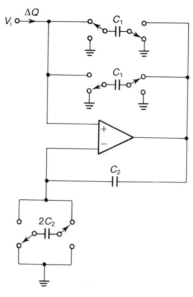

Fig. 6.8  Spahlinger's bilinear lossless grounded inductance (adapted from [6.9],
© AEU 1981).

lossless floating inductance, as shown in Figure 6.9. It can also be shown that this
circuit realizes an inductance of value $T^2/4C$.

## 6.1.6  Temes–Jahanbegloo and Nossek realizations

The circuits discussed thus far use two-phase clocking. Consider the circuit of
Figure 6.10(a) due to Temes and Jahanbegloo [6.10] . This circuit employs a four-
phase clock. It is balanced (or operates using ground) within and floating at the
input terminals, and uses three capacitors.

The capacitors $C_0$ and $C_2$ sample the input voltage in various time slots. These
sampled capacitor voltages are integrated on $C_a$. The voltage on $C_a$ is stored on $C_1$
and used to 'simulate' the charge–voltage relationships required for realizing a
bilinear lossless floating inductance. The charge conservation equations are as
follows. At phase 1:

$$\Delta Q(z) = C_0 V_m + (C_1 + C_2) V_m - C_1 V_o z^{-1/2} \tag{6.17a}$$

At phase 2:

$$C_0 V_m z^{-1/4} + C_2 V_m z^{-3/4} = C_a V_o (1 - z^{-1/2}) \tag{6.17b}$$

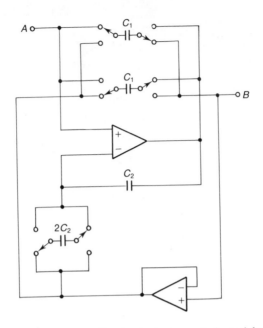

Fig. 6.9  Spahlinger's bilinear lossless floating inductance *(adapted from [6.9],*
© AEU 1981).

At phases 3 and 4, the charge conservation equation is the same as equation (6.17b). Thus, from equations (6.17) we obtain, after eliminating $V_o$,

$$\Delta Q(z) = \left[ (C_0 + C_1 + C_2) + \frac{C_1}{C_a} \frac{(C_0 z^{-1/2} + C_2 z^{-1})}{(1 - z^{-1/2})} \right] V_m \qquad (6.18)$$

It is thus clear that a charge (given by the second term) is made to flow into the input terminals in addition to the present charge $(C_0 + C_1 + C_2) V_m$. Since a bilinear loss-less floating inductance is required, equation (6.18) has to be identified with equation (6.11b) (noting that the sampling frequency is doubled in this case). Thus, we get the following design equations for capacitor ratios:

$$C_0 + C_1 + C_2 = \frac{T^2}{4L}$$

$$\frac{C_0 C_1}{C_a} = \frac{3T^2}{4L} \qquad (6.19a)$$

$$C_0 = 3C_2$$

(a)

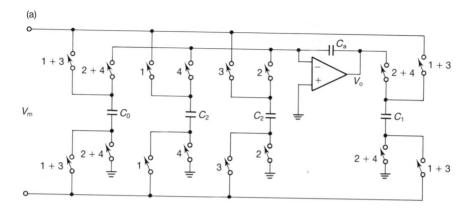

(b)

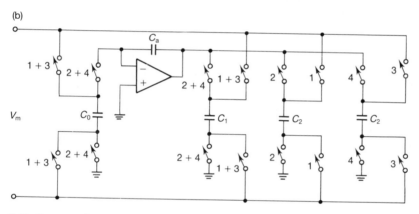

Fig. 6.10  Temes–Jahanbegloo bilinear lossless floating inductances *(adapted from [6.10],*
© 1979 IEE).

It can be shown that the spread is minimal when $C_0 = C_1$ is chosen. Then

$$C_0 = C_1 = \frac{3T^2}{28L}$$

$$C_2 = \frac{C_0}{3}$$

$$C_a = \frac{C_0}{7}$$

(6.19b)

with the resulting spread as 7.

An alternative lossless floating inductor is shown in Figure 6.10(b), which realizes the same inductance value as in the case of Figure 6.10(a). The proof is left to the reader. Note that the circuits of Figure 6.10 can be represented in the form given in Figure 6.11. The capacitors $C_0$, $C_1$ and $C_2$ are charged to different voltages $V_0$, $V_1$ and $V_2$ and then they share their charges to realize the charge–voltage relation required for a bilinear lossless inductance. It can be seen that the Brugger–Hosticka lossy inductance [6.3] of Figure 6.2 is obtained when we do not use the block $T_2$, the capacitor $C_2$ and the associated switches, and use a two-phase clock.

The lossless floating inductances of Figure 6.10 require a minimum capacitance $C_a$ of $3T^2/196L$ which, if the clock rate is high, may be a prohibitively small capacitor value. In view of this limitation, Nossek and Temes have proposed alternative realizations [6.11, 6.12].

For the circuit of Figure 6.12(a) [6.12], it can be shown that

$$\Delta Q(z) = C_0 \left[ 1 + \frac{C_0(C_0 z^{-1} + C_1 z^{-2})}{C_2(C_0 + C_1)(1 - z^{-1})} \right] V_m \qquad (6.20)$$

Equating this with the $\Delta Q(z)-V$ relationship of equation (6.11b), we obtain capacitance values as follows:

$$C_0 = 3C_1 = 4C_2 = T^2/4L_m \qquad (6.21)$$

Evidently, the spread in capacitor values is only 4 and the minimum capacitance is $T^2/16L_m$, as against 7 and $3T^2/196L_m$ for the circuits of Figure 6.10. These circuits also use fewer switches. An alternative realization is shown in Figure 6.12(b). The proof that this circuit realizes a lossless floating inductance is left to the reader.

An alternative realization with reduced capacitor spread (of 2) is shown in Figure 6.13. This circuit is slightly different from that of Figure 6.12(a), and realizes only a grounded lossless inductance.

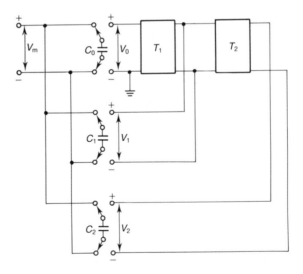

Fig. 6.11 Block diagram of SC lossless floating inductance realization (adapted from [6.10], © 1979 IEE).

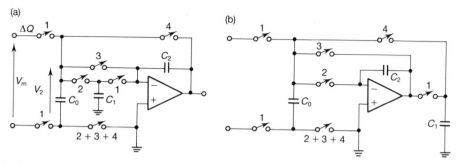

Fig. 6.12 Nossek's bilinear losses floating inductance realizations. *Adapted from [6.12]*, © 1980 IEE.

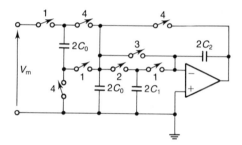

Fig. 6.13 Nossek and Temes grounded bilinear lossless inductance with capacitor spread of 2 *(adapted from [6.11]*, © 1980 IEEE).

## LC resonators

Using the method of Figure 6.11, floating LC resonators can also be realized. The circuit of Figure 6.10(a) requires the following design equations to realize an LC resonator:

$$\frac{C_1C_2}{C_a} = C_0 + C_1 + C_2 = C + \frac{T^2}{4L}$$

(6.22)

$$\frac{C_2}{C_0} = \frac{4LC + T^2}{3T^2 - 4LC}$$

For realizability, the sampling frequency should be such that $3T^2 > 4LC$. It can be shown that, for the circuit of Figure 6.12(a) also, the same condition is required for realizability. A modification is possible as shown in Figure 6.14, if $3T^2 < 4LC$. The design equations in this case can be derived as follows:

$$C_0 = C + \frac{T^2}{4L}$$

$$C_2 = \frac{C_0^2}{2C - \dfrac{T^2}{4L}} \tag{6.23}$$

$$C_1 = \frac{C_0^2}{C - \dfrac{3T^2}{4L}}$$

An alternative method is to choose $C = (3T^2/4L)$ and then, since it is required that $C > 3T^2/4L$, the remaining $C$ [i.e., $(C - (3T^2/4L))$] can be connected externally. A circuit which uses equal capacitors is shown in Figure 6.15. The capacitor $C_0$ has value $T^2/L$.

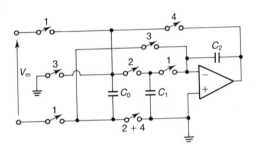

Fig. 6.14  Nossek and Temes lossless resonator ($C > 3T^2/4L$)  (adapted from [6.11], © 1980 IEEE).

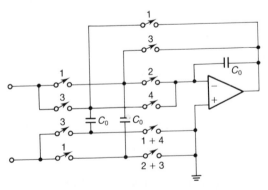

Fig. 6.15  Lossless resonator ($C = 3T^2/4L$)  (adapted from [6.11], © 1980 IEEE).

This completes the discussion on lossless floating inductance and resonator realizations using the scheme of Figure 6.11 and four-phase clocking. In the next section, the simulation of capacitance and resistance elements corresponding to LDI and bilinear transformations will be considered.

## 6.2 Capacitance simulation using SC elements [1.18]

We follow a procedure similar to that used in Example 6.1 to derive the $\Delta Q(z)-V$ relationships needed for a capacitor. In the $s$-domain, the current through a capacitor and the voltage across it are related by

$$I(s) = sCV(s)$$

or the charge in a capacitor is expressed as

$$Q(s) = \frac{I(s)}{s} = CV(s) \tag{6.24}$$

Thus, the incremental charge flowing in the $n$th interval in the $z$-domain is

$$\Delta Q(z) = (1 - z^{-1})Q(z) = C(1 - z^{-1})V_d \tag{6.25}$$

Hence

$$\frac{\Delta Q(z)}{V_d} = C(1 - z^{-1}) \tag{6.26}$$

This relationship is evidently *independent of the transformation used*. It indicates, in physical terms, the 'memory' of a capacitor, i.e., its ability to store the charge. Thus, in simulating ladder networks, the capacitors can be retained as they are, since they inherently satisfy the desired relationship of equation (6.26).

## 6.3 Resistance simulation for SC ladders using the component simulation method [1.18]

Resistance elements are used in the source and load terminations of LCR ladder filters and these 'doubly terminated' LC ladders are attractive in view of their low sensitivity in the pass band. We shall first derive the incremental charge–voltage relationships of these resistance elements using both LDI and bilinear transformations.

*Example 6.2*
Derive the incremental charge–voltage relationships in the $z$-domain for a resistor.

From Ohm's law, for resistance elements, we have

$$I_s = \frac{V(s)}{R} \quad \text{or} \quad Q_s = \frac{V(s)}{sR}$$

Thus, for LDI and bilinear transformations,

$$Q(z)_{\text{LDI}} = \frac{Tz^{-1/2}}{(1 - z^{-1})R} V_{\text{d}}$$

$$Q(z)_{\text{BT}} = \frac{T(1 + z^{-1})}{2R(1 - z^{-1})} V_{\text{d}}$$

Hence, using $\Delta Q(z) = (1 - z^{-1})Q(z)$ using equation (2.44), we obtain the charge–voltage relationships for resistors:

$$\Delta Q(z)_{\text{LDI}} = \left(\frac{Tz^{-1/2}}{R}\right) V_{\text{d}} \tag{6.27a}$$

$$\Delta Q(z)_{\text{BT}} = \frac{T(1 + z^{-1})}{2R} V_{\text{d}} \tag{6.27b}$$

Similarly, for BEI or $p$-transformation, we obtain, since $s \rightarrow (1 - z^{-1})/T$,

$$\Delta Q(z)_{\text{BEI}} = \frac{T}{R} V_{\text{d}} \tag{6.27c}$$

and, for FEI where $s \rightarrow (1 - z^{-1})/Tz^{-1}$,

$$\Delta Q(z)_{\text{FEI}} = \frac{Tz^{-1}}{R} V_{\text{d}} \tag{6.27d}$$

Thus, one should be able to realize equations (6.27a)–(6.27d) as desired, to 'suit' the other simulated components satisfying a given transformation from $s$-domain to $z$-domain. Also, two termination resistances, one for the source and another for the load, will be needed. Recall that some simulated inductances operate in one clock phase only, and that in the other phase the inputs are virtually disconnected from the inductances internally (Figures 6.1–6.4, 6.7, 6.10–6.15), while some simulated inductances operate in both the clock phases (Figures 6.8 and 6.9). For LP RLC filters of all-pole type, as shown in Figure 6.16(a), at each node of the floating inductance, capacitors to ground always exist; hence the termination resistances can make use of the fact that the capacitors retain their previous half-cycle voltages and, as such, the circuits simulating the termination resistance can be simplified. For HP RLC filters, such a facility does not exist (see Figure 6.16(b)). Hence, in general, one can conclude from the foregoing that four types of termination resistances are available. Their suitability for any type of SC ladder simulation needs to be examined.

*Example 6.3*
Examine the use of LDI transformation to simulate an SC resistor.

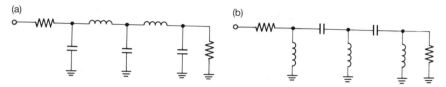

Fig. 6.16  (a) LP all-pole RLC filter;  (b) HP all-pole RLC filter.

Equation (6.27a) gives the incremental charge–voltage relationship for a resistor. Consider, in general, a block box as in Figure 6.17(a) containing the SC circuits to be realized satisfying equation (6.27a) in order to simulate an SC resistor. The voltage across the input terminals is $V_R$. From previous experience in simulation of SC inductances, certain methods of making an incremental charge flow into or out of the input terminals are possible. The voltage $V_R$ can be used to make a charge flow instantaneously into a capacitor, which is discharged in every cycle once, as shown in Figure 6.17(b). Secondly, the voltage $V_R$ can be sampled at the present instant, stored and then used to make a charge flow in the input terminals at the 'next instant', when $V_R$ is 'again' present at the input terminals. This operation requires a full clock-cycle delay. In other words, it takes a full cycle for the charge to go 'around' the termination and come back. Hence, there is no simple half-cycle delay realization in practice. Thus, during the course of this discussion, it has been noted that one can only realize a resistance using charge transfer with either no delay or a clock-cycle delay. It is simple to see that resistances can also be realized using a combination of the above two types. Thus, only the types of termination given by equations (6.27b)–(6.27d) can be achieved. Thus, we have derived the result that *LDI terminations are not realizable* and that, as such, they have to be approximated

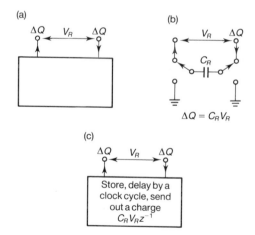

Fig. 6.17  Simulation of termination resistance:  (a) general schematic representation; (b) simulation of resistance $R$;  (c) simulation of resistance $Rz^{-1}$.

by one of the three above-mentioned terminations. Naturally, the resulting filters have some errors. If one desires bilinear transformation to be used to alleviate these problems, some direct and indirect solutions do exist, which will be discussed in due course.

The termination resistance realization for all-pole LP filters will now be discussed (see Figure 6.16(a)). Lee has suggested the circuit of Figure 6.18. This requires a unity-gain amplifier, which in some cases can be made a part of the inductance simulation scheme, thus saving an OA.

Note that $V_1$ may be a voltage source or ground for source and load terminations, respectively. It can be derived that the incremental charge flow is given by

$$\Delta Q_2(z) = (C_1 + C_2)z^{-1/2}V_1 - (C_1 + C_2z^{-1})V_2 \tag{6.28}$$

Hence, for suitable choices of $C_1$ and $C_2$, the following $\Delta Q(z) - V$ relations for the three load terminations shown in Figure 6.19 (i.e., with $V_1 = 0$) are obtained:

(1) $C_1 = 0$          $\Delta Q_2(z) = -C_2 z^{-1}V_2$

(2) $C_2 = 0$          $\Delta Q_2(z) = -C_1 V_2$          (6.29)

(3) $C_1 = C_2 = \dfrac{C_R}{2}$    $\Delta Q_2(z) = \dfrac{-C_R(1 + z^{-1})}{2} V_2$

Note that for case 2, the buffer in Figure 6.18 is not required. Note also that in place of case 2, the termination of Figure 6.17(b) can be used. Regarding source termination, it can be seen from equation (6.28) that, with $C_1 = 0$, $C_2$ present, and $C_2 = 0$, $C_1$ present, we obtain, respectively,

$$\Delta Q_2(z) = C_2 z^{-1/2}V_1 - C_2 z^{-1}V_2 \tag{6.30a}$$

and

$$\Delta Q_2(z) = C_1 z^{-1/2}V_1 - C_1 V_2 \tag{6.30b}$$

Ideally, equations (6.30a) and (6.30b) should be $C_2(V_1 - V_2)z^{-1}$ and $C_1(V_1 - V_2)$, respectively, but adding or subtracting a half-cycle delay to the input does not affect the overall frequency response of the filter. However, it can be observed from equation (6.28) that the circuit of Figure 6.18 cannot realize the source termination of the third type, $[R(1 + z^{-1})]/2$. An alternative termination is shown in

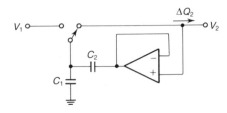

Fig. 6.18   Lee's general termination circuit   (adapted from [6.8], © 1980 IEE).

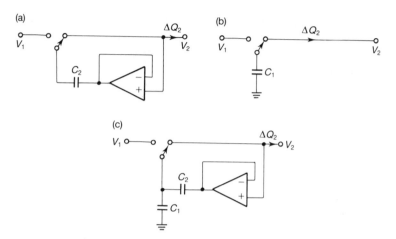

Fig. 6.19 SC terminations to realize resistances of three types: (a) $Rz^{-1}$; (b) $R$; (c) $R(1+z^{-1})/2$.

Figure 6.20, which can be used for the source as well as the load. The charge conservation equation for this circuit is

$$\Delta Q(z) = C_R[(V_{1e} + V_{1o}z^{-1/2}) - V_{2e}(1 + z^{-1})] \qquad (6.31)$$

For load termination $V_1 = 0$, and equation (6.31) realizes a bilinear termination as desired, whereas with $V_{1o} = V_{1e}z^{-1/2}$, a bilinear source termination is realized. Equation (6.31) thus requires the assumption that a capacitor precedes or follows the terminations to ensure that the output is held over a clock period. Hence, these circuits are suitable only for SC LP ladder filters.

For other types of SC ladder filter, when an inductor may be in series with (or parallel to) the termination, Lee's circuit may not be attractive. Hence, an alternative circuit is required, which also should inject or (receive) charge in only one clock phase. This is because LDI-type simulated inductances are 'connected' to the ladder in one phase only. The circuit of Figure 6.21(a) [6.13] precisely performs this function and is a simple modification of the Lee's floating lossless inductance realization of Figure 6.3. The integrator in the inductance realization of Figure 6.3 is

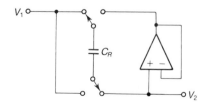

Fig. 6.20 A bilinear source or load termination.

modified as a voltage amplifier. Furthermore, since the terminal $V_1$ is a source or ground (depending on whether source or load termination is required), the buffer $A_3$ and the associated switches and capacitor can be removed. Also OAs $A_1$ and $A_2$ can be multiplexed to give OA realization of the general termination circuit, in which case it is a simple modification of the lossless grounded inductance of Figure 6.4(b) due to Viswanathan *et al.* [6.6]. This multiplexed version of the general termination circuit is shown in Figure 6.21(b).

The incremental charge–voltage relationship of this termination can be obtained as

$$\Delta Q(z) = \left( C_1 + \frac{C_2 C_4}{C_3} z^{-1} \right) (V_1 - V_2) \tag{6.32}$$

Accordingly, with $C_2 = C_3$, by choosing $C_1 = 0$ or $C_4 = 0$ or $C_1 = C_4$, the three required types of termination are realized.

It should be thus clear that the circuits of Figure 6.21 can be used in any general SC ladder filter as source or load terminations. For LP SC ladder filters, of course, the circuit of Figure 6.18 due to Lee is simpler than the circuits of Figure 6.21.

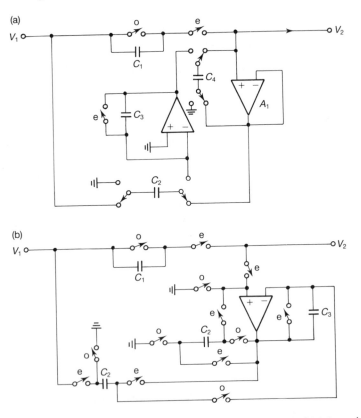

Fig. 6.21   (a) General termination circuit;  (b) multiplexed version of (a) Ananda Mohan, Ramachandran and Swamy *(adapted from [6.13], © 1983 IEE).*

## Other bilinear terminations

For ladder filters using bilinear simulated inductances working at double the clock frequency, the terminations are rather simple and the periodically reverse-switched capacitor studied earlier can be used directly [6.14]. For such a termination, shown in Figure 6.22, the net charge flowing in a clock cycle is given by

$$\Delta Q_e(z) + \Delta Q_o(z) = C_1 [(V_{1e} + V_{1o}) - (V_{2e} + V_{2o})] (1 + z^{-1/2}) \qquad (6.33)$$

and hence, the circuit of Figure 6.22 represents a bilinear termination working at double the clock frequency. Terminal 1 or 2 can be connected to source or ground irrespective of whether a capacitor or inductor precedes or follows the termination.

It is thus seen that any type of RLC ladder filter can be simulated exactly for bilinear inductance realizations, but only approximately for LDI designs. For LDI type SC ladder filters, it has to be examined whether any one type of termination is better than the others among the three non-ideal ones available. Consider the following example, which examines the nature of the three terminations.

*Example 6.4*
Evaluate the three termination resistances with respect to LDI transformation.

The incremental charge–voltage relationships that can be realized are

$$\Delta Q(z) = C_R V$$

$$\text{or} \qquad \Delta Q(z) = C_R V z^{-1}$$

$$\text{or} \qquad \Delta Q(z) = C_R V \frac{(1 + z^{-1})}{2}$$

These correspond to resistances of

$$R z^{1/2}, \ R z^{-1/2} \qquad \text{and} \qquad 2R(z^{1/2} + z^{-1/2})^{-1}$$

respectively. (In order to verify this, follow the procedure of Example 6.2 and use LDI transformation.) The three realized terminations can be rewritten as

$$R\left(\cos \frac{\omega T}{2} + j \sin \frac{\omega T}{2}\right)$$

$$R\left(\cos \frac{\omega T}{2} - j \sin \frac{\omega T}{2}\right)$$

$$R \sec \frac{\omega T}{2}$$

It is evident that the first two are complex terminations, i.e., they have series reactances, while the third termination is real. All are, however, frequency-dependent. When these non-ideal terminations are used, the one which results in the least amount of error in the pass or stop band response for the SC filter is chosen.

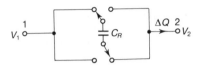

*Fig. 6.22* Bilinear source or load termination using a periodically reverse-switched capacitor.

Computer simulation of fifth-order SC LP ladder filters using the three types of termination has typically shown that the bilinear termination (which realizes real terminations) is the best in the sense that the least error in the magnitude response in the pass band is observed [6.15]. The effect of non-ideal terminations can be examined in another way. Since it is desired that an ideal resistance, $R$, be LDI transformed, which is not physically realizable, only approximate terminations can be used. It is of interest, then, to consider the simulated $L$ and $C$ element values for the chosen termination resistance [1.17]. Consider the following example.

*Example 6.5*
Examine the nature of the $L$ and $C$ elements with respect to given terminations, viz., $Rz^{1/2}$, $Rz^{-1/2}$ and $2R(z^{1/2}+z^{-1/2})^{-1}$, using LDI transformation.

Since the ideal resistance, $R$, is scaled to, say, $Rz^{1/2}$, the original $L$ and $C$ values also have to be scaled to $Lz^{1/2}$ and $Cz^{-1/2}$, respectively. Hence, the $L$ and $C$ values become complex. Using

$$\omega \rightarrow \frac{2}{T}\sin\frac{\omega T}{2}$$

the inductance and capacitance are scaled such that their new reactances are as given below for the three terminations:

*Termination $Rz^{1/2}$*

$$\text{Inductive reactance} = L\frac{2}{T}\left[j\sin\frac{\omega T}{2}\cos\frac{\omega T}{2}-\sin^2\frac{\omega T}{2}\right]$$

$$\text{Capacitive reactance} = \left[C\frac{2}{T}\left(j\cos\frac{\omega T}{2}\sin\frac{\omega T}{2}+\sin^2\frac{\omega T}{2}\right)\right]^{-1}$$

(6.34)

*Termination $Rz^{-1/2}$*

$$\text{Inductive reactance} = L\frac{2}{T}\left[j\sin\frac{\omega T}{2}\cos\frac{\omega T}{2}+\sin^2\frac{\omega T}{2}\right]$$

$$\text{Capacitive reactance} = \left[C\frac{2}{T}\left(j\cos\frac{\omega T}{2}\sin\frac{\omega T}{2}-\sin^2\frac{\omega T}{2}\right)\right]^{-1}$$

(6.35)

*Termination* $2R(z^{1/2} + z^{-1/2})^{-1}$

$$\text{Inductive reactance} = jL\,\frac{2}{T}\,\sin\frac{\omega T}{2}\,\cos\frac{\omega T}{2}$$

$$\text{Capacitive reactance} = \left(jC\,\frac{2}{T}\,\tan\frac{\omega T}{2}\right)^{-1}$$

(6.36)

Equations (6.34)–(6.36) represent the scaled reactances to suit the chosen termination.

The effect of different terminations is seen to modify the inductance and capacitance values. This effect can be seen to be twofold. One is the element value variation, and the other a dissipative effect. Since doubly terminated RLC filters are least sensitive to element value variation, the first effect does not influence the performance of the filters to a great extent. The dissipation introduced for the elements can be interpreted in terms of the $Q$-factor (i.e., reactive impedance/resistive impedance) for the three terminations, as given below:

| | $R \to Rz^{1/2}$ | $R \to Rz^{-1/2}$ | $R \to 2R(z^{1/2} + z^{-1/2})^{-1}$ |
|---|---|---|---|
| $Q_L$ | $-\cot\dfrac{\omega T}{2}$ | $\cot\dfrac{\omega T}{2}$ | $\infty$ |
| $Q_C$ | $\cot\dfrac{\omega T}{2}$ | $-\cot\dfrac{\omega T}{2}$ | $\infty$ |

(6.37)

Thus the $Q$ factors for the third transformation type resistor are infinite. This means that for real terminations the LC elements are lossless. In a practical situation, one has to evaluate the frequency response and then choose the proper termination. This concludes the discussion on the resistance simulation for SC ladder filters.

In the above analysis, it is seen that LDI-type SC ladder filters cannot be realized exactly. However, LDI designs can be 'converted' into exact bilinear designs by a simple method due to Lee and Chang [6.15] which will be considered next.

## The Lee–Chang method for realizing bilinear SC ladder filters using LDI-type simulated elements

First the incremental charge–voltage relationship of an inductance using the LDI and bilinear transformations will be examined. The inductance in the LDI domain has a $\Delta Q(z)-V$ relationship (see equation (6.11)) as follows:

$$\frac{\Delta Q(z)}{V} = \frac{T^2}{L}\,\frac{z^{-1}}{1-z^{-1}} = \frac{T^2}{4L}\,\frac{(1+z^{-1})^2}{(1-z^{-1})} - \frac{T^2}{4L}(1-z^{-1})$$

(6.38)

The first term on the right is the $\Delta Q(z)-V$ relationship for a bilinear inductance. Hence, an inductor of LDI transformation type is equivalent to a parallel combination of the same inductance realizing the bilinear transformation and a negative capacitor. Lee and Chang have, therefore, suggested that an exact bilinear SC ladder filter design can proceed with an LDI design using bilinear terminations, where each inductance is then shunted by a capacitance of $T^2/4L$ to cancel the negative capacitance. We will next consider a few examples to illustrate the design of SC ladder filters using the component simulation technique.

## 6.4  Design examples

It must be clear that by interconnection of the simulated components in exactly the same manner as for the RLC prototype, LDI or bilinear SC ladder filters are obtained. These are illustrated in the design examples that follow.

*Example 6.6*
Design an LDI inductor to resonate with a capacitor of 1 pF at 1 kHz with a sampling frequency of 8 kHz.

The prewarped resonant frequency is obtained from $\omega_c \rightarrow (2/T) \sin(\omega T/2)$ as 6122.9349 rad s$^{-1}$. Thus the inductance value required to resonate with 1 pF at this frequency is 26 673.543 H. Choosing the configuration of Figure 6.4(a), the inductance realized is

$$L = \frac{C_2 T^2}{C_1 C_3}$$

whence

$$\frac{C_2}{C_1 C_3} = \frac{L}{T^2} = \frac{1}{0.585\ 786\ 45 \times 10^{-12}}$$

In order to minimize the capacitor spread, let $C_1 = C_3$. Then, by choosing $C_2 = 1$ pF, $C_1 = C_3 = 0.765\ 366\ 8$ pF. Thus the required capacitor spread is only 1.3.

*Example 6.7*
Design a doubly terminated LP RLC filter for a Butterworth-type response with a cutoff frequency of 1000 Hz and sampling frequency of 8000 Hz. Use LDI transformation.

First the cutoff frequency is prewarped to 6122.9349 rad s$^{-1}$, using the formula $\omega_c \rightarrow (2/T) \sin(\omega T/2)$. From filter design tables, the normalized values of components for a second-order Butterworth filter are $R_1 = R_2 = 1$, $L = \sqrt{2}$ H, $C = \sqrt{2}$ F. Denormalizing to obtain the required cutoff frequency, the $L$ and $C$ values are $L = \sqrt{2}/6122.9349$ H, $C = \sqrt{2}/6122.9349$ F with $R_1 = R_2 = 1\ \Omega$. Using LDI transformation, since resistors become $R_1 = T/C_R$, where $T$ is the sampling period, $R_1$ has to be scaled to realize practical values of capacitors. Without scaling, $C_R = 1/8000$ F. To make $C_R$ 1 pF, we need to scale $R_1$, $R_2$ by a factor of $10^{12}/8000$.

Thus, scaling $L$ by the same factor and scaling $C$ by the reciprocal of the factor, $8000/10^{12}$, we obtain

$$L = 28\ 871.235\ \text{H}$$

$$C = 1.847\ 759\ 1\ \text{pF}$$

The LDI lossless inductance of Figure 6.3, with $C_1 = C_3$, shall be used for minimal spread. Since

$$L = \frac{C_2 T^2}{C_1 C_3} = \frac{C_2 T^2}{C_1^2}$$

and choosing $C_2 = 1$ pF, we have

$$C_1 = C_3 = 0.735\ 660\ 3\ \text{pF}$$

The final design values are:

$$C_1 = C_3 = 0.735\ 660\ 3\ \text{pF}$$

$$C_2 = C_R = 1\ \text{pF}$$

$$C = 1.847\ 759\ 1\ \text{pF}$$

The next step is to choose a 'proper' termination, depending on the error that can be tolerated.

*Example 6.8*
Derive the transfer functions of LDI filters of second-order LP type with the three terminations. (Consider a doubly terminated filter with equal termination resistances.)

The doubly terminated filter is shown in Figure 6.23(a). It is known that each element has an incremental charge–voltage relationship, which can be interpreted as each element having a 'capacitance' of $\Delta Q(z)/V$. Therefore, the network of Figure 6.23(b) contains exclusively capacitances. The output can be analyzed to be

$$\frac{V_o}{V_i} = \frac{C_R \dfrac{T^2}{L} z^{-1}}{\left(C_R C - \dfrac{T^2 C}{L}\right)(1 - z^{-1})^2 + \left(C_R^2 + \dfrac{T^2 C}{L} - \dfrac{2 C_R T^2}{L}\right)(1 - z^{-1}) + \dfrac{2 C_R T^2}{L}} \qquad (6.39a)$$

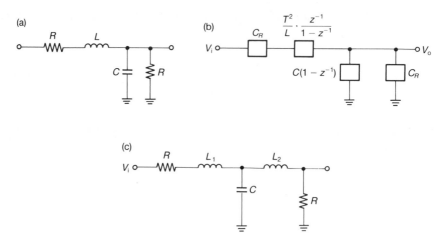

Fig. 6.23   (a) Second-order doubly terminated LP LC filter;   (b) z-domain equivalent;
(c) third-order doubly terminated LP LC filter.

Thus equation (6.39a) gives the transfer function for the resistor termination of the
type $\Delta Q(z) = C_R V$. For $\Delta Q(z) = C_R V z^{-1}$ and $\Delta Q(z) = C_R V(1 + z^{-1})/2$, we obtain
the transfer functions by substituting for $C_R$, $C_R z^{-1}$ and $C_R(1 + z^{-1})/2$ in equation
(6.39a). Thus we obtain

$$\frac{V_o}{V_i} = \frac{C_R \dfrac{T^2}{L} z^{-1}}{C_R(C - C_R)(1 - z^{-1})^2 + \left(\dfrac{T^2 C}{L} - \dfrac{2C_R T^2}{L} + C_R^2\right)(1 - z^{-1}) + \dfrac{2C_R T^2}{L}} \qquad (6.39b)$$

$$\frac{V_o}{V_i} = \frac{C_R \dfrac{T^2}{L} \dfrac{z^{-1}(1 + z^{-1})}{2}}{D} \qquad (6.39c)$$

where

$$D = \frac{C_R}{2}\left[\frac{C_R}{2} - C\right](1 - z^{-1})^3 + \left[CC_R - \frac{T^2 C}{L} - \frac{C_R^2}{2} + \frac{C_R T^2}{L}\right](1 - z^{-1})^2$$

$$+ \left[C_R^2 + \frac{T^2 C}{L} - \frac{3C_R T^2}{L}\right](1 - z^{-1}) + \frac{2C_R T^2}{L}$$

respectively. Note that the bilinear transformation type termination makes the
second-order transfer function a third-order one. The reader is urged, however, to
plot the frequency responses of equations (6.39) and compare them with the desired
analog response, whose LDI version is to be implemented in SC form. Note also
that, in general, designs using LDI inductances in series with the load and source
terminations of bilinear transformation type lead to z-domain transfer functions of

order 2 higher than the actual order of the prototype, e.g., for a third-order LP prototype as in Figure 6.23(c). Hence, in such cases the use of the dual of this network with grounded capacitors facing the load and source terminations is recommended. Note also that, otherwise, the resulting network (i.e., based on Figure 6.23(c)) is unstable and hence not usable [6.14].

*Example 6.9*
Design a fifth-order elliptic SC LP filter with prewarped specifications as follows:

- pass-band edge: 3.235 kHz,
- pass-band maximum reflection coefficient: 12%
- loss pole frequencies: 4.502 kHz and 6.567 kHz

Use a sampling frequency of 128 kHz.

Using the prewarped specifications given above, from filter design tables, the RLC filter of Figure 6.24 can be obtained, with the following values:

$$R_s = R_L = 1 \ \Omega, \qquad L_2 = 0.579 \ 717 \ 4, \qquad L_4 = 0.391 \ 319 \ 9, \qquad C_1 = 1.770 \ 911 \ 9,$$

$$C_2 = 0.418 \ 512 \ 8, \qquad C_3 = 2.904 \ 092 \ 2, \qquad C_4 = 1.320 \ 025 \ 4, \qquad C_5 = 1.207 \ 370 \ 6$$

Denormalizing for a cutoff frequency of 3.235 kHz, and choosing $R_s$ and $R_L$ such that $C_R = T/R_s = 0.5$ pF, we obtain $R_s = R_L = 6.25 \times 10^7 \ \Omega$ and

$$L_2 = 1782.55 \ \text{H}, \qquad L_4 = 1203.2552 \ \text{H}, \qquad C_1 = 1.394 \ \text{pF}, \qquad C_2 = 0.329 \ 438 \ 7 \ \text{pF},$$

$$C_3 = 2.286 \ \text{pF}, \qquad C_4 = 1.039 \ 077 \ 8 \ \text{pF}, \qquad C_5 = 0.9504 \ \text{pF}$$

The next step is to design simulated inductances to realize $L_2$ and $L_4$, using the formula $L_i = T^2 C_{2i}/C_{1i}C_{3i}$. Choosing $C_{22} = 1$ pF and $C_{1i} = C_{3i}$, we obtain $C_{12} = C_{32} = 0.740 \ 165 \ 7$ pF. Similarly, for the inductance $L_4$, we get $C_{14} = C_{34} = 0.900 \ 888 \ 7$ pF and $C_{24} = 1$ pF. Thus, all the capacitor values have been determined. The terminations can be of bilinear type for good performance.

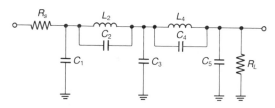

*Fig. 6.24* Fifth-order LP elliptic filter prototype.

Note that by the simple relationship suggested in equation (6.38), a bilinear design can be obtained. The new capacitor values for $C_2$ and $C_4$ in this case are, respectively,

$$C_2' = C_2 + \frac{T^2}{4L_2}$$

$$C_4' = C_4 + \frac{T^2}{4L_4}$$

which are 0.4664 pF and 1.242 pF, respectively. Thus, a bilinear design has been achieved. The reader may note that, due to the frequency warping effect of LDI transformation, the cutoff frequency is actually 3.292 kHz.

In the above sections, the simulation of each inductance in a ladder filter using switched capacitors and OAs has been considered. In some cases one OA will suffice, and in some cases two-phase clocking can realize simulated inductances. An alternative approach is considered next.

## 6.5  SC ladder filter design using switched capacitor immittance converters [6.15–6.19]

In this method, a network of inductances is simulated from a topologically similar network of capacitances using *switched capacitor immittance converters* (SCICs). The number of SCICs used is equal to the number of nodes in the $L$ network. Once the $L$ network is thus simulated at the inputs of SCICs, the SC ladder filter is obtained by connecting external capacitors and termination resistances.

Before discussing the use of immittance converters for realizing $n$-port inductive networks, the use of immittance converters for realizing just one inductance is studied first. The representation of such a scheme is shown in Figure 6.25, where output port 2 is terminated by a capacitor $C_R$ and a grounded inductive input impedance is realized.

### 6.5.1  SCIC based on two-integrator loop

Consider the SCIC of Figure 6.26(a). This is derived by Inoue and Ueno [6.15] from

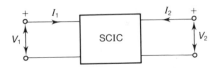

*Fig. 6.25*  Schematic representation of an SCIC.

the leap-frog active RC ladder filter of Figure 6.26(b), for which Bruton [6.20] has shown that the input impedance is an inductance. In the circuit of Figure 6.26(a), the charge flows out of the SCIC terminals during the even phase only. The reader can verify that the following relationship can be derived from the circuit of Figure 6.26(a):

$$
\begin{bmatrix} V_{1e} \\ \Delta Q_1 \end{bmatrix} = \begin{bmatrix} 1 & 0 \\ 0 & \dfrac{-z^{-1}C_F}{C_I(1-z^{-1})} \end{bmatrix} \begin{bmatrix} V_{2e} \\ \Delta Q_2 \end{bmatrix} \tag{6.40}
$$

[Note that the sign convention, well known, is not followed in writing equation (6.40).] Thus, the transmission matrix of the network of Figure 6.26(a) is the same as that of a general immittance converter (GIC) [1.29]. However, one difference

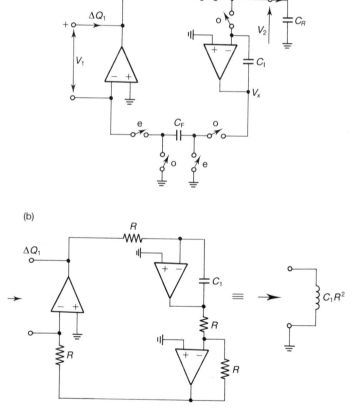

Fig. 6.26   (a) Inoue and Ueno's SCIC (adapted from [6.15], © 1980 IEE);   (b) active RC leap-frog structure used to derive (a).

between the conventional GIC and the circuit of Figure 6.26(a) is that the output terminals in the present case are loaded by capacitors. From equation (6.40), the input impedance with a load capacitance of $C_R$ is found to be

$$\frac{\Delta Q_1(z)}{V_1} = \frac{C_F C_R z^{-1}}{C_I(1 - z^{-1})} \tag{6.41}$$

thus simulating the charge–voltage relationship of a grounded LDI inductor.

## 6.5.2  A bilinear SCIC

Inoue and Ueno have proposed another scheme, shown in Figure 6.27, which simulates a bilinear grounded lossless inductor. A series-switched capacitor $C_L$ is added to realize the desired incremental charge–voltage relationships for bilinear lossless inductances. This circuit also has charges flowing in and out during the even phase only. The transmission matrix for the network excluding the switched capacitor at terminal $c$ is found to be

$$
\begin{bmatrix} V_{1e} \\ \Delta Q_1(z) \end{bmatrix} =
\begin{bmatrix} -\left(\dfrac{1 - z^{-1}}{3 + z^{-1}}\right) \dfrac{C_I}{C_R} & 0 \\ 0 & -\dfrac{C_F z^{-1}}{C_h} \end{bmatrix}
\begin{bmatrix} V_{2e} \\ \Delta Q_2(z) \end{bmatrix}
\tag{6.42}
$$

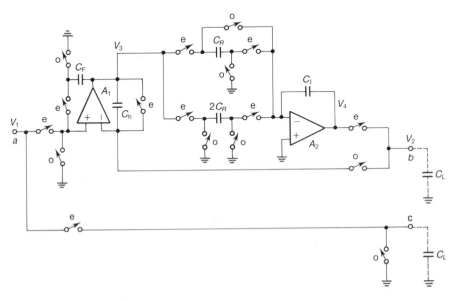

Fig. 6.27  Bilinear SCIC due to Inoue and Ueno  (adapted from [6.15], © 1980 IEE).

We obtain the total charge flowing into terminal $a$ (including the path between $a$ and $c$), as

$$\Delta Q_e(z) = \left\{ \frac{C_R C_L C_F}{C_h C_I} \frac{(3z^{-1} + z^{-2})}{(1 - z^{-1})} + C_L \right\} V_{1e} \tag{6.43}$$

Thus, when $C_R C_F = C_h C_I$, a bilinear lossless inductance is realizable at the input terminal $a$, when ports $b$ and $c$ are loaded by equal capacitors $C_L$.

## 6.5.3   LDI-type SCICs using one OA

We next consider the SCIC using a single OA, shown in Figure 6.28(a). Note that

$$V_{1e} = V_{2e} \tag{6.44a}$$

$$\Delta Q_1(z) = \frac{\Delta Q_2(z) C_f z^{-1}}{C_I(1 - z^{-1})} \tag{6.44b}$$

Thus the transmission matrix can be obtained as

$$\begin{bmatrix} V_{1e} \\ \Delta Q_1(z) \end{bmatrix} = \begin{bmatrix} 1 & 0 \\ 0 & \dfrac{C_f z^{-1}}{C_I(1 - z^{-1})} \end{bmatrix} \begin{bmatrix} V_{2e} \\ \Delta Q_2(z) \end{bmatrix} \tag{6.45}$$

Hence, when the output terminals are loaded by a capacitor $C_L$, an LDI inductance is obtained at the input terminals. The circuit of Figure 6.28(a) uses a floating OA. Another SCIC which requires a grounded OA is shown in Figure 6.28(b). The circuit is described by the same relationships as given in equation (6.45). Note that terminal $1'$ is at virtual ground and terminal 1 is at a low impedance (output of an OA).

## 6.5.4   A general definition of the SCIC and its properties

It is seen in the above that terminating an SCIC by a capacitor (or two capacitors in case of Figure 6.27) realizes a grounded inductance at the input terminals. Also, the SCIC is represented by its transmission matrix. A general definition of the SCIC given by Inoue and Ueno will now be presented. They use the hybrid matrix to characterize an SC two-port as in Figure 6.25, which uses the port currents and port voltages in both the phases. Thus, $I_{1e}, I_{1o}, V_{1e}, V_{1o}$ characterize the input port

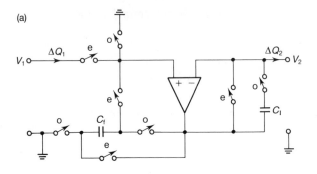

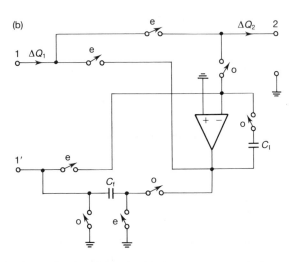

*Fig. 6.28* (a) SCIC using a floating OA; (b) SCIC using a grounded OA *(adapted from [6.15], © 1980 IEE)*.

whereas $I_{2e}, I_{2o}, V_{2e}, V_{2o}$ describe the output port. Thus, the hybrid matrix can be written as

$$\begin{bmatrix} I_1 \\ V_2 \end{bmatrix} = \begin{bmatrix} G_{11} & G_{12} \\ G_{21} & G_{22} \end{bmatrix} \begin{bmatrix} V_1 \\ I_2 \end{bmatrix} \tag{6.46}$$

where

$$I_i = \begin{bmatrix} I_{ie} \\ I_{io} \end{bmatrix} \quad \text{and} \quad V_i = \begin{bmatrix} V_{ie} \\ V_{io} \end{bmatrix}, \quad i = 1, 2$$

Accordingly, $G_{11}, G_{12}, G_{21}$ and $G_{22}$ are $2 \times 2$ matrices.

An even SCIC is defined as having

$$G_{11} = G_{22} = 0 \tag{6.47}$$

$$G_{12} = \begin{bmatrix} G_{12}^{ee} & G_{12}^{eo} \\ 0 & 0 \end{bmatrix}$$

$$G_{21} = \begin{bmatrix} G_{21}^{ee} & 0 \\ G_{21}^{oe} & 0 \end{bmatrix}$$

Thus, for an even SCIC, equation (6.46) becomes

$$\begin{bmatrix} I_{1e} \\ I_{1o} \\ V_{2e} \\ V_{2o} \end{bmatrix} = \begin{bmatrix} 0 & 0 & G_{12}^{ee} & G_{12}^{eo} \\ 0 & 0 & 0 & 0 \\ G_{21}^{ee} & 0 & 0 & 0 \\ G_{21}^{oe} & 0 & 0 & 0 \end{bmatrix} \begin{bmatrix} V_{1e} \\ V_{1o} \\ I_{2e} \\ I_{2o} \end{bmatrix} \tag{6.48}$$

From equation (6.48), the input current is zero in the odd phase. If such an SCIC is terminated at the output port by a load capacitor $C_L$, then we can write for the load capacitor

$$-I_2 = Y'V_2 \tag{6.49}$$

where $I_2$ and $V_2$ are as before and $Y'$ is a $2 \times 2$ matrix given by

$$Y' = C_L \begin{bmatrix} 1 & -z^{-1/2} \\ -z^{-1/2} & 1 \end{bmatrix} \tag{6.50}$$

which indicates the 'memory' of the capacitor.

It will now be shown that by using $n$ SCICs terminated by an $n$-port capacitive network, we obtain at the input an identical $n$-port LDI inductive network and discuss its limitations and applications in filter design. Refer to the circuit of Figure 6.29, where $n$ SCICs are terminated by an $(n + 1)$-terminal capacitor network of which one terminal is ground. In a manner similar to the previous case, we can write the hybrid matrix for the $n$ SCICs as follows:

$$\begin{bmatrix} \mathbf{I} \\ \mathbf{V}' \end{bmatrix} = \begin{bmatrix} \mathbf{0}_n & \mathbf{0}_n & G_{12}^{ee}\mathbf{1}_n & G_{12}^{eo}\mathbf{1}_n \\ \mathbf{0}_n & \mathbf{0}_n & \mathbf{0}_n & \mathbf{0}_n \\ G_{21}^{ee}\mathbf{1}_n & \mathbf{0}_n & \mathbf{0}_n & \mathbf{0}_n \\ G_{21}^{oe}\mathbf{1}_n & \mathbf{0}_n & \mathbf{0}_n & \mathbf{0}_n \end{bmatrix} \begin{bmatrix} \mathbf{V} \\ \mathbf{I}' \end{bmatrix} \tag{6.51}$$

Note that

$$\mathbf{I} = \begin{bmatrix} \mathbf{I}_e \\ \mathbf{I}_o \end{bmatrix}, \quad \mathbf{I}' = \begin{bmatrix} \mathbf{I}'_e \\ \mathbf{I}'_o \end{bmatrix}, \quad \mathbf{V} = \begin{bmatrix} \mathbf{V}_e \\ \mathbf{V}_o \end{bmatrix}, \quad \mathbf{V}' = \begin{bmatrix} \mathbf{V}'_e \\ \mathbf{V}'_o \end{bmatrix}$$

where, for example, $\mathbf{I}_e$ and $\mathbf{I}_o$ are column vectors of length $n$ corresponding to the $n$ SCICs. The matrices $\mathbf{1}_n$ and $\mathbf{0}_n$ are $n \times n$ unit matrices and $n \times n$ null matrices

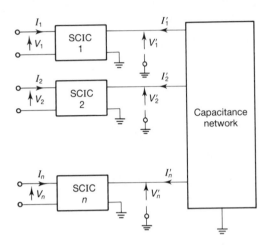

*Fig. 6.29* LDI inductance network simulation scheme using SCICs *(adapted from [6.15],*
© 1980 IEE).

respectively. The reader can verify the validity of equation (6.51). For the $(n + 1)$
terminal capacitance network used as the load termination of the $n$ SCICs, we
obtain

$$-\mathbf{I}' = \begin{bmatrix} [C_L] & -z^{-1/2}[C_L] \\ -z^{-1/2}[C_L] & [C_L] \end{bmatrix} \mathbf{V}' \tag{6.52}$$

Defining the input admittance of the complete network of Figure 6.29 as $[Y]$ as in

$$[I] = [Y][V] \tag{6.53}$$

we get, from equations (6.51)–(6.53), after simplification

$$Y = - \begin{bmatrix} G_{12}^{ee}\mathbf{1}_n & G_{12}^{eo}\mathbf{1}_n \\ \mathbf{0}_n & \mathbf{0}_n \end{bmatrix} \begin{bmatrix} [C_L] & -z^{-1/2}[C_L] \\ -z^{-1/2}[C_L] & [C_L] \end{bmatrix} \begin{bmatrix} G_{21}^{ee}\mathbf{1}_n & \mathbf{0}_n \\ G_{21}^{oe}\mathbf{1}_n & \mathbf{0}_n \end{bmatrix}$$

$$= \begin{bmatrix} F[C_L] & \mathbf{0}_n \\ \mathbf{0}_n & \mathbf{0}_n \end{bmatrix} \tag{6.54}$$

with $F$, the conversion function, given by

$$F = -\{G_{12}^{ee}G_{21}^{ee} + G_{12}^{eo}G_{21}^{oe} - (G_{12}^{ee}G_{21}^{oe} + G_{12}^{eo}G_{21}^{ee})z^{-1/2}\} \tag{6.55a}$$

or

$$F = \frac{C_f}{C_I} \frac{z^{-1}}{1 - z^{-1}} \tag{6.55b}$$

Then, equation (6.54) yields

$$[Y] = \frac{C_f}{C_I} \frac{z^{-1}}{1 - z^{-1}} [C_L] \tag{6.56}$$

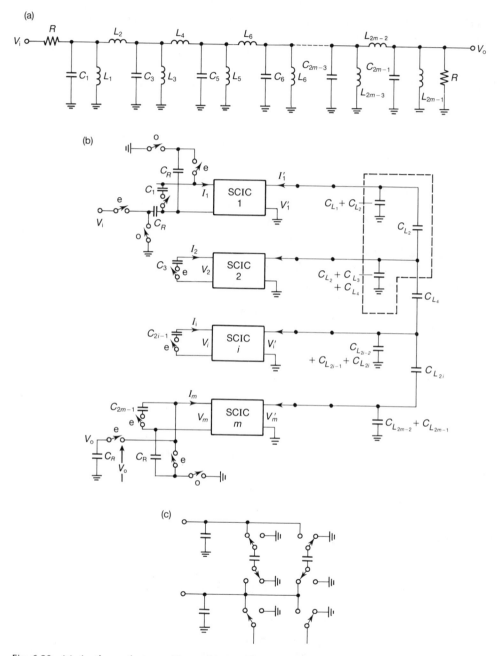

Fig. 6.30 (a) Analog reference filter; (b) SC filter obtained from (a) using SCICs; (c) technique of eliminating bottom-plate parasitic capacitance of floating capacitors in (b).

Thus, the capacitance matrix is converted into an LDI-type inductance matrix by the arrangement of Figure 6.29.

## 6.5.5  Application of SCICs to simulate SC ladder filters

Some observations are now made on the use of Figure 6.29 in the derivation of an SC ladder filter from a given analog reference filter (ARF). Referring to Figure 6.28(b), the following features are seen:

1. In the reference filters, there should be no floating capacitors. (This is because the top left-hand terminals of all SCICs are outputs of OAs and hence capacitors should not exist between outputs of OAs.)
2. Every floating terminal of the ARF must have a capacitance to ground. (This is because a feedback capacitor is required for each OA to stabilize the OA.)
3. Every floating node of the ARF must have an inductance to ground so as to absorb the parasitic capacitances present in the simulated SC network.

These restrictions limit the application of the technique of Figure 6.29 to only certain types of SC filter, especially inductive coupled resonator filters (Figure 6.30(a)).

The ARF and the corresponding SC filter obtained using SCICs are shown in Figures 6.30(a) and 6.30(b). Evidently, this scheme requires the use of floating capacitors to simulate floating inductors present in the prototype RLC circuit. Consequently, the top- and bottom-plate parasitic capacitances affect the performance of the circuit. A simple modification to make the circuit realization insensitive to bottom-plate parasitic capacitances [6.21] is shown in Figure 6.30(c). Note that the grounded capacitor values need to be increased, whereas the remaining coupling capacitors are stray-insensitive.

In the above sections, the design of high-order SC filters by the component simulation method based on LDI or bilinear transformations has been discussed. Another method of designing high-order SC filters, which can be considered to belong to the above class of filters, will be studied next.

## 6.6  SC filter design using voltage inverter switches

This method uses SC simulation of RLC components together with a device called a *voltage inverter switch* (VIS) [6.22, 6.23]. These SC filters realize a bilinear transformation so that there is no restriction on the sampling frequency being used, other than that required by the sampling theorem. Before introducing the details of the VIS, the equivalent circuits of certain basic SC structures in the $z$-domain will be considered [6.24].

## 6.6.1 Equivalent circuits of some passive SC elements

Consider a floating capacitor $C$ (see Figure 6.31(a)). The incremental charge flowing through $C$ is given by

$$\Delta q(nT) = C(v(nT) - v(n-1)T) \tag{6.57}$$

$$\Delta Q(z) = C(1 - z^{-1})V \tag{6.58}$$

The incremental charge–voltage expression in equation (6.58) can be converted into a current–voltage relationship to obtain the impedance of the capacitor:

$$I(z) = \frac{\Delta Q(z)}{T} = \frac{C}{T}(1 - z^{-1})V \tag{6.59}$$

or

$$\frac{V}{I(z)} = \frac{T}{C(1 - z^{-1})} = \frac{T}{2C}\frac{1 + z^{-1}}{1 - z^{-1}} + \frac{T}{2C} \tag{6.60}$$

Thus in the domain

$$\psi = \frac{1 - z^{-1}}{1 + z^{-1}} \tag{6.61}$$

which corresponds to bilinear $s \to z$ transformation, the impedance of a capacitor is as shown in Figure 6.31(a). Thus a lossy capacitor is simulated.

Consider next a periodically reverse-switched capacitor (Figure 6.31(b)). The simulated impedance in the $\psi$-domain is derived as in the previous case:

$$\Delta Q(z) = C(1 + z^{-1})V$$

$$I(z) = \frac{\Delta Q(z)}{T} = \frac{C}{T}(1 + z^{-1})V \tag{6.62}$$

$$\frac{V}{I(z)} = \frac{T}{C(1 + z^{-1})} = \frac{T\psi}{2C} + \frac{T}{2C}$$

The equivalents of the SC networks of Figures 6.31(c) and 6.31(d) used to simulate source and load resistances can be similarly derived. The common feature of all these equivalent circuits is that they are lossy. Hence, to obtain an exact lossless ladder simulation, these losses have to be compensated in some manner. This is accomplished by realizing a shunting negative resistance which exactly cancels the positive resistance. The negative resistance is realized by a VIS as discussed below.

## 6.6.2 VIS realization

A simple VIS realization is shown in Figure 6.32 [6.22]. In phase 1, the output voltage of the capacitive network is sampled and held on a capacitor $C_H$ at the

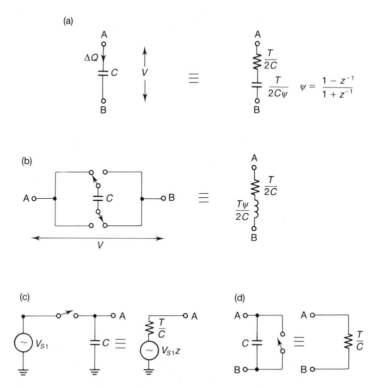

Fig. 6.31 Bilinear equivalents of four SC elements: (a) capacitor; (b) periodically reverse-switched capacitor; (c) source resistance simulating SC; (d) load resistance simulating SC.

output of OA. In phase 2, the capacitor voltage $v_{C_H}$ is available, after inversion at the output of OA, for application to the capacitive network. Thus, the VIS samples the voltage across $C$ and then applies it to the network after inverting it. The VIS makes $v(nT) = -v((n-1)T)$ so that corresponding to equation (6.57), we obtain

$$\Delta q(nT) = +2Cv(nT)$$

or

$$I(z) = \left(\frac{2C}{T}\right)V \qquad (6.63)$$

It is possible to visualize this behaviour, as though a negative resistance $(T/2C)$ is shunted across the capacitor, since the charge $\Delta q(nT)$ flows 'out' of the VIS.

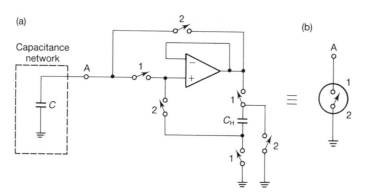

*Fig. 6.32* (a) A capacitive network connected to a VIS; (b) the symbol for a VIS. *(Adapted from [6.22], © AEU 1979.)*

## 6.6.3 The principle of lossless charge transfer using a VIS

The fact that lossless charge transfer takes place in a capacitor network closed through a VIS can also be seen by considering the charge transfer between two capacitors connected as shown in Figure 6.33 [6.25]. It can be seen that the energy in the capacitors *before* the VIS is switched on is $(1/2)(C_1 C_2/(C_1 + C_2)) \cdot (v_S)^2$. Next, the VIS switch senses the voltage $v_S$, inverts it, and applies the voltage $-v_S$ to the series combination of $C_1$ and $C_2$. It is clear that this results in no loss of charge. It is seen from the foregoing discussion that the losses present in an SC network consisting of elements simulating source and load resistances, inductors and capacitors, all of which are lossy, can be compensated by suitable negative resistances included in the various loops. Such a negative resistance is realized by using a VIS.

## 6.6.4 Theory of VIS based on the definition of 'equivalent voltage'

Fettweis has developed the theory of VIS-based SC filters in a slightly different manner [6.22, 6.23]. It is instructive to study it for further development of the

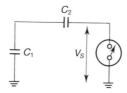

*Fig. 6.33* Capacitor loop with a VIS.

principles of these SC networks. As has been mentioned, the charge transfer in an SC network takes place because of the periodic occurrence of pulse currents. We denote the voltage across a capacitor before the arrival of such a pulse as $v_{bn}$ and after the arrival of the pulse as $v_{an}$. Thus the charge passing through the capacitor is (see Figure 6.34(a))

$$q(t_n) = C(v_{an} - v_{bn}) \tag{6.64}$$

where the suffix $n$ corresponds to the time instant under consideration. In the steady state, we can write

$$q(t_n) = Qe^{st_n}$$
$$v_a(t_n) = V_a e^{st_n} \tag{6.65}$$
$$v_b(t_n) = V_b e^{st_n}$$

where $Q$, $V_a$ and $V_b$ are complex constants. We now define an equivalent voltage $U = (V_a + V_b)/2$, and an equivalent current $J = Q/T$.

Noting that $v_b(t_n) = v_a(t_{n-1})$, we obtain

$$V_b = V_a z^{-1} \tag{6.66a}$$

and

$$U = \frac{V_a(1 + z^{-1})}{2} \tag{6.66b}$$

From equations (6.64) and (6.66a) we get:

$$Q = CV_a(1 - z^{-1}) \tag{6.67}$$

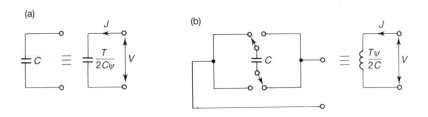

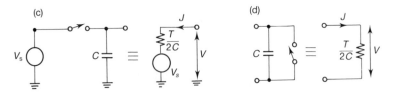

Fig. 6.34  Circuits using VIS to simulate: (a) a capacitor; (b) an inductor; (c) source resistance; (d) load resistance.

From equations (6.66b) and (6.67), we next derive

$$U = \frac{T}{2C}\left(\frac{1+z^{-1}}{1-z^{-1}}\right)J \qquad (6.68a)$$

or

$$U = \frac{T}{2C\psi}\,J \qquad (6.68b)$$

Thus a lossless capacitor in the $\psi$-domain of value $T/2C$ is realized.

It can be similarly shown that the circuits of Figures 6.34(b)–6.34(d) realize bilinear lossless inductance, source and load resistances. The above discussion centred on the definition of the equivalent voltage $U$, and also on the fact that the charge transfer should occur only once in a clock cycle. The definition of $U$ is easily satisfied, as explained next.

Consider a VIS incorporated in a loop containing capacitors, for instance, as in Figure 6.35. The VIS senses the voltage $v_{Ib}$ in the loop, inverts it and then applies $v_{Ia} = -v_{Ib}$ to the capacitive network thereby supplying the charge pulses. Thus, applying KVL before and after the occurrence of these charge pulses at time instant $n$, we obtain [1.21],

$$v_{1bn} + v_{2bn} + v_{3bn} + \cdots + v_{mbn} + v_{Ibn} = 0 \qquad (6.69a)$$

$$v_{1an} + v_{2an} + v_{3an} + \cdots + v_{man} + v_{Ian} = 0 \qquad (6.69b)$$

Applying these two equations (noting that due to the action of the VIS, $v_{Ian} = -v_{Ibn}$) we get, through the definition of equivalent voltage in equation (6.66a),

$$U_{1bn} + U_{2bn} + U_{3bn} + \cdots + U_{mbn} = 0 \qquad (6.69c)$$

This equation corresponds to KVL for the original RLC loop and further, by virtue of our earlier development, the branch impedance relations $U = F(z) \cdot J$ are also satisfied. Also, at the internal nodes, KCL holds true, i.e., $\Sigma J = 0$. In summary, the equations of the RLC network and its simulated version are identical. Thus, a lossless ladder filter is simulated through the use of VISs. The reader will have noted that in the simulated arrangement $v_a$s and $v_b$s are present, but not $U$s. The voltages

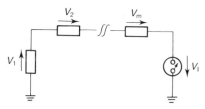

Fig. 6.35  An SC loop containing a VIS.

*U* are introduced for convenience only. Note that, since the presence of a VIS in the loop satisfies the KVL, every independent loop must have a VIS.

## 6.6.5  Design examples

Using the above-mentioned principles, we now examine how an SC circuit can be obtained corresponding to prototype RLC filters.

*Example 6.10*
Derive an SC filter using the VIS concept based on a first-order LP RC network.

Consider the first-order passive RC LP network of Figure 6.36(a). Replacing the source resistor and capacitor by their equivalent circuits derived earlier, the SC network of Figure 6.36(b) is obtained. It is important to note that the clock signals for the various switches in the circuit have to be appropriately determined. By choosing the various clock phases as shown in Figure 6.36(c), proper operation is achieved. In every clock cycle, at the beginning of instant, i.e., phase 1, the 'memory' of the capacitor simulating the source resistor is destroyed by charging it to the new input sample. This voltage is thus available in series with the voltage across $C_2$. The voltage across $C_2$ at the present instant is the voltage across it resulting in the previous cycle, because of the operations described next. The total voltage in the loop is sensed by the VIS in phase 2, held on its hold capacitor $C_H$,

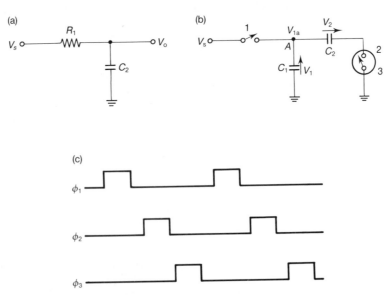

Fig. 6.36  (a) First-order LP RC network;  (b) SC implementation using a VIS; (c) switching waveforms.

and in phase 3 reapplied to the loop after inversion. Thus, a new charge division takes place resulting in the 'after' voltages. We shall evaluate the output voltage next. Similar to the prototype circuit, the output must be available at the junction of the resistor-simulating element and the capacitor.

Following equations (6.69), we write the equations characterizing the circuit of Figure 6.36(b). We denote the voltages across $C_1$ and $C_2$ as $v_1$ and $v_2$. Then, applying KVL,

$$U_1 + U_2 = 0$$

or

$$v_{1a} + v_{s1} + v_{2a} + v_{2b} = 0 \tag{6.70a}$$

Applying KCL at node A,

$$C_1(v_{1a} - v_{s1}) = C_2(v_{2a} - v_{2b}) \tag{6.70b}$$

Taking the $z$-transform of equations (6.70a) and (6.70b), we obtain

$$V_{1a} + V_{s1} + V_2(1 + z^{-1}) = 0 \tag{6.70c}$$

$$C_1(V_{1a} - V_{s1}) = C_2 V_2(1 - z^{-1}) \tag{6.70d}$$

Solving for the output $V_{1a}$, we have

$$\frac{V_{1a}}{V_{s1}} = \frac{C_1 - C_2\psi}{C_1 + C_2\psi} \tag{6.71}$$

while by substituting $R_1 = T/2C_1$ in a first-order LP transfer function and using the bilinear transformation $(s \rightarrow (2/T)\psi)$, we get

$$\frac{V_o}{V_{s1}} = \frac{1}{1 + \dfrac{C_2}{C_1}\psi} \tag{6.72}$$

The discrepancy between equations (6.71) and (6.72) should have been apparent to the reader. In the simulated network, the equivalent voltages $U$ correspond to the bilinear transformed voltages in the prototype, whereas at node $A$ in the present example, we are only considering $V_{1a}$ and not $(V_{1a} + V_{s1})/2$:

$$\frac{U_1}{V_{s1}} = \frac{\frac{1}{2}(V_{1a} + V_{s1})}{V_{s1}} = \frac{1}{1 + \dfrac{C_2}{C_1}\psi} \tag{6.73}$$

Note, however, that $V_{1a}$ is the 'reflected' voltage given by

$$\frac{V_{1a}}{V_{1b}} = S_{11} = \frac{Z_1 - R_1}{Z_1 + R_1} \tag{6.74}$$

where $R_1$ is the source resistance and $Z_1$ is the input impedance of the remaining network (in this case $C_2$).

In the above example, we could realize the reflected voltage, but not the actual transfer function. In the case of doubly terminated RLC filter simulation, however, such a problem will not exist and the exact output is available across the termination. This is demonstrated next.

*Example 6.11*
Design a second-order SC BP filter using VIS concepts [6.26].

The prototype RLC BP filter is shown in Figure 6.37(a). Using the equivalent SC elements for various passive elements in the prototype, the SC network of Figure 6.37(b) is obtained. The timing diagram is shown in Figure 6.37(c). Since there is only one loop, one VIS is required. When the switches associated with $C_2$ simulating the 'inductor' are in position $a$, the loop voltage is sampled and inverted. The operation repeats again, when $C_2$ terminals are reversed in phase $b$. Before the VIS makes measurement of the voltage in the loop to simulate the load resistance, $C_R$ must be discharged. The capacitor values are related to the prototype element values as before. It remains to be seen as to how the output is obtained. Since the load termination capacitor $C_R$ is discharged in one phase, the voltage available across $C_R$ is related to the equivalent voltage in the prototype by:

$$U_{C_R} = \frac{V_{C_{Ra}} + V_{C_{Rb}}}{2} = \frac{V_{C_{Ra}}}{2} \tag{6.75}$$

Thus, twice the equivalent voltage $U_{C_R}$ is available as $V_{C_{Ra}}$. This voltage can conveniently be sampled in phase 1, for which purpose the lower plate of the

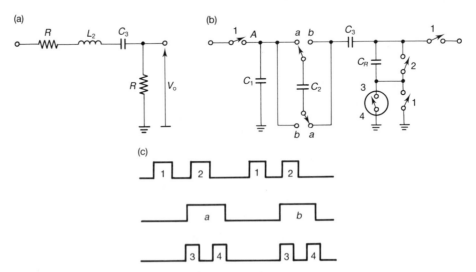

*Fig. 6.37*  (a) A second-order RLC BP filter;  (b) SC implementation using a VIS; (c) switching waveforms.

capacitor needs to be grounded. During this time, the VIS is inoperative. It may also be noted that, as in the case of the previous example, the reflected voltage is available at node $A$.

The method described in the above examples could be used to arrive at SC ladder filters corresponding to any prototype RLC ladder filter. The rules to be observed are as follows:

1. The KVL must be satisfied in each independent loop.
2. All charge transfers must be VIS-controlled.
3. The prototype must not have any loops in the series elements because simulation of such a circuit using SCs requires a *floating* VIS in that loop. Floating VISs are difficult to realize, if not impossible [6.27].

A high-order filter design is further illustrated in the next example.

*Example 6.12*
Design an SC filter using VIS concept with a maximum loss of 0.1 dB in the pass band 0–3.4 kHz, a minimum loss of 40 dB in the stop band (4.6–12 kHz) and a clock rate of 24 kHz [6.28].

The specifications are first prewarped using the formula $\Omega = \tan(\omega T/2)$. Thus, the pass-band and stop-band edges after prewarping are, respectively,

$$\Omega_c = 0.476\ 975\ 5$$

$$\Omega_s = 0.687\ 281$$

and hence their ratio $(\Omega_s/\Omega_c)$ is 1.440 914 6. For this pass-band–stop-band edge ratio, and the given ripple of 0.1 dB in the pass band, filter design tables will yield a prototype fifth-order filter configuration as shown in Figure 6.38(a). This design is chosen because it does not have floating parallel resonant circuits, thereby avoiding floating VISs.

The SC realization then follows readily, as shown in Figure 6.38(b). The component values of the SC network are evaluated from the component values of the prototype as follows. Dividing all the prototype $L$ and $C$ values by the prewarped cutoff frequency $\Omega_c$, the new component values are obtained. Next, using the equivalences

$$R_i \rightarrow \frac{T}{2C_i}$$

$$sl_i \rightarrow \psi \frac{T}{2C_i}$$

$$\frac{1}{sc_i} \rightarrow \frac{T}{2C_i\psi}$$

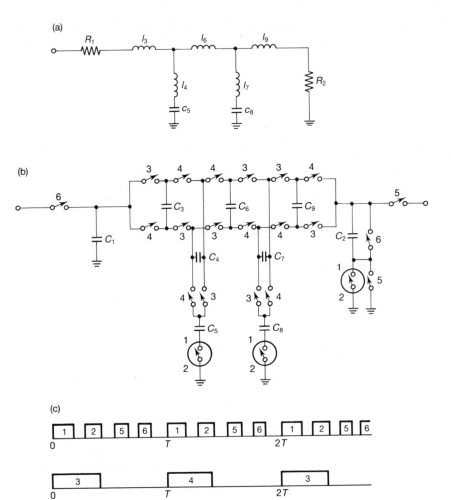

Fig. 6.38  (a) A fifth-order LP filter;  (b) SC realization;  (c) switching waveforms.
(Adapted from [6.28], © 1980 IEEE.)

we obtain

$$C_i = \frac{T}{2R_i}$$

$$C_i = \frac{T}{2l_i} \Omega_c$$

$$C_i = \frac{Tc_i}{2\Omega_c}$$

The factor $2/T$ can be eliminated, as its effect is equivalent to scaling all capacitors by $2/T$. Thus the final design equations are

$$C_i = \frac{1}{R_i} \text{ (corresponding to resistors)}$$

$$C_i = \frac{\Omega_c}{l_i} \text{ (corresponding to inductors)}$$

$$C_i = \frac{c_i}{\Omega_c} \text{ (corresponding to capacitors)}$$

The resulting capacitor values are as follows. The prototype element values are:

$$R_1 = R_2 = 1$$

$$l_3 = 1.011\ 276$$

$$l_4 = 0.170\ 181$$

$$l_6 = 1.594\ 879$$

$$l_7 = 0.504\ 392$$

$$l_9 = 0.776\ 406$$

$$c_5 = 1.197\ 206$$

$$c_8 = 0.889\ 855$$

and the SC network capacitor values are:

$$C_1 = C_2 = C_u$$

$$C_3 = 0.471\ 657C_u$$

$$C_6 = 0.299\ 067C_u$$

$$C_7 = 0.945\ 645C_u$$

$$C_4 = 2.802\ 75C_u$$

$$C_5 = 2.509\ 99C_u$$

$$C_8 = 1.865\ 62C_u$$

$$C_9 = 0.614\ 338C_u$$

This completes the design. The switching waveforms are as shown in Figure 6.38(c).

The reader will have noted the simplicity in design of the above technique. However, the resulting circuit is sensitive to both bottom- and top-plate parasitic capacitances. Hence, methods of either reducing or compensating their effects will be investigated. At least the parasitic capacitances across the VIS can be eliminated, if we use a modified VIS in place of the simple VIS of Figure 6.32; this is considered next.

### 6.6.6  Integrator VIS

Consider the integrator VIS [6.28] of Figure 6.39(a). In this VIS realization phase 1 overlaps phase 3, thereby discharging the parasitic capacitance $C_p$ (dominated by the bottom-plate parasitic capacitance of the capacitor immediately facing node $A$) as well as the integrating capacitor $C_I$. In phase 1, during the remaining period after phase 3 elapses, the capacitive network $C$ discharges into $C_I$ to develop a voltage, $v_I$, at the output of the OA. This voltage is available in phase 2 as $v_a$.

It is evident that for $v_a = -v_I = -v_b$, $C_I$ must equal the capacitance of the one-port associated with the VIS. The capacitor $C_I$ is reset in the beginning of phase 1 (i.e., phase 3) to be ready for the next voltage inversion. It is thus seen that the VIS of Figure 6.39(a) is superior to that of Figure 6.32. However, the remaining top- and bottom-plate parasitics in the loop containing the VIS (see, for example, Figure 6.38(b)) still exist. This method, however, has the disadvantage that if $C_I < C$, then it introduces a negative resistance in the circuit, more than that required to compensate for the losses, resulting in a large sensitivity in the SC filter.

### 6.6.7  Methods of reducing the effect of parasitic capacitances

Another method of eliminating the parasitic capacitance of the VIS, which does not have any disadvantage as in the case of integrator VIS (i.e., the requirement of $C_I = C$) is considered next [6.26]. Consider the loop of two capacitors and a VIS shown in Figure 6.40(a). Invariably the parasitic capacitance is present across the VIS. To eliminate this, a switch is introduced in between, as shown in Figure 6.40(b); the actual realization of Figure 6.40(b) is shown in Figure 6.40(c). First, the VIS is short-circuited in phase 1 and the voltage at terminals $A$ and $B$ through two buffers is measured and sampled on a hold capacitor $C_H$. Next, in

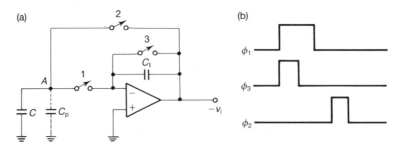

Fig. 6.39  (a) An integrator VIS;  (b) switching waveforms. *(Adapted from [6.28],* © 1980 IEEE.)

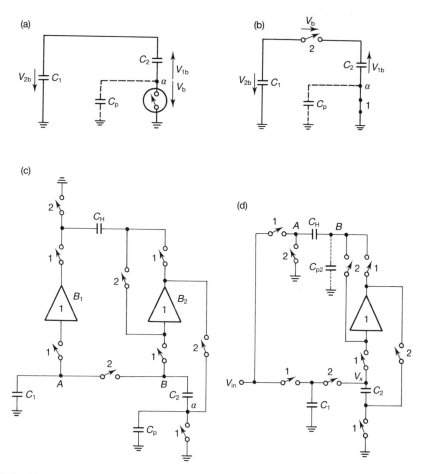

*Fig. 6.40* (a) A two-capacitor loop with a VIS;  (b) method of eliminating parasitic capacitance in (a);  (c) realization of (b);  (d) realization of source resistance and a grounded capacitor in a loop. *(Adapted from [6.26],* © 1982 IEE.)

phase 2, the voltage across $C_H$ is properly connected to the input of buffer $B_2$ and ground and is thus available after inversion for application to terminal $\alpha$ again. It is evident that no capacitor matching is required, but the circuit needs an additional OA. To realize the input circuit of the SC ladder filter, i.e., source resistance simulation, the circuit of Figure 6.40(c) can be modified as in Figure 6.40(d).

Note, however, that the parasitic top-plate capacitance of $C_H$ affects the filter transfer function in this case, as follows. The parasitic capacitance at $B$ is charged to $v_x$ while $C_H$ is charged to $v_{in} - v_x$. In phase 2, when they are connected together, the resulting voltage is fed to the bottom plate of $C_2$ and not $v_{in} - v_x$ as desired.

## 6.6.8 VIS using the inverse recharging principle

Another VIS realization can be obtained by using the principle of *inverse recharging* [6.23] described next. Consider a capacitor into which a charge is flowing to change its terminal voltage from $v_b$ to $v_a$. Then the charge required to flow into the capacitor is

$$q = C(v_a - v_b) \tag{6.76}$$

In order to realize a VIS, $v_a = -v_b$ is required; hence, the charge required to achieve this voltage inversion is

$$q = -2Cv_b \tag{6.77}$$

Now let us short-circuit the capacitance first, so that its $v_a = 0$; the charge required to do this is

$$q_0 = -Cv_b \tag{6.78}$$

Equations (6.77) and (6.78) mean that the charges which result in a voltage reversal are twice those required to reduce the voltage to zero. Thus the voltage inversion can be realized in two steps. In the first, the capacitance one-port is short-circuited, thus determining the charge $q_0$. Then, the same charge is shifted again in the same direction as before through the capacitance one-port to develop $v_a = -v_b$. This second step is what is called *inverse recharging*.

A device to perform this function is shown in Figure 6.41(a) [6.28]. In phase 1, the capacitance one-port is discharged by the OA into $C_H$, and in phase 2, $C_H$ is discharged by the OA buffer into the input port. When the VIS of Figure 6.41(a) is used, the changes occur in both the clock phases, whereas when the VIS of Figure 6.32 is used, the capacitor network is affected only in one clock phase. In the VIS of Figure 6.41(a), the OA input terminals are off the ground in one phase thus making it sensitive to parasitic capacitances. A modification to keep the OA terminals at ground in both the clock phases is shown in Figure 6.41(b). In this circuit, $C$ is the capacitance in a loop. If it is already grounded, then the ground terminal has to be removed off the ground to facilitate inverse recharging. In phase 1, the charge available in the capacitance network $C$ is transferred to $C_H$, which will be discharged into the capacitance network in phase 2. The effect of parasitic capacitances will now be considered. These are shown in dotted lines in Figure 6.41(b). Since the bottom-plate parasitic capacitance $C_B$ is always connected to ground or virtual ground, it does not affect the performance of the circuit, whereas the top-plate parasitic capacitance $C_A$ increases the value of $C$. There is no parasitic capacitance associated with $C_H$ in this VIS realization.

The VIS based on the inverse recharging concept can be used in the same way as the VIS of Figure 6.32, with a few modifications. As an illustration, the second-order BP filter of Figure 6.37(a) is considered. The resulting SC filter is shown in Figure 6.42. Note that the capacitor simulating the source-resistance is lifted off the

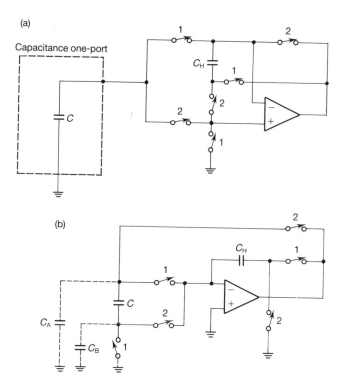

Fig. 6.41  VISs based on the inverse recharging principle  *(adapted from [6.28],*
© 1982 IEEE).

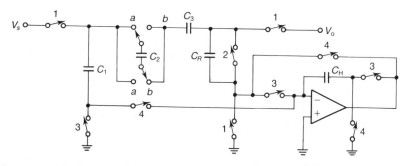

Fig. 6.42  A second-order BP SC filter based on a VIS using the inverse recharging principle.

ground to facilitate inverse recharging. The timing waveforms are the same as those
shown in Figure 6.37(c).

It is thus seen that RLC circuits can be used to derive SC filters based on VISs.
These, however, suffer from the parasitic capacitances of the bottom as well as
top plates. However, by alternative circuit arrangements VIS designs can be made
insensitive to bottom-plate parasitic capacitances.

Completely bottom-plate stray-insensitive realizations can be obtained using simulation of other circuit elements called *unit elements*. SC filter design using these is considered next.

## 6.6.9  Realization of unit elements [6.23]

Consider the SC network of Figure 6.43(a). The charges flowing when the switches $S_1$ and $S_2$ are closed alternately are expressed by the charge conservation equations:

$$\Delta Q_1(z) = C[V_{1e} - V_{2o}z^{-1/2}] \tag{6.79a}$$

$$\Delta Q_2(z) = C[V_{2o} - V_{1e}z^{-1/2}] \tag{6.79b}$$

Once again defining the equivalent voltages $U_1$ and $U_2$ as the averages of the voltages at the ports 1 and 2 before and after charge transfers, we obtain

$$U_1 = \frac{V_{1e} + V_{2o}z^{-1/2}}{2} \tag{6.80a}$$

$$U_2 = \frac{V_{2o} + V_{1e}z^{-1/2}}{2} \tag{6.80b}$$

Solving equations (6.79) for $V_{1e}$ and $V_{2o}$ and substituting in equations (6.80), we have

$$\frac{\Delta Q_1(z)}{C} = \frac{2U_1(1 + z^{-1})}{(1 - z^{-1})} - \frac{4U_2 z^{-1/2}}{1 - z^{-1}} \tag{6.81a}$$

$$\frac{\Delta Q_2(z)}{C} = \frac{-4U_1 z^{-1/2}}{(1 - z^{-1})} + \frac{2U_2(1 + z^{-1})}{(1 - z^{-1})} \tag{6.81b}$$

Introducing again the notation $I_i = \Delta Q_i/T$, we then derive the admittance matrix as

$$\begin{bmatrix} I_1 \\ I_2 \end{bmatrix} = \begin{bmatrix} \dfrac{1}{R\psi} & \dfrac{-\sqrt{1 - \psi^2}}{R\psi} \\ \dfrac{-\sqrt{1 - \psi^2}}{R\psi} & \dfrac{1}{R\psi} \end{bmatrix} \begin{bmatrix} U_1 \\ -U_2 \end{bmatrix} \tag{6.82}$$

with $R = T/2C$; the waveforms of $i_1(t)$ and $i_2(t)$ are as shown in Figure 6.43(b). A transmission matrix representation of the above two-port can be seen to be

$$\begin{bmatrix} U_1 \\ I_1 \end{bmatrix} = \begin{bmatrix} \dfrac{1}{\sqrt{1 - \psi^2}} \end{bmatrix} \begin{bmatrix} 1 & R\psi \\ \dfrac{\psi}{R} & 1 \end{bmatrix} \begin{bmatrix} U_2 \\ -I_2 \end{bmatrix} \tag{6.83}$$

Equation (6.83) represents the chain matrix of a 'unit element', a transmission line of unit length; such a transmission line is represented as shown in Figure 6.43(c).

(a)

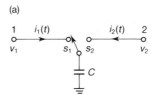

(b)

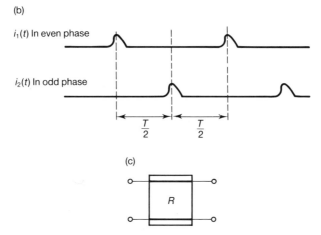

$i_1(t)$ In even phase

$i_2(t)$ In odd phase

(c)

*Fig. 6.43*  (a) SC unit element simulation;  (b) timing waveforms of $i_1(t)$ and $i_2(t)$; (c) symbol of unit element.

It has no equivalent lumped element counterpart. Such elements are useful in the synthesis of microwave filters and uniformly distributed RC networks.

A cascade of unit elements, each characterized by its $R$, the characteristic impedance, can be used to realize high-order LP filters. It is also possible to use unit elements, together with lumped inductors and capacitors, to realize high-order filters. It is thus possible to use unit-element SC simulation, together with simulated SC type RLC elements and VISs, to realize high-order SC filters. The principle of using VISs discussed before is used here too, in order to achieve lossless charge transfer. We assume in the treatment that follows that a prototype filter using all unit elements or some lumped $L$ and $C$ elements has been obtained by the synthesis procedures described by Rhodes [6.29] and Matsumoto [6.30].

A seventh-order SC filter using unit elements and VIS [6.25] is shown in Figure 6.44(b) together with the prototype in Figure 6.44(a). In each loop, the voltage across the capacitance network is sampled, inverted and reapplied. The same VIS can be thus used in the adjacent loops. Any of the VISs described in the previous discussion can be employed.

Note that the designs based on the cascade of unit elements use grounded capacitors for the source and load terminations and hence the remaining capacitors

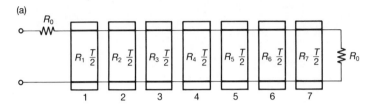

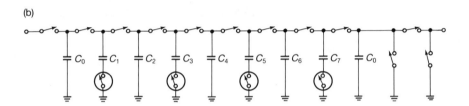

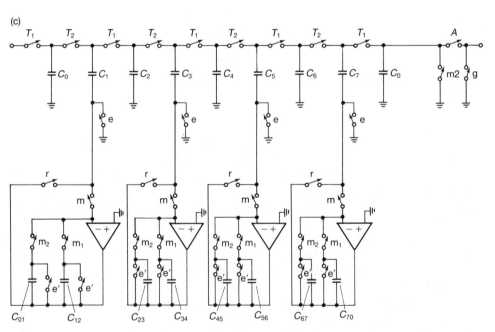

*Fig. 6.44* (a) A seventh-order unit element type LP filter; (b) SC implementation of (a); (c) practical circuit; (d) switching waveforms *(adapted from [6.25]*, © 1981 IEEE*).*

(d)

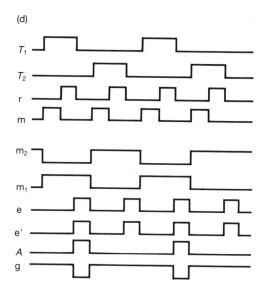

*Fig. 6.44 (continued)*

simulating the unit elements are connected to the VISs such that the bottom plates are always at virtual ground. Thus, the integrator VIS in Figure 6.39 can be employed in these circuits. However, as in the case of designs discussed before, the requirement of the integrator capacitor to be equal to the capacitance of the loop in which the VIS is used is a disadvantage. The integrator VIS type implementation and the switching waveforms are shown in Figures 6.44(c) and 6.44(d). Note also that the VIS needs two integrator capacitors, one for each loop.

## 6.6.10 Bottom-plate stray-insensitive VIS realizations and their application to the design of unit-element filters

A VIS without the matching requirement on capacitors, due to Herbst and Hosticka [6.31], is shown in Figure 6.45(a). It is especially suited for the cascade of unit elements, as will be shown next. In this VIS, in phase 1, the voltage across $C$ is sampled and held on $C_H$. In phase 2, this voltage is fed to the bottom plate of $C$. Thus, the bottom plate of $C$ is always kept at either ground or output of an OA, thereby eliminating altogether the parasitic capacitance of the bottom plates. It is easy to extend this concept to coupled loops [6.26]. Consider the SC network of Figure 6.45(b). In this network, the voltages across $C_1$, $C_2$ and $C_3$ are sampled and held on $C_{H1}$, $C_{H2}$, $C_{H3}$. Then, they are applied to the bottom plates of $C_1$, $C_2$ and $C_3$, respectively. It can be verified that the voltages between the bottom plates of $C_1$ and $C_2$, and between $C_2$ and $C_3$ are equal and opposite in sign to those before.

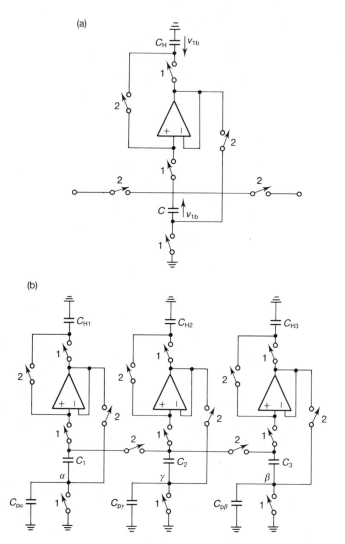

*Fig. 6.45* (a) A bottom-plate stray-insensitive VIS realization; (b) extension of (a) to coupled loops. *(Adapted from [6.31], © 1980 IEE.)*

As an illustration of the application of this method, a fifth-order SC filter using unit elements and capacitors is considered [6.26]. The prototype is shown in Figure 6.46(a) and the SC realization is as in Figure 6.46(b). It can be seen that each unit element and capacitor in the prototype needs an individual VIS for SC implementation. Note also that the source resistance simulation is different from that of Figure 6.40(d). The voltage across $C_1$ in this case is not 'inverted', since the inversion of the voltages across $C_1$ and $C_2$ requires an arrangement like that of

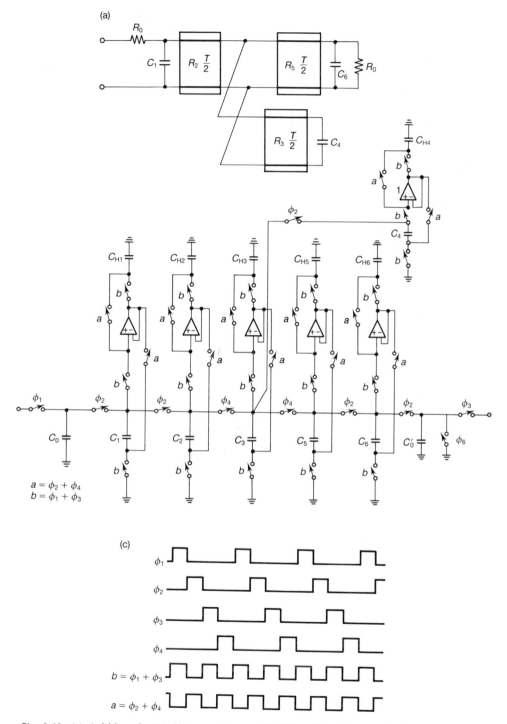

Fig. 6.46 (a) A fifth-order LP filter prototype; (b) SC realization; (c) clocking waveforms.

Figure 6.40(d), leading to a top-plate parasitic capacitance sensitivity. With this modification, the resulting SC filter will have an additional loss of 6 dB. In this method, more OAs are required than in the case of integrator VIS technique.

## Effect of parasitic capacitances due to top plates [6.25]

In the realization of bottom-plate stray-insensitive SC filters using VIS, the top-plate parasitic capacitances affect the transfer function. The effect of these is illustrated by considering a typical two-capacitor loop using a VIS. Consider the circuit of Figure 6.47. The capacitor voltages after inversion has taken place are obtained as follows. Before the VIS operates, the parasitic capacitances $C_{p1}$ and $C_{p2}$ are charged to $v_{1b}$ and $v_{2b}$, respectively. After the VIS inverts the voltage across the capacitors in the loop, the resulting voltages across the capacitors $C_1$ and $C_2$ are

$$v_{1a} = v_{1b} - A(v_{1b} - v_{2b}) \tag{6.84a}$$

$$v_{2a} = v_{2b} + B(v_{1b} - v_{2b}) \tag{6.84b}$$

where

$$A = \frac{C_2'\left(2 - \dfrac{C_{p1}}{C_1'}\right)}{C_1' + C_2'} \tag{6.84c}$$

and

$$B = \frac{2C_1' - C_{p1}}{C_1' + C_2'} \tag{6.84d}$$

with $C_1' = C_1 + C_{p1}$ and $C_2' = C_2 + C_{p2}$. For $C_{p1} = C_{p2} = 0$, $A = 2C_2/(C_1 + C_2)$ and $B = 2C_1/(C_1 + C_2)$.

It is thus seen that the effect of parasitic capacitances is to increase the capacitor values $C_1$ to $C_{p1} + C_1$ and $C_2$ to $C_{p2} + C_2$. The negative terms in equations (6.84c) and (6.84d) will cause losses and an undesirable droop in the pass band. Hence, for high-quality filters, these parasitic capacitances have to be taken into account.

In the above discussion, it is seen that by restricting the prototype elements to unit elements or grounded capacitors, bottom-plate stray-insensitive realizations are

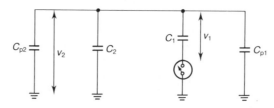

*Fig. 6.47* A two-capacitor VIS loop illustrating top-plate parasitic capacitances.

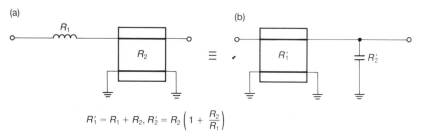

$$R'_1 = R_1 + R_2, \quad R'_2 = R_2 \left(1 + \frac{R_2}{R_1}\right)$$

*Fig. 6.48* Kuroda transformation used to convert an ungrounded unit element prototype to a grounded one.

obtained. Hence, if the prototype contains floating elements, they should be converted to grounded elements using Kuroda transformations [6.30]. As an illustration, the network of Figure 6.48(a) can be transformed into the network of Figure 6.48(b), which uses only a grounded capacitor and a unit element. The reader is referred to [6.29, 6.30] for more information on Kuroda transformations.

This completes the discussion on the VISs and their application to the simulation of lossless ladder filters. Several ICs have been fabricated using the design techniques described in this section and have been found to give excellent results [2.4, 4.3, 6.11]. As such, the practical utility of these VIS SC filters makes them comparable with other high-order SC filter design techniques. The design of SC filters using VIS concepts has been recently pursued further to develop methods of scaling them for optimal dynamic range and minimum total capacitance [6.32]. Also, stray-insensitive VISs have been proposed, which use dynamic MOS amplifiers with a tristate output capability [6.33]. These are believed to enhance the utility of this method of SC filter design. The reader is referred to the extensive list of references listed at the end of this chapter for more design information.

## 6.7  SC ladder filters based on FDNR simulation

First, we briefly review the concept of FDNR or supercapacitance due to Bruton [1.29]. Consider an RLC filter (Figure 6.49(a)) in which the impedance $Z(s)$ is transformed to $1/s(Z(s))$. Hence, the impedances corresponding to the elements $R$, $L$ and $C$ are transformed to:

$$R \to \frac{R}{s}, \qquad sL \to L, \qquad \frac{1}{sC} \to \frac{1}{s^2 C}$$

These elements corresponding to the transformed impedances may be recognized as a capacitor, a resistor and an FDNR (see Figure 6.49(b)). The impedance transformation considered above does not alter the transfer function. Further, the low-sensitivity property of doubly terminated LC ladder filters is translated to the

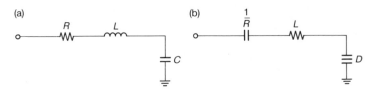

Fig. 6.49  (a) An RLC network;  (b) network obtained by FDNR transformation.

realized FDNR-R-C filters. Note that the FDNR element is denoted by $D$, whose impedance is $1/s^2D$.

## 6.7.1  Basic concepts

It is noted that the realization of SC ladder filters using FDNR concepts requires the realizations of SC FDNR blocks. As in the case of inductance simulation using SC networks, we can also derive the required $\Delta Q(z)-V$ relationships for FDNR simulation as follows.

*Example 6.13*
Derive the $\Delta Q(z)-V$ relationships of an FDNR element using various $s \rightarrow z$ transformations.

Proceeding in the same manner as in the case of inductor/resistor/capacitor simulation, we may derive the following relations for an FDNR element for both the LDI and bilinear transformations, by first writing the charge conservation equations:

$$\Delta Q(z)_{\text{LDI}} = \frac{D(1-z^{-1})^2}{Tz^{-1/2}} V \qquad (6.85)$$

$$\Delta Q(z)_{\text{BT}} = D \cdot \frac{2}{T} \frac{(1-z^{-1})^2}{(1+z^{-1})} V \qquad (6.86)$$

It thus suffices to realize the charge–voltage relationships in equation (6.86) to realize an SC FDNR. Note also that, in the case of $p$-transformation and forward Euler integration, the corresponding results are as follows:

$$\Delta Q(z)_{\text{PT}} = \frac{(1-z^{-1})^2}{T} DV$$

$$\Delta Q(z)_{\text{FEI}} = \frac{(1-z^{-1})^2}{z^{-1}T} DV$$

## 6.7.2 SC realizations

It may be noted that the Temes, Jahanbegloo and Nossek technique for floating inductance simulation can be used to realize floating FDNR realizations [6.10, 6.11] of bilinear transformation type. (It is easy to derive grounded FDNR elements by grounding one input terminal in this realization.) The block diagram of Figure 6.11 realizes a $\Delta Q(z)-V$ relationship given by

$$\Delta Q(z) = V(z)[C_0 + C_1 + C_2 - C_1 H_1(z) - C_2 H_1(z) H_2(z)] \qquad (6.87)$$

where $H_1(z)$ and $H_2(z)$ are the transfer functions of the blocks $T_1$ and $T_2$. Note also that $C_0$ discharges into the virtual ground of the OA used to realize the block $T_1$. Identifying (6.87) with the desired $\Delta Q(z)-V(z)$ relationship of equation (6.86), we note that

$$C_0 + C_1 + C_2 = \frac{2D}{T}$$

and

$$C_1 H_1(z) + C_2 H_1(z) H_2(z) = \frac{2D}{T}\left(\frac{3z^{-1} - z^{-2}}{1 + z^{-1}}\right) \qquad (6.88)$$

One possible realization is as shown in Figure 6.50. The design equations in this case are as follows:

$$C_0 + C_1 + C_2 = \frac{2D}{T}$$

$$\frac{C_1 C_0}{C_a} = \frac{6D}{T} \qquad (6.89)$$

$$C_1 = 3C_2$$

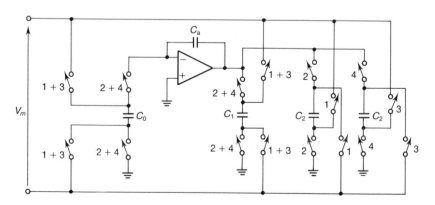

Fig. 6.50 Temes et al. bilinear FDNR realization (adapted from [6.11], © 1980 IEEE).

(Note that the OA operates during the interval between clock phase 1 and clock phase 2 in the open-loop mode and hence XY feedback is recommended.)

Hosticka and Moschytz [5.3] have proposed an SC FDNR realization of the $p$-transformation type, which requires a capacitor and a delay element $z^{-1}$ using OA buffer stages and sample-hold capacitors (see Figure 6.51). An alternative SC FDNR realization has been proposed by Faruque et al. [6.34]. This realization is based on the analog model of Figure 6.52(a). It can be shown that the input impedance realized is

$$Z_{in} = \frac{g_m}{s^2 C_3 C_4} \qquad (6.90)$$

It is thus seen that an SC transconductance is needed to simulate the current source in the analog model. The resulting SC FDNR realization is shown in Figure 6.52(b). In phase 1, the voltage across $C_3$ is sampled by $C_m$ without discharging $C_3$ (between the outputs of OA $A_1$ and buffer $A_2$). In phase 2, the buffer discharges $C_m$ into $C_4$ realizing a current source operation. The voltage across $C_4$ is available at the output of buffer $A_2$ as required in the analog model of Figure 6.52(a). Note, however, that as a result of the delays associated with sampling a voltage and making a charge to flow depending on this voltage at a later instant, the input admittance (charge domain) realized by the circuit of Figure 6.52(b) is of FEI type:

$$\Delta Q_0(z) = \frac{C_3 C_4 (1 - z^{-1})^2 V_{io}}{C_m z^{-1}} \qquad (6.91)$$

Nevertheless, the FDNR in Figure 6.52(b) is superior to the $p$-transformation type of FDNR. Consider the following example, which studies the usefulness of the SC FDNR in Figure 6.52(b).

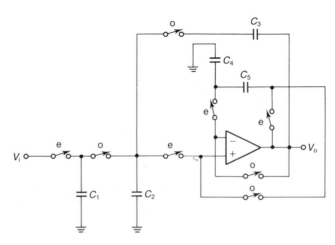

*Fig. 6.51*  SC LP filter based on FDNR realization *(adapted from [5.3], © 1980 IEEE).*

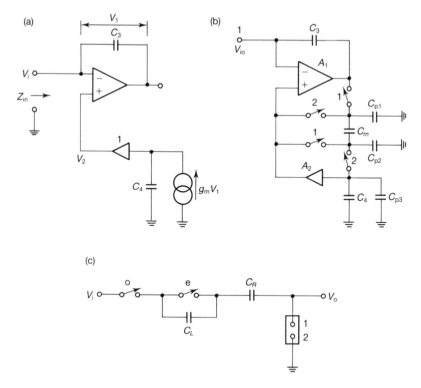

Fig. 6.52   (a) Analog model for FDNR realization;   (b) SC simulation;   (c) an SC equivalent for Figure 6.49(b). *(Adapted from [6.34], © 1982 IEEE.)*

*Example 6.14*
Derive an SC FDNR network based on the RCD network in Figure 6.49(b).

The SC FDNR of Figure 6.52(b) can be used together with an 'odd' SC resistor and a capacitor to give the network shown in Figure 6.52(c). The denominator of the transfer function for this network is given by

$$D(z) = C_3C_4(C_1 + C_2)z^2 - (C_3C_4(2C_1 + C_2) + C_1C_2C_m)z + C_1C_3C_4 \quad (6.92)$$

showing that arbitrary $Q_p$ values can be realized. Thus, while the FDNR realized is of FEI type, since the resistor and capacitor simulation may correspond to any other transformation $(s \rightarrow z)$, the resulting circuit behaviour has to be studied for each application separately. This observation is similar to that of unrealizable terminations for LDI-type SC ladder filters.

The circuit of Figure 6.52(b) is affected by parasitic capacitances of the top plates of capacitors $C_4$ and $C_m$. Faruque *et al.* have described an optimization scheme to predistort the FDNR filter specifications so as to eliminate the effect of these parasitic capacitances [6.34]. We next briefly consider the basic principle of the realization of Figure 6.52(b) from a different point of view so as to generate alternative FDNR realizations.

Consider the circuit of Figure 6.53, for which the following relation can be derived:

$$\Delta Q(z) = \left(1 - \frac{1}{H(z)}\right) C_f(z)\, V_i \tag{6.93}$$

Realizing $C_f(z)$ as a capacitor $C_1$ and $H(z)$ as a unit-delay element, we obtain

$$\Delta Q(z) = \frac{(1 - z^{-1})^2}{z^{-1}}\, C_1 V_i \tag{6.94}$$

which is the same as for the circuit in Figure 6.52(b). The reader is urged to seek alternative realizations for the delay element (perhaps stray-insensitive ones) to generate new FDNR circuits. The above network configuration is known in the active RC literature as *inverse active network* (see [6.35]).

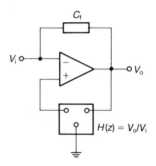

Fig. 6.53  Inverse active network.

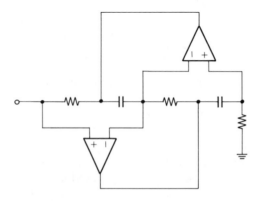

Fig. 6.54  A GIC realization.

We bring to a close the discussion of SC FDNR realizations by pointing out that the periodically reverse-switched capacitor [2.4] could be used to replace the resistors in an FDNR simulation circuit using a GIC, to realize a bilinear trans-formation type FDNR (Figure 6.54). However, since the GIC has floating terminals the parasitic capacitances at these nodes are unavoidable. Nevertheless, GIC based SC ladders have been recently realized in IC form and found to be satisfactory [6.36].

## 6.8 Conclusions

In this chapter, we have studied several design techniques for SC ladder simulation using substitution of RLC and/or distributed elements by SC networks. The resulting designs are, however, only at most bottom-plate stray-insensitive. Nevertheless, they enjoy the excellent low-sensitivity properties of RLC ladder filters. Scaling for optimal dynamic range is rarely possible in these designs. SC filters based on RLC ladder networks but with properties of stray insensitivity as well as scaling facilities will be considered in the next chapter.

## 6.9 Exercises

6.1 Derive the $z$-domain equivalent circuit of Figure 6.1 from first principles, i.e., charge conservation equations.

6.2 The Ford and Girling active RC inductance simulator is shown in Figure 6.55. By lifting the ground terminal and looking between the input and this new terminal, we obtain a floating inductance which is unilateral, meaning that the current entering terminal $B$ is zero, and the current entering terminal $A$ is

$$I_A = \frac{V_A - V_B}{R_1} + \frac{V_A - V_B}{R_2} + \frac{V_A - V_B}{sCR_1R_2}$$

as though there is a parallel combination of $R_1$, $R_2$ and inductance $CR_1R_2$. In order to make it bilateral, obviously, we have to send 'out' a current equalling $I_A$ at terminal $B$. Suggest a circuit to perform this function. Compare it with the SC circuit of Figure 6.2(a).

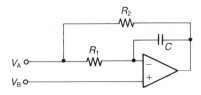

*Fig. 6.55*  Ford and Girling active RC inductance simulator.

6.3 Find the impedance between terminals 1 and 2 in the circuit described by Brugger, Hosticka and Moschytz [6.2] shown in Figure 6.56.

6.4 Riordan has proposed the circuit shown in Figure 6.57 for inductance simulation. The circuit between terminals $A$ and $B$ within dashed lines is a floating (or, to be accurate, unilateral floating) negative resistance. Evaluate the input impedance obtained between $A$ and ground. Note that we can derive the SC circuit of Figure 6.8 from the circuit of Figure 6.57 by using a periodically reversed SC in place of $R_1$, whereas the negative resistance is realized by two inverting SCs working in each phase.

6.5 In an ideal floating inductance, the current entering one terminal must leave the other, while the branch impedance satisfies the $I-V$ relationship for an inductance. Examine the effect of mismatch of $C_3$ and $C_4$ on the realized floating impedance in the circuit of Figure 6.3.

6.6 Examine the input impedance of the gyrator of Figure 6.6 loaded by a capacitor $C_L$ at the output port.

6.7 Design a second-order SC ladder filter (see Figure 6.23(b)) using Lee's LDI inductance and bilinear termination resistances for a 0.5 dB Chebyshev

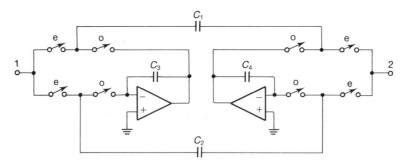

Fig. 6.56   SC inductance simulator due to Brugger et al. (Adapted from [6.2] © 1979 IEE).

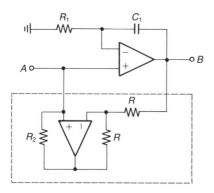

Fig. 6.57   Riordan's SC inductance simulator.

response and pass-band edge frequency of 1 kHz. For variable sampling frequency, observe the effect of non-LDI terminations by plotting the frequency responses and comparing them with the desired response. Repeat the problem for a third-order Chebyshev 0.5 dB response.

6.8 Prove that the order of the transfer function increases by 2 when a prototype of the type shown in Figure 6.23 is used to realize an SC network using LDI inductances and bilinear terminations. Verify that the resulting network yields poles outside the unit circle [6.14].

6.9 The circuit, due to Antoniou, shown in Figure 6.58(a) can work as an impedance transformer of ratio $K(s):1$, where $K(s) = Z_1 Z_3 / Z_2 Z_4$. The use of two such transformers in association with a resistor network embedded between them, as shown in Figure 6.58(b), can realize a star network of inductances for proper choices of $K(s)$. Examine the validity of this observation and study its application to RLC filter design [1.25].

6.10 Analyze the SCIC of Figure 6.59 and study whether it can be used to simulate elliptic low-pass ladder filters of the type shown in Figure 6.24.

6.11 Examine whether parasitic compensation of the capacitors $C_{L1} + C_{L2}$, etc., in Figure 6.30(b) can be realized using the methods studied in Chapter 4 [6.37].

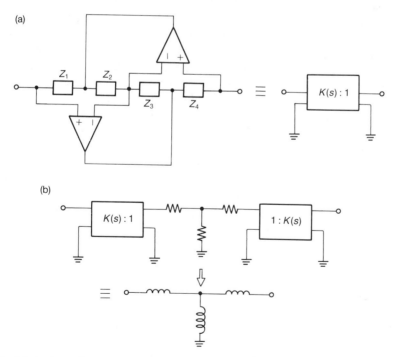

*Fig. 6.58* (a) An impedance transformer of ratio $K(s):1$, where $K(s) = Z_1 Z_3 / Z_2 Z_4$; (b) realization of a star network of inductances using two impedance transformers.

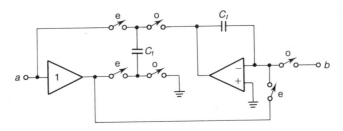

*Fig. 6.59* An SCIC circuit.

6.12 The prototype RLC filter for realizing SC ladder all-pole low-pass filters can be of minimum inductance or minimum capacitance type (Figure 6.60). Using Lee's simulated inductance, compare these two realizations. Note that the redundant buffer stages can be simplified.

6.13 Derive the transmission matrix of the SC network of Figure 6.61. Note that the shunt capacitor network is an SC transformer studied earlier. The circuit of Figure 6.61 realizes a unit element cascaded by a transformer with $N$:1 turns ratio [6.38].

6.14 Consider the inverse recharging device shown in Figure 6.62. The effect of parasitic capacitance at node $A$ can be compensated partly by the addition of switches at node $B$, and the required clock waveforms are as shown in Figure 6.62. Analyze the operation of the circuit and evaluate the charge loss in spite of the compensation arrangement. An alternative technique is to use a capacitor $C_c$ for compensation, in which case phase 0 is not necessary. Study the effect of this modification to evaluate the recharging error [6.39].

6.15 In an SC network requiring several recharging devices, it is possible to trade off the number of OAs (recharging devices) with the number of clock phases. Develop such an arrangement, called 'sequential inverse charging' in the literature, illustrating the clocking wave forms [6.40].

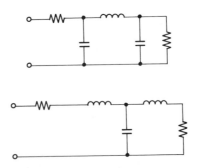

*Fig. 6.60* Prototype RLC low-pass ladder filters.

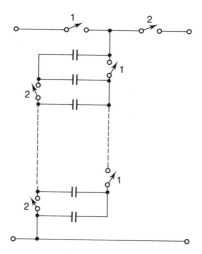

*Fig. 6.61* An SC circuit realizing a unit element cascaded by a transformer.

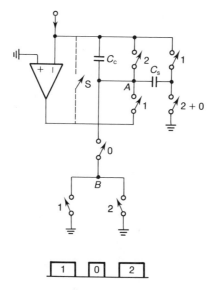

*Fig. 6.62* An inverse recharging network *(adapted from [6.39], © AEU 1983)*.

6.16 A capacitor grounded by a VIS is shown in Figure 6.63(a), where the parasitic capacitances affect the performance. Consider the configurations in Figures 6.63(b) and 6.63(c) and evaluate the recharging error [6.41].

6.17 Two floating VISs are shown in Figure 6.64. Analyze their operation [6.27].

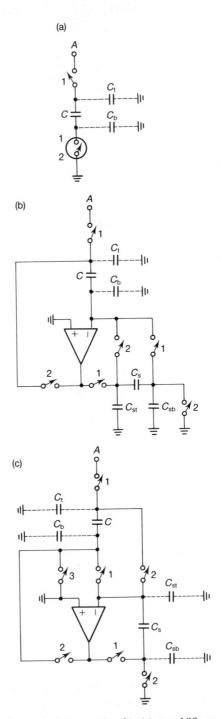

Fig. 6.63 (a) A capacitor grounded by a VIS; (b), (c) two VIS configurations for evaluation of recharging error. *(Adapted from [6.41],* © *AEU 1981.)*

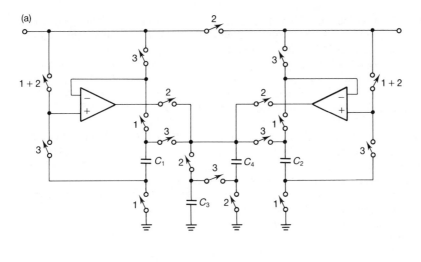

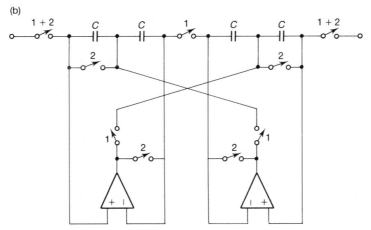

*Fig. 6.64* Two floating VIS circuits *(adapted from [6.27]*, © AEU 1979).

## 6.10  References

[6.1] B.J. Hosticka and G.S. Moschytz, Switched-capacitor simulation of grounded inductors and gyrators, *Electronics Letters*, 14, 24, 788–90, 23 November 1978.

[6.2] U.W. Brugger, B.J. Hosticka and G.S. Moschytz, Switched-capacitor simulation of floating inductors using gyrators, *Electronics Letters*, 15, 16, 494–6, 2 August 1979.

[6.3] U.W. Brugger and B.J. Hosticka, Alternative realizations of switched-capacitor floating inductors, *Electronics Letters*, 15, 21, 698–9, 11 October 1979.

[6.4] M.S. Lee, Switched-capacitor filters using floating-inductance simulation circuits, *Electronics Letters*, 15, 20, 644–5, 27 September 1979.

[6.5] T.R. Viswanathan, S.M. Faruque, K. Singhal and J. Vlach, Switched-capacitor transconductance and related building blocks, *IEEE Transactions on Circuits and Systems*, CAS-27, 6, 502–8, June 1980.

[6.6] T.R. Viswanathan, J. Vlach and K. Singhal, Switched-capacitor transconductance elements and gyrators, *Electronics Letters*, 15, 11, 318–19, 24 May 1979.

[6.7] S.M. Faruque, J. Vlach and T.R. Viswanathan, Switched-capacitor inductors and their use in LC filter simulation, *Proc. IEE*, Part *G*, 128, 4, 227–9, August 1981.

[6.8] M.S. Lee, Improved circuit elements for switched-capacitor ladder filters, *Electronics Letters*, 16, 4, 131–3, 14 February 1980.

[6.9] G. Spahlinger, Switched-capacitor simulated inductors, *AEU*, 35, 53–5, 1981.

[6.10] G.C. Temes and M. Jahanbegloo, Switched-capacitor circuits bilinearly equivalent to floating inductor or f.d.n.r., *Electronics Letters*, 15, 3, 87–8, 1 February 1979.

[6.11] J.A. Nossek and G.C. Temes, Switched-capacitor filter design using bilinear element modelling, *IEEE Transactions on Circuits and Systems*, CAS-27, 6, 481–91, June 1980.

[6.12] J.A. Nossek, Improved circuit for switched-capacitor simulation of an inductor, *Electronics Letters*, 16, 4, 141–2, 14 February 1980.

[6.13] P.V. Ananda Mohan, V. Ramachandran and M.N.S. Swamy, Terminations for component simulation type SC ladders, *Electronics Letters*, 19, 1, 29–30, 6 January 1983.

[6.14] P.V. Ananda Mohan, Use of bilinear terminations for LDI type SC ladder filters, *Electronics Letters*, 21, 14, 588–9, 4 July 1985.

[6.15] T. Inoue and F. Ueno, New switched-capacitor immittance converter using operational amplifier leap-frog structure, *Electronics Letters*, 16, 7, 266–7, 27 March 1980.

[6.16] T. Inoue and F. Ueno, New bilinear switched-capacitor immittance converter circuit, *Electronics Letters*, 16, 8, 285–6, 10 April 1980.

[6.17] T. Inoue and F. Ueno, Switched-capacitor immittance converter using a single operational amplifier, *Electronics Letters*, 16, 20, 770–1, 26 September 1980.

[6.18] T. Inoue and F. Ueno, New switched-capacitor ladder filter with reduced number of operational amplifiers, *Electronics Letters*, 17, 5, 209–10, 5 March 1981.

[6.19] T. Inoue and F. Ueno, Analysis and synthesis of switched-capacitor circuits using switched-capacitor immittance converters, *IEEE Transactions on Circuits and Systems*, CAS-29, 7, 458–66, July 1982.

[6.20] L.T. Bruton, Topological equivalence of inductorless ladder structures using integrators, *IEEE Transactions on Circuit Theory*, CT-20, 4, 434–7, July 1973.

[6.21] T. Inoue and F. Ueno, Realization of switched-capacitor immittance converters and their use in the design of active SC filters, *Transactions CAS* (Japan), 11, 61–8, 1980 (in Japanese).

[6.22] A. Fettweis, Basic principles of switched-capacitor filters using voltage inverter switches, *AEU*, 33, 13–19, 1979.

[6.23] A. Fettweis, Switched-capacitor filters using voltage inverter switches: further design principles, *AEU*, 33, 107–14, 1979.

[6.24] J.A. Nossek and H. Weinrichter, Equivalent circuits for switched capacitor networks including recharging devices, *IEEE Transactions on Circuits and Systems*, CAS-27, 6, 539–44, June 1980.

[6.25] D. Herbst, A. Fettweis, B. Hoefflinger, U. Kleine, W. Nientiedt, J. Pandel and R. Schweer, An integrated seventh-order unit element filter in VIS-SC technique, *IEEE Journal of Solid State Circuits*, SC-16, 3, 140–46, June 1981.
[6.26] D. Herbst, J. Pandel, A. Fettweis and B. Hoefflinger, VIS-SC filters with reduced influences of parasitic capacitances, *Proc. IEE*, Part G, 129, 2, 29–39, April 1982.
[6.27] A. Fettweis, D. Herbst and J.A. Nossek, Floating voltage inverter switches for switched-capacitor filters, *AEU*, 33, 376–7, 1979.
[6.28] A. Fettweis, D. Herbst, B. Hoefflinger, J. Pandel and R. Schweer, MOS switched-capacitor filters using voltage inverter switches, *IEEE Transactions on Circuits and Systems*, CAS-27, 6, 527–38, June 1980.
[6.29] J.D. Rhodes, *Theory of Electrical Filters*, Wiley, 1976.
[6.30] A. Matsumoto (ed.), *Microwave Filters and Circuits*, Academic Press, 1970.
[6.31] D. Herbst and B.J. Hosticka, Novel bottom-plate stray-insensitive voltage inverter switch, *Electronics Letters*, 16, 16, 636–7, 31 July 1980.
[6.32] J. Pandel, Scaling techniques for switched-capacitor filters employing voltage-inverter switches, *International Journal of Circuit Theory and Applications*, 11, 1, 73–96, January 1983.
[6.33] R. Schweer, B. Hoefflinger, B.J. Hosticka and U. Kleine, Novel stray-insensitive voltage inverter switches, *AEU*, 36, 270–4, 1982.
[6.34] S.M. Faruque, M. Vlach, J. Vlach, K. Singhal and T.R. Viswanathan, FDNR switched-capacitor filters insensitive to parasitic capacitances, *IEEE Trans. on Circuits and Systems*, CAS-29, 9, 589–95, September 1982.
[6.35] T.S. Rathore, Inverse active networks, *Electronics Letters*, 13, 303–4, May 1977.
[6.36] M. Cooperman and C.W. Kapral, Integrated SC FDNR filter, *IEEE Journal of Solid-State Circuits*, SC-18, 4, 378–83, August 1983.
[6.37] T. Inoue and F. Ueno, Parasitics-compensated bilinear switched-capacitor ladder filters using switched-capacitor immittance converters, *Proc. ISCAS, New Port Beach*, 582–5, 1983.
[6.38] M.K. Li, Cascade synthesis of switched-capacitor networks employing voltage inverter switches, *Electronics Letters*, 16, 10, 370–1, 9 May 1980.
[6.39] J. Pandel, Compensation of the effects of top plate parasitic capacitances in VIS-SC filters, *AEU*, 37, 65–7, 1983.
[6.40] J. Pandel, Sequential voltage-inversion principle for switched-capacitor filters, *Electronics Letters*, 15, 13, 399–400, 21 June 1979.
[6.41] J. Pandel, Switched-capacitor elements for VIS-SC filters with reduced influence of parasitic capacitances, *AEU*, 35, 121–30, 1981.

## 6.10.1 Further reading

[6.42] T. Inoue and F. Ueno, A new switched capacitor immittance converter and its exact analysis, *Transactions IECE* (Japan), 63, 231–2, 1980.
[6.43] J. Pandel and D. Herbst, VIS-SC filters for higher clock frequency applications, *Electronics Letters*, 17, 14, 504–6, 9 July 1981.
[6.44] U. Kleine, D. Herbst, B. Hoefflinger, B.J. Hosticka and R. Schweer, Real-time programmable unit element SC filter for LPC synthesis, *Electronics Letters*, 17, 17, 600–2, 20 August 1981.

[6.45] R. Raschke, Analysis of switched-capacitor filters with parasitic elements based upon the voltage inversion principle, *AEU*, 36, 119–23, 1982 (in German).

[6.46] B.J. Hosticka, D. Herbst, B. Hoefflinger, U. Kleine, J. Pandel and R. Schweer, Real-time programmable low power SC bandpass filter, *IEEE Journal of Solid State Circuits*, SC-17, 3, 499–506, June 1982.

[6.47] D. Bruckman and U. Kleine, Integrated VIS SC filter with lattice reference structure, *Electronics Letters*, 20, 15, 627–8, 19 July 1984.

[6.48] F. Montecchi, Switched capacitor recharging devices for VIS SC filters using fully differential operational amplifiers, *Electronics Letters*, 20, 6, 236–8, 15 March 1984.

[6.49] J. Pandel, Switched-capacitor filters employing voltage inverter switches – new results, *European Conference on Circuit Theory and Design Proceedings, Stuttgart*, pp. 22–4, 1983.

[6.50] U. Kleine, Design of wave SC filters using building blocks, *International Journal of Circuit Theory and Applications*, 12, 2, 69–87, April 1987.

[6.51] K. Chen and S. Eriksson, Symmetrical wave SC bandpass filter, *IEEE Transactions on Circuits and Systems*, CAS-32, 3, 301–3, March 1985.

[6.52] U. Kleine, W. Brockherde, A. Fettweis, B.J. Hosticka, J. Pandel and G. Zimmer, An integrated six-path wave SC filter, *IEEE Journal of Solid State Circuits*, SC-20, 2, 632–40, April 1985.

[6.53] A.M. Davis, Wave-variable analysis of passive switched-capacitor circuits, *IEEE Transactions on Circuits and Systems*, CAS-32, 9, 935–7, September 1985.

[6.54] J.O. Scanlan and P.J. O'Donovan, Adapter realizations for wave switched-capacitor filters with reduced capacitance spread, *Electronics Letters*, 22, 19, 969–70, 11 September 1986.

[6.55] D. Bruckmann, Design of switched-capacitor elements with voltage inverter switches for the simulation of FDNR reference structures, *International Journal of Circuit Theory and Applications*, 13, 1, 19–35, January 1985.

[6.56] F. Montecchi, Low-pass to high-pass transformations in VIS switched-capacitor filters, *Electronics Letters*, 21, 4, 146–8, 14 February 1985.

[6.57] D. Bruckmann and U. Kleine, Novel voltage inverter switches with minimum sensitivity properties, *IEEE Transactions on Circuits and Systems*, CAS-32, 7, 723–6, July 1985.

[6.58] A. Handkiewica, Switched-capacitor circuit synthesis based on gyrator-capacitor prototype, *International Journal of Circuit Theory and Applications*, 15, 3, 305–9, July 1987.

[6.59] K. Nagaraj and J. Vlach, Parasitic tolerant component simulation type switched-capacitor filter using unity-gain buffers, *IEEE Transactions on Circuits and Systems*, 35, 1, 35–43, January 1988.

# CHAPTER SEVEN

# SC filters based on the operational simulation of LC ladders and on multiloop feedback concepts

Chapter 6 discussed the design of SC ladder filters using the component simulation technique. This method has two disadvantages for realizing high-performance filters. First, parasitic capacitances are unavoidable at the various nodes in the network joining the passive components in the prototype. Second, there is no facility for scaling the dynamic range. These two disadvantages can be overcome by the *operational simulation* method. The resulting SC filters enjoy the low-sensitivity properties of the doubly terminated LC networks and the extensive design information available for RLC filters can be profitably used. Further, they can be designed to be stray-insensitive and can be scaled for optimal dynamic range.

The design of SC ladder filters can be based on the application of LDI, bilinear or *p*-transformation on the analog transfer function. All these methods will be considered in what follows. The SC filter design is based largely on that of active RC filters, so we also study the design of leap-frog active RC ladder filters in this chapter.

## 7.1  All-pole low-pass ladder filters

### 7.1.1  Active RC all-pole low-pass ladder realization
[1.25, 1.29, 7.1–7.3]

Consider the passive doubly terminated fifth-order LC LP filter shown in Figure 7.1(a). The following equations can be written:

$$\frac{V_i - V_1}{(R_s/R)} = (I_1 R)$$

$$(I_2 R) = (I_1 R) - (I_3 R)$$

$$V_1 = (I_2 R) \frac{1}{sc_1 R}$$

$$(I_3 R) = \frac{(V_1 - V_2)}{sl_2/R}$$

$$(I_4 R) = (I_3 R) - (I_5 R)$$

$$V_2 = \frac{(I_4 R)}{sc_3 R}$$

$$\frac{V_2 - V_3}{sl_4/R} = (I_5 R)$$

$$V_3 = \frac{(I_6 R)}{sc_5 R}$$

$$I_6 R = I_5 R - I_7 R$$

$$I_7 R = V_3 \left(\frac{R}{R_L}\right)$$

(7.1)

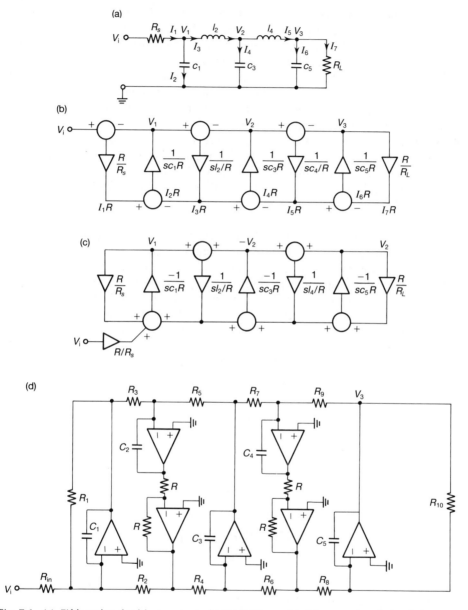

*Fig. 7.1* (a) Fifth-order doubly terminated LC LP filter; (b), (c) operational simulation block diagrams; (d) circuit implementation.

Note that the various currents are converted into voltages using an arbitrary resistance $R$. The block diagram* of Figure 7.1(b) can be constructed to realize the relationships in equations (7.1). It is evident from the block diagram that the simulation of the passive RLC ladder filter requires integrators and summers (differential summers). It is also observed that there are several two-integrator loops in the simulation obtained above. Further, each loop has a negative gain which is independent of $R$, the scaling resistor. The terminations of the ladder filter are realized in the first and last loop in Figure 7.1(b). When we attempt to realize the block diagram of Figure 7.1(b), it is observed that the two inputs to the various summers do not have the same sign. It is convenient to perform the summation at the virtual ground of the integrators, for which purpose the various inputs to be summed should have the same sign. This requirement is satisfied by a slight rearrangement of the block diagram of Figure 7.1(b) to that in Figure 7.1(c), which leads to the circuit of Figure 7.1(d).

In the block diagram of Figure 7.1(c), it is noted that some non-inverting integrators and some inverting integrators are required. The non-inverting integrators are used to realize the ladder inductances, while the inverting integrators are used to realize the ladder capacitances. Further, the outputs of the OAs represent the voltages across the capacitors and currents in the inductors in the prototype, shown at the top and bottom of Figure 7.1(d), respectively. The terminations are realized by converting the first and last lossless integrators into lossy integrators. There is an additional facility in the resistor $R_{in}$ to scale the gain of the filter, if desired.

Note that arbitrary values have been assigned for all the resistors and capacitors in Figure 7.1(d). These are related to the component values in the prototype of Figure 7.1(a) as follows:

$$C_1 R_1 = c_1 R_s$$

$$C_1 R_2 C_2 R_3 = l_2 c_1$$

$$C_2 R_5 C_3 R_4 = l_2 c_3$$

$$C_3 R_6 C_4 R_7 = l_4 c_3 \qquad (7.2)$$

$$C_4 R_9 C_5 R_8 = l_4 c_5$$

$$C_5 R_{10} = c_5 R_L$$

$$R_{in} = R_1$$

Thus from the component values in the prototype, the various component values in the practical realization can be evaluated.

In the above discussion, the design of odd-order active RC leap-frog filters has been considered. In the case of even-order filters, the prototype will be different (see Figure 7.2(a)). The resulting block diagram simulation is shown in Figure 7.2(b).

---

* This structure resembles the famous children's game 'leap-frog', and hence is called the *leap-frog* structure.

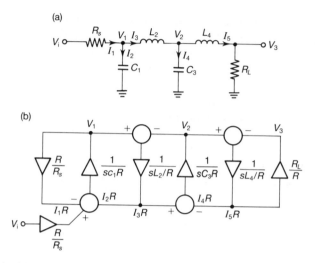

*Fig. 7.2* (a) A doubly terminated even-order filter; (b) its simulation.

Note that in this case the output $V_3$ is the current $I_5$ scaled by the load termination resistance and consequently the last multiplier is proportional to $R_L$, whereas in the case of odd-order filters the last multiplier is inversely proportional to $R_L$. This will lead to certain differences in the synthesis of SC networks, while the active RC filter design is not affected in any way.

It is also worth recalling in this connection that RLC odd-order filters can be designed with equal terminations, whereas even-order filters of equiripple response cannot be realized for equal terminations. Furthermore, the prototype RLC filter can also be of the minimum capacitance form. (Note that the configurations of Figures 7.1(a) and 7.2(a) are of minimum inductance type.) Accordingly, the operational simulation method leads to different block diagrams (see Exercise 7.2). Note, however, that the operational simulation method leads to the same circuit topology, whether minimum-inductance or minimum-capacitance prototypes are used as the starting point.

## 7.1.2  SC all-pole low-pass ladder realizations

The basic blocks required to realize an active RC ladder filter are: non-inverting and inverting lossless integrators to simulate the ladder inductances and capacitances; and lossy integrators to simulate the termination resistances. In the active RC version, it is known that non-inverting integrators are realized using two OAs, one of which provides an inversion and the other realizes an inverting integrator. The active RC integrators could be replaced with SC integrators to realize an SC ladder filter. However, as seen in Chapter 4, a non-inverting SC integrator can be realized using one OA, where the inverting switched capacitor is used to advantage. (The reader may recall that the Fleischer–Laker biquad uses a lossy integrator and a loss-less integrator, one of which is non-inverting, realized using the above principle.) Thus, it is first noted that SC all-pole LP ladder filters need as many OAs as the order of the filter.

Due to the large number of integrator configurations available, the choice of the proper integrator comes into consideration. Early SC ladder filter designs [4.2, 5.20] were based on the replacement of resistors in a two-integrator loop by parallel-switched capacitors. We first consider two such lossless integrators connected in a loop (Figure 7.3(b)) to realize the inductance and capacitance subnetwork in the LC ladder shown in Figure 7.3(a). The following equations can be written by inspection:

$$V_{1o} = \frac{-C_1 z^{-1/2}}{C_2(1 - z^{-1})} V_{2e}$$

$$V_{1e} = V_{1o} z^{-1/2} \tag{7.3}$$

$$V_{2e} = \frac{C_3 z^{-1}}{C_4(1 - z^{-1})} V_{1e}$$

The characteristic equation of the system can then be obtained:

$$D(z) = C_2 C_4 z^2 - 2C_2 C_4 z + C_1 C_3 + C_2 C_4 = 0 \tag{7.4}$$

It is easy to see that the resulting complex poles are outside the unit circle and, as such, the system is not an exact sinusoidal oscillator, but only a relaxation oscillator.

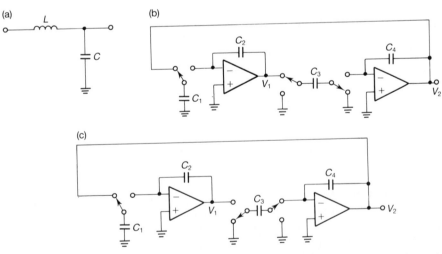

Fig. 7.3  (a) LC sub-network;  (b) SC two-integrator loop realizing right-half-plane poles; (c) SC two-integrator loop realizing $j\omega$-axis poles and LDI transformation.

It is thus noted that the arrangement of Figure 7.3(b) does not simulate exactly the LC network of Figure 7.3(a).

It is interesting, however, to see that, by changing the phasing of the switches associated with $C_3$ in Figure 7.3(b) to that shown in Figure 7.3(c), the following characteristic equation is obtained as

$$D(z) = C_2 C_4 z^2 - (2C_2 C_4 - C_1 C_3)z + C_2 C_4 = 0 \qquad (7.5)$$

showing that a pair of complex conjugate poles on the unit circle in the $z$-domain can be realized. In this case, the two transfer functions of the integrators realize an LDI transformation:

$$\frac{V_{1o}}{V_{2e}} = \frac{-C_1 z^{-1/2}}{C_2(1 - z^{-1})} \qquad (7.6a)$$

$$\frac{V_{2e}}{V_{1o}} = \frac{-C_3 z^{-1/2}}{C_4(1 - z^{-1})} \qquad (7.6b)$$

It can thus be seen that the arrangement of Figure 7.3(c) realizes exactly, based on the LDI transformation, the LC network of Figure 7.3(a) [4.2, 5.20].

Referring once again to the block diagram of Figure 7.1(b), it is noted that several differential integrators are required in the implementation. This requirement for SC designs is satisfied by a simple modification of the SC network of Figure 7.3(c) to that shown in Figure 7.4(a). Thus, an SC equivalent can be constructed from the block diagram representation of Figure 7.1(b) in a straightforward manner. There are, however, two further aspects to be considered.

The first is concerned with the parasitic capacitances that are invariably present in the configurations of Figures 7.3(b), 7.3(c) and 7.4(a). These can be avoided by using a pair of stray-insensitive integrators, as shown in Figure 7.4(b). The design of SC ladder filters using LDI transformation and with complete insensitivity to parasitic capacitances is therefore based on the arrangement of Figure 7.4(b).

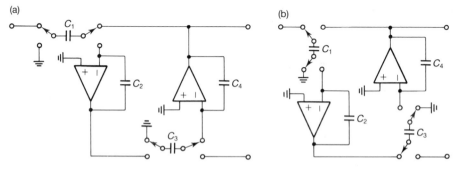

Fig. 7.4  Two-integrator loops realizing LDI transformation:  (a) top-plate sensitive type; (b) stray-insensitive realization. (The circuits are also differential summers.) (Adapted from [4.2], © 1978 IEEE.)

The second aspect to be considered is the realization of the termination resistance. In Chapter 6, it was noted that for ladder filters based on LDI transformation, the termination resistances can only be approximated and, as such, errors will arise due to these approximations. Since it is required to maintain stray insensitivity, only the series-switched capacitor can be used in the feedback path of the lossy integrator. Consequently, series-switched capacitors operating in either phase can be used in the realization of lossy integrators simulating the terminations, as discussed in Section 6.2. These are shown in Figure 7.5. We have thus arrived at the basic structure of SC ladder filters based on LDI transformation. As an illustration, the SC realization of a third-order all-pole LP filter based on the prototype of Figure 7.6(a) is shown in Figure 7.6(b). It employs series-switched capacitors in the terminations. The details of the design are considered in Example 7.1.

## *Stability of LDI-transformed filters* [1.19]

Certain observations on ladder filter realizations based on LDI transformation will now be made. Using LDI integrators in the SC version corresponding to the various integrator blocks in the block diagram of Figure 7.1(b), we obtain the following digital transfer function:

$$H_d(z) = \frac{1}{\frac{a_0}{T^n} (z^{1/2} - z^{-1/2})^n + \frac{a_1}{T^{n-1}} (z^{1/2} - z^{-1/2})^{n-1} + \cdots + 1} \qquad (7.7a)$$

corresponding to an $H_a(s)$ given by

$$H_a(s) = \frac{1}{a_0 s^n + a_1 s^{n-1} + \cdots + 1} \qquad (7.7b)$$

$H_d(z)$ is evidently of order $2n$ in $z^{1/2}$ and consequently has $2n$ roots. Further, since substitution of $z^{1/2}$ by $-z^{-1/2}$ does not change the nature of $H_d(z)$, it follows that $n$ of the $2n$ roots lie within the unit circle and $n$ lie outside the unit circle. Consequently, LDI-type mapping results in instability. Fortunately, this disadvantage is overcome by the impossibility of simulating exact terminating resistances. The denominators of the transfer functions of the lossy integrators

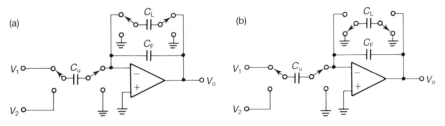

*Fig. 7.5* SC lossy integrators: (a) capacitor simulating termination ($C_L$) switched in phase with $C_u$, where $C_u$ is a unit capacitor; (b) $C_L$ switched in anti-phase with $C_u$. *(Adapted from [4.14], © 1981 IEEE.)*

simulating the termination no longer have $z^{-1/2}$ terms so that the loop gains of all two-integrator loops are rational functions of $z^{-1}$. Consequently, the order of the system reduces from $2n$ to $n$.

## Design of LDI-transformed filters

The next task before us is to design an LDI-type filter with non-ideal terminations for the given specifications [4.3]. For high clock frequencies, the termination resistances may be considered to be ideal. Thus from the prewarped $s$-domain transfer function, the element values of the prototype can be determined. Thus, the capacitor ratios can be obtained in the SC version [4.2, 5.20].

For moderate clock frequencies, it is desirable to have an exact design. One design procedure [1.19] starts with mapping of the prewarped $s$-domain transfer function. The poles of the prototype network are first mapped into the $z$-domain, according to the transformation

$$s \rightarrow \frac{z^{1/2} - z^{-1/2}}{T} \tag{7.8}$$

Evidently, two roots in the $z$-domain are obtained, of which the one within the unit circle is chosen. Thus, the $z$-domain transfer function can be obtained as discussed in Chapter 1. This will then be matched with the actual $z$-domain transfer function of the SC network. The resulting non-linear equations in capacitor ratios can be solved to obtain the various capacitor ratios, using the Newton–Raphson method, with approximate values of the capacitor ratios as the starting point. These approximate values are obtained directly from the RLC prototype element values, used to realize the prewarped specifications.

The following example illustrates the design of a third-order LP Chebyshev SC ladder filter based on LDI transformation.

*Example 7.1*
Design an LDI-type LP ladder filter to realize a 0.1 dB Chebyshev response and a cutoff frequency of 11 kHz. Choose a sampling frequency of 100 kHz.

The prewarped cutoff frequency of the filter is

$$\omega_c = \frac{2}{T} \sin \frac{\omega_d T}{2} = 67\ 747.6\ \text{rad s}^{-1}$$

(or $f_c = 10\ 782.36$ Hz). The third-order LP Chebyshev prototype shown in Figure 7.6(a) can be denormalized for this cutoff frequency first. Next, we can evaluate the various capacitor ratios in the SC ladder filter of Figure 7.6(b). Considering

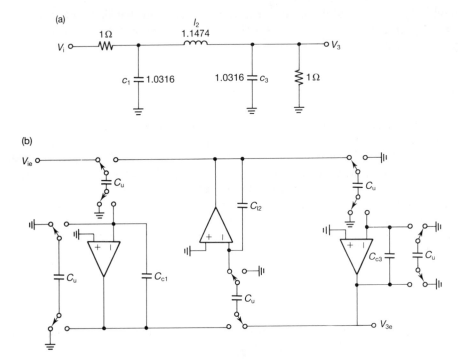

*Fig. 7.6* (a) Prototype LP filter; (b) SC version based on LDI transformation.

the input capacitors for all integrators as $C_u$, the termination capacitors are obtained as follows:

$$\frac{C_{c1}}{C_u} = \frac{c_1 f_s}{\omega_c} = 1.5227$$

$$\frac{C_{l2}}{C_u} = \frac{l_2 f_s}{\omega_c} = 1.6936$$

$$\frac{C_{c3}}{C_u} = \frac{c_3 f_s}{\omega_c} = 1.5227$$

The resulting $z$-domain transfer function is

$$\frac{V_{3e}}{V_{ie}} = \frac{z^{3/2}}{\left[1 + \dfrac{C_{c1}}{C_u}(1 - z^{-1})\right]\left[\dfrac{C_{l2}}{C_u}(1 - z^{-1})\left\{1 + \dfrac{C_{c3}}{C_u}(1 - z^{-1})\right\} + z^{-1}\right] + z^{-1}\left[(1 - z^{-1})\dfrac{C_{c3}}{C_u} + 1\right]}$$

$$= \frac{0.2546 z^{3/2}}{2.744\,753 z^3 - 4.773\,347 z^2 + 3.537\,921\,5 z - 1} \tag{7.9}$$

The desired $z$-domain transfer function is, however, obtained by mapping the poles in the $s$-domain into the $z$-domain according to the transformation in expression (7.8). Thus, we obtain the LDI-transformed $z$-domain poles as $z_1 = 0.5244$, $z_{2,3} = 0.4732 \pm j0.5157$, corresponding to the $s$-domain poles $-65\ 676$, $-32\ 838 \pm j81\ 716$. From these $z$-domain roots, we obtain the actual desired transfer function:

$$H_d(z) = \frac{0.5031 z^{3/2}}{3.8929 z^3 - 5.7254 z^2 + 3.8388 z - 1} \tag{7.10}$$

Equations (7.9) and (7.10) have to be matched to obtain the various capacitor ratios. The three non-linear equations resulting in this process can be solved to give the capacitor ratios

$$C_{c1} = 1.691\ 761\ 1 C_u$$

$$C_{l2} = 1.699\ 524\ 1 C_u$$

$$C_{c3} = 0.691\ 276\ 1 C_u$$

This completes the design procedure.

Note that, in this example, the capacitor ratios are dependent on the coefficients of the transfer function desired and there is no freedom to optimize the dynamic range. However, by increasing the number of switches and capacitors, it is possible to scale the circuit in order to achieve optimal dynamic range. This facility is achieved by realizing the differential summing operation performed by the capacitors $C_u$ in Figure 7.6(b) through two different switched capacitors. This modified circuit is shown in Figure 7.7.

An even-order stray-insensitive LDI-type SC filter structure is shown in Figure 7.8. Notice the phasing of the termination resistances in the SC filters of

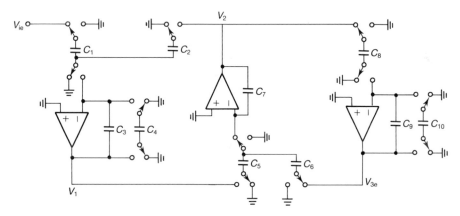

Fig. 7.7   LDI transformed filter with facility to optimize the dynamic range.

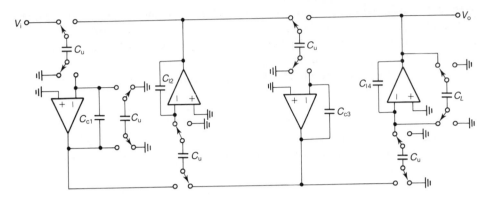

*Fig. 7.8* An even-order stray-insensitive LDI SC filter.

Figures 7.6(b) and 7.8. These are opposite, because, in both the cases, the termination resistance realized is of type $Rz^{1/2}$.

The above-discussed design technique for LDI transformation type SC ladder filters requires the solution of non-linear equations. However, a rigorous theoretical method exists for the exact synthesis of LDI filters. This will be considered next.

## Exact synthesis of LDI transformation type SC all-pole low-pass filters [7.5, 4.14]

No matter how the SC equivalent is derived, i.e., from a minimum inductance or minimum capacitance ladder, the stray-insensitive configurations are unique. The transfer functions of the SC lossless (LLI) and lossy (LI) integrators derived from Figure 7.5(a) with $C_L = 0$ and $C_L \neq 0$ are, respectively,

$$T_{LLI} = \frac{-z^{-1/2}[V_1 - V_2]}{\left(\dfrac{C_F}{C_u}\right)(1 - z^{-1})} \tag{7.11a}$$

$$T_{LI} = \frac{-z^{-1/2}[V_1 - V_2]}{\left(\dfrac{C_F}{C_u}\right)(1 - z^{-1}) + \dfrac{C_L}{C_u}}$$

which can be rewritten as

$$T_{LLI} = \frac{-[V_1 - V_2]}{2\left(\dfrac{C_F}{C_u}\right)\sinh\left(\dfrac{sT}{2}\right)} \tag{7.11b}$$

$$T_{LI} = \frac{-[V_1 - V_2]}{\left(\dfrac{2C_F + C_L}{C_u}\right)\sinh\left(\dfrac{sT}{2}\right) + \dfrac{C_L}{C_u}\cosh\left(\dfrac{sT}{2}\right)}$$

(Note that $z = \exp(sT)$ has been used in the above equations.)

Introducing the notation

$$\sinh\left(\frac{sT}{2}\right) = \gamma$$

$$\cosh\left(\frac{sT}{2}\right) = \mu \tag{7.12}$$

$$\tanh\left(\frac{sT}{2}\right) = \lambda$$

the above transfer functions of the integrators become

$$T_{\text{LLI}} = \frac{-[V_1 - V_2]}{2\dfrac{C_F}{C_u}\gamma}$$

$$T_{\text{LI}} = \frac{-[V_1 - V_2]}{\dfrac{2C_F + C_L}{C_u}\gamma + \dfrac{C_L}{C_u}\mu} \tag{7.13}$$

Using the transfer functions given by equations (7.13), we can construct the RLC equivalent circuits for the stray-insensitive LDI SC odd-order and even-order filters of Figures 7.6(b) and 7.8, as shown in Figures 7.9(a) and 7.9(b). Note that, in both cases, the SC filter equivalent circuit has frequency-dependent terminations (because of the factor $\mu$). In the even-order case, the terminations, however, are $\mu R_s$ and $R_L/\mu$, as against $\mu R_s$ and $\mu R_L$ for the odd-order case.

The component values in the equivalent circuit of odd-order filters can be scaled by the factor $\mu$ so that the terminations are purely resistive, whereas the remaining network is a reactance network in the new frequency variable $\lambda$. This new equivalent circuit is shown in Figure 7.9(c).

Note that since the desired transfer function is $(I_L/V_{ie})$ for the odd-order SC ladder filters, the scaling performed to obtain Figure 7.9(c) will change the load current $I_L$ while the voltage remains constant. Thus, the transfer functions of Figures 7.9(a) and 7.9(c) are related by

$$H'_{21} = \mu H_{21} \tag{7.14}$$

Matters are, however, different in the case of even-order LDI SC filters. Scaling by the factor $\mu$ the network of Figure 7.9(b), we obtain Figure 7.9(d) in which the termination resistance is $R_L/\mu^2$. The equivalent circuit of Figure 7.9(d) is modified to that of Figure 7.9(e) to make the load frequency-independent. This modification, however, realizes across the load $R_L$ the transfer function $H'_{21}$ $(=\mu H_{21})$ thus satisfying equation (7.14) in this case as well.

The object is now to synthesize the reactance networks in Figures 7.9(c) and 7.9(e) for the desired specifications. Note that these reactance networks are not of conventional LC type, but are similar to ladder networks encountered in microwave

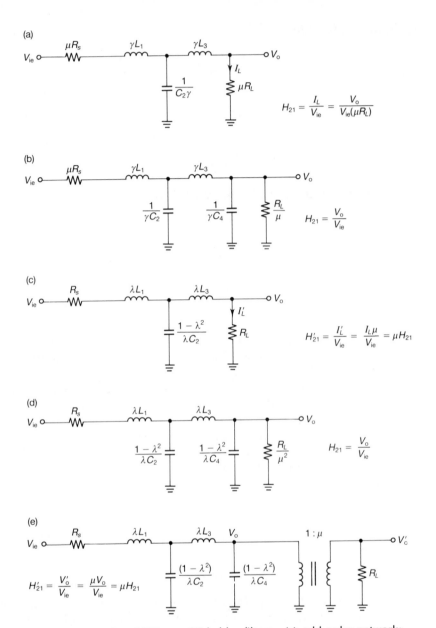

Fig. 7.9 Equivalent circuits of LDI-type SC ladder filters: (a) odd-order network; (b) even-order network; (c) network obtained from (a) by dividing all impedances by $\mu$; (d) network obtained from (b) by dividing all impedances by $\mu$; (e) modification of (d) illustrating that $H'_{21} = \mu H_{21}$.

filters. This analogy will be apparent when we invoke the concept of a 'unit element' which has an $ABCD$ matrix given by

$$\frac{1}{\sqrt{1-\lambda^2}} \begin{bmatrix} 1 & Z_0\lambda \\ Y_0\lambda & 1 \end{bmatrix} \qquad (7.15)$$

where $Z_0$ is the characteristic impedance. A cascade of two such unit elements (Figure 7.10(a)) with characteristic impedances $Z_1$ and $Z_2$ has $ABCD$ matrix given by

$$\frac{1}{1-\lambda^2} \begin{bmatrix} 1+Z_1Y_2\lambda^2 & (Z_1+Z_2)\lambda \\ (Y_1+Y_2)\lambda & 1+Z_2Y_1\lambda^2 \end{bmatrix} \qquad (7.16)$$

The two-port shown in Figure 7.10(b) has the same $ABCD$ parameters as in expression (7.16). Thus the ladder network in Figure 7.9(c) or 7.9(e) can be looked upon as a cascade of unit elements. The inductances in the series branches in this cascade of Figure 7.11(a) for odd-order filters are positive or negative, and all of them can be shifted to one end of the ladder resulting in the equivalent circuit of Figure 7.11(b). The resulting transfer function [7.4] will be of the form

$$H'_{21} = \frac{(1-\lambda^2)^{(n-1)/2}}{D_n(\lambda)} \qquad (7.17)$$

where the numerator is the result of $(n-1)$ terms, each being $(1-\lambda^2)^{1/2}$ corresponding to a unit element in the cascade of the $n-1$ unit elements. The result of the series inductance is a transmission zero at infinity. Using equation (7.14) in equation (7.17), we immediately obtain, for odd-order networks,

$$H_{21} = \frac{(1-\lambda^2)^{n/2}}{D_n(\lambda)} \qquad (7.18)$$

In the case of even-order networks, the transformer at the load end shown in Figure 7.9(e), redrawn in Figure 7.11(c), has to be taken into account. Proceeding in the same manner as in the case of odd-order networks, by substituting a cascade of two unit elements for each two-port of the type shown in Figure 7.10(b), the

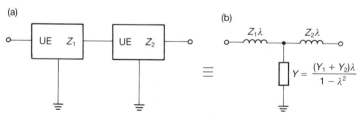

*Fig. 7.10* (a) A cascade of two unit elements; (b) equivalent of (a).

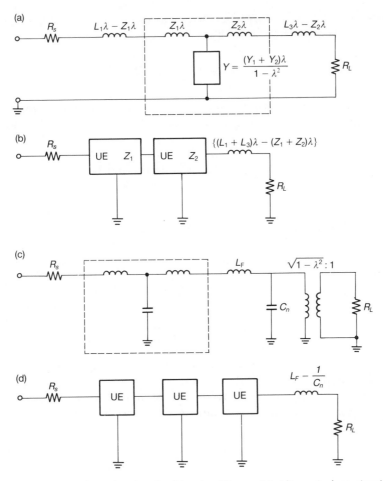

*Fig. 7.11* (a), (b) Equivalent circuits of odd-order filters;  (c), (d) equivalent circuits of even-order filters.

equivalent circuit shown in Figure 7.11(d) is obtained. The last inductance $L_F$, capacitance $C_n$ and the transformer have the overall $ABCD$ matrix

$$\frac{1}{\sqrt{1-\lambda^2}} \begin{bmatrix} 1+(L_F C_n - 1)\lambda^2 & L_F\lambda \\ C_n\lambda & 1 \end{bmatrix} \qquad (7.19)$$

which can be seen to correspond to an inductance $L_F - 1/C_n$ in cascade with a unit element of characteristic impedance $1/C_n$. Thus, once again, in the even-degree case as well, the equivalent circuit has $n-1$ unit elements with a series inductor in cascade. The next problem to be considered is the realization of the transfer function $H_{21}$ given in equations (7.11) to approximate the desired specification.

Noting that $\lambda = j \tan(\omega T/2)$, the magnitude squared function $|H_{21}|^2$ can be written from equation (7.18), after some manipulation, as

$$|H_{21}|^2 = \frac{1}{F_n\left(\sin^2 \dfrac{\omega T}{2}\right)} \qquad (7.20)$$

The denominator of $|H_{21}|^2$ in equation (7.20) can be chosen so as to realize a maximally flat or Chebyshev approximation:

$$|H_{21}|^2 = \frac{1}{1 + \left(\dfrac{\sin \theta}{\alpha}\right)^{2n}} \qquad (7.21a)$$

for the maximally flat case, and

$$|H_{21}|^2 = \frac{1}{1 + \varepsilon^2 T_n^2\left(\dfrac{\sin \theta}{\alpha}\right)} \qquad (7.21b)$$

for the Chebyshev case, where $\theta = \omega T/2$, $T_n$ is the Chebyshev polynomial of $n$th order, $\varepsilon$ is the pass-band ripple and $\alpha$ is chosen so as to specify the 3 dB point at the desired frequency.

Now the usual procedure used in the synthesis of doubly terminated reactance networks is followed. First, $\hat{H}_{21}$ is evaluated using equation (7.14). Denoting the transducer power gain as $S_{21}$, we obtain for the networks of Figures 7.9(c) and 7.9(e),[*]

$$|S_{21}|^2 = |\hat{H}_{21}|^2 4R_L R_g, \qquad \text{for } n \text{ odd} \qquad (7.22a)$$

$$|S_{21}|^2 = |\hat{H}_{21}|^2 \frac{4R_g}{R_L}, \qquad \text{for } n \text{ even} \qquad (7.22b)$$

(Note that in the relationship obtained above for odd-order networks, $H_{21}$ is defined as $I_L/V_{ie}$ and hence the difference.) Thus the input reflection coefficient is now obtained from

$$|S_{11}(\lambda)|^2 = 1 - |S_{21}(\lambda)|^2 \qquad (7.23)$$

Selecting the left-half $\lambda$-plane zeros, we obtain

$$S_{11} = \frac{N(\lambda)}{D(\lambda)} \qquad (7.24)$$

[*] The transducer power gain is the ratio of power available to maximum power that can be obtained. Thus for input voltage $V_{in}$ and load voltage $V_2$,

$$|S_{21}|^2 = \left(\frac{V_2^2}{R_L}\right) \Big/ \left(\frac{V_{in}^2}{4R_g}\right) = \left(\frac{V_2}{V_{in}}\right)^2 \cdot \frac{4R_g}{R_L}$$

It is next an easy matter to determine the input impedance from the reflection coefficient through the formula

$$\hat{Z}_{in} = \frac{1 + S_{11}(\lambda)}{1 - S_{11}(\lambda)} \tag{7.25}$$

It is observed that, by virtue of impedance scaling performed in deriving the equivalent circuits of Figures 7.9(a)–7.9(e), the input impedance has been scaled down by $\mu$. Therefore, the actual input impedance is obtained as

$$Z_{in} = \mu\hat{Z}_{in} = \frac{\mu[1 + S_{11}(\lambda)]}{[1 - S_{11}(\lambda)]} \tag{7.26}$$

The last part of the solution is to synthesize the network for the above $Z_{in}$.

Note that $Z_{in}$ in equation (7.26) is a function of $\lambda$ and $\mu$. By using the relationships

$$\mu^2 = (1 + \gamma^2)$$
$$\lambda = \gamma/\mu \tag{7.27}$$

it is possible to obtain $Z_{in}$ for the odd- and even-order cases, respectively, as

$$Z_{in(odd)} = \frac{\mu A_{n-1}(\gamma) + B_n(\gamma)}{\mu C_{n-2}(\gamma) + D_{n-1}(\gamma)} \tag{7.28a}$$

$$Z_{in(even)} = \frac{A_n(\gamma) + \mu B_{n-1}(\gamma)}{C_{n-1}(\gamma) + \mu D_{n-2}(\gamma)} \tag{7.28b}$$

where $A$, $B$, $C$, $D$ are the entries of the chain matrix of the reactive two-ports in Figures 7.9(a) and 7.9(b). Equations (7.28) show that $Z_{in(odd)}$ is realized as a ladder network terminated in a resistance $R_L\mu$, whereas $Z_{in(even)}$ can be realized as a reactance ladder terminated in a resistance $R_L/\mu$. Thus, by performing the continued fraction expansion of $Z_{in}$ about $\gamma$ in equations (7.28), we obtain the element values of the ladders in Figures 7.9(c) and 7.9(e). The next step is to obtain the capacitor ratios from these $L$ and $C$ values. Baher and Scanlan have presented an algorithm to obtain equations (7.28) from equation (7.26); the reader is referred to [7.5]. The following example demonstrates the ideas developed in the above method.

*Example 7.2*
Design a third-order Butterworth filter with a cutoff frequency of 1 kHz and using a sampling frequency of 16 kHz to realize LDI transformation.

We first obtain $\theta_0 = \omega_0 T/2$ as $\pi/16$. Thus the desired transfer function is obtained from equation (7.21a) as

$$|H_{21}|^2 = \frac{0.25}{1 + \left(\dfrac{\sin\theta}{\sin\theta_0}\right)^6} = \frac{0.25}{1 + 18\,138\,\sin^6\theta}$$

From equations (7.14) and (7.22a), we obtain

$$|S_{21}|^2 = \frac{\cos^2\theta}{1 + 18\ 138\ \sin^6\theta}$$

Using equation (7.23), we next obtain

$$|S_{11}|^2 = 1 - |S_{21}|^2 = \frac{\sin^2\theta + 18\ 138\ \sin^6\theta}{1 + 18\ 138\ \sin^6\theta}$$

At this stage, the relationship

$$\sin^2\theta = -\frac{\lambda^2}{1 - \lambda^2}$$

obtained from equations (7.12), is introduced. Thus

$$|S_{11}|^2 = S_{11}(\lambda)S_{11}(-\lambda) = \frac{-\lambda^2(1-\lambda^2)^2 - 18\ 138\lambda^6}{(1-\lambda^2)^3 - 18\ 138\lambda^6}$$

Factorizing and choosing the left half-plane poles in the $\lambda$-domain, we get

$$S_{11}(\lambda) = \frac{\lambda(134.68\lambda^2 + 16.473\lambda + 1)}{(5.2221\lambda + 1)(25.788\lambda^2 + 5.2252\lambda + 1)}$$

Hence $\hat{Z}_{\text{in}}$ is obtained as

$$\hat{Z}_{\text{in}} = \frac{269.34\lambda^3 + 69.547\lambda^2 + 11.447\lambda + 1}{36.602\lambda^2 + 9.4473\lambda + 1}$$

Then $Z_{\text{in}}$ is formed as $\mu\hat{Z}_{\text{in}}$ according to equation (7.26) and manipulated using equation (7.27) to obtain an expression in $\gamma$ and $\mu$:

$$Z_{\text{in}} = \mu\hat{Z}_{\text{in}} = \frac{\{280.79\gamma^3 + 11.447\gamma\} + \mu\{1 + 70.547\gamma^2\}}{\{37.602\gamma^2 + 1\} + \mu\{9.4473\gamma\}}$$

$Z_{\text{in}}$ can be synthesized, as shown in Figure 7.12, by a continued fraction expansion in $\gamma$:

$$Z_{\text{in}} = 7.4675\gamma + \cfrac{1}{9.4482\gamma + \cfrac{1}{3.9798\gamma + \mu}}$$

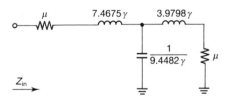

Fig. 7.12  Synthesized equivalent circuit for Example 7.2.

Note that the source and load resistances are measured in microhms. Identifying the equivalent circuit of Figure 7.12 with the third-order LDI SC filter of Figure 7.6(b) to obtain the capacitor ratios, the design values are as follows:

$$C_s = C_L = C_u$$

$$C_{c1} = 3.2337C_u$$

$$C_{l2} = 4.7241C_u$$

$$C_{c3} = 1.4899C_u$$

This completes the design of the desired SC ladder filter using LDI transformation.

The circuit obtained in Example 7.2 can be designed for optimal dynamic range also using the circuit arrangement of Figure 7.7 and then evaluating the outputs of the immediate OAs and scaling the component values to equalize the various maxima. This is left as an exercise to the reader.

In this section, the design of all-pole LP filters has been considered. Rafat and Mavor [7.6] have described experimental verification of the results of Scanlan's design procedure, by fabricating a monolithic SC LP filter based on LDI transformation. We consider elliptic-type LP filter design in the next section.

## 7.2  Low-pass elliptic ladder filters

### 7.2.1  Active RC realizations

We first consider a third-order doubly terminated elliptic LP network as shown in Figure 7.13(a), which can be modified as shown in Figure 7.13(b) [4.2, 5.20]. The resulting block diagram and a practical circuit are shown in Figures 7.13(c) and 7.13(d). In the practical situation, the summing operations must be performed at the virtual ground of the OAs, as shown in Figure 7.13(d). Notice that the cross-connection of capacitors $C_4$ realizes the function of $C_4$ in the original circuit.

The design values can be obtained from the component values of the RLC prototype. The outputs $V_1$ and $V_2$ of the OAs in the active RC realization correspond to the node voltages in the prototype. This procedure can be extended to higher order filters.

In the next section, we consider the design of SC elliptic LP ladder filters.

### 7.2.2  LDI-type realizations [7.7, 7.8]

The SC equivalents of LDI-type elliptic filters are obtained from their RLC prototypes, noting that the terminations are frequency-dependent in $\mu$, while the remaining is a reactance network in $\gamma$. The transfer functions of these networks can be obtained as rational functions in $\gamma$ and $\mu$ as in the all-pole LP case, but with finite transmission zeros in the $\gamma$-plane. The squared magnitude transfer function is first obtained as a real rational function in $\lambda^2$. The synthesis procedure then follows along the same lines as in the all-pole case.

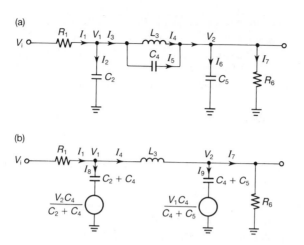

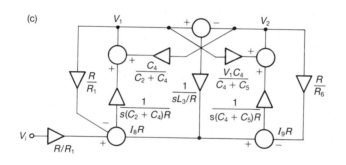

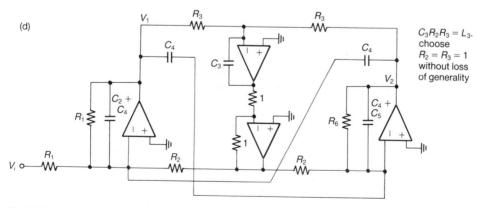

*Fig. 7.13*  (a) Third-order elliptic prototype filter;  (b) modification to eliminate $C_4$ in (a) using dependent voltage sources;  (c) simulation of block diagram of (b);  (d) practical active RC realization of (c).

In order to illustrate the procedure, the SC equivalent of the arrangement of Figure 7.13(d) is considered. The resulting SC LP elliptic ladder filter is shown in Figure 7.14. In the third-order SC filter, both inverting and non-inverting integrators are used; hence, the feeding arrangement to realize the transmission zeros is similar to that in the active RC case. The circuit can only be designed approximately.

The above procedure is illustrated in detail by Example 7.3.

*Example 7.3*
Design an SC elliptic LP filter of third order using LDI transformation for the following specifications:

- ratio of stop-band edge to pass-band edge frequencies = 2;
- ratio of sampling frequency to pass-band edge = 9.25;
- ripple = 0.28 dB (reflection coefficient $\rho = 25\%$);
- stop-band attenuation = 28.6 dB

The required $s$-domain transfer function can be obtained from filter design tables as

$$H_a(s) = \frac{0.208\ 685\ 4(s^2 + 5.153\ 208\ 7)}{s^3 + 1.453\ 809\ s^2 + 1.811\ 208s + 1.075\ 399\ 5}$$

The squared magnitude function in the $s$-domain is next obtained as

$$H_a(s)H_a(-s) = \frac{0.043\ 549\ 6(s^4 + 10.306\ 417s^2 + 26.555\ 56)}{-s^6 - 1.508\ 881s^4 - 0.153\ 669\ 8s^2 + 1.156\ 484\ 1}$$

Since synthesis is carried out in the $\gamma$-domain directly, replacing $s = j\gamma$, we get

$$|H_a(j\gamma)|^2 = \frac{0.043\ 549\ 6(\gamma^4 - 10.306\ 417\gamma^2 + 26.555\ 56)}{\gamma^6 - 1.508\ 881\gamma^4 + 0.153\ 669\ 8\gamma^2 + 1.156\ 484\ 1}$$

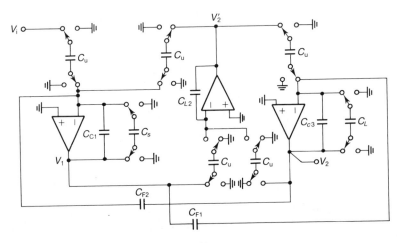

*Fig. 7.14*  A third-order elliptic SC LP ladder filter.

Denormalizing for the required ratio of sampling frequency to pass-band edge of 9.25, i.e. $\gamma_0 = \sin(\omega_0 T/2) = \sin[2\pi/(2 \times 9.25)] = 0.333\ 139\ 8$, we obtain

$$|H_a(j\gamma)|^2 = \frac{12.2264\gamma^4 - 13.984\ 945\gamma^2 + 4}{2530.2043\gamma^6 - 423.71\gamma^4 + 4.789\ 17\gamma^2 + 4}$$

Using the identity $\gamma^2 = -\lambda^2/(1 - \lambda^2)$ which can be obtained from equation (7.12), we obtain

$$|H_a(j\lambda)|^2 = \frac{(1 - \lambda^2)(2.241\ 455\lambda^4 + 5.984\ 945\lambda^2 + 4)}{-2115.2835\lambda^6 - 402.131\ 66\lambda^4 - 16.789\ 17\lambda^2 + 4}$$

The transducer power gain is obtained next by multiplying $|H_a(j\lambda)|^2$, obtained above, by $\mu^2$. (Note that the factor $4R_L/R_G$ equals unity since we have not taken the attenuation as 0.5 at the starting point of the design example). Then, since $\mu^2 = 1/(1 - \lambda^2)$,

$$|S_{21}(\lambda)|^2 = \frac{2.241\ 455\lambda^4 + 5.984\ 945\lambda^2 + 4}{-2115.2835\lambda^6 - 402.131\ 66\lambda^4 - 16.789\ 17\lambda^2 + 4}$$

The next step is to evaluate $|S_{11}|^2$ as $(1 - |S_{21}(\lambda)|^2)$, thus obtaining

$$|S_{11}(\lambda)|^2 = \frac{-\lambda^2(2115.2835\lambda^4 + 404.373\ 12\lambda^2 + 22.774\ 115)}{-2115.2835\lambda^6 - 402.131\ 66\lambda^4 - 16.789\ 17\lambda^2 + 4}$$

Factorizing $|S_{11}(\lambda)|^2$ and choosing the left-half $\lambda$-plane zeros gives

$$S_{11}(\lambda) = \frac{\lambda(45.909\ 521\lambda^2 + 5.880\ 557\lambda + 4.7702)}{45.909\ 521\lambda^3 + 23.932\ 403\lambda^2 + 10.6072\lambda + 2}$$

Next, the input impedance is obtained:

$$\hat{Z}_{in}(\lambda) = \frac{1 + S_{11}(\lambda)}{1 - S_{11}(\lambda)} = \frac{45.905\ 21\lambda^3 + 14.906\ 48\lambda^2 + 7.6887\lambda + 1}{9.025\ 923\lambda^2 + 2.9185\lambda + 1}$$

The actual input impedance is $\mu\hat{Z}_{in}(\lambda)$; hence, manipulating using equation (7.27), we get

$$Z_{in} = \frac{53.598\ 222\ 1\gamma^3 + 15.906\ 48\mu\gamma^2 + 7.6887\gamma + \mu}{10.025\ 923\gamma^2 + 2.9185\mu\gamma + 1}$$

The equivalent network in the $(\gamma, \mu)$-domain can now be synthesized as shown in Figure 7.15(a), where a transmission zero is required at $\gamma = \pm j0.756\ 701\ 1$. It is further noted that the network of Figure 7.15(b) is equivalent to that of Figure 7.15(a).

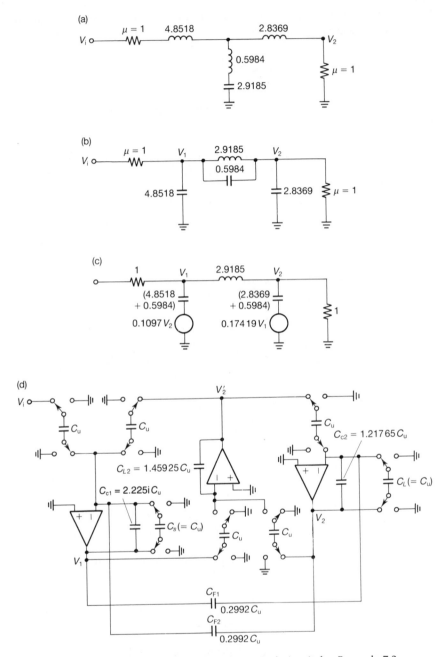

Fig. 7.15  Various steps in the development of practical circuit for Example 7.3.

The next step is to obtain capacitor ratios for the SC network. From Figure 7.15(b), the equivalent network of Figure 7.15(c) is obtained. The SC network of Figure 7.15(d) realizes the three respective transfer functions (due to non-ideal terminations) as

$$V_{1e} = - \frac{\left\{ V_{io} + V'_{2o} + V_{2e} \dfrac{C_{F2} 2\gamma}{C_u} \right\}}{\left( \dfrac{2C_{c1} + C_s}{C_u} \right) \gamma + \left( \dfrac{C_s}{C_u} \right) \mu}$$

$$V'_{2o} = \frac{(V_{1e} + V_{2e})}{\left( \dfrac{2C_{L2}}{C_u} \right) \gamma}$$

and

$$V_{2e} = \frac{- \left\{ V'_{2o} + V_{1e} \left( \dfrac{2C_{F1}}{C_u} \right) \gamma \right\}}{\gamma \left( \dfrac{2C_{c3} + C_L}{C_u} \right) + \left( \dfrac{C_L}{C_u} \right) \mu}$$

Identifying with the network of Figure 7.15(c), we obtain the following capacitor ratios:

$$C_s = C_L = C_u$$

$$C_{L2} = 1.459\,25 C_u$$

$$C_{c1} = 2.2251 C_u$$

$$C_{c3} = 1.217\,65 C_u$$

$$C_{F1} = C_{F2} = 0.2992 C_u$$

This completes the design. The designer can optimize the circuit for optimal dynamic range, if desired.

## 7.3 High-pass ladder filters

### 7.3.1 Active RC high-pass filters

Consider the doubly terminated third-order HP filter of Figure 7.16(a). The following equations can immediately be written by inspection:

$$\frac{V_i - V_1}{R_1} - \frac{V_1}{sL_1} = I_3$$

$$V_1 - V_2 = \frac{I_3}{sC_2} \tag{7.29}$$

$$V_2 = I_3 \left[ \frac{R_2 s L_3}{R_2 + s L_3} + \frac{1}{sC_4} \right]$$

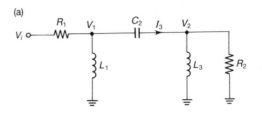

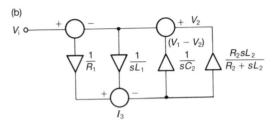

*Fig. 7.16* Active RC simulation of HP ladder filters.

The corresponding simulation block diagram is shown in Figure 7.16(b), wherein we have used only integrators throughout.

The design of HP elliptic ladder filters also starts with block diagram simulation of the RLC prototypes. For example, consider the third-order passive ladder HP filter of Figure 7.17(a), for which the corresponding simulations are shown in Figures 7.17(b)–7.17(d).

We next consider the SC simulation of all-pole as well as elliptic HP filters.

## 7.3.2  SC high-pass filters

To illustrate the procedure, we first consider the realization of a third-order SC all-pole HP filter corresponding to the prototype shown in Figure 7.16(a). This realization [7.9] is obtained from the corresponding block diagram of Figure 7.16(b) and is shown in Figure 7.18(a). The realized transfer function is given by

$$\frac{V_{20}}{V_{io}} = \frac{C_1 C_5 C_6 (1 - z^{-1})^3}{\begin{aligned} C_2 C_5 C_9 (1 - z^{-1})^3 + C_3 C_5 C_9 (1 - z^{-1})^2 \\ + [C_8(1 - z^{-1}) + C_7][C_1 C_{10}(1 - z^{-1})^2 + C_2 C_4(1 - z^{-1}) + C_3 C_4 z^{-1}] \end{aligned}}$$

$$(7.30)$$

which is an HP function. Thus, from the given specifications, the $z$-domain transfer

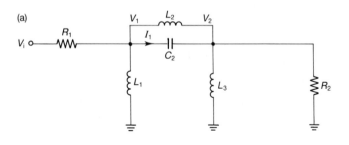

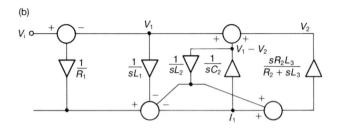

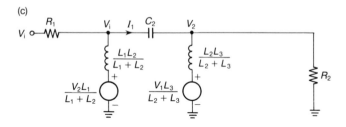

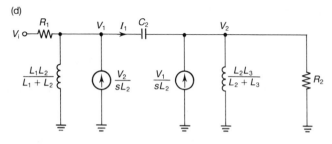

Fig. 7.17  (a)–(d) HP elliptic ladder filter and simulation.

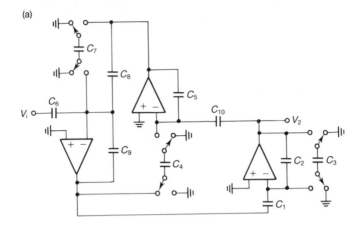

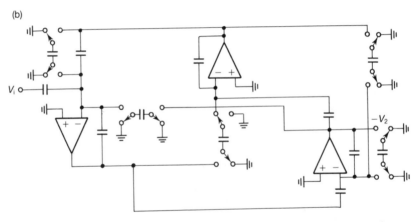

*Fig. 7.18* (a) A third-order SC HP structure corresponding to the prototype of Figure 7.16(a); (b) a third-order elliptic HP filter.

function obtained through any technique can be matched with equation (7.30) to yield the capacitor ratios. Alternatively, the exact synthesis procedure can be used, as in the synthesis of LP filters. The exact procedures will not involve solution of non-linear equations, which otherwise is inevitable. These are left as exercises to the reader.

The above technique *cannot* be extended to higher-order all-pole HP ladder filters without increasing the number of OAs.

We next consider the third-order elliptic HP realization [7.8]. In order to realize the block diagram of Figure 7.17(b), the SC all-pole HP realization can be easily modified to obtain the elliptic realization of Figure 7.18(b). Unfortunately, the results cannot be extended easily to the high-order cases without additional OAs, and so we terminate the discussion here. In the next section, we consider the simulation of active RC and SC BP filters.


## 7.4  Band-pass ladder filters

### 7.4.1  Active RC band-pass ladder filters

Active RC high-order BP filters can be obtained in two ways. In one method, the BP filters are obtained from the active RC LP filters through LP to BP transformation. It is only possible to obtain geometrically symmetric BP filters in this manner. In the second method, the leap-frog simulation of a prototype BP filter is carried out. The prototype in this latter case may be obtained in any manner from the given specifications. We will consider the first method [1.25]. In this technique, from the BP specifications, specifications of an LP prototype are first evaluated. The LP prototype is then designed from filter design tables. Then, using LP to BP transformation, namely,

$$s \rightarrow \frac{s^2 + \omega_0^2}{sB} \tag{7.31}$$

where $\omega_0$ and $B$ are the centre frequency and bandwidth of the BP filter, the BP realization is obtained. In the transformation given by expression (7.31), $\omega_0$ is obtained using the formula

$$\omega_0 = \sqrt{\omega_1 \omega_2} \tag{7.32}$$

where $\omega_1$, $\omega_2$ are the lower and upper pass-band edges. The bandwidth $B$ is $\omega_2 - \omega_1$. The design procedure is illustrated in the following example.

*Example 7.4*
Design a sixth-order BP filter based on a third-order Chebyshev 1 dB ripple LP prototype. The centre frequency is 10 kHz and the bandwidth is 1 kHz.

The simulation of the LP filter with the specifications given above is shown in Figure 7.19(a). The filter has a d.c. gain of unity. Using the LP to BP transformation given in expression (7.31), the sixth-order BP block diagram simulation shown in Figure 7.19(b) is obtained. Evidently, the simulation needs inverting and non-inverting second-order BP filters, one of which has an infinite $Q$ factor. Any active RC biquad circuit can be used in this realization. As an illustration, the Tow–Thomas biquads are used, resulting in the final circuit of Figure 7.19(c). Note that the time constants of the integrators are chosen to be equal in order to optimize the dynamic range within the two integrator loops. The reader is urged to verify the design values given in Figure 7.19(c), obtained using equal capacitor values.

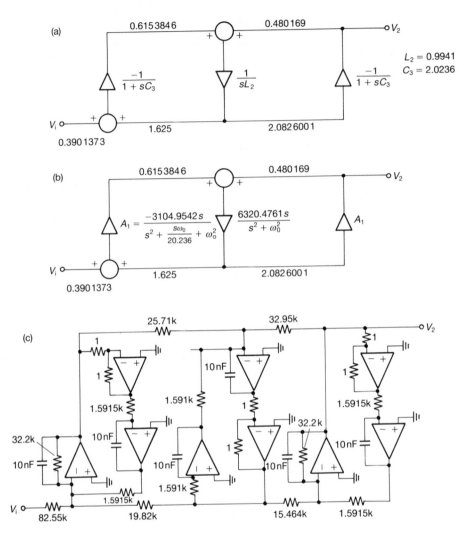

*Fig. 7.19* Design of BP active RC filter for Example 7.4.

The second method of obtaining active RC BP ladder filters starts with direct simulation of RLC BP prototypes [1.25, 7.2]. As an illustration, a fourth-order BP filter is shown in Figure 7.20(a). A block diagram simulation, presented in Figure 7.20(b), can be implemented as shown in Figure 7.20(c). Note that each series-resonant or parallel-resonant circuit is implemented by a two-integrator loop. The reader may compare the realization of Figure 7.20(c) with the realization obtained from an LP active RC filter by LP to BP transformation, and should find that they are the same.

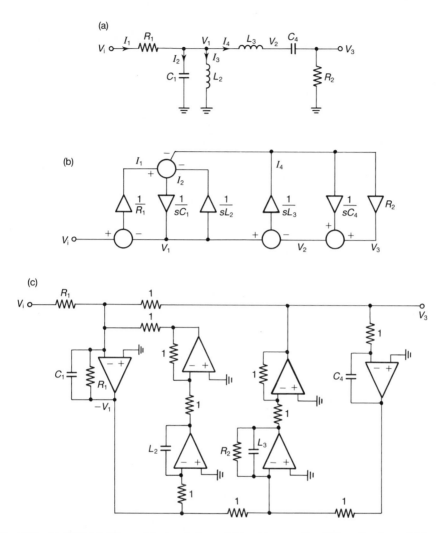

*Fig. 7.20* (a) RLC BP filter; (b) simulation of (a); (c) an active RC realization of (b).

The above techniques can be applied to BP RLC filters derived from low-pass prototypes. These are left as exercises to the reader. We will next deal with the SC BP ladder realizations.

## 7.4.2 SC band-pass ladder filters based on LDI transformation

A straightforward implementation of the simulation block diagram of Figure 7.20(b) yields the SC ladder filter of Figure 7.21(a) [4.2]. Once again the terminations are

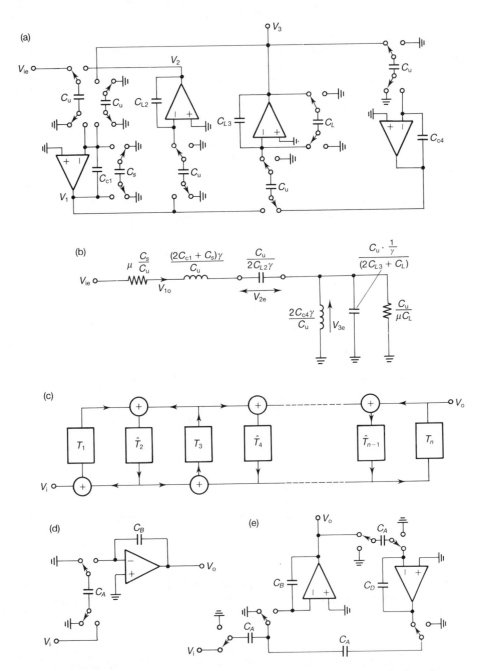

*Fig. 7.21* (a) A fourth-order LDI-type BP filter; (b) equivalent of (a); (c) high-order BP/HP simulation block diagram; (d) first-order block useful in (c); (e) second-order block useful in (c). (d) and (e) *(both adapted from [7.10], © 1984 IEEE).*

non-ideal, whereas the remaining network is reactive. An equivalent circuit can be readily obtained from the above equations, as shown in Figure 7.21(b). The synthesis of the LDI transformation type SC BP filters can be thus carried out easily in a manner similar to that of LP ladder filters. This method, independently advanced by Tawfik *et al.* [4.15] and Baher and Scanlan [7.10], is considered briefly in what follows.

The synthesis of BP ladder filters requires the evaluation of the transfer function from the given specifications. The given specifications are the upper pass-band edge, $\omega_2$, and lower pass-band edge, $\omega_1$, and the sampling frequency, $f_s$. Since the squared magnitude transfer functions are rational functions in the variable $\sin^2 \theta$, where $\theta = \pi f / f_s$, we first obtain the $\sin \theta$ values corresponding to the band edges. Thus, the bandwidth and centre frequency (geometrically symmetric, see equation (7.32)) in the $\sin \theta$ domain are obtained as

$$\beta = \left( \sin \frac{\pi f_1}{f_s} - \sin \frac{\pi f_2}{f_s} \right) \tag{7.33a}$$

$$\sin \theta_0 = \sqrt{\left( \sin \frac{\pi f_1}{f_s} \right) \left( \sin \frac{\pi f_2}{f_s} \right)} \tag{7.33b}$$

Noting that the maximally flat LP transfer function of an LDI-type network is of the form given by

$$|H_{21}|^2 = \frac{1}{1 + \sin^{2n} \theta} \tag{7.34}$$

where $n$ is the order of the filter, the BP transfer function can be obtained by LP to BP transformation in the $\sin \theta$ domain:

$$(\sin \theta)_{\text{LP}} \rightarrow \frac{\sin^2 \theta - \sin^2 \theta_0}{\beta \sin \theta} \tag{7.35}$$

Note the similarity between expressions (7.35) and (7.31). Thus, equation (7.34) becomes

$$|H_{21}|^2 = \frac{1}{1 + \left( \dfrac{\sin^2 \theta - \sin^2 \theta_0}{\beta \sin \theta} \right)^{2n}} \tag{7.36}$$

The synthesis procedure is then exactly similar to that in the case of SC LP ladder filters. In the case of Chebyshev filters, in place of $\sin^{2n} \theta$ in equation (7.34), we have $T_n^2(\sin \theta)$, where $T_n$ is the Chebyshev polynomial.

It can be shown that the transfer function realized by the above method will be of the form [7.10]

$$H_{21}(\lambda) = \frac{K\lambda^m (1 - \lambda^2)^{m/2}}{P_{2m}(\lambda)} \tag{7.37a}$$

It is evident that the response at $f_s/2$ is non-zero, because of the mapping property of the LDI transformation. Baher and Scanlan suggested an alternative stray-insensitive SC structure, which is in fact a cascade of a $k$th-order LDI-type LP network and a $2m$th-order LDI-type BP network studied in the previous sections. This structure [7.10], shown in Figure 7.21(c), employs first- and second-order blocks presented in Figures 7.21(d) and 7.21(e), and realizes a transfer function given by

$$H_{21}(\lambda) = \frac{K\lambda^m (1 - \lambda^2)^{(m+k)/2}}{P_{2m+k}(\lambda)} \qquad (7.37b)$$

Note that the LP section contributes to the additional zeros (i.e., the factor $(1 - \lambda^2)^{k/2}$ in the numerator). Note also that $H_{21}(\lambda)$ in equation (7.37b) can realize either a BP or HP response [7.10, 7.11]. The synthesis, however, requires the use of generalized Chebyshev functions and is rather involved. The reader is referred to Baher [7.11] and Rhodes [6.29] for an introduction to this subject and to the work of Baher and Scanlan [7.10, 7.11] for the synthesis of the above LDI-type SC BP and HP filters.

In the above sections, the design of leap-frog filters based on LDI transformation has been considered in detail. The advantages of the resulting SC filters are that they are stray-insensitive and that they can be designed exactly. The number of OAs required is equal to the order of the filter for LP, elliptic LP and BP filters. HP filter realization by itself is not suitable for operational simulation, whence SC realizations too are not economical. The design procedure is based on an exact synthesis of an RLC filter with non-ideal terminations and is rather involved. However, developing computer-aided design procedures may alleviate this problem of computational complexity. The main drawback of the LDI-type designs is that the capacitor ratios in the SC ladder filter cannot be obtained directly from the prototype RLC filter element values. An added drawback is that though the circuits using LDI transformation can be scaled for optimal dynamic range, this requires evaluation of the $z$-domain transfer functions from the designed SC network after design (at the outputs of OAs) and then scaling them to be equal. It is important to note that the information of nodal voltages and/or branch currents in the prototype should correspond to outputs of OAs in a 'easy to design' configuration, so that when we obtain an SC filter, by bilinear transformation, the scaling of the networks for optimal dynamic range is straightforward. Fortunately, some methods exist for realizing 'bilinear-transformation' type SC ladder filters. One of these is designated the *node-voltage simulation* method, and will be considered next.

## 7.5 Bilinear transformation type SC ladder filters based on nodal voltage simulation

### 7.5.1 Low-pass SC filters

This method has been advanced by Lee, Chang, Temes and Ghaderi [7.12–7.14], and we illustrate it by considering the third-order elliptic LP ladder filter prototype,

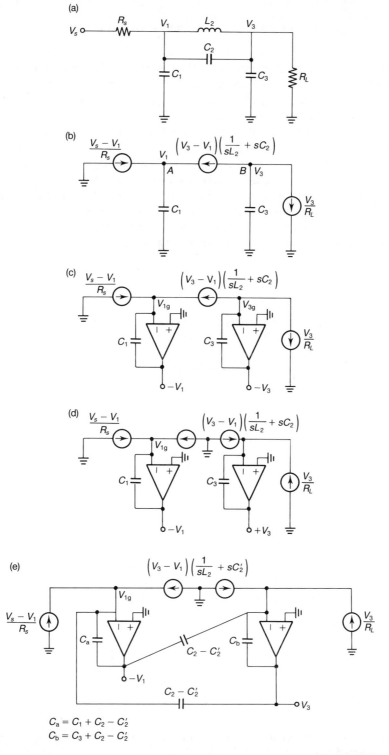

Fig. 7.22 Development of bilinear SC equivalent of a prototype RLC elliptic LP filter (adapted from [7.14], © 1981 IEEE).

shown in Figure 7.22(a). A circuit equivalent to that in Figure 7.22(a) is shown in Figure 7.22(b), where the voltages at nodes $A$ and $B$ are the same as those in the prototype. In the next step, a circuit transformation is introduced to change nodes $A$ and $B$ to virtual grounds and to develop voltages $-V_1$ and $-V_3$ (see Figure 7.22(c)). In order to avoid using inverters requiring additional OAs, $V_3$ is realized instead of $-V_3$, requiring a modification as shown in Figure 7.22(d), where the directions of the current sources associated with the right OA are reversed. An alternative realization of Figure 7.22(c) is shown in Figure 7.22(e).

Next the SC circuit based on Figure 7.22(d) is realized. This SC circuit must realize 'bilinear transformation' type transfer functions. Hence, we first investigate the design of various current sources in Figure 7.22(d) realizing bilinear transformation. Considering the current source simulating source resistance $R_s$, it is required to obtain

$$I(s) = \frac{V_s - V_1}{R_s} \tag{7.38}$$

Recalling the methods of obtaining charge–voltage relationships in the $z$-domain, $Q(s)$ is obtained as $I(s)/s$ from which, using bilinear transformation, we obtain

$$Q(z) = \frac{(V_s - V_1)}{R_s} \cdot \frac{T}{2} \cdot \frac{(1 + z^{-1})}{(1 - z^{-1})} \tag{7.39}$$

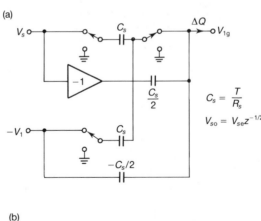

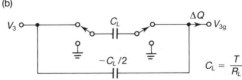

Fig. 7.23 (a) Circuit for simulating source current; (b) circuit realizing load current source. (Adapted from [7.14], © 1981 IEEE.)

Next, the incremental charge flow is obtained as

$$\Delta Q(z) = Q(z)(1 - z^{-1}) = \frac{(V_s - V_1)}{R_s} \frac{T}{2} (1 + z^{-1}) \qquad (7.40)$$

The circuit realization to simulate the incremental charge flow equations given in expression (7.40) is shown in Figure 7.23(a). It is known that $V_s$ and $-V_1$ are sources. Further, $V_1$ and $V_3$ are held over a clock period. Also, $V_s$ should be held over a clock period, necessitating a sample and hold stage at the input. The circuit of Figure 7.23(a) requires an additional OA and is stray-insensitive.

We next consider the load current source for which the incremental charge–voltage relationship is derived in a manner similar to that in equation (7.40), and the circuit implementation is shown in Figure 7.23(b). The remaining task is to synthesize the current sources simulating the LC tank circuit in the prototype. Consider for this purpose the SC network of Figure 7.24 whose incremental charge–voltage relationship is given by

$$\Delta Q(z) = \frac{z^{-1}}{1 - z^{-1}} \frac{C_a C_c}{C_b} [V_3(z) - V_1(z)] \qquad (7.41)$$

Using the identity

$$\frac{z^{-1}}{1 - z^{-1}} = \frac{1}{4} \left[ \frac{(1 + z^{-1})^2}{(1 - z^{-1})} - (1 - z^{-1}) \right] \qquad (7.42)$$

the corresponding impedance realized in the $s$-domain is

$$Z(s) = \frac{\dfrac{C_b}{C_a C_c}}{\left(\dfrac{1}{sT^2} - \dfrac{s}{4}\right)} - \frac{1}{\left(\dfrac{C_a C_c}{C_b T^2} \cdot \dfrac{1}{s} - \dfrac{s}{4} \dfrac{C_a C_c}{C_b}\right)} \qquad (7.43)$$

Thus a parallel combination of an inductance $T^2 C_b / C_a C_c$ and a negative capacitor $-(C_a C_c / 4 C_b)$ is realized. In other words, a current source of value

$$\left(sC_2' + \frac{1}{sL_2}\right)(V_3 - V_1)$$

with

$$C_2' = \frac{-T^2}{4L_2}$$

$$L_2 = \frac{T^2 C_b}{C_a C_c}$$

$$(7.44)$$

is realized.

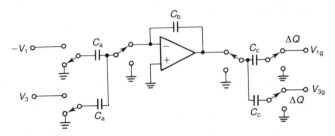

Fig. 7.24  Circuit for realizing current sources simulating LC resonant circuit *(adapted from [7.14], © 1981 IEEE)*.

Noting that

$$\left(sC_2 + \frac{1}{sL_2}\right)(V_3 - V_1) = s(C_2 - C_2')(V_3 - V_1) + \left(sC_2' + \frac{1}{sL_2}\right)(V_3 - V_1) \qquad (7.45)$$

we realize the current sources using the circuit of Figure 7.24 together with that shown in Figure 7.22(e). This completes the design of bilinear SC LP elliptic filters. For the all-pole LP case, since $C_2 = 0$, the results are immediately applicable. The complete realization of a third-order elliptic SC LP filter is shown in Figure 7.25,

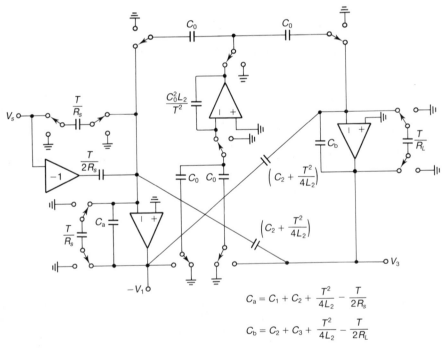

$$C_a = C_1 + C_2 + \frac{T^2}{4L_2} - \frac{T}{2R_s}$$

$$C_b = C_2 + C_3 + \frac{T^2}{4L_2} - \frac{T}{2R_L}$$

Fig. 7.25  A third-order bilinear SC elliptic LP filter *(adapted from [7.14], © 1981 IEEE)*.

where the design of SC filter is straightforward from the component values in the prototype.

We will next consider, as an illustration, the third-order elliptic LP filter considered in Example 7.3.

*Example 7.5*

Design a bilinear SC LP elliptic ladder filter for a clock frequency/BP edge ratio of 9.25, 0.28 dB ripple, and ratio of stop-band edge and pass-band edge frequencies of edges $\Omega_s = 2$.

The prototype doubly terminated elliptic LP filter is shown in Figure 7.26(a), with

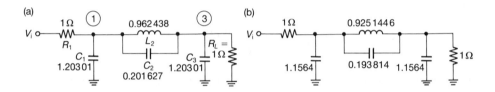

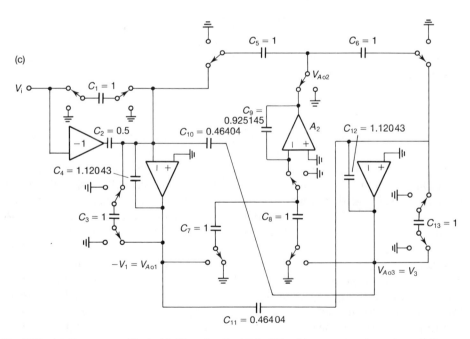

*Fig. 7.26* (a) Prototype elliptic LP filter for Example 7.5; (b) prewarped version of (a); (c) SC implementation of (b).

an angular cutoff frequency of $1 \text{ rad s}^{-1}$. The prewarped cutoff frequency for the given specifications can be calculated (normalized) to be

$$\omega_c = \frac{2}{T} \tan \frac{\omega_c T}{2} = 1.040\ 310\ 8 \text{ rad s}^{-1}$$

Then, the frequency-scaled prototype is derived as shown in Figure 7.26(b) whose SC realization is required. From Figure 7.25, the various capacitor values as shown in Figure 7.26(c) are obtained. Without loss of generality, $T = 1$ s and $C_o = 1$ pF have been chosen in order to simplify the calculations.

The next step is to evaluate the maxima of the outputs of all three OAs. Recall that the voltages $-V_1$ and $V_3$ are the same as those in the prototype at nodes 1 and 3 (see Figure 7.26(a)). From the SC network of Figure 7.26(c), the output of OA $A_2$ is

$$V_{Ao2}(z) = \frac{C_8(V_3 - V_1)}{C_1} \frac{z^{-1}}{(1 - z^{-1})} \tag{7.46}$$

Using the bilinear transformation with $T = 1$ s, equation (7.46) is rewritten as

$$V_{Ao2}(s) = \frac{V_3 - V_1}{C_9} \left( \frac{1 - \dfrac{s}{2}}{s} \right) \tag{7.47}$$

Thus evaluating $V_3/V_i$ and $V_1/V_i$ from the prototype and substituting in equation (7.47), we obtain, using the given component values,

$$\frac{V_{Ao2}(s)}{V_i(s)} = \frac{(1 + 1.203\ 011s)\left(1 - \dfrac{s}{2}\right)}{1.859\ 771s^3 + 2.703\ 75s^2 + 3.368\ 46s + 2}$$

The maximum of $V_{Ao2}$ is evaluated to be 1.072 69, and the maxima of $V_{Ao1}$ (i.e., $-V_1$) and $V_{Ao3}$ (i.e., $V_3$) to be 0.722 98 and 0.5, respectively. Using this information, the scaling for optimal dynamic range is carried out next.

To scale $V_{Ao2}$ to $\mu V_{Ao2}$, we need to divide the values of the capacitors connected to the output $V_{Ao2}$ by $\mu$. Thus, we get the scaling factors for $V_{Ao2}$ as $\mu_2 = V_{Ao3(max)}/V_{Ao2(max)} = 0.466\ 117$. Scaling the output $V_{Ao1(max)}$ also to 0.5 (i.e., $V_{Ao3(max)}$), we get $\mu_1 = 0.691\ 582\ 1$. The above scaling procedure yields the capacitance values given in Table 7.1. The scaling of capacitors to make the minimum capacitance unity has to be carried out next. For this purpose, all the capacitors connected to the virtual ground of each OA are grouped. For the circuit of Figure 7.26(c), the capacitors $\{C_1, C_2, C_3, C_4, C_5, C_{10}\}$ are grouped together, as are $\{C_7, C_8, C_9\}$ and $\{C_6, C_{11}, C_{12}, C_{13}\}$. The minimum capacitance in each

**Table 7.1** Scaling of capacitors for optimal dynamic range (circuit in Figure 7.26(c)).

| Capacitor | Original | Values when $V_{Ao2} = V_{Ao3} = V_{Ao1(max)}$ | Values after scaling for minimum capacitance $= 1$ |
|---|---|---|---|
| $C_1$ | 1 | 1 | $2.154\ 98 C_u$ |
| $C_2$ | 0.5 | 0.5 | $1.077\ 49 C_u$ |
| $C_3$ | 1 | 1.445 96 | $3.116\ 01 C_u$ |
| $C_4$ | 1.120 43 | 1.620 1 | $3.491\ 28 C_u$ |
| $C_5$ | 1 | 2.145 38 | $4.623\ 27 C_u$ |
| $C_6$ | 1 | 2.145 38 | $3.197\ 359\ 6 C_u$ |
| $C_7$ | 1 | 1.445 96 | $1.446 C_u$ |
| $C_8$ | 1 | 1 | $C_u$ |
| $C_9$ | 0.925 145 | 1.984 79 | $1.984\ 79 C_u$ |
| $C_{10}$ | 0.464 04 | 0.464 04 | $C_u$ |
| $C_{11}$ | 0.464 04 | 0.670 983 | $C_u$ |
| $C_{12}$ | 1.120 43 | 1.120 43 | $1.669\ 83 C_u$ |
| $C_{13}$ | 1 | 1 | $1.490\ 35 C_u$ |

group shall be unity. The resulting scaled capacitor values are shown in Table 7.1 in the last column. The total capacitance is readily found to be $27.2485 C_u$. The circuit has been optimized for dynamic range and minimum total capacitance. This completes the design.

It is worth mentioning that, for dynamic range optimization, the prototype filter can be analyzed for maxima of all nodal voltages, and additionally for the remaining amplifier outputs as given by equation (7.47). This computation can be done in the $s$-domain itself, which is an advantage. In the next sub-section, we consider the design of SC HP and BP ladder filters.

## 7.5.2 Band-pass SC filters [7.14]

A prototype RLC network suitable for realizing all-pole or elliptic BP filters is shown in Figure 7.27(a). As in the case of LP filters discussed previously, the realization suitable for SC modelling is obtained as in Figure 7.27(b). It may be seen that the current source $I_{L1}$ and $I_{L2}$ have different values. The realization of $I_{L1}$ and $I_{L2}$ is similar to the current sources in Figure 7.25 except that the input capacitors have different values and that at the output one capacitor will suffice to feed the virtual ground input of the respective OAs. The resulting complete SC realization is shown in Figure 7.27(c). The scaling for dynamic range and minimal total capacitance can be carried out as before.

## 7.5.3 High-pass SC filters [7.14]

In the case of HP filters, the choice of $C_1 = C_3 = 0$ in the prototype of Figure 7.27(a) cannot be useful. This is because the bilinear source resistance feeds a charge

$$\Delta Q(z) = \frac{T}{2R_s} (1 + z^{-1}) V_s(z) \qquad (7.48)$$

which at $z^{-1} = -1$ makes $\Delta Q(z) = 0$. In contrast, the HP filter should have the output non-zero at a frequency $f_s/2$, whence the above realization of Figure 7.27(a) with $C_1 = C_3 = 0$ cannot be useful to realize HP filters. A solution to this problem is to use the impedance transformation of the prototype by dividing all the impedances by $s$. The resulting passive circuit needs resistors, capacitors and FDNRs [1.29]. The passive HP network and its realization schematic are shown in Figures 7.28(a) and (b). The capacitors and resistors can be realized using the methods discussed before. Regarding the realization of the current source

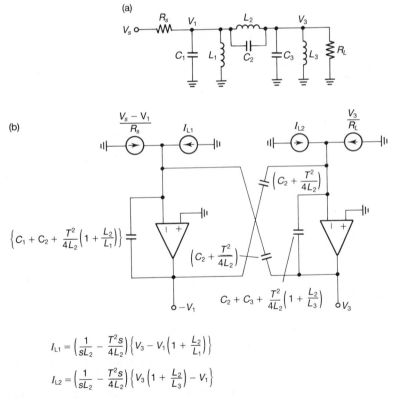

$$I_{L1} = \left( \frac{1}{sL_2} - \frac{T^2 s}{4L_2} \right) \left\{ V_3 - V_1 \left( 1 + \frac{L_2}{L_1} \right) \right\}$$

$$I_{L2} = \left( \frac{1}{sL_2} - \frac{T^2 s}{4L_2} \right) \left\{ V_3 \left( 1 + \frac{L_2}{L_3} \right) - V_1 \right\}$$

Fig. 7.27   (a) Prototype SC BP filter;   (b) equivalent of (a);   (c) SC implementation. *(Adapted from [7.14], © 1981 IEEE.)*

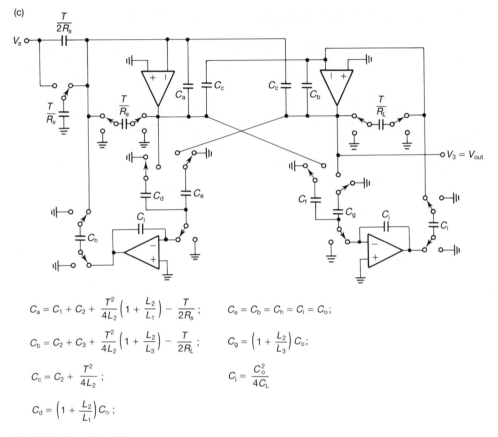

$$C_a = C_1 + C_2 + \frac{T^2}{4L_2}\left(1 + \frac{L_2}{L_1}\right) - \frac{T}{2R_s};$$

$$C_b = C_2 + C_3 + \frac{T^2}{4L_2}\left(1 + \frac{L_2}{L_3}\right) - \frac{T}{2R_L};$$

$$C_c = C_2 + \frac{T^2}{4L_2};$$

$$C_d = \left(1 + \frac{L_2}{L_1}\right)C_o;$$

$$C_e = C_b = C_h = C_i = C_o;$$

$$C_g = \left(1 + \frac{L_2}{L_3}\right)C_o;$$

$$C_j = \frac{C_o^2}{4C_L}$$

Fig. 7.27  (continued)

$s^2C_2(V_3 - V_1)$, it is useful to examine the necessary incremental charge–voltage relationship, which from equation (6.86) is

$$\Delta Q(z) = \frac{2C_2}{T}(V_3 - V_1)\left[(1 + z^{-1}) - \frac{4z^{-1}}{1 + z^{-1}}\right] \tag{7.49}$$

The first term on the right side of equation (7.49) may be realized by a resistor (see equation (6.27b)) of value $T^2/4C_2$. By choosing $C_{o1}C_{o3}/C_o = 8C_2/T$ in the circuit of Figure 7.28(c), whose transfer function is given by

$$\Delta Q(z) = \frac{-C_{o1}C_{o3}V_{3e}z^{-1}}{C_o(1 + z^{-1})} \tag{7.50}$$

the second term on the right side of equation (7.49) may be realized. The resulting SC realization is shown in Figure 7.28(d), which is only bottom-plate stray-insensitive.

## 7.5.4  Circuits using a four-phase clock [7.12, 7.14]

In the above discussion, we have studied bilinear SC ladder filter design using nodal voltage simulation method and a two-phase clock. The stray-insensitive two-phase clock based configurations require two additional amplifiers, one for the inverter in the source termination and another for the complete circuit to provide $V_{so} = V_{se}z^{-1/2}$. Using a four-phase clock, however, it is possible to dispense with these additional amplifiers. In this case, the circuits for current sources simulating the load termination and the LC resonant circuits in the series path are the same as in the case of two-phase circuits, but they operate twice in a clock cycle. As an illustration, the SC current source of Figure 7.24 after such modification is shown in Figure 7.29(a). The current source used to simulate the source resistor is realized as shown in Figure 7.29(b). For this circuit, it is easy to see that charge is transferred in phases 2 and 4. These charges are

$$-\Delta Q_2(z) = \frac{C_s}{2} V_{s1}z^{-1/4} + \frac{C_s}{2} V_{s3}z^{-3/4} \qquad (7.51a)$$

$$-\Delta Q_4(z) = \frac{C_s}{2} V_{s3}z^{-1/4} + \frac{C_s}{2} V_{s1}z^{-3/4} \qquad (7.51b)$$

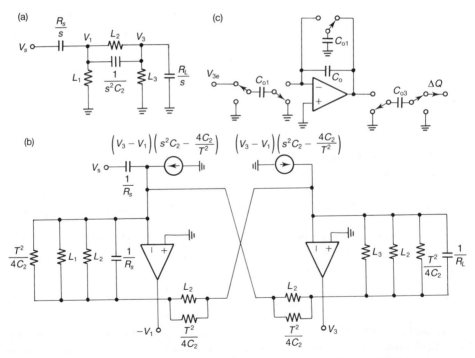

*Fig. 7.28*  Design of SC HP bilinear filters *(adapted from [7.14], © 1981 IEEE).*

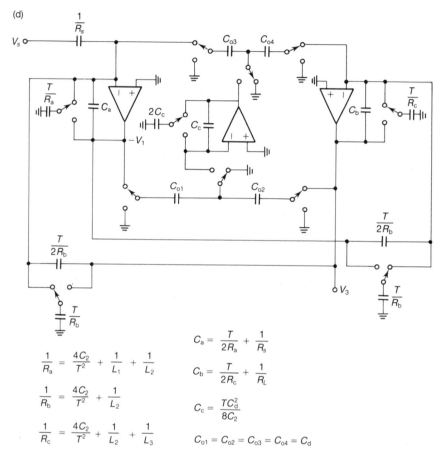

(d)

$$C_a = \frac{T}{2R_a} + \frac{1}{R_s}$$

$$\frac{1}{R_a} = \frac{4C_2}{T^2} + \frac{1}{L_1} + \frac{1}{L_2}$$

$$C_b = \frac{T}{2R_c} + \frac{1}{R_L}$$

$$\frac{1}{R_b} = \frac{4C_2}{T^2} + \frac{1}{L_2}$$

$$\frac{1}{R_c} = \frac{4C_2}{T^2} + \frac{1}{L_2} + \frac{1}{L_3}$$

$$C_c = \frac{TC_d^2}{8C_2}$$

$$C_{o1} = C_{o2} = C_{o3} = C_{o4} = C_d$$

*Fig. 7.28  (continued)*

Assuming that the clock frequency is doubled, $V_{s3}$ becomes $V_{s1}$ in the above equations so that we have

$$\Delta Q(z) = \frac{C_s}{2} V_s z^{-1/2}(1 + z^{-1}) \tag{7.52}$$

Thus the sample and hold function as well as realization of a bilinear resistor simulation are achieved. The realized resistance value is $T/C_s$, according to equation (7.40). Thus the design of bilinear SC filters with a four-phase clock and a reduced number of OAs is straightforward using the source and load current sources and inductance–capacitance simulation circuits of Figures 7.23 and 7.24.

In the Lee *et al.* method [7.14], the nodal voltages are simulated using the virtual ground property of the OAs. Bilinear SC ladder filters could be designed with either

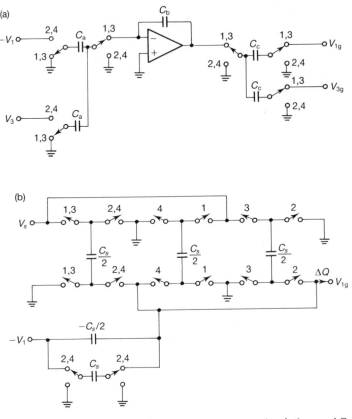

*Fig. 7.29* (a) A four-phase circuit to realize a current source simulating an LC resonant circuit; (b) a stray-insensitive four-phase current source for source resistor simulation which also performs a sample and hold function. *(Adapted from [7.12], © 1980 IEE.)*

two- or four-phase clocking schemes using this method. Other design methods to realize bilinear SC ladder filters do exist and these will be considered next.

## 7.5.5 The Martin–Sedra method of designing bilinear SC ladder filters

Two approaches have been discussed by Martin and Sedra [5.13] in order to realize high-order bilinear stray-insensitive SC filters. In the former approach, the starting point is the RLC LP filter obtained from the prewarped specifications. The LP–BP transformation then results in a leap-frog structure containing second-order inverting and non-inverting structures, as shown in Example 7.4 (see Figure 7.19)). In order to realize a bilinear SC ladder filter, bilinear stray-insensitive second-order inverting and non-inverting SC BP networks are required. For elliptic BP realizations, the LP–BP transformation can be used on the block diagram simulation scheme of Figure 7.13(b). The resulting biquads need BP as well as notch transfer functions. Thus, for this application, a stray-insensitive biquad SC capable of realizing an inverting notch or inverting BP transfer function is essential.

The general biquad of Figure 5.4(a) can be used to realize inverting bilinear transfer functions. Its transfer function is given by

$$\frac{V_o(z)}{V_i(z)} = \frac{-[k_3 z^2 + (-2k_3 + k_1 k_5 + k_2 k_5)z + k_3 - k_2 k_5]}{z^2 + (-2 + k_4 k_5 + k_5 k_6)z + 1 - k_5 k_6} \qquad (7.53)$$

For a desired $s$-domain second-order function given by

$$\frac{-(a_2 s^2 + a_1 s + a_0)}{(s^2 + b_1 s + b_0)} \qquad (7.54)$$

the corresponding bilinear $z$-domain transfer function can be obtained through bilinear $s \to z$ transformation. Matching this with equation (7.53), we obtain the design values for inverting BP and notch transfer functions to be respectively

$$k_1 = 0, \qquad k_2 k_5 = 2ma_1, \qquad k_3 = ma_1 \qquad (7.55a)$$

$$k_2 = 0, \qquad k_1 k_5 = 4ma_0, \qquad k_3 = m(a_0 + a_2) \qquad (7.55b)$$

where $m = (1 + b_0 + b_1)^{-1}$.

The non-inverting BP function is realized using the SC BP filter of Figure 5.4(b). The transfer function can be obtained as

$$\frac{V_o(z)}{V_i(z)} = \frac{(k_2(k_5 + k_6) - k_3)z^2 + (2k_3 - k_2(k_5 + 2k_6))z + k_2 k_6 - k_3}{z^2 + (-2 + k_4(k_5 + k_6))z + 1 - k_4 k_6} \qquad (7.56)$$

The design equations for realizing the bilinear BP transfer function are as follows:

$$k_4 k_5 = 4mb_0, \qquad k_4 k_6 = 2mb_1, \qquad k_2 k_5 = 2ma_1 \qquad (7.57)$$

Thus, any general biquad can be realized by an appropriate combination of BP and notch transfer functions [5.13]. The reader is urged to design SC filters using this technique.

In the second method proposed by Martin and Sedra, general-parameter BP filters can be designed from the desired prototype filter obtained using filter synthesis programs. Then, using a nodal voltage simulation method proposed by Yoshihoro *et al.* [7.15], second-order blocks can be used to realize a complete simulation of the RLC filter. As an example, we consider the third-order elliptic LP filter of

Figure 7.30(a). The node voltages $V_1$ and $V_3$ can be expressed in terms of the adjacent node voltages (i.e., nodal voltages of left and right nodes) as

$$V_1(s) = \frac{\dfrac{s}{(C_1 + C_2)R_s}}{s^2 + \dfrac{s}{(C_1 + C_2)R_s} + \dfrac{1}{L_2(C_1 + C_2)}} V_s(s)$$

$$+ \frac{\dfrac{C_2}{C_1 + C_2}\left(s^2 + \dfrac{1}{L_2 C_2}\right)}{s^2 + \dfrac{s}{(C_1 + C_2)R_s} + \dfrac{1}{L_2(C_1 + C_2)}} V_3(s) \qquad (7.58)$$

$$V_3(s) = \frac{\dfrac{C_2}{C_2 + C_3}\left(s^2 + \dfrac{1}{L_2 C_2}\right)}{s^2 + \dfrac{s}{(C_2 + C_3)R_L} + \dfrac{1}{L_2(C_2 + C_3)}} V_1(s)$$

It is thus noted that the block diagram simulation is as shown in Figure 7.30(b). Since only notch and BP second-order transfer functions are required to simulate the RLC filter of Figure 7.30(a), bilinear stray-insensitive SC filters can be obtained by replacing the blocks with the SC filter of Figure 5.4(a). Note that inverting biquadratic filter sections can serve this purpose exclusively. The SC version of a third-order elliptic LP realization is shown in Figure 7.30(c), requiring four OAs. The outputs of the two biquads correspond exactly to node voltages $V_1$ and $V_3$ in the prototype filter. Hence, the scaling can be performed easily. As in the case of geometrically symmetric filters discussed previously, the other OA outputs have to be evaluated to scale their outputs also in order to achieve optimal dynamic range. The following example illustrates the Martin–Sedra method for a third-order filter.

*Example 7.6*
Design a third-order SC LP bilinear ladder filter with the same specifications as in Example 7.5 using the Martin–Sedra method.

The prewarped RLC prototype is shown in Figure 7.26(b). From the component values depicted there, the transfer functions $T_{12}$, $T_{23}$, $T_{32}$ are evaluated first, using equation (7.58). These are, respectively,

$$T_{12} = \frac{0.740\ 626s}{s^2 + 0.740\ 626s + 0.800\ 551}$$

$$T_{23} = T_{32} = \frac{0.143\ 544(s^2 + 5.7705)}{s^2 + 0.740\ 626s + 0.800\ 551}$$

Next $T_{12}$, $T_{23}$ and $T_{32}$ are compared with (7.54) to identify the values of $a_0$, $a_1$, $a_2$, $b_0$ and $b_1$. The next step is to calculate the capacitor ratios of the resulting SC filter

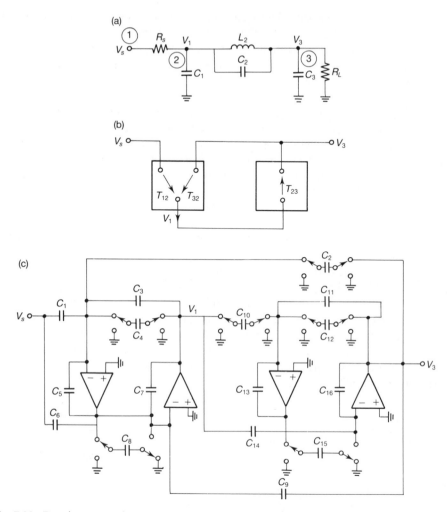

Fig. 7.30 Development of stray-insensitive SC filter design due to Martin and Sedra (adapted from [7.15], © 1977 IEEE).

using equations (7.55) and assuming $k_4 = k_5$ for optimal dynamic range. The results are obtained as follows (note that the second biquad used to realize $T_{23}$ is the same as that used to realize $T_{32}$):

(a) band-pass realization:

$$k_1 = 0, \qquad k_2 = k_6 = 0.519\ 262, \qquad k_3 = 0.291\ 45, \qquad k_4 = k_5 = 1.122\ 55$$

(b) notch realization:

$$k_1 = k_4 = k_5 = 1.122\ 55, \qquad k_2 = 0, \qquad k_3 = 0.371\ 519, \qquad k_6 = 0.519\ 262$$

The resulting SC network is shown in Figure 7.30(c). The maxima of $V_1$ and $V_3$ are estimated in Example 7.5 as 0.722 98 and 0.5, respectively. Hence, it is required to scale down $V_1$ to 0.5. The last step is to scale such that minimum capacitance is unity. The actual values and scaled values are shown in Table 7.2. This completes the design.

It is thus shown how to generate SC filters based on general-parameter RLC filters or geometrically symmetric RLC filters. In the Martin–Sedra method, based on active RC simulation method by Yoshihoro et al., it is evident that only the nodal voltages are simulated and not the internal working of the ladder filters. However, the resulting SC filters have been found to exhibit low-sensitivity properties and as such are recommended.

## 7.5.6 The equivalence of Martin–Sedra and Lee et al. methods

Wellekens [7.16] has demonstrated that the nodal voltage simulation methods for designing SC filters developed by Martin and Sedra [5.13] and Lee, Temes, Chang and Ghaderi [7.12–7.14] are equivalent and hence possess identical properties. This aspect will be considered in what follows. The Lee et al. realization needs source and load current sources and another current source to simulate the series branch of the analog reference filter. We have considered some current source designs in Figures 7.23 and 7.24. Alternative current sources have been identified by Wellekens in Martin–Sedra SC structures.

**Table 7.2** Design values for Example 7.6.

| Capacitor | Values when $V_1 = V_{3(\max)}$ | After scaling for minimum capacitance = 1 |
|---|---|---|
| $C_1$ | 0.519 262 | 1 |
| $C_2$ | 0.776 338 | 1.495 06 |
| $C_3$ | 0.519 262 | 1 |
| $C_4$ | 1.122 55 | 2.161 82 |
| $C_5$ | 1 | 1.925 81 |
| $C_6$ | 0.291 45 | 1.134 33 |
| $C_7$ | 1 | 3.892 02 |
| $C_8$ | 1.122 55 | 4.369 01 |
| $C_9$ | 0.256 936 | 1 |
| $C_{10}$ | 1.623 17 | 3.125 91 |
| $C_{11}$ | 0.519 262 | 1 |
| $C_{12}$ | 1.122 55 | 2.161 82 |
| $C_{13}$ | 1 | 1.925 81 |
| $C_{14}$ | 0.537 201 | 1 |
| $C_{15}$ | 1.122 55 | 2.089 63 |
| $C_{16}$ | 1 | 1.861 50 |

Consider the source or load current source realization shown in Figure 7.31(a) which, under the condition $C_2C_4 = 2C_1C_3$, realizes a bilinear source termination. Adding an additional feed-in capacitor to input $-V_1$ realizes the source current $(V_s - V_1)/R_s$. This addition is also shown in Figure 7.31(b). The current source of Figure 7.31(a) can be used also as a load current source. The two cross-connecting capacitors remain as before in the case of the Lee *et al.* method. Thus the inductor remains to be simulated. Once again using Lee *et al.*'s method, this current source can be realized as shown in Figure 7.31(c). Two such current sources have to feed the virtual ground of each OA, simulating the nodal voltages $-V_1$ and $V_3$.

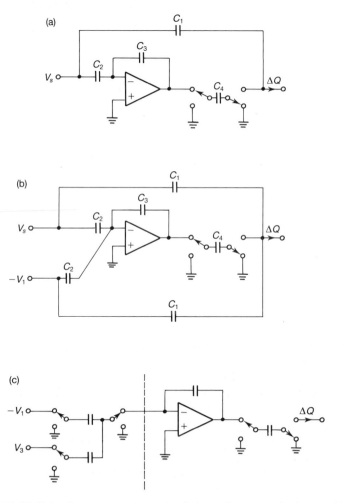

*Fig. 7.31* (a)–(c) Welleken's current source simulation schemes; (d) their application to simulate bilinear SC ladder filters as for the Martin–Sedra and Lee *et al.* methods. (a) and (c) *(adapted from [7.16],* © *1982 IEE).*

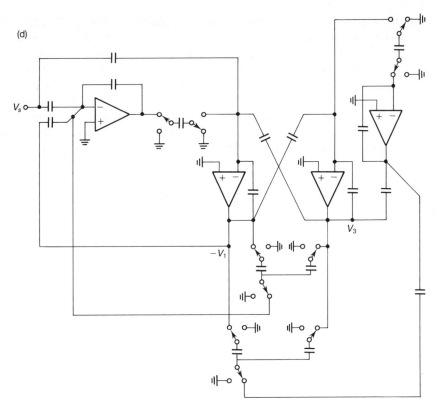

*Fig. 7.31   (continued)*

Since the circuit to the right of the dotted line in Figure 7.31(c) already exists in the source and load current sources, it suffices to add the remaining switched capacitors. The result is the Martin–Sedra SC network shown in Figure 7.31(d). Thus, even though the approaches are different, the two methods realize the same filter topology. However, as illustrated in Examples 7.5 and 7.6, the design procedures are different and the user can choose either method as convenient. It may be of interest to note that, as pointed out earlier, in the Lee *et al.* method, exact scaling for optimal dynamic range is possible, because the outputs of the OA realizing the current sources simulating the series branch are evaluated. The comparison discussed above also leads to the useful conclusion that the Lee *et al.* method simulates only the interaction between nodal voltages. In contrast, the LDI transformation type filters simulate the internal working of the doubly terminated RLC prototypes.

## 7.5.7  The Datar–Sedra method

The study of LDI and bilinear transformation type SC ladder filters has thus demonstrated that the resulting SC structures are similar in topology except for the input-switched capacitors. They use integrators, preferably of the stray-insensitive type, and enjoy low sensitivities. Datar and Sedra have presented a unified approach for the design of stray-insensitive SC ladder filters [7.17]. As against the nodal voltage simulation technique to derive bilinear SC ladder filters from doubly terminated prototype RLC networks, their method is based on rigorous synthesis procedures akin to passive RLC ladder synthesis and is therefore very systematic. It serves to integrate the results obtained by Baher and Scanlan [4.14, 7.5], Lee et al. [7.14] and Jacobs et al. [4.2]. We briefly study their technique in what follows.

Consider the realization of odd-order SC ladder filters having $(N-1)/2$ pairs of transmission zeros at $\lambda = \pm 1$ and one transmission zero at $f_s/2$. Such a network has a transfer function $H(\lambda)$ such that

$$H(\lambda) = \frac{(1 - \lambda^2)^{(N-1)/2}}{D(\lambda)} = \frac{P(\lambda)}{E(\lambda)} \qquad (7.59)$$

Thus, we have

$$P(\lambda)P(-\lambda) = (1 - \lambda^2)^{(N-1)} \qquad (7.60)$$

Also, by Feldtkeller's equation (see [1.25] for further details)

$$E(\lambda)E(-\lambda) = P(\lambda)P(-\lambda) + F(\lambda)F(-\lambda) \qquad (7.61)$$

Thus, choosing $F(\lambda)F(-\lambda)$ to obtain the desired pass-band response (maximally flat, Chebyshev, etc.) and following the conventional synthesis procedures, $H(\lambda)$ is determined.

Datar and Sedra next observe that $H(\lambda)$ as expressed in equation (7.59) can be realized by the network shown in Figure 7.32(a) where the transmission zeros are realized by parallel resonant circuits. Then, using a computer-aided synthesis technique that can remove transmission zeros at $\lambda = \pm 1$ sequentially, they obtain values for the components of the circuit in Figure 7.32(a). Note that this network is termed an auxiliary network by Baher and Scanlan [7.5] (see Figure 7.9(c)). Next, scaling all the impedances in Figure 7.32(a) by $\mu$, the network of Figure 7.32(b) is obtained, which realizes the desired transfer function. The equations describing this ladder are as follows:

$$V_1 = \frac{(1/R_s)\mu V_i - (\mu I_2)}{\gamma c_1 + \mu(1/R_s)}$$

$$(\mu I_2) = \frac{V_1 - V_3}{\gamma l_2}$$

$$V_3 = \frac{(\mu I_2) - (\mu I_4)}{\gamma c_3} \qquad (7.62a)$$

$$(\mu I_4) = \frac{V_3 - V_5}{\gamma l_4}$$

$$V_o = V_5 = \frac{(\mu I_4)}{\gamma c_5 + \mu(1/R_L)}$$

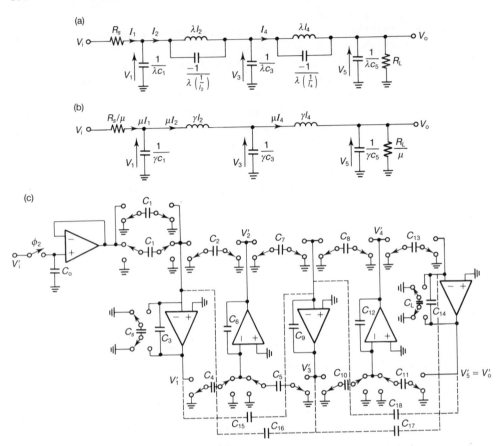

*Fig. 7.32* Synthesis of fifth-order SC ladder filter with two pairs of transmission zeros at $\lambda = \pm 1$: (a) prototype network; (b) network obtained by dividing all impedances in (a) by $\mu$; (c) SC realization of (b). *(Adapted from [7.17], © 1983 IEEE.)*

An SC network that precisely realizes the above equations is shown in Figure 7.32(c), for which the following relations hold:

$$V_1' = \frac{-2C_1\mu V_i' - C_2 z^{-1/2} V_2'}{\gamma(2C_3 + C_s) + \mu C_s}$$

$$V_2' = \frac{C_4 z^{-1/2} V_1' + C_5 z^{-1/2} V_3'}{\gamma(2C_6)}$$

$$V_3' = \frac{-C_7 z^{-1/2} V_2' - C_8 z^{-1/2} V_4'}{\gamma(2C_9)} \qquad (7.62b)$$

$$V_4' = \frac{C_{10} z^{-1/2} V_3' + C_{11} z^{-1/2} V_5'}{\gamma(2C_{12})}$$

$$V_o' = V_5' = \frac{-C_{13} z^{1/2} V_4'}{\gamma(2C_{14} + C_L) + \mu C_L}$$

Comparison of equations (7.62a) and (7.62b) shows the correspondence between OA output voltages in the SC network and capacitor voltages and inductor currents in the ladder network of Figure 7.32(b). We also can obtain the following equalities:

$$\frac{2C_3 + C_s}{C_s} = c_1 R_s$$

$$\frac{2C_1}{C_s} = K, \qquad \frac{2C_3 + C_s}{C_2} = c_1$$

$$\frac{2C_6}{C_4} = \frac{2C_6}{C_5} = l_2, \qquad \frac{2C_9}{C_7} = \frac{2C_9}{C_8} = c_3 \qquad (7.63)$$

$$\frac{2C_{12}}{C_{10}} = \frac{2C_{12}}{C_{11}} = l_4, \qquad \frac{2C_{14} + C_L}{C_{13}} = c_5$$

$$\frac{2C_{14} + C_L}{C_L} = c_5 R_L$$

Note that $K$ is the ratio of the required gain to the ladder gain. There are 11 equations and 16 unknown capacitor values. Thus any 5 capacitors can be assumed and the rest evaluated. The design steps for dynamic range scaling and scaling for minimum total capacitance then follow as usual.

The reader will have noted that by replacing the input feed-in branch by a simple series-switched capacitor, we obtain a transfer function given by equation (7.18). This operation amounts to multiplying $H(\lambda)$ in equation (7.59) by $(1 - \lambda^2)^{1/2} = \mu$. This network has been synthesized using the rigorous procedure of Baher and Scanlan in Section 7.1.2. However, Datar and Sedra have suggested an alternative technique, which is briefly considered next.

For the case under consideration, as mentioned before, we have

$$P(\lambda)P(-\lambda) = (1 - \lambda^2)^N \qquad (7.64)$$

(by virtue of multiplying $P(\lambda)$ by $(1 - \lambda^2)^{1/2}$ due to the series-switched capacitor feeding the input). Thus equation (7.61) can be solved for a given $F(\lambda)F(-\lambda)$ depending on the passband requirement and $H(\lambda)$ can be obtained. However, this $H(\lambda)$ cannot be realized by the network of Figure 7.32(a), which must have a transmission zero at $\lambda = \infty$. A solution to this problem is to choose $P'(\lambda)$, a new function such that

$$P'(\lambda)P'(-\lambda) = (1 - \lambda^2)^{N-1} \qquad (7.65)$$

while using the $E(\lambda)$ as that obtained before by solving Feldtkeller's equation. Thus, the new $H'(\lambda)$ becomes $P'(\lambda)/E(\lambda)$. The network can then be synthesized. When a series-switched capacitor is used as feed-in, we realize

$$H(\lambda) = \frac{\mu P'(\lambda)}{E(\lambda)} = \frac{P(\lambda)}{E(\lambda)} \qquad (7.66)$$

thus realizing the required specifications.

We next consider SC filter realization with finite transmission zeros. In this case, the λ-plane prototype satisfying the required specifications is as shown in Figure 7.33(a). A modified version, equivalent to this is shown in Figure 7.33(b). (Recall that LDI inductance has been shown in a previous section, to be equivalent to a bilinear inductance in parallel with a negative capacitance, which is the basis for this step.) Thus it remains to obtain the simulation of capacitances $C_2'$, $C_4'$ in the SC version, noting that the remaining SC network can be developed as in the case of Figure 7.32(a). We next borrow the results of Jacobs *et al.* [4.2] (see Section 7.2) to realize the function of capacitors $C_2'$ and $C_4'$ using voltage-controlled current sources. The equivalent circuit thus obtained is as shown in Figure 7.33(c). Scaling all the components by the frequency variable $(1/\mu)$, we obtain the network of Figure 7.33(d), which can be realized by the network of Figure 7.32(c) using the additional four capacitors shown in dotted lines. The above technique can be extended to the case with no transmission zero at $f_s/2$. The design procedure is the same as in the LP case discussed above.

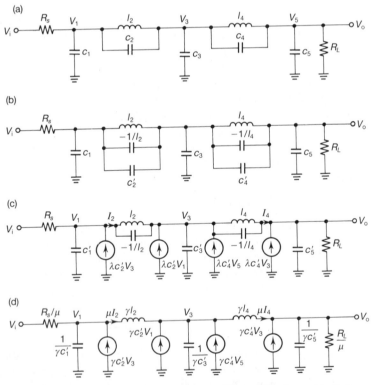

Fig. 7.33 Various stages in the development of SC filter realization with finite transmission zeros: (a) prototype; (b) modified version of (a); (c) realization of capacitors across inductors by dependent current sources; (d) network obtained from (c) by scaling all impedances by $\mu$. (Adapted from [7.17], © 1978 IEEE.)

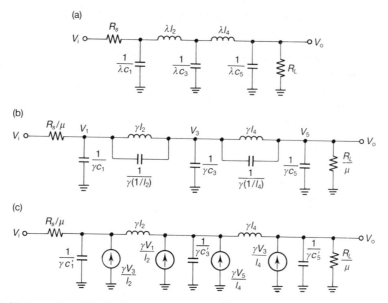

Fig. 7.34 Various stages in the development of true all-pole SC filters: (a) $\lambda$-domain prototype; (b) network obtained from (a) by dividing all impedances by $\mu$; (c) circuit for eliminating the capacitors across inductors in (a) by using dependent current sources. (Adapted from [7.17], © 1978 IEEE.)

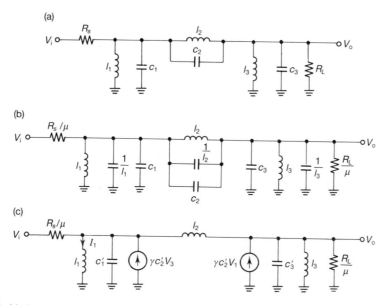

Fig. 7.35 Various stages in the development of bilinear BP SC filters: (a) prototype; (b) scaled version of (a) ($\hat{Z}_i = Z_i/\mu$); (c) modification of (b). (Adapted from [7.17], © 1978 IEEE.)

Now consider the realization of another class of LP filters that may be designated as true all-pole transfer functions, the reason being that all the transmission zeros of this class of filters are at $\omega = f_s\pi$. The design procedure starts, as before, with a prototype LP ladder as shown in Figure 7.34(a). After scaling by $1/\mu$, we have the network of Figure 7.34(b). Next, the use of voltage controlled current sources leads to the realization in Figure 7.34(c), which can be implemented in SC form.

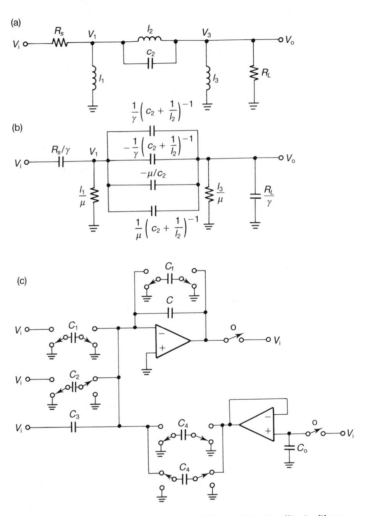

Fig. 7.36 Various stages in the development of bilinear SC HP elliptic filters: (a) prototype network; (b) scaled network ($\hat{Z}_i \to Z_i/\gamma$); (c), (d) blocks used to realize equations (7.67); (e) final structure. (Adapted from [7.19], © 1983 IEE.)

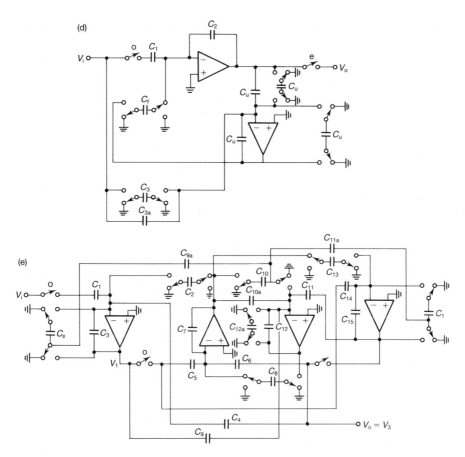

Fig. 7.36  (continued)

We finally consider a BP example as in Figure 7.35(a). The two steps involved are as shown in Figures 7.35(b) and (c). It is thus easy to see the systematic manner in which the Datar–Sedra method leads to SC networks from prototype RLC networks. Remember, however, that in the synthesis procedure, only design aids that have the facility for extracting poles at $\lambda = \pm 1$ will be helpful. (The FILTOR program due to Snelgrove and Sedra [7.18] has this facility.)

Datar and Sedra have extended the above procedure for SC HP filter realization as well [7.19]. For illustration we consider the prototype of Figure 7.36(a), which is a ladder in the $\lambda$-domain (where $\lambda$ is the bilinear transformation variable $\lambda = \tanh(sT/2)$). Next, scaling all the impedances down by a factor $\gamma = (z^{1/2} - z^{-1/2})/2$, we obtain the equivalent network of Figure 7.36(b). Note that we have manipulated the series branch impedances to suit SC realization, as will be explained next. The equations describing the network of Figure 7.36(b) can be written as

$$V_1 = \frac{\dfrac{\gamma}{R_s} \cdot V_i + \gamma\left(c_2 + \dfrac{1}{l_2}\right)V_3 - I_2}{\gamma\left(\dfrac{1}{R_s} + c_2 + \dfrac{1}{l_2}\right) + \dfrac{\mu}{l_1}}$$

$$-I_2 = \frac{\left[\left(c_2 + \dfrac{1}{l_2}\right)\gamma(\mu - \gamma) - \dfrac{1}{l_2}\right](V_1 - V_3)}{\mu} \tag{7.67}$$

$$V_0 = V_3 = \frac{\gamma\left(c_2 + \dfrac{1}{l_2}\right)V_1 + I_2}{\gamma\left(\dfrac{1}{R_L} + c_2 + \dfrac{1}{l_2}\right) + \dfrac{\mu}{l_3}}$$

where $\mu = \cosh(sT/2)$. Two SC networks which can be used to realize the above equations are shown in Figures 7.36(c) and (d). The respective transfer functions can be derived as

$$\frac{V_0}{V_i} = \frac{-C_1 z^{1/2} + C_2 z^{-1/2} - 2C_3\gamma - 2C_4\mu}{\gamma(2C + C_f) + \mu C_f} \tag{7.68a}$$

$$\frac{V_{oe}}{V_{io}} = -\left[\frac{C_1\gamma(\mu - \gamma) + C_3 C_f/2}{C_2\mu}\right] \tag{7.68b}$$

where $C_f = 2C_2$ in equation (7.68b). Using these two blocks, equations (7.67) can be realized to yield the complete SC network in a straightforward manner, as in Figure 7.36(e). The two OAs in the middle realize $I_2$ while the OAs at either end realize the outputs $V_1$ and $V_3$ as desired.

Note that all-pole HP filters can be obtained by deleting $C_9$, $C_{9a}$, $C_{11}$ and $C_{11a}$ in Figure 7.36(e). Further, the above design method can be readily extended to the arbitrarily high-order case.

This completes the discussion of bilinear SC ladder filters. We next study a high-order filter design technique for active RC and SC cases known as *multiloop feedback design*, which has the advantage of low sensitivity.

## 7.6  Multiloop feedback type high-order filters

### 7.6.1  Active RC topologies

In this method, several second-order (or first-order) active filters will be used in various configurations to realize high-order filters. Two such configurations have been studied earlier, namely, the leap-frog and coupled biquad types. The others are known as the *follow-the-leader feedback* (FLF) topology, the *inverse follow-the-leader feedback* (IFLF) topology, and the *minimum sensitivity feedback* (MSF) topology, respectively. Other topologies described in the literature can be considered to be particular cases of the above topologies. The reader is referred to [7.20] for a comprehensive treatment on this subject. We consider these topologies only briefly to highlight their specific features.

Historically, the primary resonator block [1.25] method was the first multiloop feedback technique described in the literature. This method first realizes an LP active RC filter using identical first-order LP blocks, as shown in Figure 7.37(a). Note the absence of the first feedback loop in this configuration. Each first-order block in Figure 7.37(a) has a transfer function given by

$$T_i(s) = \frac{k}{s+k} \tag{7.69}$$

The transfer function of the complete network is given by

$$\frac{V_{out}}{V_{in}} = \frac{\prod\limits_{i=1}^{n} T_i(s)}{1 + \sum\limits_{j=2}^{n} F_j \prod\limits_{i=1}^{j} T_i(s)}$$

$$= \frac{k^n}{(s+k)^n + \sum\limits_{j=2}^{n} F_j k^j (s+k)^{n-j}} \tag{7.70}$$

Suppose it is desired to realize an LP transfer function given by

$$T(s) = \frac{1}{s^n + a_{n-1}s^{n-1} + \cdots + a_0} \tag{7.71}$$

Matching the denominator coefficients of equations (7.70) and (7.71), we obtain the following design equations:

$$k = \frac{a_{n-1}}{n}$$

$$F_2 = \frac{a_{n-2}}{k^2} - \frac{n(n-1)}{2!} \tag{7.72}$$

$$F_i = \frac{a_{n-i}}{k^i} - \frac{n!}{(n-i)!i!} - \frac{1}{(n-1)!} \sum_{l=2}^{i-1} F_l \frac{(n-l)!}{(i-l)!}, \qquad i = 3, \ldots, n$$

Thus $k$ and all the feedback coefficients can be determined. The factor $k$ is the 'shift' of the pole at the origin corresponding to a lossless integrator used in the direct-form state variable structure, hence this topology has been called the *shifted-companion form*.

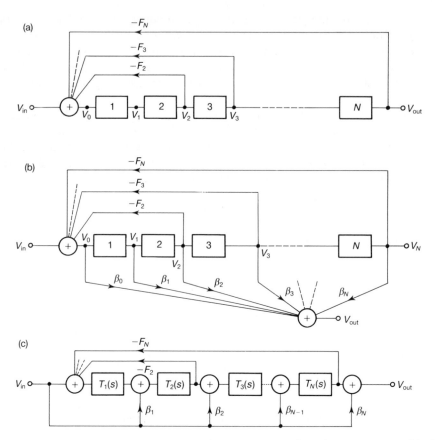

Fig. 7.37 (a) Basic multiloop feedback configuration using the primary resonator block technique; (b) realization of transmission zeros by summation; (c) realization of transmission zeros by feedforward technique.

Once the feedback factors and $k$ are determined, the next step is to use LP–BP transformation on the first-order blocks. Thus, a second-order BP filter needs to be used for each block, so that a geometrically symmetric type of BP response of $2n$th order, where $n$ is the order of the LP transfer function, can be obtained. The next step in the design will be to scale the circuit for optimal dynamic range, which can be carried out on the prototype LP network itself. During synthesis, if negative terms are obtained for any $F_i$, an additional inverter is used.

If a transfer function other than the BP type is desired, there are two ways of using the feedforward technique. In one method, an additional OA can be used to sum the outputs of the various blocks and the input (see Figure 7.37(b)). Alternatively, each block can be given a feedforward input as in Figure 7.37(c). In either case, design equations can be developed. The reader is referred to literature for details of these steps.

In the above method, the first feedback loop is omitted, which facilitates easy evaluation of $k$ and other coefficients $F_i$. In the FLF technique, the first feedback loop also exists. The resulting system can be solved by matrix methods, which will not be considered here.

In the generalized FLF case shown in Figure 7.38(a), each block $T_i$ can be a general biquad. Thus the zeros of these biquads are used to derive the zeros of the complete transfer function. However, the zeros of the individual biquads also affect the poles of the overall system. The IFLF structure is shown in Figure 7.38(b). The resulting transfer function is

$$\frac{V_o(s)}{V_i(s)} = \frac{-\alpha \prod_{j=1}^{N} T_j(s)}{1 + \sum_{k=1}^{N} F_{kN} \prod_{j=k}^{N} T_j(s)} \tag{7.73a}$$

where

$$\alpha = \prod_{i=1}^{N} \frac{G_i}{G_{Ti}}, \qquad G_{Ti} = G_i + G_{fi} + G_{i,\text{in}}$$

$$F_{iN} = \frac{G_{fi}}{G_i} \prod_{j=1}^{N} \frac{G_j}{G_{Tj}} \qquad i = 1, ..., N \tag{7.73b}$$

Note that $G_{i,\text{in}}$ is the input conductance of the $i$th block $T_i$. It may be shown that IFLF and FLF structures can be related easily. However, the synthesis of IFLF structures is somewhat difficult as compared to that of FLF structures.

We finally mention the MSF topology shown in Figure 7.38(c), which is the most general of all the multiloop feedback topologies. This employs all types of feedback, and by choosing these feedback paths 'properly' it has been shown that low sensitivity can be achieved. We conclude the discussion of these active RC topologies with a remark that experience has shown multiple feedback topologies to be superior to cascade designs as regards sensitivity. In addition, the modularity of these topologies retaining the advantages of biquads makes them a viable alternative to leap-frog ladder simulation methods discussed extensively in this chapter. As such, we consider SC topologies using multiloop feedback next.

## 7.6.2 SC topologies

Attaie and El-Masry have proposed high-order SC-filter realization techniques using multiloop feedback [7.21, 7.22]. They describe IFLF and FLF topologies. The former structure is shown in Figure 7.39(a). The various $T_{jk}$s are of the form

$$T_{jk}(z) = \frac{-a_{jk} + b_{jk}z^{-1}}{1 - z^{-1}} = \frac{N_{jk}(z)}{D_{jk}(z)} \tag{7.74}$$

This particular choice guarantees the realization using stray-insensitive SC blocks.

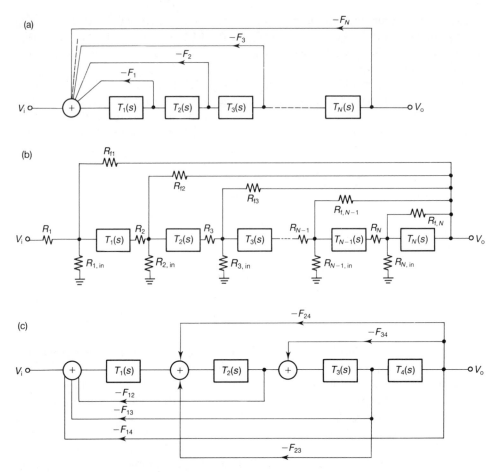

Fig. 7.38 (a) Follow-the-leader feedback (FLF) method; (b) inverse follow-the-leader feedback (IFLF) method (adapted from [7.20], © 1979 IEEE); (c) general multiple feedback technique (minimum sensitivity feedback (MSF)).

Specifically, the circuit of Figure 7.39(b) realizes the $T_{jk}(z)$ given in equation (7.74). The complete transfer function of the SC network of Figure 7.39(a) can be obtained. Note, however, without loss of generality, that the blocks $T_{11}$ to $T_{1,n-1}$ can be designed with the respective $a$ values as zero and $b$ values as unity in equation (7.74). The completely general circuit will be as shown in Figure 7.39(c). The design procedure starts by matching the transfer function of the circuit of Figure 7.39(c) with the desired $H(z)$:

$$H(z) = \frac{\displaystyle\sum_{i=0}^{N} \alpha_i z^{-i}}{\displaystyle\sum_{j=0}^{N} \beta_j z^{-j}} \qquad (7.75)$$

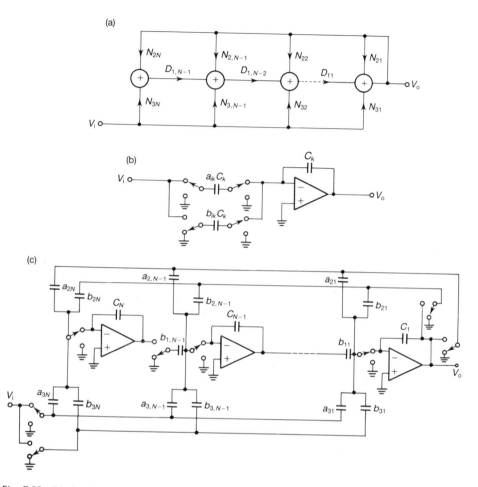

Fig. 7.39 (a) An IFLF structure; (b) block for realizing summer in (a), showing $T_{ij}$
realization; (c) complete IFLF SC structure. (*Adapted from [7.21], © 1983 IEEE.*)

Note that there are several extra degrees of freedom available in the topology of
Figure 7.39(c); these are chosen so as to make the resulting values $a_{jk}$, $b_{jk}$ positive
to facilitate easy realization (i.e., without requiring extra OAs realizing inverters).
A closed-form solution for the above design problem can be obtained, for which the
reader is referred to [7.22].

The sensitivities of the resulting SC structures have been examined by Monte
Carlo simulation, and Attaie and El-Masry have observed that the resulting designs
are superior to cascade designs, as is also the case with active RC realizations.

It is noted that direct-form digital filter structures can be considered to be
state–space realizations. For example, in the direct-form structure of Figure 7.40
(which incidentally is an FLF structure), the outputs of all the delay elements can
be considered as state variables. Also, the transmission zeros are realized using

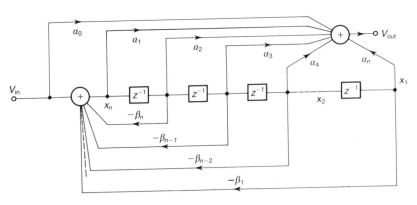

*Fig. 7.40* A direct-form digital filter structure.

feedforward inputs to all the intermediate outputs (state variables) and the input, as mentioned before. However, the direct-form structure is known to be highly sensitive to coefficient inaccuracies. Hence, suitable modifications of the elementary state–space direct-form structures are necessary in order to derive low-sensitivity realizations.

El-Masry [5.16, 7.23, 7.24] has described an elegant state–space approach for the design of stray-insensitive multiloop feedback SC filters. This approach is applicable to both IFLF and FLF structures, and the reader is referred to his work for details.

## 7.7 High-order switched capacitor filters based on *p*-transformation [5.23]

An active RC high-order filter structure due to Bach [7.25] is shown in Figure 7.41(a). The circuit realizes an LP transfer function given by

$$\frac{V_{out}}{V_{in}} = \frac{1}{C_n R_n \ldots C_2 R_2 C_1 R_1 s^n + \cdots + C_2 R_2 C_1 R_1 s^2 + C_1 R_1 s + 1} \tag{7.76}$$

An interesting property of this circuit is that the time constants can be determined in an iterative manner, i.e., $C_1 R_1$ is first identified, then $C_2 R_2$, and so on. (Second-order versions of these networks were studied in Chapter 5.) The circuit uses unity-gain amplifiers and requires the minimum number of components. However, one limitation compared to other high-order filter structures is the lack of facility for optimizing the dynamic range. This is because no degrees of freedom are available to equalize the outputs of the various OAs. Similar to the modification suggested in the case of second-order filters, the dynamic range optimization is possible through the use of lossy integrators/differentiators.

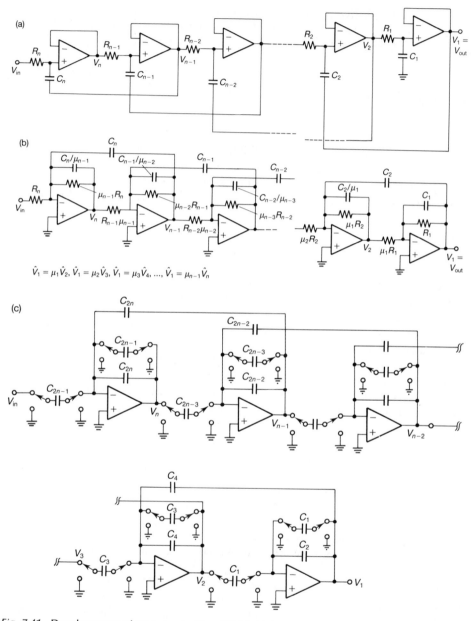

$$\hat{V}_1 = \mu_1\hat{V}_2, \; \hat{V}_1 = \mu_2\hat{V}_3, \; \hat{V}_1 = \mu_3\hat{V}_4, \; ..., \; \hat{V}_1 = \mu_{n-1}\hat{V}_n$$

*Fig. 7.41* Development of stray-insensitive SC LP high-order filters based on Bach's active RC configuration of (a) Amanda Mohan, Ramachandran and Swamy *(adapted from [5.23],* © 1983 IEEE). (The ˆ symbol denotes maxima.)

The equivalent active RC network for this purpose is shown in Figure 7.41(b). Note that when $\mu_i = 1$, $i = 1, 2, \ldots, n - 1$, the circuit realizes the same transfer function as that of Figure 7.41(a). The various outputs of the other OAs are given by

$$\frac{V_j}{V_{in}} = \frac{C_{j-1}R_{j-1} \ldots C_1 R_1 s^{j-1} + C_{j-2}R_{j-2} \ldots C_1 R_1 s^{j-2} + \cdots + C_1 R_1 s + 1}{C_n R_n \ldots C_1 R_1 s^n + C_{n-1} R_{n-1} \ldots C_1 R_1 s^{n-1} + \cdots + C_2 R_2 C_1 R_1 s^2 + C_1 R_1 s + 1},$$

(7.77)

However, the choice of $\mu_i \neq 1$ will facilitate scaling for optimal dynamic range. It may be remembered, however, that this modification to Bach's network results in large $Q_p$ sensitivity, as in the case of second-order active RC networks. The maxima of the various transfer functions may be evaluated using equation (7.77). Defining

$$\frac{(V_{i+1})_{max}}{(V_1)_{max}} = \frac{1}{\mu_i}$$

(7.78)

the scaling can be done as illustrated in Figure 7.41(b). This completes the design.

Low-sensitivity high-order SC filters can be obtained from Figure 7.41(a) by replacing resistors with parallel- or series-switched capacitors. The use of the parallel-switched capacitors, however, does not preserve the advantage of Bach's filter, namely, iterative evaluation of the time constants, i.e., capacitor ratios, since the resulting $z$-domain transfer functions are not related to the $s$-domain transfer functions in any manner. However, the use of series-switched capacitor preserves this property. The resulting design, however, is sensitive to top-plate parasitic capacitances as well as parasitic capacitances at the non-inverting inputs of OAs. Parasitic compensation techniques such as those discussed in the second-order case can be used at the expense of almost duplicating the passive part of the network (i.e., switches and capacitors). Alternatively, stray-insensitive realizations can be derived from the active RC filter of Figure 7.41(b) by $p$-transformation. The resulting SC filter (assuming $\mu_i = 1$) is illustrated in Figure 7.41(c). The high $Q_p$ sensitivity is a limitation of this design, as in the case of the active RC filter of Figure 7.41(a). Nevertheless, the simplicity in design in the $p$-domain, availability of scaling facility and the capability of being 'exactly' designed are some advantages of this topology.

The $z$-domain transfer functions can be obtained by substituting $s \rightarrow (1 - z^{-1})/T$ in equations (7.76) and (7.77) and $R_i = T/C_i$. $T$ can be chosen as 1 s without loss of generality. These SC networks can realize all-pole LP transfer functions. Using matched-$z$ transformation, the $z$-domain transfer function $H(z)$ can be obtained from the $s$-domain transfer function. Then, rewriting $H(z)$ as a polynomial in $1 - z^{-1}$ for the denominator and identifying with equation (7.76) yields the capacitor ratios directly. Next, the internal transfer functions at the outputs of other OAs can be determined and their maxima evaluated to facilitate subsequent scaling for optimal dynamic range and minimum total capacitance. The design procedure is illustrated in the next example.

*Example 7.7*

Design a third-order Butterworth SC filter for a pole frequency of 1 kHz, using a sampling frequency of 16 kHz.

The $s$-domain transfer function is

$$H_a(s) = \frac{1}{(s+1)(s^2+s+1)}$$

Using matched-$z$ transformation, for the given clock frequency and $\omega_p$, the $z$-domain transfer function can be obtained directly in the '$p$-domain' itself, i.e., as a polynomial in $1 - z^{-1}$. The result is

$$H_d(z) = \frac{1}{11.149\ 956p^3 + 8.646\ 528p^2 + 3.658\ 491\ 3p + 1}$$

Identifying $H_d(z)$ with the transfer function of the third-order SC network shown in Figure 7.42, i.e.

$$\frac{V_{out}}{V_{in}} = \frac{V_1}{V_{in}} = \frac{1}{\dfrac{C_6}{C_5}\dfrac{C_4}{C_3}\dfrac{C_2}{C_1}p^3 + \dfrac{C_4}{C_3}\dfrac{C_2}{C_1}p^2 + \dfrac{C_2}{C_1}p + 1}$$

we obtain

$$\frac{C_2}{C_1} = 3.658\ 491\ 3$$

$$\frac{C_4}{C_3} = 2.363\ 413\ 6$$

$$\frac{C_6}{C_5} = 1.289\ 529\ 9$$

Now the outputs of other OAs can immediately be written as

$$\frac{V_2}{V_{in}} = \frac{3.658\ 491\ 3p + 1}{11.149\ 956p^3 + 8.646\ 528p^2 + 3.658\ 491\ 3p + 1}$$

$$\frac{V_3}{V_{in}} = \frac{8.646\ 528p^2 + 3.658\ 491\ 3p + 1}{11.149\ 956p^3 + 8.646\ 528p^2 + 3.658\ 491\ 3p + 1}$$

The maxima of $V_2/V_{in}$ and $V_3/V_{in}$ are next computed, using $z = \exp(j\omega T)$ or $p = 2s/(1+s)$ to evaluate:

$$\frac{V_{2(max)}}{V_{1(max)}} = \frac{1}{\mu_1} = 1.4722$$

$$\frac{V_{3(max)}}{V_{1(max)}} = \frac{1}{\mu_2} = 1.3729$$

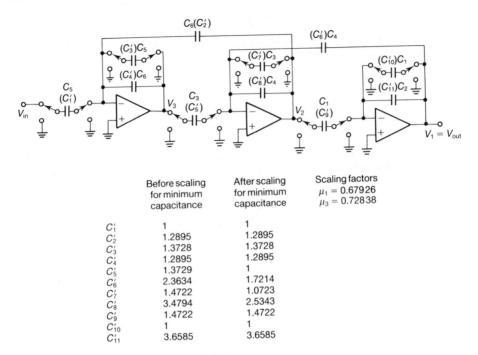

| | Before scaling for minimum capacitance | After scaling for minimum capacitance | Scaling factors $\mu_1 = 0.67926$ $\mu_3 = 0.72838$ |
|---|---|---|---|
| $C_1'$ | 1 | 1 | |
| $C_2'$ | 1.2895 | 1.2895 | |
| $C_3'$ | 1.3728 | 1.3728 | |
| $C_4'$ | 1.2895 | 1.2895 | |
| $C_5'$ | 1.3729 | 1 | |
| $C_6'$ | 2.3634 | 1.7214 | |
| $C_7'$ | 1.4722 | 1.0723 | |
| $C_8'$ | 3.4794 | 2.5343 | |
| $C_9'$ | 1.4722 | 1.4722 | |
| $C_{10}'$ | 1 | 1 | |
| $C_{11}'$ | 3.6585 | 3.6585 | |

*Fig. 7.42* SC filter for design of Example 7.7.

The resulting SC network, after scaling for optimum dynamic range, is shown in Figure 7.42 where the capacitors now are those shown in the paranthesis, with values shown. The circuit can now be scaled for minimum total capacitance, and the corresponding capacitor values are also given in Figure 7.42. The total capacitance is $17.4105C_u$ and the capacitance spread is 3.6585. Note that several switches can be combined in the SC network to simplify it further.

The above discussion deals with LP filters. SC HP filters can be obtained in a similar manner. The transfer functions of the active RC HP network of Figure 7.43(a) are

$$\frac{V_{out}}{V_{in}} = \frac{C_n R_n \dots C_1 R_1 s^n}{C_n R_n \dots C_2 R_2 C_1 R_1 s^n + \dots + C_2 R_2 C_1 R_1 s^2 + C_1 R_1 s + 1} \tag{7.79a}$$

$$\frac{V_j}{V_{in}} = \frac{C_n R_n \dots C_1 R_1 s^n + \dots + C_j R_j \dots C_1 R_1 s^j}{C_n R_n \dots C_2 R_2 C_1 R_1 s^n + \dots + C_2 R_2 C_1 R_1 s^2 + C_1 R_1 s + 1} \tag{7.79b}$$

The z-domain transfer functions of the SC network of Figure 7.43(b) are obtained by substituting $s \to (1 - z^{-1})/T$ and $R_j = T/C_j$. The design procedure and scaling rules are similar to those of LP filters. The resulting SC filters can be designed also using bilinear transformation, since the numerator is of the form $(1 - z^{-1})^n$ for an $n$th-order SC network.

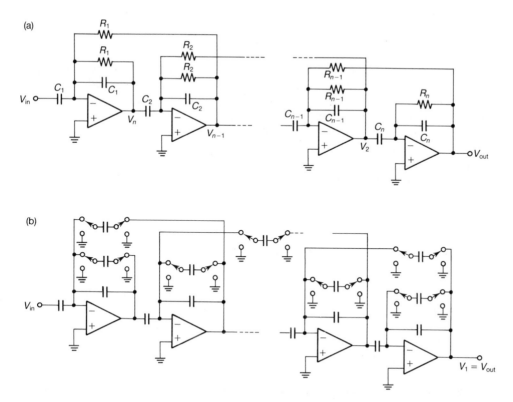

*Fig. 7.43* (a) Active RC and (b) stray-insensitive SC HP filters derived from Bach's active RC filters. *(Adapted from [5.23], © 1983 IEEE.)*

The designs of Figures 7.41 and 7.43 have one restriction, however. This is concerned with the sampling frequency that should be used to realize a given specification. This restriction arises because of the mapping property of the *p*-transformation itself; maximum $Q_p$ of the complex conjugate pair should satisfy the restriction

$$\frac{f_s}{\omega_p} > Q_p \qquad (7.80)$$

where $\omega_p$ is the bilinearly prewarped radian frequency, $f_s$ is the sampling frequency. As an illustration, for a third-order Butterworth characteristic, the maximum $Q_p$ required is 1, since the quadratic factor associated with this pole pair is $s^2 + s + 1$. Thus the constraint in inequality (7.80) implies that $f_s$ must at least be $2\pi$ times the pole frequency. The second limitation is the large $Q_p$ sensitivity of these structures. It may also be noted that other types of transfer function cannot be realized using these topologies.

## 7.8  Conclusions

This chapter has considered SC high-order designs using LDI, bilinear and $p$-transformations. These enjoy several advantages. They can be designed exactly as stray-insensitive. The suitability of a particular circuit for a given application thus proceeds with the evaluation of all the available techniques, and then choosing the circuit with minimum chip area. While the circuits described in this chapter have been extensively useful in industry, in some applications where very narrow band responses are required, an alternative technique based on simple modification of SC ladder filter topology is useful. These are known as SC $N$-path filters and will be treated in the next chapter in some detail.

## 7.9  Exercises

7.1 Derive an operational simulation block diagram of the ladder network of Figure 7.2(a).

7.2 The prototype of RLC LP filter can also be as shown in Figure 7.44 for the even- and odd-order cases. Derive the operational simulation block diagrams for these networks. What are the differences between these and those of Figures 7.1(a) and 7.2(a)?

7.3 A second-order doubly terminated LP filter is to be realized. Sketch the two possible circuit configurations and discuss the restrictions on source and load resistance values.

7.4 Stray-insensitive SC versions of the second-order doubly terminated LP filters can be derived using two lossy integrators. What are all the possible practical realizations? Derive the transfer functions evaluating the effect of terminations switched in phase and out of phase. Discuss the results.

7.5 Assuming that a high sampling frequency is being used, develop approximate design formulae for the LDI-type elliptic LP ladder filter (choose order 5).

7.6 Using a computer program, analyze the effect of using either of the three terminations on a third-order SC LP LDI-type filter. Study the effect of terminations switched in anti-phase.

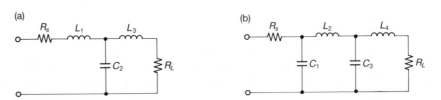

Fig. 7.44  Alternative prototype of RLC LP filter:  (a) odd-order case;  (b) even-order case.

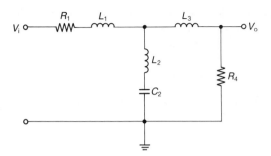

*Fig. 7.45*  Prototype RLC LP elliptic filter.

7.7 Using feedforward inputs to the inputs of the various OAs in a leap-frog active RC filter, it is possible to realize HP, LP elliptic and HP elliptic realizations. Demonstrate this observation for a third-order transfer function. Compare these realizations with those obtained using operational simulation discussed in this chapter.

7.8 Develop the design procedure for the LDI-type SC HP filter shown in Figure 7.17(e), using the procedure illustrated for LP filters.

7.9 Assuming that a bilinear transformation type of LP response is desired, evaluate the sensitivities of a third-order SC LP filter obtained using the Lee *et al.* method [7.14]. Compare these with the results of the doubly terminated RLC prototype. Give reasons for the differences observed, if any.

7.10 Derive an SC HP third-order filter using the Yoshihoro *et al.* approach [7.15]. Study the similarity between the resulting design and the Datar–Sedra and Hokenek–Moschytz [7.26] designs.

7.11 A prototype RLC LP elliptic filter can be of the type shown in Figure 7.45. Derive a leap-frog simulation systematically.

7.12 Investigate whether the Datar–Sedra technique can be applied to even-order bilinear LP filter design. Discuss the disadvantages, if any.

## 7.10  References

[7.1] F.E.J. Girling and E.F. Good, Active filters 12 – The leap-frog or active ladder synthesis; Active filters 13 – Applications of the active ladder synthesis, *Wireless World*, 76, 341–5, 445–50, July, September 1970.

[7.2] G. Daryanani, *Principles of Active Network Synthesis and Design*, Wiley, 1976.

[7.3] W.E. Heinlein and W.H. Holmes, *Active Filters for Integrated Circuits*, R. Oldenbourg Verlag, 1974.

[7.4] H.J. Carlin, Distributed circuit design with transmission line elements, *Proc. IEEE*, 59, 7, 1059–81, July 1971.

[7.5] H. Baher and S.O. Scanlan, Stability and exact synthesis of low-pass switched-capacitor filters, *IEEE Transactions on Circuits and Systems*, CAS-29, 7, 488–92, July 1982.

[7.6] M.A. Rafat and J. Mavor, Experimental validation of exact design of switched-capacitor ladder filters, *Electronics Letters*, 17, 7, 275–6, 2 April 1981.

[7.7] J. Taylor, Stability test for switched-capacitor filters, *Electronics Letters*, 19, 3, 89–91, 3 February 1983.

[7.8] J. Taylor, Exact design of elliptic switched-capacitor filters by synthesis, *Electronics Letters*, 18, 19, 807–9, 16 September 1982.

[7.9] T. Hui and D.J. Allstot, MOS switched capacitor high-pass/notch ladder filters, *Proc. ISCAS*, 309–12, 1980.

[7.10] H.O. Baher and S.O. Scanlan, Exact synthesis of band-pass switched capacitor ladder filters, *IEEE Transactions on Circuits and Systems*, CAS-31, 4, 342–9, April 1984.

[7.11] H. Baher, Synthesis of high-pass switched-capacitor LDI ladder filters, *Electronics Letters*, 21, 2, 79–80, 17 January 1985.

[7.12] M.S. Lee, Parasitics-insensitive switched-capacitor ladder filters, *Electronics Letters*, 16, 12, 472–3, 5 June 1980.

[7.13] G.C. Temes and M.B. Ghaderi, Bilinear switched-capacitor ladder filter with reduced number of amplifiers, *Electronics Letters*, 16, 11, 412–13, 22 May 1980.

[7.14] M.S. Lee, G.C. Temes, C. Chang and M.B. Ghaderi, Bilinear switched-capacitor ladder filters, *IEEE Transactions on Circuits and Systems*, CAS-28, 8, 811–22, August 1981.

[7.15] M. Yoshihoro, A. Nishihara and T. Yanagisawa, Low sensitivity active and digital filters based on the node voltage simulation of LC ladder structures, *Proc. ISCAS, Phoenix, Arizona*, 446–9, 1977.

[7.16] C.J. Wellekens, Equivalence of two designs of bilinear switched-capacitor ladder filters, *Electronics Letters*, 18, 6, 246–7, March 1982.

[7.17] R.B. Datar and A.S. Sedra, Exact design of stray-insensitive switched-capacitor ladder filters, *IEEE Transactions on Circuits and Systems*, CAS-30, 12, 888–98, December 1983.

[7.18] W. Snelgrove and A.S. Sedra, FILTOR-2: A computer-aided filter design program, 1980, Dept. of Elec. Engr., University of Toronto, Toronto, Canada.

[7.19] R.B. Datar and A.S. Sedra, Exact design of stray-insensitive switched-capacitor high-pass ladder filters, *Electronics Letters*, 19, 24, 1010–12, 24 November 1983.

[7.20] K.R. Laker, R. Schaumann and M.S. Ghausi, Multiple-loop feedback topologies for the design of low-sensitivity active filters, *IEEE Transactions on Circuits and Systems*, CAS-26, 1, 1–20, January 1979.

[7.21] N. Attaie and E.I. El-Masry, Multiple-loop feedback switched-capacitor structures, *IEEE Transactions on Circuits and Systems*, CAS-30, 12, 865–72, December 1983.

[7.22] N. Attaie and E.I. El-Masry, Synthesis of switched-capacitor filters in the multiple-input follow-the-leader feedback topology, *Proc. ISCAS*, 175–8, 1981.

[7.23] E.I. El-Masry, Synthesis of a follow-the-leader feedback switched-capacitor structure, *Proc. IEE*, Part *G*, 132, 18–23, 1985.

[7.24] E.I. El-Masry, Design of switched-capacitor filters in the biquadratic state-space form, *Proc. ISCAS*, 179–82, 1981.

[7.25] R.E. Bach, Jr, Selecting $R-C$ values for active filters, *Electronics*, 33, 82–5, 1960.

[7.26] E. Hokenek and G.S. Moschytz, Design of parasitic insensitive bilinear-transformed admittance-scaled (BITAS) SC ladder filters, *IEEE Transactions on Circuits and Systems*, CAS-30, 12, 873–88, December 1983.

## 7.10.1  Further reading

[7.27] F. Montecchi, Time-shared switched-capacitor ladder filters insensitive to parasitic effects, *IEEE Transactions on Circuits and Systems*, CAS-31, 4, 349–54, April 1984.

[7.28] H.G. Dimopoulos, New switched-capacitor ladder filters, *Electronics Letters*, 21, 4, 152–4, 14 February 1985.

[7.29] G.J. Smolka, Synthesis of switched-capacitor circuits simulating canonical reactance sections, *IEEE Transactions on Circuits and Systems*, CAS-32, 6, 513–21, June 1985.

[7.30] J.T. Inoue and F. Ueno, On robustness of the stability of multiphase switched-capacitor two-ports, *IEEE Transactions on Circuits and Systems*, CAS-32, 6, 522–9, June 1985.

[7.31] A.M. Davis, Flow-graph synthesis of Darlington–Cauer switched-capacitor filters, *IEEE Transactions on Circuits and Systems*, CAS-32, 7, 727–32, July 1985.

[7.32] G. Fischer and G.S. Moschytz, SC leapfrog filters without additional adder stages, *IEEE Transactions on Circuits and Systems*, CAS-32, 8, 853–6, August 1985.

[7.33] T. Inoue and F. Ueno, Switched-capacitor ladder filters with zero worst-case sensitivity at reflection zeros, *Electronics Letters*, 21, 21, 1000–1, 10 October 1985.

[7.34] S. Eriksson, Offset compensated bilinear SC filter of leapfrog type using a two-phase clock, *Electronics Letters*, 21, 22, 1043–4, 24 October 1985.

[7.35] H. Baher, Selective linear phase switched-capacitor filters modelled on classical structures, *IEEE Transactions on Circuits and Systems*, CAS-33, 2, 141–9, February 1986.

[7.36] G. Roberts, D.G. Nairn and A.S. Sedra, On the implementation of fully-differential switched-capacitor ladder filters, *IEEE Transactions on Circuits and Systems*, CAS-33, 4, 452–5, April 1986.

[7.37] J. Taylor and J. Mavor, Exact design of stray-insensitive switched-capacitor LDI ladder filters from unit-element prototypes, *IEEE Transactions on Circuits and Systems*, CAS-33, 6, 613–22, June 1986.

[7.38] P.J.O. Donavan, Parasitic-insensitive realization of switched-capacitor leapfrog ladder filters with reduced number of operational amplifiers, *Electronics Letters*, 22, 20, 1085–6, 25 September 1986.

[7.39] H. Baher, Transfer functions for switched-capacitor and wave digital filters, *IEEE Transactions on Circuits and Systems*, CAS-33, 11, 1138–42, November 1986.

[7.40] D.B. Ribner and M.A. Copeland, On the implementation of fully differential switched-capacitor ladder filters, *IEEE Transactions on Circuits and Systems*, CAS-33, 11, 1152, November 1986.

[7.41] J. Inoue and F. Ueno, Design of very low-sensitivity low-pass switched-capacitor filters, *IEEE Transactions on Circuits and Systems*, CAS-34, 5, 524–32, May 1987.

[7.42] E.I. El-Masry and H.L. Lee, Low-sensitivity realization of switched-capacitor filters, *IEEE Transactions on Circuits and Systems*, CAS-34, 5, 510–23, May 1987.

[7.43] P. Landau, D. Michel and D. Melnik, A reduced capacitor spread algorithm for elliptic band-pass SC filters, *IEEE Journal of Solid-State Circuits*, SC-22, 4, 624–6, August 1987.

[7.44] T.S. Rathore and B.B. Bhattacharyya, A systematic approach to the design of stray-insensitive SC circuits from active RL- or RLC prototypes, *International Journal of Circuit Theory and Applications*, 15, 4, 371–89, October 1987.

[7.45] L. Ping and J. Sewell, The LUD approach to switched-capacitor filter design, *IEEE Transactions on Circuits and Systems*, CAS-34, 12, 1611–14, December 1987.

[7.46] D.G. Nairn and A.S. Sedra, Auto-SC, an automated switched-capacitor ladder filter design program, *IEEE Circuits and Devices Magazine*, 4, 2, 5–9, March 1988.

[7.47] L. Ping and J.I. Sewell, Filter realization of passive network simulation, *Proc. IEE, Part G, ECS*, 135, 4, 167–76, August 1988.

[7.48] P.V. Ananda Mohan, Exact design of LDI type of SC ladder filters, *Proc. ISCAS, Portland*, pp. 1701–4, 1989.

[7.49] P.V. Ananda Mohan, Operational simulation of RLC high-pass filters, *Electronics Letters*, 24, 13, 779–80, 23 June 1988.

[7.50] T. Curran and M. Collier, Sensitivity properties of SC filters derived from LC ladder prototypes, *IEEE Transactions on Circuits and Systems*, CAS-37, 12, 1544–6, December 1990.

[7.51] H.M. Sandler and A.S. Sedra, Programmable switched-capacitor low-pass ladder filters, *IEEE Journal of Solid-State Circuits*, SC-21, 6, 1109–19, December 1986.

[7.52] L. Ping and J.I. Sewell, The TWINTOR in band-stop switched-capacitor ladder filter realization, *IEEE Transactions on Circuits and Systems*, CAS-36, 7, 1041–4, July 1989.

[7.53] C. Ikeda, Y. Horio and S. Mori, SC bilinear band-elimination ladder filters with extended biquad, *Electronics Letters*, 24, 432–4, 31 March 1988.

[7.54] D.J. Allstot and K.S. Tan, Simplified MOS switched-capacitor ladder filter structures, *IEEE Journal of Solid-State Circuits*, SC-16, 6, 724–9, December 1981.

[7.55] B.S. Song and P.R. Gray, Switched-capacitor high-$Q$ band-pass filters for IF applications, *IEEE Journal of Solid-State Circuits*, SC-21, 6, 924–33, December 1986.

[7.56] M.F. Fahmy, M. Abo-Zahhad and M.I. Sobhy, Design of selective linear phase band-pass switched-capacitor filters with equiripple pass-band amplitude response, *IEEE Transactions on Circuits and Systems*, CAS-35, 10, 1220–9, October 1988.

# Switched capacitor *N*-path filters

In the previous chapters, SC filter realizations using cascaded first- and second-order blocks or using high-order configurations based on doubly terminated RLC networks have been extensively studied. While realizing very narrow BP filters using either of these two approaches, the sensitivity of the transfer function can become appreciably large. This difficulty can be overcome by using the *N*-path filtering principle originally introduced by Franks and Sandberg [8.1].

## 8.1 Principles of *N*-path filters

The basic *N*-path filter structure is shown in Figure 8.1. It consists of *N* identical paths, each having multipliers at the input and output. A continuous-time network with a transfer function $H_a(s)$ is embedded between these multipliers. The second input to the multiplier can be a phase-shifted sinusoid, a rectangular waveform or any arbitrary waveform. The various paths require these inputs to be suitably phased, so that each operates at a different time. The output summer sums the outputs of all the *N* paths to provide the desired frequency response characteristics.

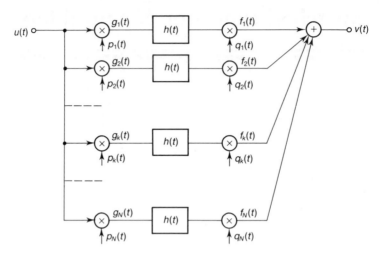

*Fig. 8.1* The N-path configuration.

The networks realizing $H_a(s)$ can have different types of transfer function such as LP, BP, notch, etc., which will determine the response at the system output.

The N-path filter structure in Figure 8.1 can use any type of multiplying waveforms, as mentioned before. However, we consider for simplicity the case when $p(t)$ and $q(t)$ are sinusoids. Since any arbitrary waveform (having the same periodicity) can be expressed as a Fourier series expansion, the results are immediately applicable [8.2]. Denoting the frequency of the multiplying sinusoid as $\omega_0$, the various $p(t)$, $q(t)$ waveforms are given by $p_n(t) = q_n(t) = \cos(\omega_0 t + 2\pi n/N)$, where $n$ is the number of the path under consideration and $N$ the total number of paths. The Laplace transform of $p_n(t)$ is obtained immediately:

$$P_n(s) = \tfrac{1}{2}\,[\delta(s - j\omega_0)e^{-j2\pi n/N} + \delta(s + j\omega_0)e^{j2\pi n/N}] \tag{8.1}$$

where the delta function is the unit impulse. The output of the $n$th multieplier is given by

$$g_n(t) = u(t)p_n(t) \tag{8.2}$$

Since multiplication in the time domain corresponds to convolution in the frequency domain, we have, from equation (8.2),

$$G_n(s) = U(s) * P_n(s) \tag{8.3}$$

Alternatively, by definition of the convolution integral, we obtain

$$G_n(s) = \int_{-\infty}^{\infty} U(p)P_n(s - p)\,\mathrm{d}p \tag{8.4}$$

Substituting for $P_n(s)$ from equation (8.1) in equation (8.4), we obtain

$$G_n(s) = \tfrac{1}{2} \left[ U(s - j\omega_0)e^{-j2\pi n/N} + U(s + j\omega_0)e^{j2\pi n/N} \right] \tag{8.5}$$

The output of the $n$th path network with the impulse response $h(t)$ in Figure 8.1 has Laplace transform

$$L_n(s) = G_n(s)H_a(s) \tag{8.6a}$$

Thus,

$$L_n(s) = \tfrac{1}{2} \left[ U(s - j\omega_0)H_a(s)e^{-j2\pi n/N} + U(s + j\omega_0)H_a(s)e^{j2\pi n/N} \right] \tag{8.6b}$$

The output multiplier performs a time–domain multiplication, in a manner similar to the input multiplier, resulting in a convolution in the frequency domain. We have, therefore,

$$\begin{aligned} F_n(s) &= \int_{-\infty}^{\infty} L_n(p)P_n(s-p)\, dp \\ &= \tfrac{1}{4} \left[ U(s)H_a(s - j\omega_0) + U(s)H_a(s + j\omega_0) \right] \\ &\quad + \tfrac{1}{4} \left[ U(s - 2j\omega_0)H_a(s - j\omega_0)e^{-j4\pi n/N} + U(s + 2j\omega_0)H_a(s + j\omega_0)e^{j4\pi n/N} \right] \end{aligned} \tag{8.7}$$

The output signal is the sum of the outputs of all output multipliers, i.e.,

$$v(t) = \sum_{n=1}^{N} f_n(t) \tag{8.8a}$$

or

$$V(s) = \sum_{n=1}^{N} F_n(s) \tag{8.8b}$$

The result is

$$\begin{aligned} V(s) &= \frac{U(s)}{4} \left[ H_a(s - j\omega_0) + H_a(s + j\omega_0) \right] + \tfrac{1}{4}U(s - 2j\omega_0)H_a(s - j\omega_0) \sum_{n=1}^{N} e^{-j4\pi n/N} \\ &\quad + \tfrac{1}{4}U(s + 2j\omega_0)H_a(s + j\omega_0) \sum_{n=1}^{N} e^{j4\pi n/N} \end{aligned} \tag{8.9}$$

We observe from equation (8.9) that, for $N > 3$, the last two terms cancel vectorially, and

$$V(s) = \frac{U(s)}{4} \left[ H_a(s - j\omega_0) + H_a(s + j\omega_0) \right] \tag{8.10}$$

Thus the transfer function of the $N$-path filter of Figure 8.1 with sinusoids as multipling waveforms is

$$T(s) = \tfrac{1}{4} \left[ H_a(s - j\omega_0) + H_a(s + j\omega_0) \right] \tag{8.11}$$

It is thus seen that if the $H_a(s)$ used in the $N$-path configuration of Figure 8.1 is an LP filter, the output of the $N$-path filter is a BP response centred at $\omega_0$. In essence, a frequency translation or arithmetic shift has been realized. A look at equation (8.7) shows that at the output of each path additional frequency components exist, which are cancelled vectorially, when the outputs of all such paths are added, which is the essential principle of the $N$-path filter.

These results can be extended to the case using arbitrary periodic waveforms for $p(t)$ and $q(t)$, considering that these are made up of several harmonically related sinusoids. Thus, the frequency translation of the LP filter response $H_a(s)$ to each of these harmonic frequencies is realized. The result is a *comb* response. Further, there is no restriction that $p(t)$ and $q(t)$ shall be identical. They can be, in general, different periodic waveforms. Considering that they are expressible as the Fourier series expansions,

$$p(t) = \sum_{m=-\infty}^{\infty} P_m e^{j2\pi mt/T} \qquad (8.12a)$$

$$q(t) = \sum_{l=-\infty}^{\infty} Q_l e^{j2\pi lt/T} \qquad (8.12b)$$

the $N$-path filter can be characterized by the overall impulse response by analyzing the configuration of Figure 8.1 in the time domain. The transfer function can also be derived. These details are omitted here; the reader is referred to [7.3, 8.1, 8.3].

The $N$-path configuration that has been studied briefly above consists of $H_a(s)$ in the continuous-time domain, while the modulating functions can be even switching waveforms. Hence, to facilitate analysis of such configurations, time-domain formulation is convenient. In contrast, when the 'path network' is itself a discrete-time network and the modulating waveforms are also switching waveforms (ON and OFF types), there is, in general, no need to follow the time-domain approaches used in $N$-path filter characterization. Instead, the complete system of Figure 8.1 with $H_d(z)$ as the path network and with $p(t)$, $q(t)$ replaced by switches can be conveniently analyzed in the $z$-domain itself. We will follow this approach, due to Lee and Chang [8.4], in this chapter. The basic principles, however, remain the same regarding the usage and functioning of the $N$-path systems.

## 8.2  SC N-path passive networks

The basic $N$-path filter structure, using pulse modulation, realized by an $N$-phase clock operating the switches, is shown in Figure 8.2(a). It consists of $N$ identical 'cells' which are cyclically connected between the input and output at a 'commutation rate' of $f_c$. Note that the switches at the input and output perform the modulation function. A summer at the output is not required, because the output is tapped 'all the time'. In the time domain, the output of each path is available successively and hence, the output is considered all the time and the summing

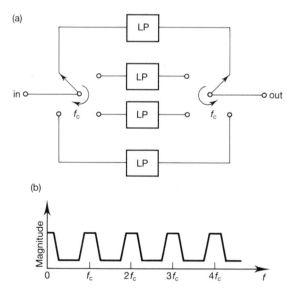

Fig. 8.2  (a) N-path filter structure;  (b) frequency response of (a).

function will be realized. Assuming an LP sub-network in each path, the resulting response is as shown in Figure 8.2(b). A simple BP response can be realized by passing the output of Figure 8.2(a) through a BP filter which rejects the LP response and responses at harmonics of $f_c$, viz., $2f_c, 3f_c, \ldots$. Alternatively, if a comb filter response is desired, band-limiting filters can be used at the input and at the output to limit the frequencies to $Nf_c/2$. This is explained by the fact that the overall sampling rate is $Nf_c$.

A first-order LP cell is shown in Figure 8.3(a). Its transfer function can be derived as

$$H_{LP}(z) = \frac{az^{-1}}{1 - bz^{-1}} \tag{8.13}$$

where $a = C_1/(C_1 + C_2)$ and $b = C_2/(C_1 + C_2)$. This network uses a two-phase clock. A four-path filter can be realized by using four such networks appropriately connected as shown in Figure 8.3(b). Note that, in each phase, the input is sampled and output is available. The transfer function of the four-path filter is

$$H(z) = \frac{az^{-4}}{1 - bz^{-4}} \tag{8.14}$$

noting that the sampling frequency in this case is increased four times. It can be verified that the response of the LP cell is preserved at zero frequency and repeats symmetrically on either side of $f_s$, $2f_s$ and $3f_s$. The response at d.c., $2f_s$ and $3f_s$ can be removed in several ways, if a single pass-band around $f_s$ is only desired.

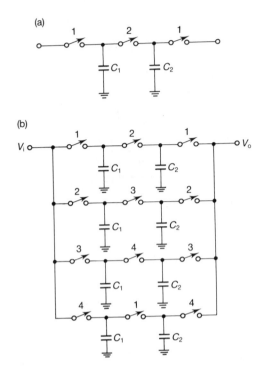

Fig. 8.3    (a) A first-order SC LP cell; (b) *N*-path filter ($N = 4$) based on (a).

In one method due to von Grunigen *et al.* [8.5], a simple FIR filter is used as a prefilter to introduce 'zeros' at even multiples of $f_s$. It realizes a transfer function

$$H(z) = \tfrac{1}{2}(1 - z^{-2})$$  (8.15)

This can be realized by another SC network or combined with the hardware required to realize the four-path filter.

In an alternative method due to Faruque [8.6], the LP cell is designed to have zeros at even multiples of $f_s$. This circuit is shown in Figure 8.4(a). The transfer function of this LP cell is

$$H(z) = \frac{A(1 - z^{-1/2})}{(1 - Bz^{-1})}$$  (8.16)

where $A = C_1/(C_1 + C_2)$ and $B = C_2/(C_1 + C_2)$. Thus, this circuit can be considered to be a cascade connection of a FIR filter and first-order SC LP network. A four-path filter using this cell is shown in Figure 8.4(b). The above two methods realize zeros at d.c. and even multiples of clock frequency. But we need, in general, circuits for suppressing all components above $f_s$ and at zero. This is achieved by using the frequency sampling filter approach due to von Grunigen *et al.* [8.7].

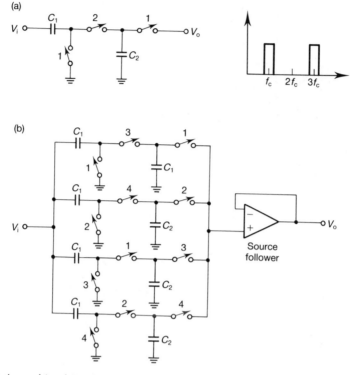

Fig. 8.4  (a) A combined FIR filter/LP cell introducing zeros at even clock-frequency multiples;  (b) N-path filter ($N = 4$) using (a).

A frequency sampling filter introduces zeros at equal intervals on the unit circle at $n$ points. A structure realizing a frequency sampling filter is presented in Figure 8.5 [1.33, 8.8]. The coefficients

$$H_i = H(e^{-j2\pi i/n}) \tag{8.17}$$

are the desired frequency response $H(j\omega)$ sampled at $n$ points on the unit circle, hence the name *frequency sampling filter*. Note that in the structure in Figure 8.5(a), zeros are introduced at desired points by suitably choosing the coefficients $H_i$. As an illustration for $n = 8$, by introducing zeros at $0, 2f_c, 3f_c, 4f_c, 5f_c, 6f_c$, a filter that passes only the frequency components $f_c$ and $7f_c$ is realized. Thus, only $H_1$ and $H_7$ exist and all other $H_i$s are zero. Note that $H_1$ and $H_7$ can be summed to yield a realizable second-order function. The resulting frequency response is shown in Figure 8.5(b).

It may also be shown that the overall transfer function of the structure of Figure 8.5(a) is that of an FIR filter. Cascading this FIR filter with the N-path filter, the side-bands at d.c., $2f_c, 3f_c$, etc., can be suppressed. It is thus seen that, by cascading N-path filters appropriately, either a filter with BP response at $f_s$ or a

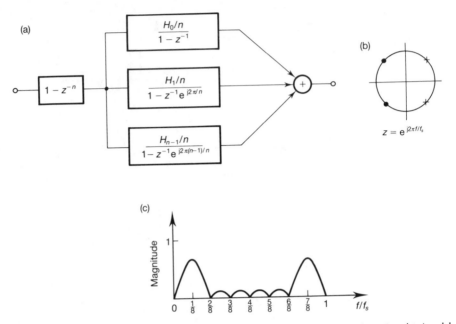

Fig. 8.5 (a) A frequency sampling filter; (b) pole–zero plot in the z-domain obtained by (a); (c) frequency response of (a).

comb filter can be realized. Any of the stray-insensitive realizations discussed earlier can be used for realizing the FIR filters.

The passive cell of the N-path filter considered in Figure 8.3(a) can be replaced by any other such LP network. A particularly interesting LP cell is shown in Figure 8.6(a) [8.9]. This is based on the LP cell of Figure 8.3(a). It may be noted, however, that in the three phases of the clock cycle, successive transfer of charge from the capacitor $C/\alpha$ to each $C$ takes place. The resulting transfer function for each path is $[H(z)]^3$, where $H(z)$ is given by

$$H(z)|_{LP} = \frac{(C/\alpha)z^{-1/2}}{\dfrac{C}{\alpha} + C(1 - z^{-1})} \tag{8.18}$$

The overall transfer function of the N-path filter is obtained by replacing $z^{-1}$ with $z^{-4}$.

Hence, the transfer function of the N-path filter is

$$H(z)|_N = \frac{z^6}{[z^4(1+\alpha) - \alpha]^3} \tag{8.19}$$

It may be noted that the circuit of Figure 8.6(a) is equivalent to a three-stage passive SC network isolated by buffers between every two stages. It is possible to activate

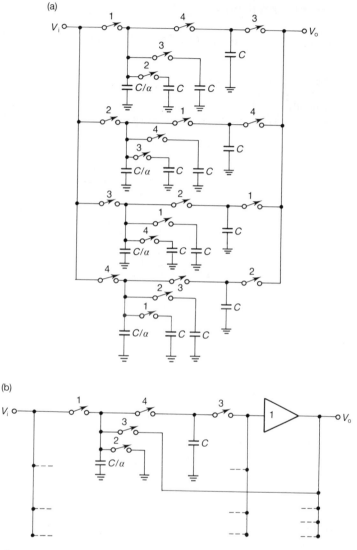

Fig. 8.6 (a) A passive N-path filter; (b) active filter derived from (a). (Adapted from [8.9], © 1982 IEEE.)

the SC network by using a unity-gain amplifier, as shown in Figure 8.6(b). The reader is urged to derive the transfer function of the SC filter of Figure 8.6(b).

In the above discussion, we have seen that the $z$-domain 'overall' transfer functions obtained for the $N$-path filters are related to the 'path $z$-domain transfer function' through the substitution $z \to z^N$. We will next consider methods of implementing this transformation in any two-phase SC network.

## 8.3 $z \to z^N$ transformation

Consider the SC network of Figure 8.7(a). The transfer function of this SC network, assuming $V_{3o} = V_{3e}z^{-1/2}$, is given by

$$C_0(1 - z^{-1})V_{0e} = -V_{2e}C_2 - V_{3e}C_3z^{-1} - V_{1e}C_1(1 - z^{-1}) \qquad (8.20a)$$

In order to implement the $z \to z^N$ transformation each switched/unswitched capacitor in Figure 8.7(a) can be substituted by new SC branches. The resulting SC

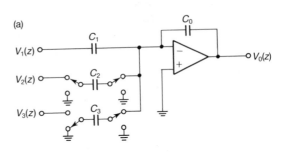

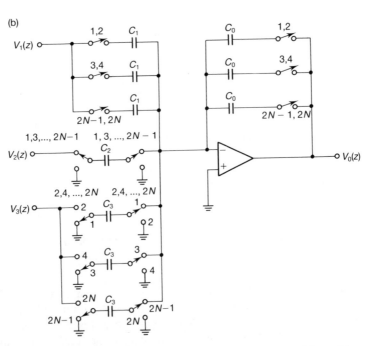

Fig. 8.7   (a) A two-phase SC network;   (b) N-phase network derived from (a) implementing $z \to z^N$ transformation. *Adapted from [8.10],* © 1980 IEEE.

network of Figure 8.7(b) can be checked to have the transfer function (in any phase),

$$C_0(1 - z^{-N})V_0 = -V_2C_2 - V_3C_3z^{-N} - V_1C_1(1 - z^{-N}) \qquad (8.20b)$$

The reader may have noted that the series-switched capacitor $C_2$ need not be replaced by $N$ series-switched capacitors working in each phase. This concept can, in general, be applied to any SC filter. Allstot and Tan first described such $N$-path filters, called *stop-go N-path filters* [8.10] because at any given time one path will be in operating (go) condition, while others are in hold (stop) condition. In Allstot and Tan's method, each memory element (capacitor) is replaced by $N$ capacitors, while leaving other elements unchanged (see Figure 8.8(a)). However, the clock phasing of the input capacitors should be as shown in Figure 8.8(b). Thus, the input capacitors can be saved at the expense of a complex clocking scheme. Since each path is operating at $f_s$, the clock feedthrough reaches the output and is present in the BP response around $f_s, 2f_s, \ldots$.

The $N$-path switched capacitor filter operation can be explained in a simple manner. Considering each path, it is noted that the frequency of operation is $f_s$. Hence, the prototype filter, e.g., LP, will have a response as shown in Figure 8.9(a). By keeping the sampling frequency the same, and using $z \rightarrow z^N$ transformation, we obtain the $N$-path SC filter with response shown in Figure 8.9(b). The bandwidth is decreased by $N$ times, while the response repeats at intervals of $f_s/N$. This is explained by considering the first-order network (entry 7, Table 4.1), with $H(z)$ of the resulting $N$-path filter given by

$$H(z) = \frac{-C_1}{C_3 + C_2(1 - z^{-N})} \qquad (8.21)$$

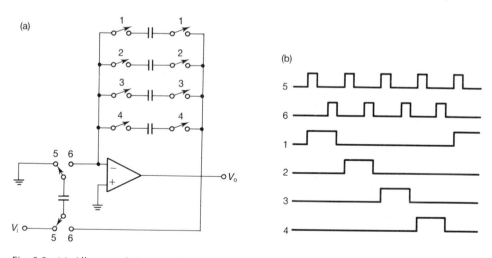

Fig. 8.8 (a) Allstot and Tan 4-path integrator; (b) clocking wave forms. *Adapted from* [8.10], © 1980 IEEE.

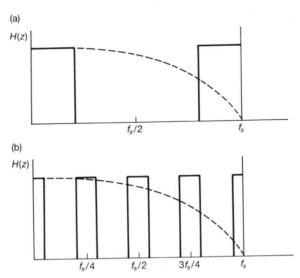

Fig. 8.9 (a) A discrete-time frequency response; (b) the corresponding response after the $z \rightarrow z^N$ transformation.

The resulting LP cutoff frequency is given by

$$\omega_c = \frac{f_s}{N} \cos^{-1} \left[ \frac{2C_2^2 + 2C_2C_3 - C_3^2}{2C_2(C_2 + C_3)} \right] \qquad (8.22)$$

It is thus evident that the base-band as well as the comb responses have bandwidths decreased proportional to the number of paths $N$.

A limitation of the above discussed $N$-path filters becomes obvious, if it is noted that the $N$ paths may not be identical. In this case, the cancellation of the image frequency components as required in the manner explained in Section 8.1 may not be perfect and the resulting spurious responses may degrade the performance of the $N$-path filter. A solution to this so-called *balancing problem* in $N$-path filters has been suggested by Wupper and Fettweis [8.11]. This new method leads to *pseudo N-path filters*, which will be considered next.

## 8.4  Pseudo *N*-path filters  [8.9, 8.11, 8.12]

In this new technique, all the $N$ paths are made identical by using an analog shift register in place of the bank of $N$ commutated capacitors. The SC inverting integrator of Figure 8.10(a) can be used to realize an $N$-path version ($N = 3$) as shown in Figure 8.10(b). In this circuit, in phase 1, the feedback capacitor $C$ receives charge from the input capacitor $C_0$ as well as the storage capacitor $C_3$. In phase 2, the charge on $C_2$ is transferred to $C_3$, and in phase 3 the charge on $C_1$ is transferred to

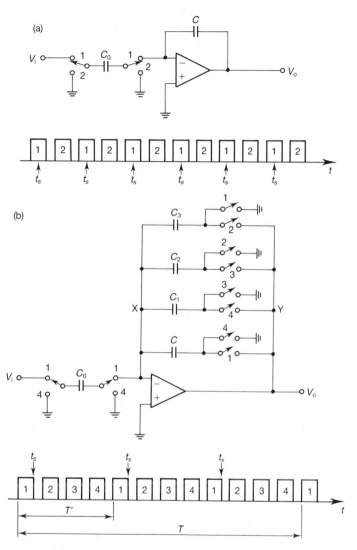

*Fig. 8.10* (a) An SC integrator; (b) *N*-path version realizing $z \to z^N$ transformation ($N = 3$). *(Adapted from [8.9], © 1982 IEEE.)*

$C_2$. In phase 4, the charge on $C$ is transferred to $C_1$ and the input capacitor is discharged. This operation repeats in a similar manner thereafter. Thus, if the original two-phase SC network is assumed to operate at a rate of $f_s$ Hz, the 3-path filter operates at a rate $12 f_s$ Hz. This is explained as follows. In the circuit of Figure 8.10(a), in each set of four phases, the charge on $C_1$ would have moved one step forward in the shift register. Thus, after the four clock phases period discussed above, the updated charge on $C$ has just moved to $C_1$. Thus, 12 clock phases are required to move the updated charge on $C_1$ to $C_3$.

The resulting $N$-path filter is stray-insensitive. The charge flows through the same path, while realizing all individual 'path-transfer functions'. There is no loss of charge in the process of charge transfer from one capacitor to another, assuming a high gain for the OA.

The transfer function of the prototype SC integrator of Figure 8.10(a) is transformed to

$$H(z) = \frac{-C_0/C}{1 - z^{-3}} \qquad (8.23)$$

as desired. The first pass-band (other than that at d.c.) is located at $f_0 = 1/T$ (see Figure 8.10(b)). Note that all charge packets follow the same path through the circuit. Hence, the lowest clock feedthrough frequency is $1/T'$ — it is far away from the pass band. The right-hand side switch of $C_0$ can be interchanged to give a non-inverting $N$-path integrator realizing

$$H(z) = \left( \frac{z^{-3}}{1 - z^{-3}} \right) \frac{C_0}{C} \qquad (8.24)$$

It is a straightforward matter to extend the above method to realize high-order $N$-path SC filters. The $N$-path concept can be applied to low-sensitivity realizations based on the leap-frog ladder simulation method in a straightforward manner so

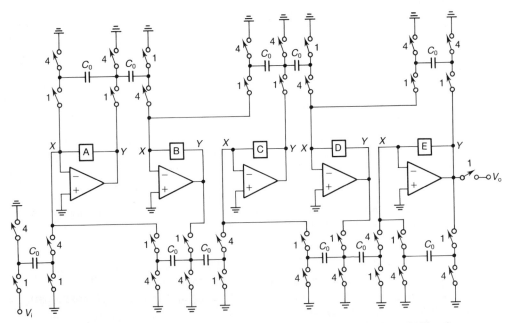

*Fig. 8.11* An SC all-pole LP ladder filter suitable for conversion as an $N$-path filter by replacing the blocks denoted A–E by a circulating shift register comprising $N+1$ capacitors *(adapted from [8.9], © 1982 IEEE)*.

that the resulting responses have the advantages of *N*-path filters as well as of SC ladder filters [8.11]. As has been mentioned before, the $z \to z^N$ transformation can be realized together with the pseudo *N*-path concept. Each memory-possessing element (integrating capacitor) can be replaced by an analog shift register of the kind shown in Figure 8.10(b). Thus an LDI-type SC LP ladder filter operating with a two-phase clock (studied in Chapter 7) can be, for instance, converted to an *N*-path filter as shown in Figure 8.11. The blocks labelled A–E in Figure 8.11 are replaced by the shift register comprising $C_1$, $C_2$, $C_3$ and $C$. It is thus seen that the SC LP filter can be modified to realize an *N*-path filter.

The above method considered realization of *N*-path filters based on all-pole LP ladder filters. However, when the starting point is an elliptic LP ladder filter, some unswitched capacitors interlinking the outputs and inputs of OAs appropriately need replacement by the capacitor banks. However, this is not necessary, if we note that for sampled and held input conditions, a capacitor $C_1$ as in Figure 8.12(a) can be replaced by the switched capacitor $C_1$ as in Figure 8.12(b). Thus, only the integrating capacitors need be modified.

The pseudo *N*-path filters considered above have the disadvantage that a large number of clock phases are required for operation, resulting in the requirement that

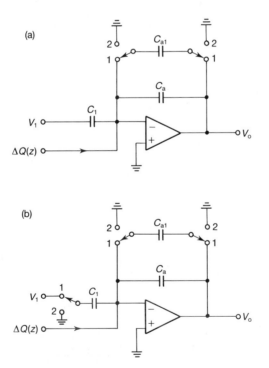

*Fig. 8.12* (a) First-order SC network occurring in an elliptic LP realization; (b) modification avoiding replication of $C_1$ by a shift register. *(Adapted from [8.9], © 1981 IEEE.)*

OAs operate fast. The number of clock phases to perform a full cycle of operation can be reduced to 8 by using a random access memory (RAM) in place of the circulating shift register of Figure 8.10 [8.9]. Such an arrangement is shown in Figure 8.13(a), together with the clocking waveforms in Figure 8.13(b). In phase 1, the input charge as well as the charge on $C_1$ is transferred to $C$ (since phase 3 is synchronous with phase 1 at this instant). In phase 2, the updated charge on $C$ is transferred back to $C_1$. This charge stored on $C_1$ remains until the next phase 3 period. Similarly, in phases 1 and 5, the charge on $C_0$ and $C_2$ is transferred to $C$ and stored back on $C_2$. The operation is similar in phases 7 and 8. Thus each charge packet is stored for an interval $T$ $(=3T')$ rather than $T'$. The circuit of Figure 8.13(a) can be used to replace the integrators in SC ladder filters to realize $N$-path structures. This is left as an exercise to the reader. The RAM method needs a more complex clocking scheme than the circulating shift register method. The

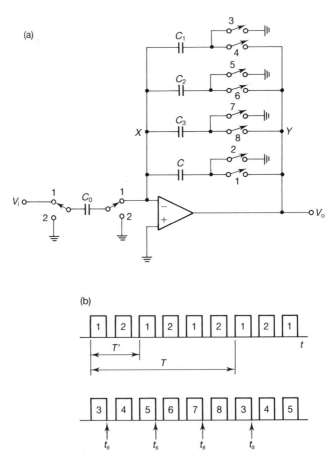

Fig. 8.13  An SC N-path integrator using a RAM *(adapted from [8.9], © 1982 IEEE).*

OAs, however, need work at only half the speed of the circulating shift register type $N$-path filters. Further, the RAM method is not immune to clock feedthrough noise in the centre of the pass band, since charge packets belonging to different paths are stored on different $C_i$. A complete comparison of the experimental performance of these pseudo $N$-path filters can be found in [8.9].

It may be noted that, in the $N$-path filters discussed above, the prototype LP filter response is mapped to d.c. and multiples of $f_s$. If we desire to have the pass bands around d.c., $2f_s, 3f_s, 4f_s$, etc., suppressed, additional BP filters have to be cascaded with the $N$-path filters. (Note that we have already studied methods of suppressing unwanted pass bands in Section 8.2.) Patangia and Cartinhour have suggested that by providing an inversion in the path updating the capacitor $C$, (e.g., the last path $C_3$ can incorporate an inversion as shown in the modified two-path filter in Figure 8.14, employing a circulating shift register), two adjacent pass bands may be separated by twice the centre frequency of the lowest pass band and that around d.c. eliminated. Note that the input capacitor $C_i$ has been used to provide a first-order LP function as well as to feed the input [8.13, 8.14].

Another possibility is to realize $z \rightarrow -z^N$ transformation (see exercise 8.5). As an illustration, for $N = 3$, we have, for the network (entry 3, Table 4.1)

$$H(z) = \frac{z^{-3}C_1}{C_3 + C_2(1 + z^{-3})} \tag{8.25}$$

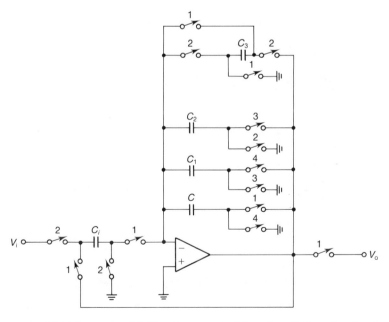

*Fig. 8.14* A modified first-order SC $N$-path network. (Note that $C_3$ has circuitry for inversion of charge.)

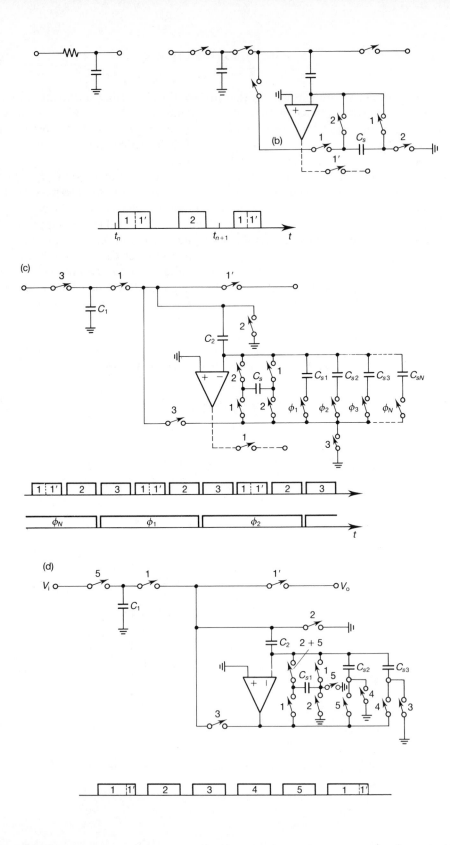

(Note the change in sign of the $z^{-3}$ term in the denominator.) The reader is urged to plot the response given by equation (8.25) in order to observe the effect of this modification. It may be mentioned that the case of $N = 2$ has been considered in [8.15].

## 8.4.1   Pseudo N-path SC filters based on VIS SC ladder filters
[6.52, 8.16–8.18]

The $z \rightarrow z^N$ transformation can also be applied to SC ladder filters based on the VIS concept. Pandel has extensively investigated this aspect, which will be outlined briefly in what follows.

Consider the simulation of the first-order RC network shown in Figure 8.15(a) using a VIS of inverse recharging type. The resulting 1-path circuit is as in Figure 8.15(b). Using a RAM for the memory capacitor $C_s$ in Figure 8.15(b), we obtain the N-path filter of Figure 8.15(c), where the clocking waveforms are also shown. The registration step is the same in both the 1- and N-path cases of Figures 8.15(b) and 8.15(c) (i.e., in phase 1). Prior to this phase, in phase 3, the charge on the appropriate memory capacitor $C_{si}$ is 'read' into $C_2$. In phase 2, the remaining charge on $C_2$ as well as the 'inverted' charge on $C_s$ are transferred to the RAM. Note that an HP output is available as shown in Figure 8.15(c).

An N-path filter obtained using a circulating shift-register is shown in Figure 8.15(d). This circuit, however, requires a five-phase clock. The reader is urged to verify its operation. Pandel has applied this concept to realize high-order wave SC filters, for which the reader is referred to [6.52]. A 6-path wave SC filter has also been fabricated successfully, thus demonstrating the realizability of N-path SC filters for narrow-band applications [6.52].

## 8.5   Conclusions

In this chapter, application of N-path principles to SC networks has been studied in detail. Very narrow BP filters can be successfully realized using this technique, where the prototype can, in general, be chosen to be any of the types of SC structure detailed in Chapters 4–7. We conclude the discussion on SC network structures in this book here and proceed in the next chapter to a detailed study of practical considerations in SC network design as well as applications of SC networks.

---

Fig. 8.15   (a) A first-order passive LP RC network;   (b) an SC version of (a) using an inverse recharging type of VIS;   (c) realization of N-path version of (b) using a RAM; (d) realization of N-path version of (b) using a circulating shift register. (b) (c) (d) *(adapted from [8.16], © AEU 1982)*.

## 8.6 Exercises

8.1 Integrated frequency selective amplifiers have been built in IC form using concepts similar to those of *N*-path filters. A 2-path structure is shown in Figure 8.16 [8.19]. Analyze the circuit and suggest a method that can be used to realize a BP response.

8.2 Is a 2-path SC network possible while using $z \to z^N$ transformation? Use a first-order LP network and an integrator to assist your investigation.

8.3 Prove that $z \to -z$ transformation converts LP responses to HP responses [8.15].

8.4 An integrated SC filter based on *N*-path and frequency sampling principles has been described by von Grunigen *et al.* [8.20]. It uses a Sallen–Key second-order LP network. Figure 8.17 contains a modification at the input terminals (switched capacitors $C_1$ and $C'$). Discuss the purpose of this modification. Derive the transfer function of this network.

8.5 The $z \to -z^N$ transformation can be used on a bilinear transformation type first-order LP network or other types of first-order LP network (i.e., forward Euler or backward Euler or LDI types). Discuss the differences between the realizations obtained. Describe a practical realization for the bilinear transformation type network.

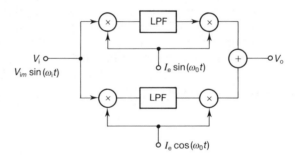

*Fig. 8.16* An integrated frequency selective amplifier using a two-path structure.

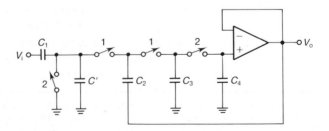

*Fig. 8.17* A modification of the integrated SC filter described by Van Grunigen *et al.*

# 8.7 References

[8.1] L.E. Franks and I.W. Sandberg, An alternative approach to the realization of network transfer functions: the N-path filter, *Bell System Technical Journal*, 39, 1321–50, 1960.

[8.2] A. Acampora, The generalized transfer function and pole-zero migrations in switched networks, *RCA Review*, 27, 2, 245–62, June 1966.

[8.3] L.E. Franks, N-path filters, in G.C. Temes and S.K. Mitra (eds), *Modern Filter Theory and Design*, 465–503, Wiley, 1973.

[8.4] M.S. Lee and C. Chang, Exact synthesis of N-path switched-capacitor filters, *Proc. ISCAS*, 166–9, 1981.

[8.5] D.C. von Grunigen, U.W. Brugger, G.S. Moschytz and W. Vollenweider, Combined switched-capacitor FIR N-path filter using only grounded capacitors, *Electronics Letters*, 17, 21, 788–90, 15 October 1981.

[8.6] S.M. Faruque, Switched-capacitor FIR cell for N-path filters, *Electronics Letters*, 18, 10, 431–2, 13 May 1982.

[8.7] D.C. von Grunigen, U.W. Brugger and G.S. Moschytz, Switched-capacitor frequency-sampling N-path filter, *Proc. ISCAS*, 209–12, May 1982.

[8.8] A.V. Oppenheim and R.W. Schafer, *Digital Signal Processing*, Prentice Hall Inc., 1975.

[8.9] M.B. Ghaderi, J.A. Nossek and G.C. Temes, Narrow-band switched-capacitor band-pass filters, *IEEE Transactions on Circuits and Systems*, CAS-29, 8, 557–71, August 1982.

[8.10] D.J. Allstot and K.S. Tan, A switched-capacitor N-path filter, *Proc. ISCAS*, 313–16, April 1980.

[8.11] A. Fettweis and H. Wupper, A solution to the balancing problem in N-path filters, *IEEE Transactions on Circuit Theory*, CT-18, 3, 403–5, May 1971.

[8.12] M.B. Ghaderi, G.C. Temes and J.A. Nossek, Switched-capacitor pseudo-N-path filters, *Proc. ISCAS*, 519–22, 1981.

[8.13] H.C. Patangia and J. Cartinhour, A tunable switched-capacitor N-path filter, *Proc. 24th Midwest Symposium on Circuits and Systems*, 400–4, June 1981.

[8.14] H.C. Patangia and J. Cartinhour, 2-phase circulating shift register for switched-capacitor N-path filters, *Electronics Letters*, 18, 11, 444–5, 27 May 1982.

[8.15] A.G. Constantinides, Spectral transformation for digital filters, *Proc. IEE*, 117, 8, 1585–90, August 1970.

[8.16] J. Pandel, Principles of pseudo N-path switched-capacitor filters using recharging devices, *AEU*, 36, 5, 177–87, May 1982.

[8.17] U. Kleine and J. Pandel, Novel switched-capacitor pseudo N-path filter, *Electronics Letters*, 18, 2, 66–8, 21 January 1982.

[8.18] J. Pandel, Design of higher-order pseudo-N-path filters and frequency symmetrical filters in VIS-SC technique, *AEU*, 37, 7/8, 251–60, July/August 1983.

[8.19] G.A. Rigby, Integrated selective amplifiers using frequency translation, *IEEE Journal on Solid-State Circuits*, SC-1, 1, 39–44, September 1966.

[8.20] D.C. von Grunigen, R.P. Sigg, J. Schmid, G.S. Moschytz and H. Melchoir, An integrated CMOS switched-capacitor band-pass filter based on N-path and frequency sampling principles, *IEEE Journal on Solid-State Circuits*, SC-18, 6, 753–61, December 1983.

## 8.7.1 Further reading

[8.21] J.E. Da Franca, A single-path frequency-translated switched-capacitor band-pass filter system, *IEEE Transactions on Circuits and Systems*, CAS-32, 9, 938–44, September 1985.

[8.22] J.E. Da Franca and D.G. Haigh, Design and applications of single-path frequency-translated switched-capacitor systems, *IEEE Transactions on Circuits and Systems*, 35, 4, 394–408, April 1988.

[8.23] J.L. Abdon Garcia-Vazquez and E. Sanchez-Sinencio, Finite gain-bandwidth product effects on a pair of pseudo-N-path SC filters, *IEEE Transactions on Circuits and Systems*, CAS-31, 6, 583–4, June 1984.

[8.24] T.H. Hsu and G.C. Temes, An improved circuit for pseudo-N-path switched-capacitor filters, *IEEE Transactions on Circuits and Systems*, CAS-32, 10, 1071–3, October 1985.

[8.25] J. Pandel, D. Bruckmann, A. Fettweis, B.J. Hosticka, U. Kleine, R. Schweer and G. Zimmer, Integrated 18th-order pseudo-N-path filter in VIS-SC technique, *IEEE Transactions on Circuits and Systems*, CAS-35, 2, 158–66, February 1986.

[8.26] J.C. Lin and J.H. Nevin, Differential charge-domain bilinear-z switched-capacitor pseudo-N-path filters, *IEEE Transactions on Circuits and Systems*, 35, 4, 409–15, April 1988.

[8.27] J.E. Da Franca, Nonrecursive polyphase switched-capacitor decimators and interpolators, *IEEE Transactions on Circuits and Systems*, CAS-32, 9, 877–87, September 1985.

[8.28] J.E. Da Franca and S. Santos, FIR switched-capacitor decimators with active delayed block polyphase structures, *IEEE Transactions on Circuits and Systems*, 35, 8, 1033–7, August 1988.

# CHAPTER NINE

# Practical considerations in the design of switched capacitor networks and their applications

The preceding chapters have discussed analysis and design methods for SC networks at great length. Numerous design approaches are thus shown to be available. The important factors to be considered in choosing a particular design are the stray insensitivity (depending on the stringency of the requirements), total capacitance required (deciding the area), the number of OAs, number of switches, ease of design (transformations used in the design, scaling facilities, etc.). In all the above analyses, the OA has been considered to have infinite gain and infinite bandwidth. In practice, the OAs are non-ideal. Furthermore, the switches have their non-idealities, especially finite on-resistance and finite capacitance. The wiring on the chip required to interconnect the various components has its own contribution regarding coupling, crosstalk, noise, etc. Thus, to satisfy stringent requirements in a practical design, it is imperative that the chip designer takes into account all the relevant factors. For this purpose, modelling or characterization of these non-idealities is essential in the analysis of SC networks. A great deal of the literature tackles these various problems. These aspects are discussed here.

## 9.1  MOS operational amplifiers

### 9.1.1  Performance characteristics of MOS OAs

Two types of OA, with different design requirements, are usually needed on an IC chip realizing analog functions: the first type is used to drive an on-chip capacitive load of a few picofarads; and the second type to interface the sub-system being integrated on the IC with the outside world. In the second case, there may be a necessity to drive a resistive load or a capacitive load. The first type is denoted as an *internal amplifier*, and the second as an *output buffer*.

The important parameters of an OA are the following:

1. open-loop voltage gain, $A_0$;
2. output resistance;
3. bandwidth, $B$;
4. response time or slew rate, $S$ (volts per second);
5. power supply rejection ratio (PSRR);
6. common mode rejection ratio (CMRR);
7. noise input voltage, $v_{ni}$;

8. input offset voltage, $v_{io}$;
9. input offset current, $I_{io}$.

We briefly explain the significance of all these parameters.

For MOS OAs, the input current is negligible and input resistance is infinite. These need not be considered here, as they are invariably obtained in MOS technology. The *input offset voltage* is the voltage required between the input terminals of the OA, to force the output voltage to be zero. The *input offset current* is the difference between bias currents at the two inputs of OA with the output voltage zero. The *common-mode rejection ratio* is the ratio of differential-mode voltage gain to common-mode voltage gain. The *slew rate* is the maximum rate of change of output voltage, $(dv_0/dt)_{max}$, for a step input voltage.

*Noise.* The OA has several internal active components (passive components like resistors are also simulated in MOS OAs by MOS transistors). It is possible to characterize the overall noise of an OA by the noise model of Figure 9.1. It consists of a noise voltage source and a noise-free OA. There are two contributions to the OA noise: the *thermal* noise and the *flicker* or $1/f$ noise.

The thermal noise arises because of the finite ohmic resistance of the channel of the transistors.[*] The equivalent input thermal noise voltage of an MOS FET is

$$\overline{V_{eq}}^2 \propto 4kT\left(\frac{2}{3g_m}\right)\Delta f \qquad (9.1)$$

where $g_m$ is the transconductance of the device and $\Delta f$ is the bandwidth used for measurement. The equivalent input flicker noise voltage resulting from random fluctuations in the density of charge carriers in the channel of MOS transistors is

$$\overline{V_{eq}}^2 \propto \frac{1}{C_0}\frac{\Delta f}{f} \qquad (9.2)$$

where $C_0$ is the gate capacitance of the device. In both expressions (9.1) and (9.2), the proportionality constants depend on the geometry of the device. In voice frequency applications, the OA noise is mainly considered to be of the $1/f$ type. Because of the large gain of the input stage of OAs, the noise contribution of the input stage

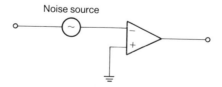

Noise source

*Fig. 9.1* Representation of non-ideal amplifier by a noise source and noise-free amplifier.

---

[*] Shot noise does not occur in MOS transistors, because the current flow is through drift or ohmic conduction.

is the most important in deciding the overall noise of an OA and hence the design of the OA must be such as to minimize the noise of the input transistors.

*Open-circuit voltage gain and bandwidth.* MOS OAs are usually realized as two-stage amplifiers in order to obtain open-circuit d.c. voltage gains of the order of 1000 to 20 000. A schematic representation of a two-stage OA is given in Figure 9.2. Each stage provides a gain of the order of 50. The second stage provides the frequency compensation by virtue of the feedback capacitor $C_c$. The last stage provides a low output impedance. In the design of the OA, it is useful to realize as large a gain as possible in the first stage itself so that the noise generated in the first stage is the main contributor to the overall noise.

The bandwidth and slew rate of the OA are decided by $C_c$, which is referred to as a *pole-splitting capacitor*. The capacitor $C_c$ is magnified by the second-stage gain (Miller effect) and serves as the load for the first stage. Thus a pole at a low frequency is realized. The high-frequency pole is realized by the second stage load capacitor as well as $C_c$. This pole is usually at a much higher frequency than the unity-gain bandwidth of the OA. The design of the OA should be such that capacitive loading of the OA should not cause instability. This topic on frequency compensation will be studied in detail at a later stage.

*Power dissipation.* For use in SC filters, the OA must be able to charge the load capacitance within the allowed time. This may impose a requirement on the standing current through the output stage, thus deciding the power dissipation. Power dissipation can, however, be reduced by using dynamic amplifiers. In these, the quiescent current in the absence of signals is allowed to decay to zero. No d.c. paths exist in these circuits for the current to flow from the supply.

*Offset voltage.* The offset voltage occurs because of the mismatch of channel lengths, widths and threshold voltages of the MOS transistors in the differential amplifier stages.

*Power supply rejection ratio.* This is the ratio of the voltage gain from input to output to that from the supply to the output (under open-loop condition). This factor is important because, in a complex IC containing digital and analog blocks, the supply noise due to high-frequency switching waveforms would couple into the analog signal paths. Thus the overall signal to noise ratio would be degraded.

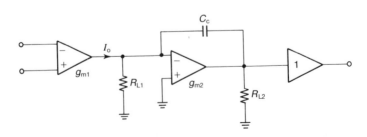

*Fig. 9.2*  Simplified model of a two-stage OA.

By suitable design, all the above factors can be taken into account and MOS OAs can be designed for good performance.

Either the NMOS or the CMOS technology can be used in the design of MOS OAs. Several NMOS OA designs have been described in the literature and have been employed in practical designs, which we shall consider in some detail in the next subsection, while the CMOS designs are discussed in Section 9.1.3. Also, these OAs tend to dissipate power continuously. Designs which require power only when they are performing some operation such as integration, termed *dynamic amplifiers*, are also of interest. Some work done in this area is also briefly considered in a subsequent section.

## 9.1.2  NMOS amplifiers

The differential amplifier or long-tailed pair is basic in the building of all OAs and is used at the first as well as intermediate stages. The earliest NMOS amplifier due to Fry [9.1] is presented in Figure 9.3(a). Interestingly, it has incorporated offset correction as well to reduce the OA offset voltage. The first stage is a differential amplifier formed by $T_1$–$T_5$, of which $T_5$ is the current source; $T_3$ and $T_4$ serve as loads for $T_1$ and $T_2$. The differential signal between the drains of $T_1$ and $T_2$ is converted to a single-ended output by $T_6$ and $T_7$. The transistors $T_6$–$T_8$ also achieve another useful function, that of increasing CMRR by giving comon-mode feedback to $T_5$. A source follower is realized by $T_{12}$ and $T_{13}$. The next eight transistors serve to amplify the signal further. The overall gain realized by the circuit is 65 dB.

The offset correction circuit (Figure 9.3(b)) functions as follows. When $\phi_c$ is negative, $T_9$, $T_{10}$ and $T_{11}$ are turned on and the OA offset voltage is stored on $C$. Next, when $\phi_{op}$ is on and $\phi_c$ is off, $V_1$ is directly applied to $A$, while the voltage stored on $C$ due to offset sampled earlier given as $(e_oA/(1 + A))$ is added to $V_2$, the second input of the OA. This voltage opposes the offset voltage $e_o$, thus yielding a net offset of

$$e_o - \frac{e_o A}{1 + A} = \frac{e_o}{1 + A} \tag{9.3}$$

which is very small. The OA in Figure 9.3(a) may be frequency-compensated using a capacitor from $E$ to ground.

It is interesting to study in greater detail the differential input stage in Figure 9.3(a). Since the gain of an amplifier is given as $g_m R_L$, where $g_m$ pertains to the driver transistor and $R_L$ to the load, for large gains, both $g_m$ and $R_L$ will be as large as possible. The requirement of large $g_m$ implies that either one or both the quantities $W$ and $L$ shall be large. A large $W$ (gate width) implies a large gate area and hence large parasitic capacitances, while a large value of $L$ results in large power dissipation or large power supply voltages. Consequently, the gain obtainable from single channel MOS stages is limited [9.2].

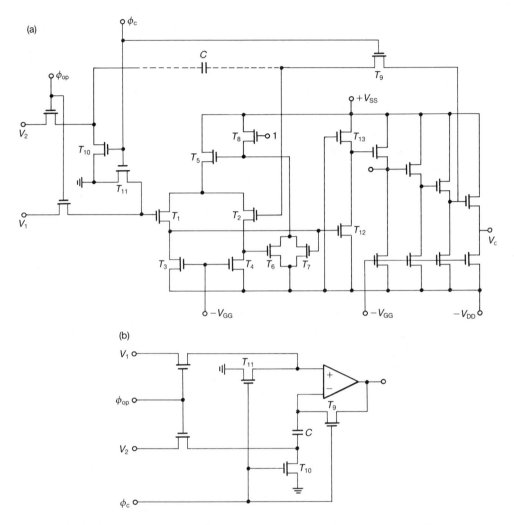

*Fig. 9.3* (a) Fry's NMOS OA; (b) offset-cancellation circuit. *(Adapted from [9.1],* © 1969 IEEE).

The gain of the differential stage formed by $T_1-T_5$ is given by

$$G_{dm} = \frac{g_{m1}}{g_{m3}}$$    (9.4a)

As an illustration, using $W/L$ ratios for driver and load devices of 432/12 and 42/24, a gain as low as 5 can be obtained from a differential stage. The CMRR of a differential stage is given by

$$G_{cm} = \frac{1}{2R_{08}g_{m3}}$$    (9.4b)

Thus, by using a long channel for the current source transistor, $R_{08}$ can be increased, thus reducing the common-mode gain.

An internally compensated NMOS OA has been described by Tsividis and Gray [9.2]. This design, presented in Figure 9.4, uses an input differential stage which is the same as that considered above, followed by differential to single-ended converter $(T_4, T_5, T_{11}, T_{12})$, a cascode stage $(T_{18}-T_{20})$, an output stage driver $(T_{21}, T_{22})$ and an output stage $(T_{23}-T_{26})$. The differential to single-ended converter passes one of its inputs through an inverting path (source follower $T_4$, $T_5$ and inverter formed by $T_{11}$ and $T_{12}$) and another through a non-inverting path (source follower $T_{11}-T_{12}$). Both these in-phase signals are thus added at the drain of $T_{12}$. The differential gain of this stage is 0.95. The output at the drain of $T_{12}$ is passed through the cascode stage formed by $(T_{18}-T_{20})$. The purpose is to reduce the Miller capacitance at the gate of $T_{20}$, which will load the previous stage. The gain of the cascode stage is 11. The source follower $T_{21}-T_{22}$ shifts the d.c. level of the voltage at the drain of $T_{19}$ as well as providing isolation between the last output stage and the cascode stage. The output stage employs shunt–shunt feedback through $T_{23}$ in order to lower the output resistance so that capacitive loads of up to 70 pF can be driven. Also, the standing current in $T_{25}-T_{26}$ is chosen large to enable driving the capacitive load. The gain of this stage is about 6.6. Thus the total gain in the Tsividis–Gray design is 51 dB. It may be noted that the strings of transistors $T_1-T_3$ and $T_{15}-T_{17}$ for voltage dividers and the voltages across $T_3$, $T_{16}$ and $T_{17}$ are used to bias the various current sources as well as the cascode stage. For a detailed discussion on the choice of $W/L$ for realizing quiescent d.c. conditions, the reader is referred to [9.2]. The

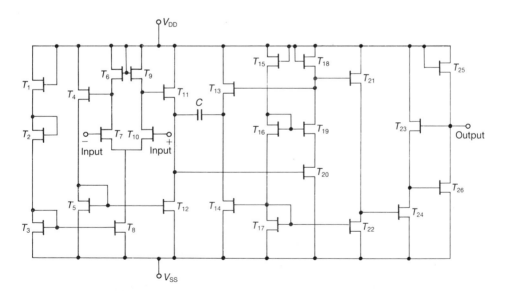

Fig. 9.4  Tsividis and Gray's NMOS OA *(adapted from [9.2], © 1976 IEEE).*

important point to be noted is that the large threshold voltage ($V_T$) variation of MOS transistors shall be taken into account in the design so that the biasing of all transistors is unconditionally stabilized.

The first OA in SC filters was described by Hosticka *et al.* [4.1], and is presented in Figure 9.5(a). It uses depletion load devices as against enhancement load devices employed in the previous designs. The main advantage due to this modification is the availability of increased gain per stage. The input stage is a single-ended differential amplifier. Transistors $T_8$ and $T_9$ perform the level shift function (source follower). A cascode stage is realized using $T_{10}$, $T_{13}$ and $T_{14}$, the output of which (available at the drain of $T_{13}$) is buffered by $T_{17}$ and $T_{18}$. Note that additional buffer stage using $T_{15}$ and $T_{16}$ is useful for frequency compensation. Furthermore, the

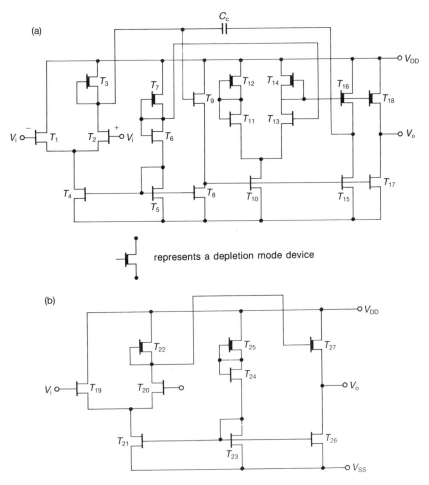

Fig. 9.5  Hosticka *et al.* designs using depletion-mode devices:  (a) high-gain NMOS OA; (b) low-gain NMOS OA. *(Adapted from [4.1]*, © 1977 IEEE.)

output stage $(T_{17}-T_{18})$ is a push-pull driver, dominated by the source follower section of $T_{18}$. The overall gain of this stage is 75 dB. If an OA with low gain is desired, Hosticka *et al* [4.1] have recommended a simple stage, shown in Figure 9.5(b). The gain of this stage is 32 dB.

It may be noted that the maximum practical voltage gain available with a depletion-mode load device is about 100. The formulas for voltage gain for an amplifier using a depletion-mode load device and an enhancement-mode load device are, respectively [9.3],

$$A_{VD} = -\frac{2}{\gamma}\sqrt{\beta_R(V_o + V_{BB} + 2\phi_F)} \tag{9.5a}$$

$$A_{VE} = -\sqrt{\beta_R}\ \cfrac{1}{1 + \cfrac{\gamma}{2\sqrt{(V_o + V_{BB} + 2\phi_F)}}} \tag{9.5b}$$

where $\beta_R$ is the ratio of the die area of active device to load device, $\gamma$ is the body factor, $\phi_F$ is the substrate built-in potential, $V_{BB}$ is the body bias, and $V_o$ is the output voltage.

Thus the gain advantage for a given $\beta_R$ is

$$\frac{A_{VD}}{A_{VE}} = \frac{2}{\gamma}\sqrt{(V_o + V_{BB} + 2\phi_F)} \tag{9.6}$$

For typical values of $V_{DD} = 10$ V, $V_{BB} = 0$, $\gamma = 0.4$, $\phi_F = 0.3$, 13 times more gain is obtained with depletion-mode load. Thus, later designs employed only depletion-mode load devices.

The NMOS OA due to Senderowicz *et al.* [9.3] is considered next. The basic concept used to realize this OA is shown in Figure 9.6. Transistors $T_1-T_7$ perform the function of the basic differential amplifier with depletion-mode load devices and the bias network for the current source. The next stage formed by the floating batteries and current mirror realized by devices $T_{17}$ and $T_{18}$ achieves differential to single-ended conversion using a technique known as *replica biasing*. This stage has high output impedance. The batteries $V_{LS}$ shall be chosen so as to yield maximum common-mode voltage range. This requires that $V_{LS}$ tracks the changes in $V_{DD} + V_{TD}$, where $V_{TD}$ is threshold voltage required to keep $T_3$ and $T_4$ in conduction.

The level shifter circuit, i.e., the circuit realizing the floating batteries $V_{LS}$, is shown in Figure 9.6(b). This shunt feedback circuit does not conduct until the voltage across and source drain of $Q_2$ exceeds $V_{T1} + V_{T2} + (2I_x(W/L)/\mu C_o)^{1/2}$, where $I_x$ is the biasing current. The incremental resistance of this battery is $1/g_{m2}$, which is quite small. The realization of the complete level shifter is presented in Figure 9.6(c). The bias current, $I_x$, for the battery is a 'replica' of that through $Q_3$. The currents through $Q_1$ and $Q_{1A}$ are identical because of current mirror action of $Q_3$, $Q_4$ and $Q_5$. However, the currents through $Q_2$ and $Q_{2A}$ are not necessarily equal,

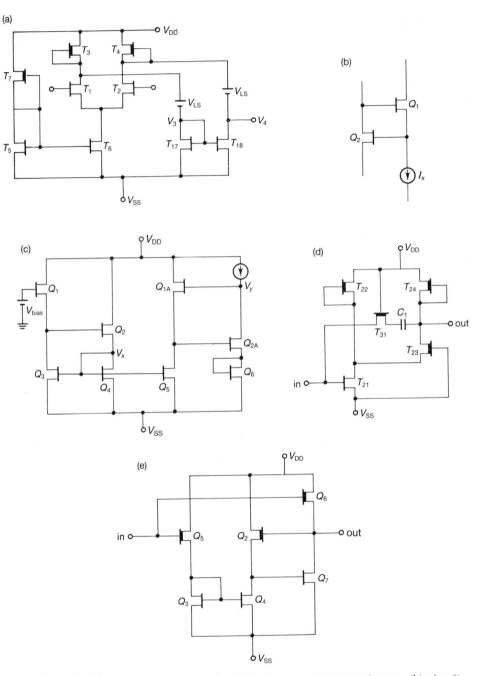

*Fig. 9.6* Senderowicz *et al.* design using replica biasing: (a) basic scheme; (b) circuit realizing floating batteries; (c) level shifter using (b); (d) frequency compensation stage; (e) output stage. *(Adapted from [9.3], © 1978 IEEE.)*

but, because of the large transconductance of these devices, $V_y$ is only slightly different from $V_{bias}$.

The frequency compensation is accomplished using the next stage (Figure 9.6(d)), which is a cascode stage with the $g_m$ of transistor $T_{21}$ increased because of the current source $T_{22}$, as explained above in the discussion of the Hosticka *et al.* design [4.1]. A capacitor and a resistor simulated by $T_{31}$ are connected between the input and output of this stage; this type of frequency compensation is studied later in some detail. The last stage is a broadband unity-gain stage (Figure 9.6(e)) with low output impedance. When the input signal swings in the positive direction with resistive load connected at the output, the transistors $Q_5$ and $Q_6$ conduct. The transistor $Q_5$ can be made large so as to deliver large load currents. In the case of negative swing of the input voltage, the currents in $Q_3$, $Q_4$ decrease, thereby increasing the $V_{GS}$ of $Q_7$. Thus, $Q_7$ sinks the load current. Thus the output stage can operate in class AB mode. Note also that the output stage is a modification of the output stage in the Tsividis–Gray design [9.2]. The overall gain achieved is 1000 and the bandwidth is 3 MHz.

An NMOS OA family described by Gray *et al.* [9.4] has been employed in Intel's PCM filters. These use Senderowicz *et al.* design [9.3] as the basic cell. A low-noise amplifier is realized by adding an additional low-noise input stage to this basic cell. It contains a common-mode feedback loop as in Fry's design [9.1]. The third type of OA is useful for driving resistive loads, e.g., a transformer (300 Ω). Hence, a high-current output stage has been added. The only difference between the output

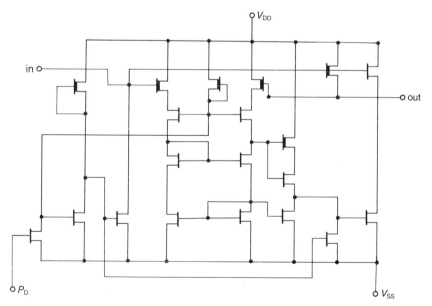

*Fig. 9.7* An output stage capable of driving resistive loads by Gray *et al.* (adapted from [9.4], © 1979 IEEE).

stage in the circuit of Figure 9.6(e) and this design is that a Wilson current source has been employed, as shown in Figure 9.7.

Young has described an NMOS OA (with entirely enhancement-mode transistors) [9.5]. The reader should be able to identify the various blocks in the schematic representation of this design (Figure 9.8). The second stage, formed by $T_{11}$ and $T_{12}$, provides gain and enables the use of feedforward compensation. The last stage is the same as in the Tsividis–Gray design [9.2]. The biasing has been chosen so as to make the design threshold insensitive. The overall voltage gain realized is 2200. It also employs common-mode feedback as in Fry's design and a split-load inverter to reduce area (with long channel devices).

In an inverter using a depletion load device, the gain is limited by the body effect. Consider the depletion-mode load device shown in Figure 9.9(a). When the output signal changes, the source to body voltage varies, thus reducing the effective load impedance presented by the load device. Toy [9.6] has suggested a modification to alleviate this problem, which is illustrated in Figure 9.9(b). In this circuit, $T_2$ acts as a voltage-controlled voltage source to bias the gate of $T_1$, while $T_3$ is a current source which biases $T_2$. The resulting design, employing a level shifter as previously considered, a cascode stage with improved transconductance and an output stage, is illustrated in Figure 9.9(c). Toy has realized a gain of 30 000 for the NMOS OA using this technique.

In a conventional differential amplifier using enhancement-type loads, the variations in the threshold voltages of the devices change the operating point as well

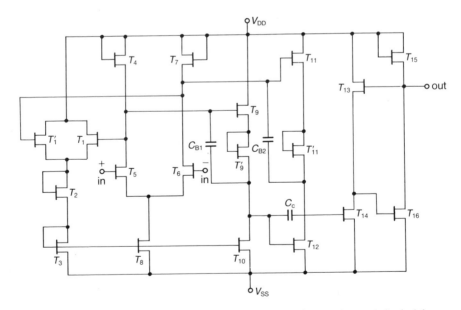

Fig. 9.8 Young's NMOS amplifier using enhancement-mode transistors (adapted from [9.5], © 1979 IEEE).

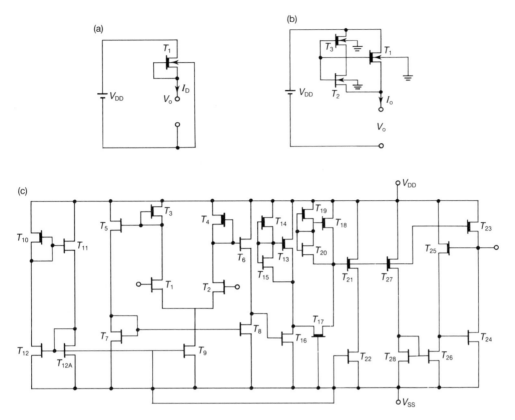

*Fig. 9.9* (a) A depletion-mode load device; (b) Toy's modification for body effect compensation; (c) NMOS OA using (b). *(Adapted from [9.6], © 1979 IEEE.)*

as the common-mode range. This problem has been solved by Tsividis *et al.* [9.7], using additional devices in the differential amplifier stage. In their circuit, shown in Figure 9.10, the required quiescent values are set through negative feedback, as in Fry's design [9.1]. The common mode feedback loop is from node 7 to node 6 through $T_5$, $T_6$ and $T_7$. The voltages at nodes 7, 8 and 9 track that of node 3. The required device dimensions are chosen for achieving $I_{D6} = I_{D7} = I_{D19} = I_{D20}$, $V_{GS5} = V_{GS7}$ and $V_{GS6} = V_{GS19}$ in Figure 9.10. The remaining stages can easily be identified by the reader.

Bosshart [4.17] has proposed another NMOS OA, which uses a cascode differential OA and employs Toy's method for load device body effect compensation and common-mode feedback as shown in Figure 9.11. Device $T_{11}$ clamps the differential output of the first stage, which avoids building up large voltages across the first-stage output during positive slewing. The second stage is similar to the cascode-type inverter due to Hosticka *et al.* [4.1], but with a small difference. The drain of $T_{27}$

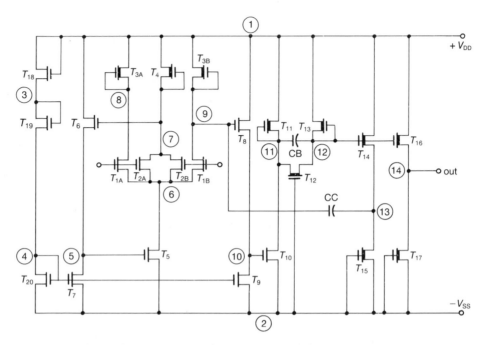

Fig. 9.10 Tsividis *et al.'s* NMOS OA *(adapted from [9.7], © 1980 IEEE).*

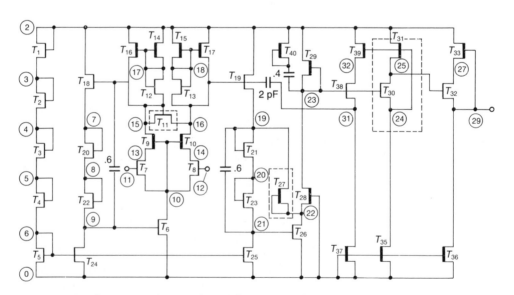

Fig. 9.11 Bosshart's NMOS OA *(adapted from [4.17], © 1980 IEEE).*

is not connected to the supply voltage but to the output of the previous stage. Thus the current through $T_{19}$ is increased considerably. The output stage contains a cascode doubler buffer formed by $T_{30}$ and $T_{31}$. The d.c. gain realized in this design is 4200 and the unity-gain frequency is 4.7 MHz.

Young has described another NMOS structure for OAs for use in PCM filter ICs [9.8]. His design is illustrated in Figure 9.12(a). Transistors $T_1$–$T_5$ realize a

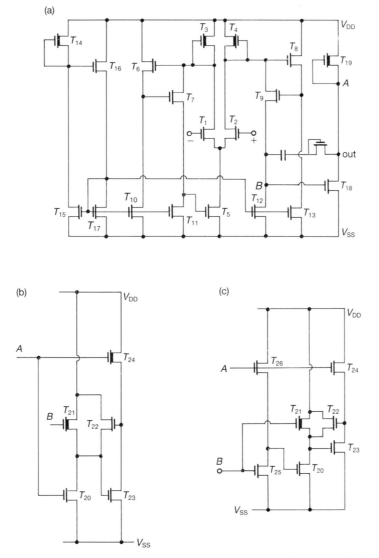

Fig. 9.12   (a) Young's NMOS amplifier;   (b), (c) modified output stages of class AB type. (Adapted from [9.8],.© 1980 IEEE.)

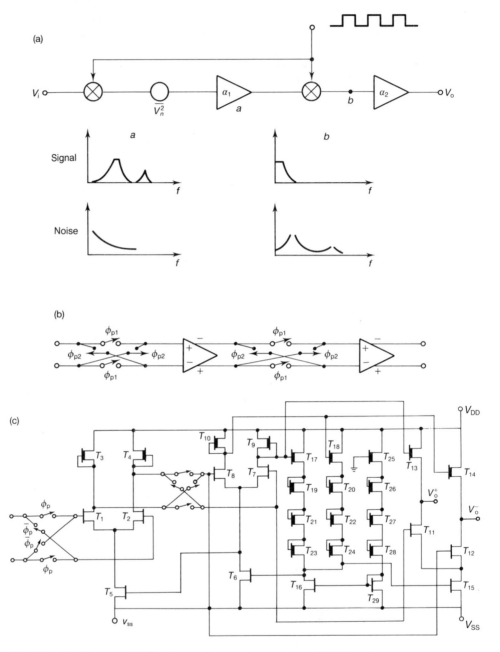

*Fig. 9.13* (a) Chopper stabilization technique for reducing OA 1/ f noise;
(b) implementation of (a) using switches and OAs; (c) practical circuit *(adapted from
[9.9]*, © 1981 IEEE).

differential amplifier, while $T_6$, $T_7$, $T_{10}$ and $T_{11}$ realize a differential to single-ended converter, which can also achieve a good CMRR. Transistors $T_8$, $T_9$, $T_{12}$ and $T_{13}$ perform the level shift function. The output stage is a simple inverter formed by $T_{18}$ and $T_{19}$. By using additional output stages, as in Figures 9.12(b) and 9.12(c), OAs capable of driving resistive loads can be designed. Note that these two output stages are of class AB type.

A chopper stabilization technique has been advanced by Hsieh *et al.* [9.9], which helps to reduce the noise considerably. In this method, conceptually illustrated in Figure 9.13(a), the input is multiplied by a square wave, thus sampling the signal. A post-multiplication operation enables recovery of the signal spectrum. The modulation operation shifts the noise spectrum ($1/f$ noise) to the harmonics of the chopping frequency. Thus the low-frequency component is due to the folding back of the harmonic frequencies. If the chopping frequency is much larger than the

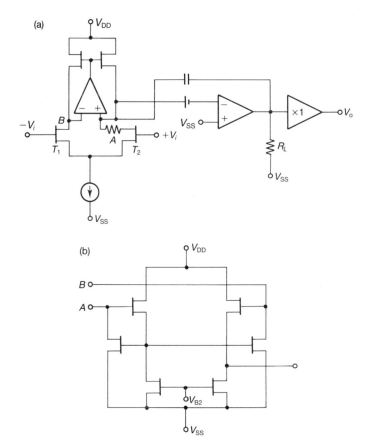

Fig. 9.14   (a) Bootstrapping techniques due to Senderowicz and Huggins *(adapted from [9.10], © 1982 IEEE)*;   (b) replica biasing technique using only one floating battery.

signal bandwidth, then the $1/f$ noise in the signal band will be greatly reduced. The modulation operation can be realized using cross-coupled switches, as shown in Figure 9.13(b). Note that the circuit whose implementation is shown in Figure 9.13(c) is fully differential. In this circuit, $T_1-T_5$ and $T_6-T_{10}$ realize differential amplifiers. The operating point of the input stage is biased through common-mode feedback through the second stage. The common mode d.c. output voltage and the operating points of the second stage are set by the common-mode feedback circuit $T_{16}-T_{24}$. Transistors $T_{25}-T_{29}$ form a bias string. The second stage has pole-splitting frequency compensation. Transistors $T_{11}-T_{15}$ realize the output stage. Such OAs can be employed in a straightforward manner in an SC circuit.

Senderowicz and Huggins [9.10] have considered further low-noise realizations for OAs. They observe that maximum front stage gain is realized to improve the noise performance. An impedance bootstrapping technique, illustrated in Figure 9.14(a), can be employed to reduce the body effect of load devices. The result is the effective increase of load impedance at the collector of $T_2$. The next stage uses replica biasing, however, in a modified form. Note that in the circuit shown in Figure 9.14(b), only one floating 'battery' is employed.

Goto et al. [9.11] have described an NMOS OA which can drive capacitive loads up to 200 pF. This circuit uses the blocks described above: a differential amplifier, a differential to single-ended converter, a Hosticka et al. type transconductance boosted gain stage with Toy's body effect compensated load devices, level shifter, and body effect compensation type inverter. These can be identified by the reader in the circuit of Figure 9.15.

An NMOS output stage with output impedance of about 100 Ω has been described by Calzolari et al. [9.12]. This circuit is presented in Figure 9.16. Devices $T_{10}$ and $T_{11}$ form a level shifter required in order properly to bias the differential stage $T_3-T_7$. Transistors $T_1$ and $T_2$ form a source-follower stage. Transistors $T_8$ and

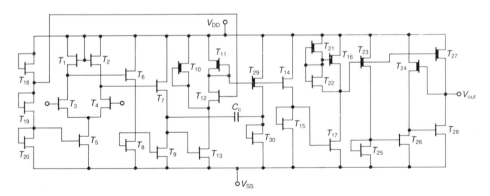

*Fig. 9.15*  Goto *et al.*'s NMOS amplifier *(adapted from [9.11], © 1982 IEEE).*

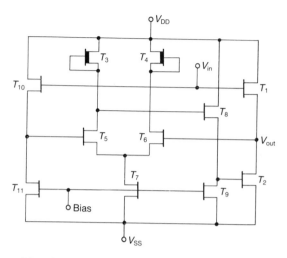

*Fig. 9.16* NMOS amplifier due to Calzolori *et al.* (adapted from [9.12], © 1981 IEE).

$T_9$ form a level shifter stage. The operation of the circuit is as follows. The gate voltage of $T_2$, $V_{g2}$, is realized as

$$V_{g2} = h_1 V_{out} - h_2 V_{in} \tag{9.7}$$

By proper choice of $h_1$ and $h_2$, low output impedance can be realized.

In this subsection, a variety of NMOS OAs have been presented, which have been integrated and used in many practical ICs available. It will have been observed that high-performance designs require great care in design — especially regarding biasing and removal of non-ideal effects of the devices. The use of CMOS technology has been shown to reduce the complexity of OA design (less components used), as well as providing larger gains than are obtainable in NMOS designs. Since the power consumption associated with the digital logic using CMOS technology is also lower, state-of-the-art SC circuit design employs CMOS designs extensively. We consider these next in some detail.

## 9.1.3 CMOS OAs

A CMOS OA due to Smarandoiu *et al.* [9.13] is presented in Figure 9.17. It uses p-channel driver transistors $T_1$, $T_2$ in the differential amplifier stage (with gain 40) biased by current source $T_5$. The single-ended output of $T_2$ is amplified (about 30 times) by $T_8$, which has a current source load formed by $T_9$. Transistors $T_6$ and $T_7$ realize a source follower useful for providing frequency compensation through $C_c$. Devices $T_{10}$ and $T_{11}$ form a voltage divider, biasing all the current sources. This simple design provides a d.c. open loop gain of 62 dB and unity-gain bandwidth of

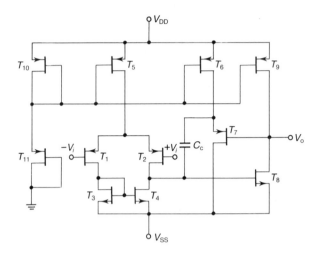

Fig. 9.17  CMOS OA due to Smarandoiu *et al.* (adapted from [9.13], © 1978 IEEE).

3.2 MHz. Transistors $T_3$ and $T_4$ form a current mirror load for differential amplifier, thus enabling realization of high gain. Since n-channel transistors fabricated in p-wells exhibit large body effect, in this design, the driver transistors in the differential amplifier were chosen to be p-channel transistors. However, if the source and body of the n-channel input transistors are connected together, the body effect can be overcome. Such a design has been described by Steinhagen and Engl [9.14]. It comprises a differential amplifier followed by a source follower.

Haque *et al.* have described a CMOS OA with high output impedance [9.15], which is presented in Figure 9.18(a). Note that the first stage uses n-channel input transistors and is a differential amplifier with single-ended output ($T_3$–$T_7$). Transistors $T_8$ and $T_9$ realize a level shifter, while $T_1$ and $T_2$ realize the output buffer and gain stage. The frequency compensation is achieved through a resistor (simulated by $T_{10}$ and $T_{11}$) and series capacitor $C_1$. The capacitor $C_2$ will improve the PSRR, as explained next. At low frequencies, the noise induced on $V_{DD}$ appears at node 1 unattenuated. Since the gate–source noise voltage of $T_1$ is zero, $T_1$ does not amplify this noise. However, at high frequencies, due to the rolloff introduced by $C_1$, the output at node 1 falls off, thus increasing the gate–source noise voltage of $T_1$. This noise is amplified by $T_1$ and is coupled to the output of OA. This effect can be reduced by reducing $C_1$. The use of $C_2$ facilitates such reduction. The open-loop d.c. gain realized has been about 90 dB. Haque *et al.* have described another OA capable of driving low impedance loads as shown in Figure 9.18(b). It realizes an open-loop gain of 60 dB, the power dissipation, however, being larger (about 20 mW).

Black *et al.* have considered in detail the techniques that need to be used to achieve good PSRR and low noise [9.16]. The parasitic capacitances between gate

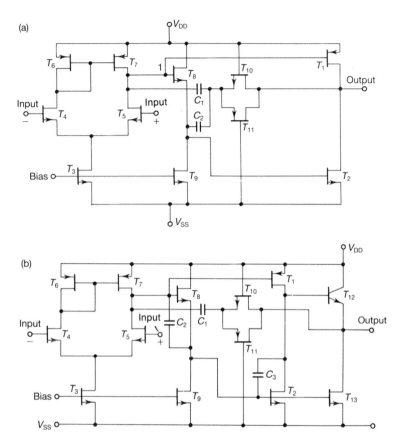

*Fig. 9.18* CMOS amplifiers for: (a) high-impedance loads; (b) low-impedance loads (adapted from [9.16], © 1980 IEEE).

and drain, and gate and source (see Figure 9.19(a)) influence the output voltage through the following approximate relationships:

$$\frac{\partial V_{\text{out}}}{\partial V_{\text{SS}}} = \frac{C_{\text{gs}}}{C_{\text{I}}} \left[ \frac{\partial I}{\partial V_{\text{SS}}} \frac{1}{2g_{\text{m1}}} + \frac{\partial V_{\text{T1}}}{\partial V_{\text{SS}}} \right] \tag{9.8a}$$

$$\frac{\partial V_{\text{out}}}{\partial V_{\text{DD}}} = \frac{-C_{\text{gd}}}{C_{\text{I}}} \left[ 1 - \frac{\partial I}{\partial V_{\text{DD}}} \frac{1}{2g_{\text{m3}}} \right] \tag{9.8b}$$

where $C_{\text{I}}$ is the integrating capacitance given that the OA is used in SC integrator. Since $C_{\text{gs}}$ is typically large, equation (9.8a) tends to be the more troublesome one. The increase in size of the input device to increase $C_{\text{gs}}$ leads to low noise as well as reducing the negative supply rejection. The first term in equation (9.8a) can be eliminated by using a supply-independent current source, while the second term can be reduced by placing the input devices in an isolated well. Black *et al.* have used

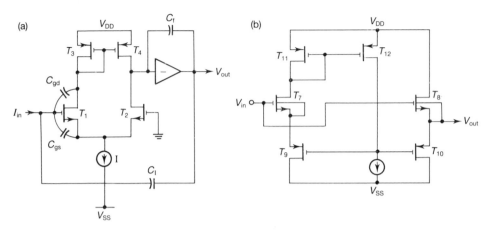

Fig. 9.19 (a) Differential amplifier showing the parasitic capacitances affecting PSRR; (b) class B output stage with improved swing. *(Adapted from [9.16],* © *1980 IEEE.)*

this technique in their designs. In view of the use of a supply-independent current source, equation (9.8b) reduces to $-C_{gd}/C_I$, which can be further reduced using a cascode circuit for $T_1$ and $T_2$.

In the OA designs, Black *et al.* have used series $R_fC_f$-type frequency compensation. However, in order to maintain a proper $R_f$ independent of temperature, power supply voltages and process parameters, a tracking RC compensation scheme has been employed, for details of which the reader is referred to their work [9.16].

The class B output stage employed in their design is presented in Figure 9.19(b). In this design, constant quiescent current is maintained, independent of the output level, by sensing the current through the dummy output pair of devices $T_7$ and $T_9$ by way of the PMOS current mirror $T_{11}$–$T_{12}$. The gate voltages of $T_9$ and $T_{10}$ are adjusted by the current through $T_{12}$ such that the current through $T_7$ and $T_9$ remains constant independent of the threshold voltage of PMOS devices.

Saari [9.17] has described low-power high-drive CMOS OAs which use a simple modification of the basic CMOS OA, with the advantage that standing power dissipation can be quite low. His design is presented in Figure 9.20, in which the current through $T_6$ can be made low. However, as the output slews negatively, $T_1$ turns hard and $T_2$ turns off resulting in sufficient $V_{GS}$ across $T_9$, thus turning it on. This pulls up the gate of $T_6$ and thereby increases considerably the pulldown current. The resistor $R_2$ isolates the charges at the gate of $T_6$ from the input pair current source $T_7$. This design has been used extensively in Marsh *et al.* design [9.18], where PSRR has been improved by putting the input devices of the OA in p-tubs, which are tied to their sources. Further, the input stage has employed large gate areas to reduce the $1/f$ noise.

Jolly and McCharles have described a CMOS OA with low noise [9.19], where they have implemented the concepts for low-noise design developed by Bertails

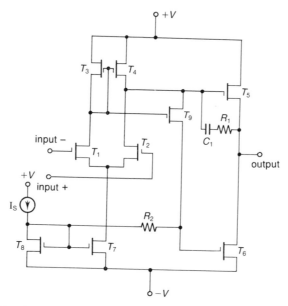

*Fig. 9.20* CMOS OA due to Marsh *et al.* *(adapted from [9.18], © 1981 IEEE).*

[9.20]. Their design, shown in Figure 9.21, uses long channels for $N_2$ and $N_3$ relative to $P_4$ and $P_5$. It also uses cascode transistors $P_6$ and $P_7$ without impairing the noise performance. The compensation technique used also is different from earlier schemes.

Choi *et al.* have considered the design of fast-settling CMOS amplifiers for high-frequency applications [9.21]. Their OA configuration is shown in Figure 9.22(a), which is known as the *folded cascode* configuration. The reason for this name can be understood by inspection of the circuit in Figure 9.22(a). The actual realization is shown in Figure 9.22(b). The cascode transistors are $T_5-T_6$. The transistors $T_7-T_{15}$ realize a common-mode feedback loop. The dynamic range realizable has been about 70 dB in this design.

Krummenacher and Zufferey [9.22] have suggested CMOS OA designs using cascoded transistors, similar to the Jolly and McCharles design [9.19] mentioned above. This design is presented in Figure 9.23(a), where $T_5$ and $T_6$ are the cascoded transistors and $T_7$, $T_8$, $T_{11}$ and $T_{13}$ realize the biasing arrangement. Transistor $T_9$ realizes a source follower. The open-loop gain realized is 80 dB. Note that no frequency compensation is required in this circuit, and that the dominant pole is decided by the load capacitance. Krummenacher [9.23] subsequently described a CMOS high-gain operational transconductance amplifier (OTA) which uses a conventional differential input stage followed by a cascoded output stage. The second stage also performs differential to single-ended conversion, as shown in Figure 9.23(b).

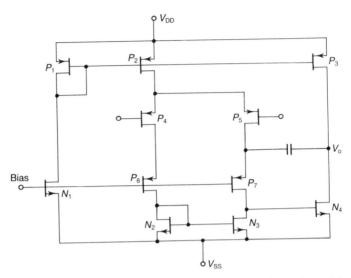

Fig. 9.21  A low-noise CMOS amplifier due to Jolly and McCharles *(adapted from [9.19],*
© 1982 IEEE).

Krummenacher [9.24] has described another OA, which uses a single stage as shown in Figure 9.23(c). In this OA, $T_3$ and $T_4$ are cascode transistors. The inverter is of push-pull type. Transistors $T_1$ and $T_5$ form a current mirror. Capacitor $C$ provides a.c. coupling for the input signals to the gate of $T_1$. However, the PSRR for this OA is rather small (about 6 dB).

Before concluding the discussion on the static OA designs available in the literature, we consider briefly the frequency-compensation techniques used in MOS OAs.

## 9.1.4  Frequency compensation techniques

The need for dominant pole compensation is well described in the literature; the reader is referred to an excellent treatment by Budak [9.25]. MOS OAs are dominant pole-compensated similar to bipolar OAs. However, certain differences exist between the techniques used for these two types of OA. These stem from the fact that MOS devices exhibit rather low $g_m$ values. This point is illustrated best by considering the equivalent circuit of a typical two-stage MOS amplifier, shown in Figure 9.24(a).

It can be shown, for this circuit, that

$$\frac{V_o(s)}{V_i(s)} = \frac{a\left(1 - \dfrac{C_c s}{g_{m2}}\right)}{1 + bs + cs^2} \tag{9.9}$$

where

$$a = g_{m1} g_{m2} R_1 R_2$$
$$b = (C_2 + C_c) R_2 + (C_1 + C_c) R_1 + g_{m2} R_1 R_2 C_c$$
$$c = R_1 R_2 (C_1 C_2 + C_c C_2 + C_c C_1)$$

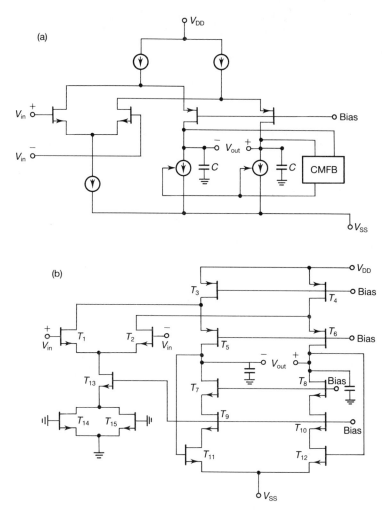

*Fig. 9.22* (a) Folded cascode configuration; (b) application to a CMOS amplifier. *(Adapted from [9.21], © 1983 IEEE).*

Assuming that the two poles are widely spaced, they can be approximated as follows:

$$P_1 = \frac{-1}{(1 + g_{m2}R_2)C_cR_1} \qquad (9.10a)$$

$$P_2 = \frac{-g_{m2}C_c}{C_1C_2 + C_cC_2 + C_cC_1} \qquad (9.10b)$$

Note that $C_c$ is multiplied by the Miller effect, thus pushing one pole much lower than the other. Thus a dominant pole can be realized. However, the zero present

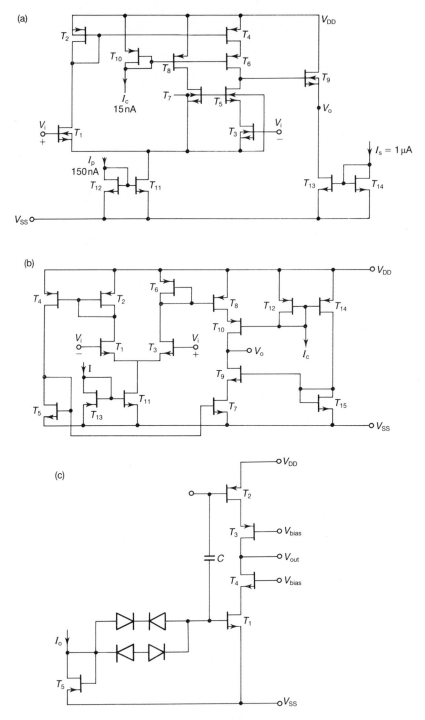

*Fig. 9.23* (a) CMOS OA due to Krummenacher and Zufferey *(adapted from [9.22], © 1980 IEE)*; (b) CMOS transconductance OA due to Krummenacher *(adapted from [9.23], © 1981 IEE)*; (c) simple CMOS OA (Krummenacher) *(adapted from [9.24], © 1982 IEEE)*.

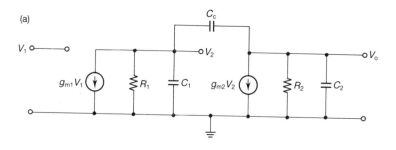

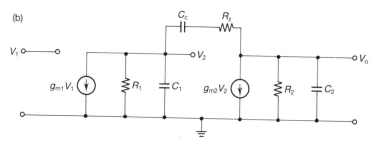

Fig. 9.24  (a) Frequency compensation in a two-stage amplifier using Miller capacitance; (b) frequency compensation which cancels the feedforward effect of $C_c$ in (a).

in equation (9.9) presents a problem causing possible instability. This is explained as follows. At high frequencies, the signal can directly flow to the output through $C_c$, thus making the second-stage transistor look like a diode of resistance $1/g_{m2}$. Thus, the output voltage appears with phase inversion while still being very large relative to the input voltage.

Tsividis and Gray have suggested the isolation of $C_c$ from the second-stage output to its input, using a source-follower stage in between them (see Figure 9.4) [9.2]. Thus, the transmission zero occurring in equation (9.9) can be eliminated. Haque *et al.* [9.15] have suggested the addition of a resistance $R_z$ in series with $C_c$ (Figure 9.24(b)). The $R_z$ can be realized using two MOS transistors connected in parallel, as shown in Figure 9.18(a). In this case, the zero realized is obtained from

$$s\left(\frac{C_c}{g_{m2}} - R_z C_c\right) = 1 \qquad (9.11)$$

The poles realized in this case are also approximately (for $R_z \ll R_1, R_2$) given by equations (9.10). Note that the zero vanishes when $R_z = 1/g_{m2}$, requiring the tracking of $R_z$ with $g_{m2}$, which can be achieved by proper choice of geometries.

For internal OAs required to drive capacitive loads, choosing

$$\frac{g_{m2}}{C_2} \gg \frac{g_{m1}}{C_c} \qquad (9.12)$$

the second-stage pole can be kept larger than that of the first stage. The advantage of the second design over the first is that the source follower stage can be avoided. Further, it helps to reduce the die size and the power dissipation [9.16].

Black *et al.* [9.16] have recommended another choice of $R_z$ which cancels the zero with a pole in equation (9.9). This requires

$$R_z = \frac{C_2 + C_c}{g_{m2}C_c} \tag{9.13}$$

The resulting condition for stability can be derived as

$$C_f \geqslant \sqrt{\frac{g_{m1}}{g_{m2}} C_1 C_2} \tag{9.14}$$

which shows that $C_f$ can be small. Many OA designs have employed any one of the three above methods.

## 9.1.5 Dynamic MOS amplifiers

The basic idea in these amplifiers is to save the static power in the amplifier and also to obtain reductions in chip area and $1/f$ noise. This is accomplished by replacing the static OAs in MOS SC integrators by a number of switches.

Consider the circuit shown in Figure 9.25. The portion of the circuit inside the dashed lines is the dynamic amplifier. This cannot act alone; the entire circuit has to be considered in order to understand its behaviour. During $\phi_1$ (that is, $S_1$ and $S_6$ are closed), $\alpha_i C$ charges to $V_{in}$ and $\alpha_o C$ transfers its previous charge to the subsequent stage. When $\phi_1$ opens, $\phi_2$ closes. This corresponds to $S_2$, $S_3$ and $S_5$ closing simultaneously. It is seen that $V_{out}$ will become equal to $V_{DD}$ while keeping both $\alpha_i C$ and $\alpha_o C$ precharged. A little later $S_3$ is opened and $S_4$ is closed, keeping $S_2$ and $S_5$ closed. $V_{out}$ starts discharging towards ground potential and hence the potential at

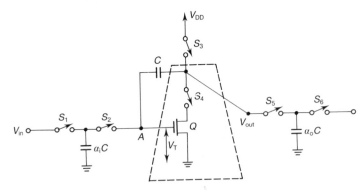

Fig. 9.25　Dynamic amplifier due to Copeland and Rabaey *(adapted from [9.26]*, © 1979 IEE).

node $A$ will be equal to $V_T$, the threshold of the MOS switch $Q$. It can be shown that, at this time,

$$V_{\text{out}} = V_T - \frac{Q_i}{C}$$

where $Q_i$ is the total charge input to the integrator summing node in this and all previous cycles. Note that $S_1$ and $S_6$ have the same switching waveforms at the respective gates, while $S_2$, $S_5$ and $S_6$ operate synchronously. Since the channel of $Q$ is empty at the end of the discharge, no $1/f$ noise will be present. Copeland and Rabaey [9.26] have observed that gains as large as 60 dB can be achieved.

Hosticka [9.27] has described a dynamic amplifier which is a simple modification of a source coupled pair. This circuit, shown in Figure 9.26(a), uses a series SC-type current source. In order to understand the operation, consider this dynamic amplifier in a typical SC integrator, as shown in Figure 9.26(b). During $\phi_1$, $C_1$ is charged to $V_i$ and $C_o$ is discharged. In $\phi_2$, the amplifier operates with the bias current $I_o$ supplied by $C_o$. The charge in $C_1$ is transferred to $C_2$, because of the feedback action. At the beginning of $\phi_2$, $I_o$ is large, since the voltage across $C_o$ is zero. Then the OA slowly enters the linear region, since $V_o$ increases gradually, thus decreasing the gate source voltages of $Q_1$ and $Q_2$. The open-loop gain thus increases till $Q_1$ and $Q_2$ enter the sub-threshold region. The rising voltage $V_o$ then cuts off the bias current, and subsequently no more power is consumed. The amplifier does not need any output stage, since the output current is zero, when $V_o$ has approached its final

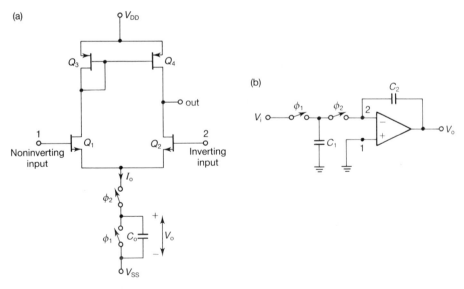

Fig. 9.26 (a) Dynamic amplifier due to Hosticka *(adapted from [9.27], © 1979 IEE)*; (b) SC integrator using the amplifier in (a).

value. The value of $C_o$ must be chosen so as to facilitate full discharge during $\phi_1$. This CMOS design, however, has the drawback that the negative output voltage swing is limited by $Q_2$.

Hosticka has described a dynamic biasing scheme, which can be used in multi-stage OA designs. The reader is referred to [9.28] for details.

## Adaptive biasing technique

In this technique, the bias current of the OA is made signal-dependent so that the power consumption can be reduced. The reason for such an adaptation can be explained briefly as follows. The slew rate of the OA is dependent on the fact that the tail current source is limited. Hence, the slew rate can be improved by using a tail current source, whose current increases for large differential input signals. This is realized by the circuit shown in Figure 9.27.

Recall that when an input voltage is applied for a differential input OA, the currents $I_1$ and $I_2$ are different. The bias current can be made signal-dependent by

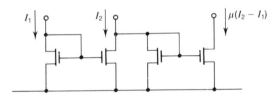

*Fig. 9.27*  Current subtractor realizing $\mu(I_1 - I_2)$.

**Table 9.1**  Typical performance, conventional two-stage CMOS internal operational amplifier ($\pm 5$ V Supply, 4 $\mu$m SI Gate CMOS)  *(adapted from [9.31],* © *1982 IEEE).*

| | |
|---|---|
| d.c. gain (capacitive load only) | 5000 |
| Settling time, 1 V step, $C_1 = 5$ pF | 500 ns |
| Equiv. input noise, 1 kHz | 100 nV Hz$^{-1/2}$ |
| PSRR, d.c. | 90 dB |
| PSRR, 1 kHz | 60 dB |
| PSRR, 50 kHz | 40 dB |
| Supply capacitance | 1 fF |
| Power dissipation | 0.5 mW |
| Unity-gain frequency | 4 MHz |
| Die area | $50 \times 10$ m$^{-9}$ m$^2$ |
| Systematic offset | 0.1 mV |
| Random offset standard deviation | 2 mV |
| CMRR | 80 dB |
| CM range | within 1 V of supply |

**Table 9.2** Typical NMOS operational amplifier performance (25 °C, 5 V supplies) *(adapted from [9.4], © 1979 IEEE).*

| | Internal amplifier | Low-noise amplifier | Transformer driver amplifier |
|---|---|---|---|
| Open-loop gain | 1000 | 5000 | 1000 |
| Common-mode range | ±3 V | ±2.5 V | ±3 V |
| Output swing | ±3.5 V at 10 kΩ | ±3.5 V at 10 kΩ | ±3.0 V at 300 Ω |
| PSRR (d.c.) | 60 dB | 65 dB | 60 dB |
| CMRR (d.c.) | 70 dB | 70 dB | 70 dB |
| Equivalent input noise at 1 kHz | $300 \, \mathrm{nV \, Hz^{-1/2}}$ | $120 \, \mathrm{nV \, Hz^{-1/2}}$ | $300 \, \mathrm{nV \, Hz^{-1/2}}$ |
| Area | $160 \times 10^{-9} \, \mathrm{m^2}$ | $320 \times 10^{-9} \, \mathrm{m^2}$ | $520 \times 10^{-9} \, \mathrm{m^2}$ |
| Power dissipation | 8 mW | 12 mW | 25 mW |
| Unity-gain bandwidth | 3 MHz | 1.5 MHz | 1.5 MHz |

adding an additional current source which realizes $\mu(I_1 - I_2)$. The parameter $\mu$ can be designated as the 'current feedback factor'. This additional current source may be realized by two subtractors. The reader is referred to [9.29] for more information.

In the above discussion on OAs, we have considered various schemes useful for realizing desirable properties such as low noise, good PSRR, large gain, etc. The details regarding d.c. conditions, dimensions of devices required to achieve the d.c. conditions and all the above-mentioned properties have not been considered. Several excellent tutorial papers are available which give further design information [1.6, 9.30–9.32], and the reader is recommended to consult these. The realized specifications of typical CMOS and NMOS OAs are presented in Tables 9.1 and 9.2, respectively.

In this section we have considered the design of MOS OAs. The next important aspect to be studied is the non-ideal behaviour of the switches. This aspect is best studied by referring to specific circuits.

## 9.2 Clock-feedthrough and offset-voltage cancellation techniques

As mentioned in Chapter 1, MOS switches possess finite on-resistances, parasitic capacitances between source and substrate, drain and substrate, drain and gate, and source and gate. They also have finite junction leakage currents.

The effect of the parasitic capacitances of the switches is to cause clock feed-through. Thus, if a gate pulse of amplitude $A$ is used to control a MOS switch in an SC integrator, as shown in Figure 9.28, the resulting output of the OA is obtained

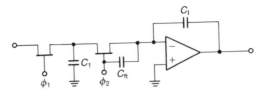

Fig. 9.28  Circuit illustrating the mechanism of clock feedthrough.

as a pulse of amplitude $AC_{ft}/C_I$, where $C_{ft}$ is the feedthrough capacitance (i.e., gate to drain capacitance of the MOS switch) and $C_I$ is the integrator feedback capacitance [3.1].

Consider the integrator of Figure 9.29. Assume first that the input SC is disconnected. It can be seen that the OA is now in self-biasing mode. A leakage current flows through $C_I$ which is integrated at the output of the OA. This voltage is sampled by $C_F$ in $\phi_2$ and is applied to the input of the OA at the beginning of $\phi_1$, thus causing an abrupt transition. The resulting ramp-like waveform has an amplitude of the order of few millivolts. If an input SC is also connected during $\phi_2$, the summing junction leakage flows through $C_I$, causing a step in the output of the OA.

Both leakage and clock feedthrough cause error voltages which are periodic and occur at the clock rate. Due to sampling at the subsequent stages, this error will appear aliased as a d.c. offset voltage. In addition to the offset introduced by the parasitic capacitances of the switches and leakage currents of capacitors, as mentioned before, the OA itself has an offset voltage, which is much larger than the above effects considered together. Some methods of cancellation of offset and clock-feedthrough effects have been described in the literature. These are briefly considered next.

## 9.2.1  Offset cancellation techniques

For purposes of analysis, the OA offset can be represented by a voltage source at the non-inverting input terminal, as shown in Figure 9.30. To analyze the effect of

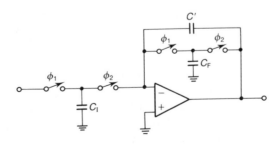

Fig. 9.29  An SC integrator.

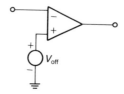

*Fig. 9.30* OA model including offset voltage.

this, we consider the gain stage of Figure 9.31(a). Routine analysis gives the charge conservation equation as

$$\alpha C\left[v_i\left(nT - \frac{T}{2}\right) - v_{\text{off}} - (-v_{\text{off}})\right] + C[0 - (v_o(nT) - v_{\text{off}})] = 0$$

or

$$v_o(nT) = \alpha v_i\left(nT - \frac{T}{2}\right) - v_{\text{off}} \qquad (9.15)$$

The offset voltage $v_{\text{off}}$ in the output can be eliminated by using the arrangement of Figure 9.31(b) [9.33]. In this modified circuit, during the even phase, the offset voltage is stored on $C$ and subtracted from the output in the odd phase. Thus the $v_{\text{off}}$ term in equation (9.15) is eliminated.

This concept can be extended to other first-order SC networks [9.33]. As an illustration, a first-order SC network and its modification to remove offset are shown in Figures 9.31(c) and (d) respectively. The capacitor $C_o$ serves to hold the output voltage in the previous half-cycle. It does not interfere with the operation of the circuit. A disadvantage of the above circuits is that the output is zero for half the clock period, whereas, for the circuit without offset compensation, the output is held over a clock period. An advantage, however, is that during the period when the input and output of the OA are shorted, the equivalent input noise (as well as the offset) is sampled on the capacitors $C$ and $\alpha_3 C$ and eliminated in the subsequent half-cycle. These techniques can be applied to biquad designs as well. This method of noise cancellation is known as *double correlated sampling* (DCS) [9.34].

This method assumes that the offset voltages (noise voltages) in adjacent half-cycles of the clock period are correlated. The resulting transfer function can be obtained as $1 - z^{-1/2}$, thus indicating that the gain at d.c. is zero. Since the $1/f$ noise of the MOS OA is predominant at low frequencies, the zero at d.c. introduced by the DCS technique effectively reduces the overall noise of the SC network [9.24].

Lam and Copeland [9.35] have suggested an alternative solution to the offset-compensation problem. An inverting integrator employing their technique is shown in Figure 9.32(a). In $\phi_1$, the OA offset voltage is sampled on $C_1$, while the SC $C_2$ is discharged. In $\phi_2$, node $A$ acts as virtual ground, thus transferring the charge completely from $C_2$ to $C_1$. The capacitor $C_1$ is a 'dummy'. Thus, this technique is an arrangement of switches and capacitors modified from that of Gregorian and Fan

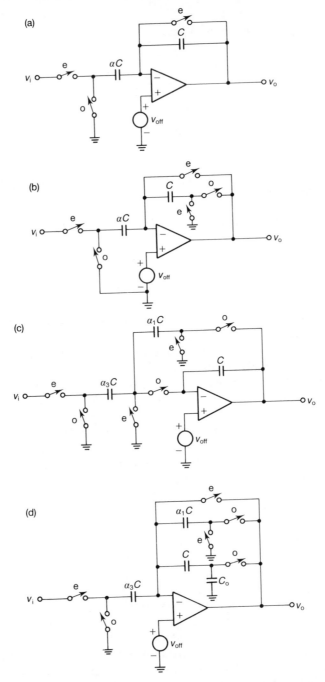

Fig. 9.31 (a) A stray-insensitive gain stage; (b) modification for offset cancellation; (c) first-order SC network; (d) modification for offset cancellation due to Gregorian and Fan. (Adapted from [9.33], © 1980 IEEE.)

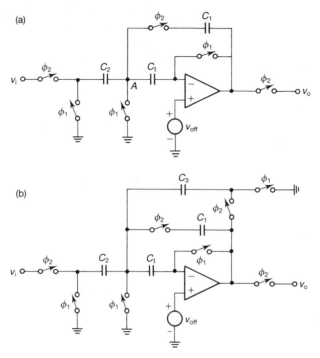

*Fig. 9.32* Lam and Copeland technique of offset compensation applied to:   (a) lossless inverting integrator;   (b) inverting first-order LP network. *(Adapted from [9.35],* © *1983 IEE.)*

in Figure 9.31(d) with $\alpha_1 = 0$. The extension to a first-order low-pass network is straightforward, as shown in Figure 9.32(b). The circuits in Figure 9.32 also suffer from the limitation that the OA has to slew in one clock phase, thus prohibiting its usefulness for high clock frequency applications.

Temes and Haug [9.36] have suggested an interesting technique to alleviate this problem. Consider their offset-free inverting integrator circuit, shown in Figure 9.33(a) $(C_1 = C_3)$. It can be shown that the circuit performs as an offset-free integrator in $\phi_1$; in $\phi_2$, however, the output is different by only $v_{\text{off}}$ from that in $\phi_1$. In this method, $C_3$ enables $C_1$ to discharge to $-v_{\text{off}}$ during $\phi_2$. The capacitor $C_3$ is precharged earlier to $(v_{o1} - v_{i1})$ to be able to absorb charge from $C_1$ without a significant change in $v_o$. An offset-free non-inverting integrator can be obtained in a similar manner, as shown in Figure 9.33(b). We obtain, in this case, in the transformed domain:

$$V_{oo} = V_{oe} z^{-1/2} + \left(1 + \frac{C_1}{C_3}\right) V_{\text{off}} \qquad (9.16)$$

thus making $V_{oo}$ only slightly different from $V_{oe}$. A damped integrator can be realized by the addition of an extra SC, as shown in broken lines in Figure 9.33(b). The above offset-free blocks can be employed to realize stray-insensitive biquads.

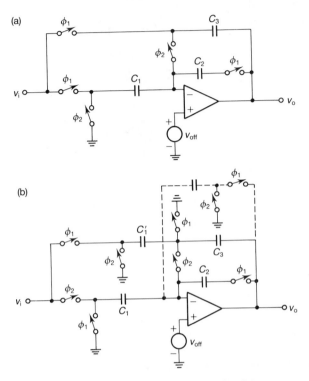

*Fig. 9.33* Temes–Haug technique of offset compensation applied to: (a) inverting integrator; (b) non-inverting integrator. *(Adapted from [9.36],* © 1984 IEE.*)*

Watanabe and Fujiwara [9.37] have described offset-compensated SC circuits similar to Temes–Haug realizations. These are shown in Figure 9.34. The SC amplifier in Figure 9.34(a) amplifies the input in $\phi_1$, thus charging $C_3$ to the output voltage. Also in $\phi_1$, $C_4$ is discharged; while in $\phi_2$, the voltage on $C_3$ is amplified by $C_3/C_4$. The choice of $C_3 = C_4$ facilitates retention of the same output voltage as in $\phi_1$, except for an increase by $(1 + C_3/C_4)v_{\text{off}}$. By interchanging the clock phases for switches $S_1$ and $S_2$ in Figure 9.34(a), a non-inverting amplifier with the same properties is obtained.

All the offset-compensation circuits above have one drawback in that the OA feedback loop can be open during the interval between $\phi_1$ and $\phi_2$. This situation can be remedied by using either XY feedback or a circuit due to Watanabe and Fujiwara [9.37]. This SC amplifier circuit is shown in Figure 9.34(b), where a continuous feedback path is provided by $C_2$. The following relationships can be derived:

$$V_{\text{oe}} = -(C_1/C_2)V_{\text{ie}} + 2V_{\text{off}} \qquad (9.17a)$$

$$V_{\text{oo}} = V_{\text{oe}}z^{-1/2} + 3V_{\text{off}} \qquad (9.17b)$$

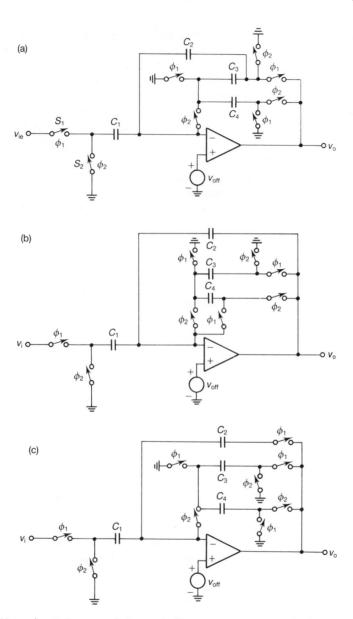

*Fig. 9.34* Watanabe–Fujiwara technique of offset compensation applied to: (a) inverting amplifier; (b) inverting amplifier with continuous feedback path; (c) inverting integrator. *(Adapted from [9.37], © 1984 IEE.)*

A non-inverting amplifier can be obtained by interchanging the phasing of the input switches in Figure 9.34(b). Note that, in these circuits, the offset compensation is achieved when $C_2 = C_4$ and $C_3 = 2C_2$. Further, the gain is independently controllable by choosing $C_1$ appropriately. A simple modification of the circuit of Figure 9.34(a) will yield an offset-free SC integrator, as shown in Figure 9.34(c).

The application of the offset-compensation technique to SC ladder filters based on operational simulation method has been studied by Eriksson and Chen [9.38]. Their circuit employs Gregorian and Fan's scheme of Figure 9.31(d). However, since the input and output sampling instants can be different in a leap-frog ladder filter in a two-integrator loop realizing the LC section, two-phase clocking will not be able to facilitate offset compensation. Therefore, three-phase clocking must be used. This additional clock phase permits a connection between inverting input and output of the OA so as to sample the offset as required. The reader should consult [9.38] for details.

We next consider clock-feedthrough cancellation techniques.

## 9.2.2  Clock-feedthrough cancellation techniques

In one technique [9.39], a dummy transistor $Q_2$ with drain and source connected together (to prevent d.c. current flow) is driven by the complement of the gate voltage of $Q_1$ (see Figure 9.35(a)). The channel area of $Q_2$ is half that of $Q_1$. The basic idea in this technique is that the charge injection due to feedthrough capacitances will be equal in magnitude and opposite in sign. However, the clock-feedthrough cancellation in this technique depends on the exact matching of the two devices $Q_1$ and $Q_2$ as well as on the clocking waveforms being exactly complementary to each other.

In another approach [9.39] a dummy capacitor also is added as shown in Figure 9.35(b). In this case also, the dummy switch has half the area of $Q_1$. By symmetry, this configuration assures that half the channel charge injection of switch $Q_1$ flowing into the dummy transistor is cancelled. We next consider another circuit design technique.

Consider for illustration the circuit of Figure 9.36(a) . When the MOS transistor implementing the switch S is turned off, a negative feedthrough error voltage is

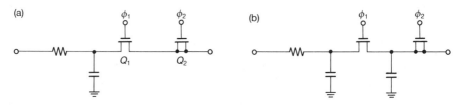

*Fig. 9.35* Clock-feedthrough cancellation techniques:  (a) using a dummy switch; (b) using a dummy capacitor and dummy switch.

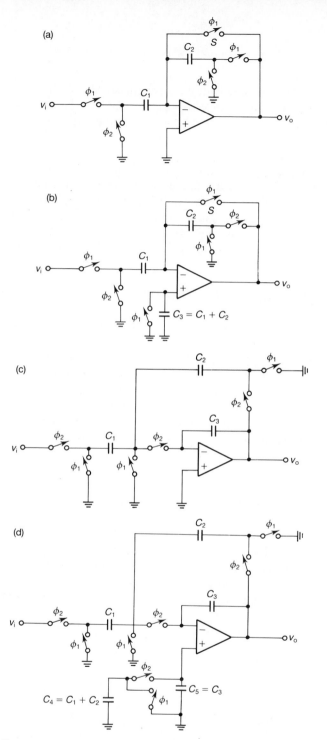

Fig. 9.36 (a) Offset-cancelled gain stage; (b) modification of (a) to cancel clock feedthrough; (c) first-order damped integrator; (d) clock-feedthrough cancellation in (c). *(Adapted from [9.40], © 1982 IEE.)*

introduced due to the device gate drain capacitance. Martin [9.40] has suggested a simple modification to cancel this error. In this method, the non-inverting input is lifted off the ground and connected to a switch and capacitor, $C_3 = C_1 + C_2$, as shown in Figure 9.36(b). The result is that at the falling edge of $\phi_1$, a negative voltage is developed across $C_3$ equal in amplitude to that of the inverting input. It is evident that a matching condition is required on $C_1$, $C_2$ and $C_3$. It is possible to extend the above method to other first-order active SC networks. As an example, a damped integrator realization is shown in Figure 9.36(c), together with the modification for clock-feedthrough cancellation in Figure 9.36(d). It is thus seen that offset-voltage and clock-feedthrough cancellation can be realized in stray-insensitive SC networks, enhancing their applicability for stringent applications.

## 9.3  Noise generated in SC networks

The dynamic range of an SC filter depends on the minimum signal level that the filter can process without being unduly affected by the noise. There are several sources of noise in SC networks. These are due to the finite on- and off-resistances of the MOS switches, which hitherto in this book have been assumed to be zero and infinite, respectively. Thermal noise is generated because of the finite on-resistance of the MOS switch. The OAs have their own contribution to the output noise. It has been mentioned before that the $1/f$ noise of the OA is predominant at low frequencies. It is thus important to evaluate the performance of SC networks, taking into account these noise sources.

### 9.3.1  Noise analysis of an SC integrator

In order to develop the basic ideas of noise evaluation, an MOS switch shunting a capacitor is considered first (Figure 9.37(a)). The switch can be either on or off. When the switch is off, the off-resistance is so large that the resulting first-order low-pass network has a cutoff frequency at very low frequencies (fractions of a hertz) and hence this noise is ignored. In the on state, the noise power generated is

$$e_n^2 = 4k\theta R_{on}\left(\frac{1}{4R_{on}C}\right) = \frac{k\theta}{C} \qquad (9.18)$$

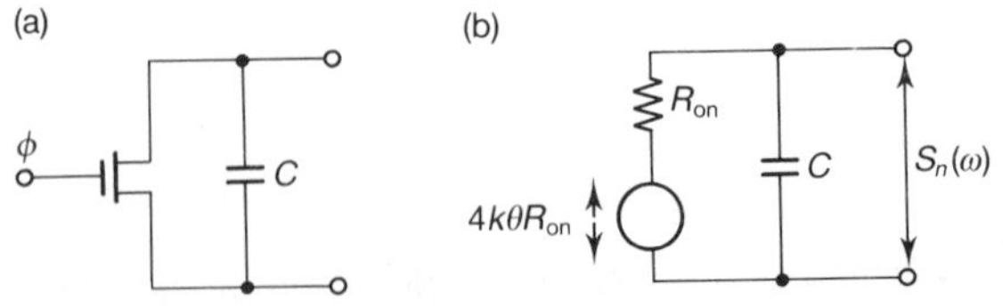

Fig. 9.37  (a) An MOS switch across a capacitor $C$; (b) noise equivalent of (a).

where $\theta$ represents the temperature in kelvin (K). The term $(1/4R_{on}C)$ is the 'noise bandwidth' of the low-pass filter. It is obtained by considering the equivalent first-order low-pass RC network of Figure 9.37(b) with the transfer function

$$H_a(s) = \frac{1}{1 + sCR_{on}}$$

and evaluating

$$\int_0^\infty |H(j\omega)|^2 \, d\omega = \frac{1}{4R_{on}C}$$

This noise, which is independent of $R_{on}$, is often called '$k\theta/C$ noise', where $k$ is Boltzmann's constant ($k = 1.38 \times 10^{-23}$ J K$^{-1}$). For current transistor sizes and capacitors of a few picofarads, the noise bandwidth is of the order of a few megahertz when the MOS transistor is conducting. With these basic ideas in mind, we now proceed to consider the noise generated in the SC integrator shown in Figure 9.38(a).

The input is grounded since we are interested in noise analysis only. During $\phi_1$, the on-resistance $R_{on1}$ of $S_1$ generates thermal noise, which is sampled by $\alpha C$. During $\phi_2$, the on-resistance $R_{on2}$ of $S_2$ and the equivalent noise of the OA are filtered by the circuit comprising the OA, $C$, $R_{on2}$ and $\alpha C$, as shown in Figure 9.38(b). The resulting output noise is designated as $S_{bb1}$. Next, during $\phi_1$, $S_2$ is off and hence the OA noise source contributes to the output noise (see Figure 9.38(c)). These two contributions to the output noise can be evaluated by taking into account the single-pole model of the OA. As an illustration, in $\phi_1$

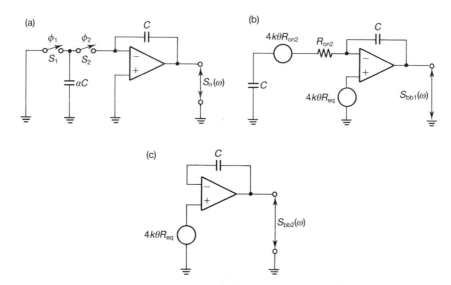

*Fig. 9.38* (a) An SC integrator; (b) continuous-time noise equivalent in $\phi_2$; (c) continuous-time noise equivalent in $\phi_1$.

when $S_2$ is off, the OA is just a unity-gain buffer and hence has transfer function given by

$$H_a(j\omega) = \left(\frac{1}{1 + j\dfrac{\omega}{B}}\right) \tag{9.19}$$

Thus the resulting output noise spectral density is the input spectral density multiplied by $|H_a(j\omega)|^2$, i.e.,

$$S_{bb2}(\omega) = \left(\frac{T - \Delta}{T}\right)(4k\theta R_{eq}) \frac{1}{1 + \dfrac{\omega^2}{B^2}} \tag{9.20}$$

Note that the factor $(T - \Delta)/T$ is included to take into account the fact that this noise is present all the time in a clock period except for a duration $\Delta$ when the switch to the right of $\alpha C$ is closed. It can be similarly shown that, in the situation depicted in Figure 9.38(b), the noise spectral density is a result of superposition of the sources $4k\theta R_{on2}$ and $4k\theta R_{eq}$ and is given by

$$S_{bb1}(\omega) = \frac{\Delta}{T} \cdot \frac{4k\theta[\alpha^2 R_{on2} + R_{eq}\{(1 + \alpha)^2 + \alpha^2 C^2 \omega^2 R_{on2}^2\}]}{1 + \dfrac{\omega^4}{B^2}\alpha^2 C^2 R_{on2}^2 + (1 + \alpha)^2 \dfrac{\omega^2}{B^2} + \alpha^2 C^2 R_{on2}^2 \omega^2 + \dfrac{2\alpha^2 C R_{on2}\omega^2}{B}} \tag{9.21}$$

In addition to the above broadband noise components, there is another source of noise. This can be described as 'sampled and held' noise, as will be explained next. In $\phi_1$, when $S_1$ is closed, as in Figure 9.39(a), the capacitor $\alpha C$ is charged to the noise voltage. This noise voltage is held on $\alpha C$ in $\phi_1$ and will be effective in $\phi_2$.

Further during $\phi_2$, the noise due to the OA noise source as well as that due to the noise of $S_2$ are 'frozen' (i.e., sampled and held) on $\alpha C$ to be transferred later to the output of the OA. The resulting circuit arrangements to enable noise calculation are shown in Figures 9.39(b) and 9.39(c). Thus the total broadband noise can be computed by summing these three contributions and multiplying the result by the frequency response of the sampled data filter.

The broadband noise due to the switches and OA has a large noise bandwidth (since the capacitances are very small in MOS SC filters and the time constants involved are very small to enable complete charge transfer) and is limited by the OA bandwidth. The sample and hold process inherent in the SC networks samples this broadband noise with a low sampling frequency. This phenomenon is called *under-sampling*. As a result, the noise sidebands resulting from the sampling process fold over to the baseband. As an illustration, consider the case where the noise bandwidth is three times as large as the sampling frequency, as shown in Figure 9.40(a). The resulting sidebands are also shown in Figure 9.40(b). The noise in these adjacent bands is uncorrelated and hence can be added. The result is the increased noise in

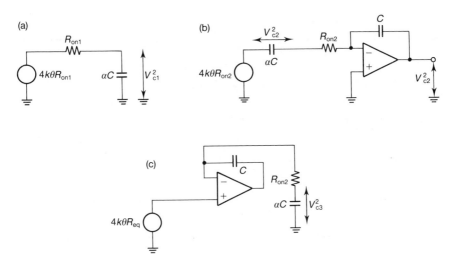

*Fig. 9.39* Sampled and held noise voltage contributions in various phases: (a) in $\phi_1$; (b), (c) in $\phi_2$.

the baseband. For a given noise bandwidth $BW_n$ and sampling frequency $f_s$, the total noise spectral density is seen to be given by [9.41]

$$\eta_T = \eta_n + \left[2\left(\frac{n}{f_s}\right) - 1\right]\eta_{sb} \tag{9.22}$$

where $\eta_n$ and $\eta_{sb}$ are the spectral densities in the baseband and sidebands, respectively. The term in the parentheses gives the number of sidebands (sb) aliasing or 'folding over' into the based band (n). Thus, for the example considered above, this factor is 5, as can be verified from Figure 9.40(b). The $1/f$ noise of the OA is dominated by the 'foldover' noise.

The foldover noise is evaluated next for the networks of Figure 9.39 as follows. The noise cutoff frequencies are first determined. They are obtained as

$$\omega_{ca} = (R_{on1}\alpha C)^{-1} \tag{9.23a}$$

$$\omega_{cb} = \omega_{cc} \simeq \left\{2R_{on2}\alpha C\sqrt{\frac{1}{4} + \frac{\alpha\omega_{on}}{B} + (\alpha+1)^2 \cdot \left(\frac{\omega_{on}}{B}\right)^2}\right\}^{-1} \tag{9.23b}$$

The integrator transfer function yields

$$|H(e^{j\omega T})|^2 = \left(\frac{\alpha}{2}\right)^2\left(\frac{1}{\sin\frac{\omega T}{2}}\right)^2 \tag{9.24}$$

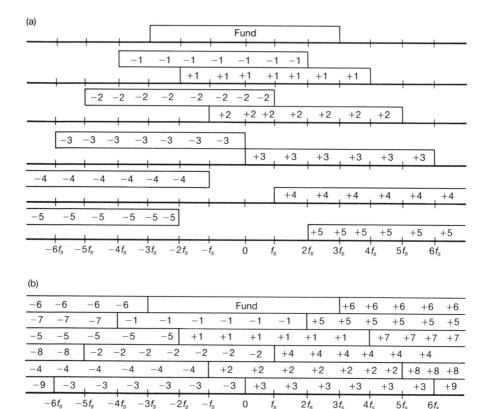

Fig. 9.40 Undersampling effect for noise bandwidth $3f_s$ and sampling frequency $f_s$ (adapted from [9.41], © 1982 IEEE).

Thus the sampled and held noise spectral density at the output of OA is given by

$$S_{sh}(t) = \frac{T}{2}\left[\omega_{ca}4k\theta R_{on1} + \omega_{cb}4k\theta(R_{on2} + R_{eq})\right] |H(e^{j\omega T})|^2 \operatorname{sinc}^2\left(\frac{\omega T}{2}\right)$$

$$= 2T\frac{k\theta}{\alpha C}\left[1 + \frac{\frac{1}{2}\left(1 + \dfrac{R_{eq}}{R_{on2}}\right)}{\sqrt{\dfrac{1}{4} + \dfrac{\alpha\omega_{on}}{B} + (\alpha + 1)^2\left(\dfrac{\omega_{on}}{B}\right)^2}}\right]\left(\frac{\alpha}{\omega T}\right)^2 \quad (9.25)$$

where sinc $x = (\sin x)/x$. The factor $T/2$ is necessary to denote the foldover noise. The sinc$^2(\omega T/2)$ term denotes the sampled and held nature of the noise under consideration.

The output noise of the integrator of Figure 9.38(a) is thus the algebraic sum of the noise spectral densities obtained in equations (9.20), (9.21) and (9.25). This completes the noise evaluation of the integrator. It can be noted from equations (9.20), (9.21) and (9.25) that the sampled and held noise dominates the other two noise contributions, since $\omega_{on}$ and $\omega_{eq}$ are very large. It thus suffices to consider the sampled and held noise in even complex SC networks. This is justified because of the assumptions that the noise of OAs is predominant at low frequencies only and that the sampled and held noise is largely due to the foldover effects. The direct noise is usually small. However, as the situation demands, the investigation of all of these noise sources can be carried out to yield accurate results.

Furrer and Guggenbuhl [9.42] have studied the noise behaviour of a two-OA stray-insensitive biquad due to Fleischer and Laker, where only the sampled and held noises are considered. As an illustration of the approach of the analysis, we briefly discuss their techniques in what follows.

## 9.3.2  Noise analysis of SC biquad

The various noise sources that are to be considered are the OA noise sources and the noise due to on-resistances of the switches in the SC network of Figure 9.41. The noise analysis can be carried out separately for the four cases: noise due to OA $A_1$; due to OA $A_2$; due to the group of switches $S_1$, $S_2$, $S_3$, $S_4$, $S_9$ and $S_{10}$; and due to another group of switches, $S_5$, $S_6$, $S_7$ and $S_8$.

The OA noise is represented by a noise source at the non-inverting input. The noise voltage of this source of OA $A_1$ is given by $4k\theta R_{eq}B_1$, where $B_1$ is the noise bandwidth of the circuit comprising OA $A_1$ and $C_1$, $C_2$, $C_5$ and $C_6$ (in $\phi_2$) given by

$$B_1 = \frac{B}{4(1 + \alpha + \gamma + \delta + \varepsilon)} \qquad (9.26)$$

where $\alpha = C_1/C_2$, $\gamma = C_5/C_2$, $\delta = C_6/C_2$ and $\varepsilon = C_7/C_4$. Thus the noise at the output of OA $A_2$ in $\phi_2$ is given by

$$S(\omega) = \frac{4k\theta R_{eq}B}{4(1 + \alpha + \gamma + \delta + \varepsilon)} H(e^{j\omega T})H(e^{-j\omega T})|\bar{H}(\omega T)|^2 \qquad (9.27)$$

where $H(e^{j\omega T})H(e^{-j\omega T})$ denotes the square of the magnitude of the transfer function of the noise source of OA $A_1$ to the output of $A_2$, and $|\bar{H}(\omega T)|^2$ is $\{T\sin(\omega T/2)/(\omega T/2)\}^2$, which takes into account the sampled and held nature of the noise. The quantity $H(z)$ can be determined using $z$-domain analysis (e.g., Laker's equivalent circuits). Note also that the noise of OA $A_1$ in $\phi_1$ does not contribute to the output of $A_2$.

The next step is to consider the noise of OA $A_2$. The transfer functions from this noise source in both phases $\phi_1$ and $\phi_2$ can be evaluated. The respective noise bandwidths of the OA $A_2$ are seen, as in the previous case, to be $B_2/[4(1 + \beta + \varepsilon)]$ in $\phi_1$, where $\beta = C_3/C_4$, and $B_2/[4(1 + \varepsilon)]$ in $\phi_2$. Thus once again, assuming that the noise sources in both phases are uncorrelated, the contribution to the output can be evaluated using expressions similar to equation (9.27).

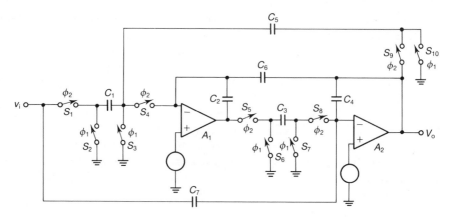

Fig. 9.41  Fleischer–Laker E-type biquad including noise sources. *(Adapted from [9.42],*
© *AEU 1983.)*

It is interesting next to study the evaluation of the noise of resistances due to switches $S_2$, $S_3$ and $S_{10}$ in $\phi_1$. The resulting network is shown in Figure 9.42. The interaction of the various noise sources results in noise being sampled on $C_1$ and $C_5$. This noise can be evaluated as follows. Suppose that we are interested in finding the noise voltage across $C_1$. This is given by

$$\bar{S}_1(\omega) = 4k\theta R_{eq} \tag{9.28}$$

where $R_{eq} = \text{Re}[Z_{C1}(j\omega)]$, the real part of the input impedance of the network of Figure 9.42 at the port $C_1$. The noise power is then obtained by integration of $\bar{S}_1(\omega)$:

$$\bar{N}_1 = \frac{1}{2\pi} \int_0^\infty \bar{S}_1(\omega) \, d\omega \tag{9.29}$$

Instead of evaluating this integral, which is a difficult process, one can use Bode's resistance integral theorem [9.43] which states that

$$\int_0^\infty \text{Re}[Z(j\omega)] \, d\omega = \frac{\pi}{2} \frac{1}{C_\infty} \tag{9.30a}$$

for minimum resistance networks and

$$\int_0^\infty \text{Re}[Z(j\omega)] \, d\omega = \frac{\pi}{2} \left( \frac{1}{C_\infty} - \frac{1}{C_0} \right) \tag{9.30b}$$

for non-minimum resistance networks, with

$$\frac{1}{C_\infty} = \lim_{s \to \infty} sZ(s), \quad \text{and} \quad \frac{1}{C_0} = \lim_{s \to 0} sZ(s) \tag{9.30c}$$

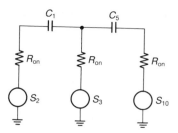

Fig. 9.42  Noise equivalent of switches $S_2$, $S_3$ and $S_{10}$ in $\phi_1$.

The resulting simplification is best illustrated with respect to Figure 9.42. The input impedance at the port across $C_1$ is

$$Z(s) = \frac{(3R_{on}sC_5 + 2)R_{on}}{3s^2C_1C_5R_{on}^2 + 2sR_{on}(C_1 + C_5) + 1} \qquad (9.31a)$$

whence

$$\int_0^\infty \text{Re}[Z(j\omega)] \ d\omega = \frac{\pi}{2}\frac{1}{C_\infty} = \frac{\pi}{2C_1} \qquad (9.31b)$$

Using equation (9.31b) in equation (9.28) and using equation (9.29), the noise powers on $C_1$ and $C_5$ are obtained as $k\theta/C_1$ and $k\theta/C_5$, respectively.

The above noise voltages stored on $C_1$ and $C_5$ in $\phi_1$, together with the noise voltage arising in $\phi_2$ because of the on-resistances of the switches $S_1$, $S_4$ and $S_9$, have to be considered. The latter contribution can be analyzed using the noise equivalent circuit of Figure 9.43. The noise current flowing into the inverting input of OA $A_1$ charges $C_2$ and affects the noise spectral density at the output of $A_2$. This noise current $i_3$ can be obtained as follows. The noise spectral density is given for the current $i_3$ as

$$\bar{S}_3(\omega) = 4k\theta \ \text{Re}[Y(j\omega)] \qquad (9.32)$$

where $Y(j\omega)$ is the admittance of the passive network formed by the on-resistances

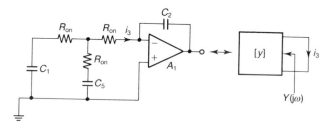

Fig. 9.43  Noise equivalent circuit in $\phi_2$ considering the effect of switches $S_1$, $S_4$ and $S_9$.

of $S_1$, $S_4$, $S_9$ and $C_1$, $C_5$. Following in a manner similar to the evaluation of the network in Figure 9.42, it is observed that the noise voltage stored on $C_2$ is

$$\bar{N}_3 = \frac{1}{2\pi} \int_0^\infty \frac{\bar{S}_3(\omega)}{(\omega C_2)^2} \, d\omega \qquad (9.33a)$$

Using Bode's resistance integral theorem, equation (9.33a) simplifies to

$$\bar{N}_3 = k\theta \left( \frac{\alpha^2}{C_1} + \frac{\gamma^2}{C_5} \right) \qquad (9.33b)$$

This noise voltage developed on $C_2$, together with the noise voltages on $C_1$ and $C_5$ in $\phi_1$, results in the overall noise at the output of $A_2$.

The noise due to the on-resistances of switches $S_5$, $S_6$, $S_7$ and $S_8$ can also be evaluated in a simple manner. In $\phi_1$, $S_6$ and $S_8$ introduce a noise with equivalent voltage given by $k\theta/C_3$; while in $\phi_2$, $S_5$ and $S_7$ result in another noise voltage of $k\theta/C_3$. Thus, we are now in a position to compute the overall noise at the output. The analysis shows that the various spectral densities corresponding to the four groups considered above are as follows:

$$\bar{S}_{Aout}(\omega T) = \frac{|\bar{H}(\omega T)|^2}{|\bar{D}(\omega T)|^2} k\theta R_{eq1} B \beta^2 \left[ \frac{(\alpha + \gamma)^2}{1 + \alpha + \beta + \delta} + 2(1 + \delta)(1 - \cos(\omega T)) \right] \qquad (9.34a)$$

$$\bar{S}_{Bout}(\omega T) = \frac{|\bar{H}(\omega T)|^2}{|\bar{D}(\omega T)|^2} k\theta R_{eq2} B \left[ \frac{2\beta^2(1 - \cos(\omega T))}{1 + \beta + \varepsilon} + 4(1 + \varepsilon)(1 - \cos(\omega T))^2 \right] \qquad (9.34b)$$

$$\bar{S}_{Cout}(\omega T) = \frac{|\bar{H}(\omega T)|^2}{|\bar{D}(\omega T)|^2} 2k\theta \beta^2 \left( \frac{\alpha^2}{C_1} + \frac{\gamma^2}{C_5} \right) \qquad (9.34c)$$

$$\bar{S}_{Dout}(\omega T) = \frac{|\bar{H}(\omega T)|^2}{|\bar{D}(\omega T)|^2} \frac{4k\theta \beta^2}{C_3} [1 - \cos(\omega T)] \qquad (9.34d)$$

with

$$|\bar{D}(\omega T)|^2 = D(z)D(z^{-1})|_{z=e^{j\omega T}}$$
$$= 4(1 - \beta\delta)\cos^2(\omega T) - 2(2 - \beta\delta)(2 - \beta\delta - \beta\gamma)\cos(\omega T)$$
$$+ [2 - \beta(\delta + \gamma)]^2 + (\beta\delta)^2 \qquad (9.34e)$$

$$|\bar{H}(\omega T)|^2 = T \left[ \frac{\sin\left(\frac{\omega T}{2}\right)}{\frac{\omega T}{2}} \right]^2 \qquad (9.34f)$$

Derivation of these results is left as an exercise for the reader. Note from equations (9.34a), (9.34b) and (9.34f) that the noise due to OAs is directly proportional to the bandwidth and the equivalent noise resistance of the OAs, and inversely proportional to the sampling frequency. A high $f_s$ implies a large $B$ due to the fact that the OA should be able to settle completely. In addition, for high $f_s$, the $1 - \cos(\omega T)$ terms are small, leading to a reduction in low-frequency noise. Thus, the noise due to OA $A_2$ is suppressed considerably.

## 9.3.3  Noise analysis using network analysis programs

It is of interest to model the noise analysis of the SC network in existing computer-aided network analysis programs. An attempt at this has been made by Fischer [9.41], who employs SPICE for noise analysis of SC networks. However, such modelling requires some approximations, the most important being the use of high sampling frequencies so that the SCs $C_i$ are treated as resistors $R_i(T/C_i)$. In this method, the OAs in the SC HP notch filter of Figure 9.44(a) are replaced by blocks containing a continuous-time output as well as switched output, as shown in Figure 9.44(b). The continuous output is connected to paths that are not switched. Note that the inversion realized by the SC $C_2$ is modelled by a voltage-controlled voltage source (VCVS) of gain $-1$. The OA is next modelled as shown in Figure 9.45. Note that the unsampled wideband and $1/f$ noise sources are included

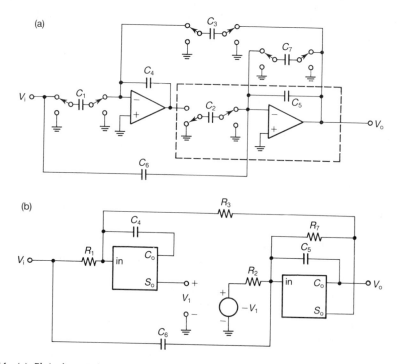

Fig. 9.44  (a) Fleischer–Laker SC HP notch filter;  (b) Fischer's noise equivalent model of (a). (Adapted from [9.41], © 1982 IEEE.)

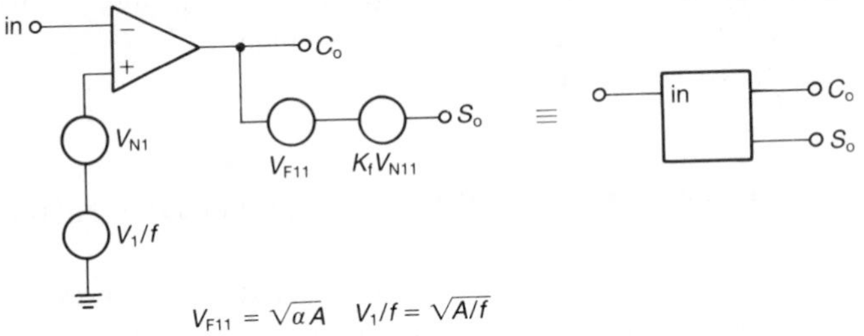

*Fig. 9.45* Frequency-domain OA noise simulation model.

at the non-inverting input. The foldover effect is represented by the switched output $S_o$ and associated wideband and $1/f$ noise sources. The factor $K_f$ is the foldover factor obtained from equation (9.22):

$$K_f = \sqrt{\left[2\left(\frac{BW_n}{f_s}\right) - 1\right]} \tag{9.35}$$

Note that $K_f V_{N11}$ represents a noise voltage source. Similarly, denoting the $1/f$ noise voltage of the OA as $(A/f)^{1/2}$ with $A$ constant, the foldover effect is to result in an overall noise with density given by

$$\eta(f) = A\left[\frac{1}{f} + \sum_{i=1}^{N}\left\{\frac{1}{(if_s - f)} + \frac{1}{(if_s + f)}\right\}\right] \tag{9.36}$$

where $N$ is the number of sidebands falling into the baseband. It may be noted that the $1/f$ noise is limited only by the bandwidth of the OA. The summation indicated in equation (9.36) is denoted as $\alpha$ in Figure 9.45. Fischer has shown that a $1/f$ noise foldover contribution above 100 kHz can usually be neglected [9.41]. It is possible to simulate the complete noise model of the OA of Figure 9.45 in SPICE. However, since SPICE does not cater for user-defined functional expressions, a device model must be used to model the unsampled $1/f$ noise. The complete SPICE OA model is shown in Figure 9.46. The enhancement-mode transistor $M_1$, biased with a current source $I_{dc}$, is used to generate $1/f$ noise. This noise voltage at the drain of $M_1$ is buffered and available as $V_1/f$. The dc voltage is blocked by $C_{BLK}$ and the noise voltage is injected into the non-inverting terminal of the OA. The resistors $R_{N1A}, R_{N1B}$ etc., generate wideband noise which can be used appropriately to simulate the foldover noise sources as well as OA input wideband noise. The resistor values are dependent on the noise source voltages that are to be simulated. This completes the discussion on modelling noise and evaluation of SC filter performance using these models.

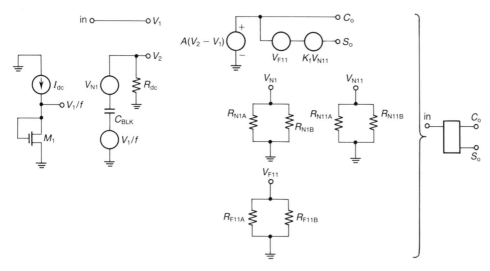

*Fig. 9.46* Complete SPICE OA noise model.

## 9.4 Effect of finite d.c. gain of the OA

In the previous chapters, SC networks using OAs have been studied assuming that the OA has infinite gain. In reality, however, MOS OAs have relatively lower gain as compared to bipolar OAs. Thus, an OA with a gain of 1000 might typically be used in practice. The effect of the finite d.c. gain is of interest in evaluating the performance of SC networks. As an illustration, consider the stray-insensitive integrator of Figure 9.47(a).

Denoting the d.c. gain of the OA as $A_o$, the transfer functions of the integrator are derived in a straightforward manner as follows:

$$\frac{V_{oe}(z)}{V_{ie}(z)} = \frac{-1}{\dfrac{C_2}{C_1}(1-z^{-1})\left(1+\dfrac{1}{A_o}\right)+\dfrac{1}{A_o}} \tag{9.37a}$$

$$V_{oo}(z) = V_{oe}(z)z^{-1/2} \tag{9.37b}$$

Thus the effect of the d.c. gain of the OA is to introduce a finite real pole. The effect of this pole is best examined, when we consider a two-integrator loop as existing in an $F$-type Fleischer–Laker biquad. For simplicity, the feedforward paths are omitted and only the feedback loop shown in Figure 9.47(b) is considered. Deriving

(a)

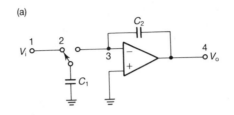

(b)

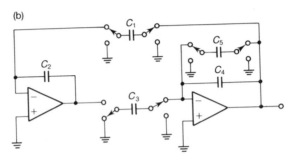

Fig. 9.47  SC networks considered for evaluation of OA gain effects:  (a) integrator;
(b) Fleischer–Laker SC biquad.

the loop gain and equating it to unity, the denominator of the transfer function is
obtained as follows:

$$D(z) \equiv \left\{ C_4 + C_5 + \frac{C_3 + C_4 + C_5}{A_o} - C_4 z^{-1} \left(1 + \frac{1}{A_o}\right) \right\}$$

$$\times \left\{ C_2 + \frac{C_1 + C_2}{A_o} - C_2 z^{-1} \left(1 + \frac{1}{A_o}\right) \right\} + C_1 C_3 z^{-1} \qquad (9.38)$$

assuming identical OAs. We simplify the analysis by considering the optimal dynamic
range design (i.e., equal time constants for both the integrators: $(C_1/C_2) = (C_3/C_4)$,
and choose $C_2 = C_4 = 1$) to give

$$D(z) \equiv z^{-2} \left(1 + \frac{2}{A_o}\right) - z^{-1} \left(2 + C_5 + \frac{2C_1 + 2C_5 + 4}{A_o} - C_1^2\right)$$

$$+ \left(1 + C_5 + \frac{2C_1 + 2C_5 + C_1 C_5 + 2}{A_o}\right) \qquad (9.39)$$

Note that we have ignored terms of order $1/A_o^2$. The resulting expressions for $\delta$ and $Q_p$ are

$$\delta = \frac{\sqrt{4 + 2C_5 - C_1^2 + \dfrac{4C_1 + 4C_5 + 8 + C_1C_5}{A_o}}}{2\sqrt{C_1^2 + \dfrac{C_1C_5}{A_o}}} \qquad (9.40a)$$

$$Q_p = \frac{\sqrt{\left(C_1^2 + \dfrac{C_1C_5}{A_o}\right)\left(4 + 2C_5 - C_1^2 + \dfrac{4C_1 + 4C_5 + 8 + C_1C_5}{A_o}\right)}}{2\left(C_5 + \dfrac{2C_1 + 2C_5 + C_1C_5}{A_o}\right)} \qquad (9.40b)$$

As an example, for desired values $\delta = 1$, $Q_p = 30$, the design values are $C_5 = 1/37$ and $C_1 = \sqrt{(30/37)}$. The resulting $\delta$ value is $1.001\,786\,2$ for an OA gain of 1000, whereas the $Q_p$ value is 28.1 as against the desired value of 30. It is thus evident that because of the relatively low d.c. gain of the OA, the realized $Q_p$ values of the SC filter deviate to a certain noticeable extent. Based on the above evaluation of the deviation in $Q_p$ value, one can choose a suitable gain OA. Note that, depending on the complexity of the OA, the gain varies. Simple designs of the OAs tend to have lower gain, as shown in Section 9.1, and for some applications these may suffice.

We next consider the methods of incorporating the finite d.c. gain of an OA into SC network analysis [9.44, 9.45]. Routine analysis can take into account the effect of finite d.c. gain of the OA, as in Laker's equivalent circuit method. For other methods, notably for computer simulation using the IAM method studied in Chapter 2, the operations performed to incorporate the OA into the IAM of the passive SC network need to be modified slightly. Recall that the differential input voltage of the OA is assumed zero in deriving the IAM in Chapter 2. We also have used the fact that the output current of the OA is not an independent variable, to delete certain rows and columns in the IAM. If a finite d.c. gain is employed in the analysis, these operations are modified to a small extent, as explained next. The voltages at the non-inverting $V_+$ and inverting $V_-$ inputs are related to the output voltage of the OA $A_o$ as follows:

$$V_+ - V_- = \frac{V_o}{A_o} \qquad (9.41)$$

This relationship evidently holds good in both the clock phases. Thus, in the IAM, the three node voltages corresponding to $V_+$, $V_-$ and $V_o$ can be reduced to two, i.e., $V_o$ and either of $V_+$ or $V_-$. Assuming, for instance, $V_+$ to be deleted, using equation (9.41), we observe that the elements $x_i$ in the column $V_+$ shall be added to the elements in the column $V_-$. Further, the elements in column $V_+$ divided by $A_o$ shall be added to the elements in the column $V_o$. The resulting IAM can be used for further analysis. The rows corresponding to the outputs of the OA can be deleted as explained in Chapter 2.

It is possible to compensate the effect of the finite d.c. gain of the OA on SC filters using two techniques. In the first method, due to Fischer and Moschytz [9.46], an additional buffer OA is required. This method is illustrated in Figure 9.48(a). The basic principle of operation is as follows. In the even phase, $C_1$ is charged to the input voltage and in the conventional circuit (that is, without the buffer shown inside the box), some charge on $C_1$ remains after the transfer of charge to $C_2$ in the odd phase because the OA input is at a finite non-zero potential. An elegant method of completely transferring the charge on $C_1$ to $C_2$ is to discharge $C_1$ using a unity-gain buffer as shown in Figure 9.48(a). Note, however, that the input differential

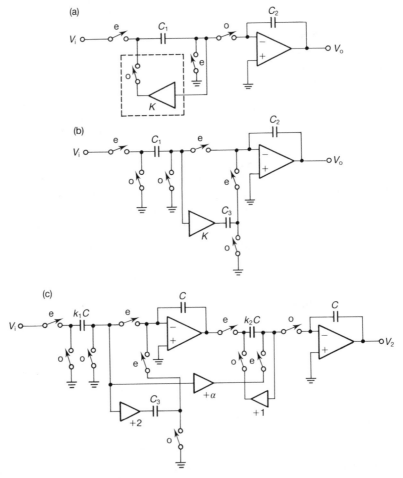

Fig. 9.48  Fischer and Moschytz's active compensation scheme for finite gain of OA applied to: (a) non-inverting integrator; (b) inverting integrator; (c) cascade of (a) and (b). (Adapted from [9.46], © 1984 IEEE.)

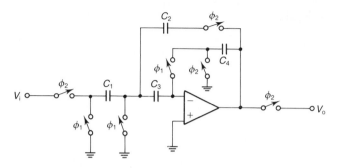

Fig. 9.49  Nagaraj et al. passive compensation technique (adapted from [9.47],
© 1985 IEE).

voltage of the OA is still $-V_o/A$. An analysis of the circuit shows that

$$\frac{V_{oo}(z)}{V_{ie}(z)} = \frac{C_1 z^{-1/2}}{C_1\left(\dfrac{K-1}{A_1}\right) + C_2\left(1 - \dfrac{1}{A_1}\right)(1 - z^{-1})} \tag{9.42}$$

where $K$ is the gain of the buffer. When $K = 1$, it is evident that the integrator time constant is modified. The results for the case when the buffer OA is not used can be obtained by substituting $K = 0$ in equation (9.42). Note that for a 'real' buffer, $K = A_2/(A_2 + 1)$, where $A_2$ is the d.c. gain of this OA. However, the error introduced is negligible to first order. The application of this technique to the inverting SC integrator is shown in Figure 9.48(b), for which we obtain

$$\frac{V_{oe}(z)}{V_{ie}(z)} = \frac{-C_1}{C_2\left(1 + \dfrac{1}{A_1}\right)(1 - z^{-1}) + \dfrac{C_1 + C_3(1 - K)}{A_1}} \tag{9.43}$$

Thus, for $K = (C_1 + C_3)/C_3$, an inverting integrator with gain error exactly opposite in magnitude to that of the non-inverting integrator can be realized. For a cascade of inverting and non-inverting integrators, the circuit of Figure 9.48(c) can be employed for compensating the effect of OA finite gain. Note that an additional OA with gain $\alpha$ has to be employed. The output voltage of this circuit is given by

$$\frac{V_{2o}(z)}{V_{ie}(z)} = \frac{-k_1 k_2 z^{-1/2}\left(1 + \dfrac{\alpha}{A_1}\right)}{\left(1 + \dfrac{1}{A_2}\right)\left(1 + \dfrac{1}{A_1}\right)(1 - z^{-1})^2} \tag{9.44}$$

Note that, for identical OAs, $\alpha = 2$, neglecting second-order effects. The reader is urged to study the application of the method to the lossy integrator case.

    We next consider the Nagaraj et al. technique [9.47] which is a simple modification of the offset compensation scheme due to Lam and Copeland studied in

Section 9.2. Their gain-compensated lossless inverting integrator circuit is shown in Figure 9.49. The transfer function of this circuit can be shown to be as follows:

$$\frac{V_o(z)}{V_i(z)} = \frac{C_1/C_2}{(1 - z^{-1}) + \dfrac{1}{A}\left(1 - z^{-1} + \dfrac{C_1}{C_2}\right)\left(1 - \dfrac{z^{-1}}{D}\right)} \tag{9.45a}$$

where

$$D = 1 + \frac{1}{A} + \frac{1}{A}\frac{C_3}{C_4}(1 - z^{-1}) \tag{9.45b}$$

Note that $D$ is approximately equal to unity, so that, from equation (9.45a), we have

$$\frac{V_o(z)}{V_i(z)} = \frac{C_1/C_2}{(1 - z^{-1})\left(1 + \dfrac{\left(1 + \dfrac{C_1}{C_2} - z^{-1}\right)}{A}\right)} \tag{9.46}$$

Thus, in addition to the pole at the origin, a parasitic pole is introduced. Note that this circuit has the advantage that the offset voltage of the OA can also be eliminated using the same circuit elements as those used to compensate the OA finite-gain effects. Nagaraj et al. [9.47] have shown that the use of these 'passively' compensated integrators reduces the errors considerably in SC ladder filters based on operational simulation.

## 9.5 Effect of finite OA bandwidth [9.45–9.48]

In the discussion so far, the OA is assumed to have infinite bandwidth. The effect of finite OA bandwidth on SC networks can be analyzed with some difficulty. Matters are complicated by the fact that the analysis has to be carried out in the time domain. The frequency-dependent open-loop gain is expressed as follows:

$$A(s) = \frac{V_o(s)}{V(s)} = \frac{-B}{s + \omega_a} \tag{9.47}$$

where $B$ is the bandwidth of the OA, $\omega_a$ is the open-loop cutoff frequency and $A_o$ is the d.c. gain, where $A_o\omega_a = B$. Assuming, for simplicity, $A_o \gg 1$, equation (9.47) can be rewritten to relate the output and input of the OA in the time domain as follows:

$$v(t) = -\frac{1}{B}\frac{dv_o(t)}{dt} \tag{9.48}$$

Thus, unless $v(t)$ contains impulses, $v_o(t)$ is a continuous function of time. The time-domain behaviour expressed in equation (9.48) should be used to evaluate the voltages in the circuits incorporating the non-ideal OAs, at various instants in time. For the purposes of analysis [9.48], we consider an inverting integrator, shown in

Figure 9.50(a). Considering the instant the switches are in $\phi_2$, the charge conservation equation can be written as follows:

$$C_1[v_i(t_{n-1}) - v(t)] = C_2[\{v(t) - v_o(t)\} - \{v(t_{n-1}) - v_o(t_{n-1})\}] \quad (9.49a)$$

or

$$C_2\{v_a(t) - v_a(t_{n-1})\} = C_1\{v(t) - v_i(t_{n-1})\} \quad (9.49b)$$

where $v_a(t)$ is the instantaneous voltage across $C_2$, $v(t)$ that across the input terminals of the OA, and we have written $t_{n-1} = (n-1)T$, $T$ being the switching period. Further, the above equations are valid only during the period $t_{n-1} \leqslant t \leqslant t_{n-1/2}$. (In other words, we have assumed that $\phi_2$ starts at the instant $t_{n-1}$.) Note that

$$v_o(t) = v_a(t) + v(t) \quad (9.50)$$

From equations (9.48)–(9.50), the time-domain behaviour of $v(t)$ and $v_o(t)$ is obtained through the equation

$$\frac{1}{KB}\frac{dv_o(t)}{dt} + v_o(t) = \hat{v}(t_{n-1}) \quad (9.51)$$

where

$$K = \frac{C_2}{C_1 + C_2}$$

$$\hat{v}(t_{n-1}) = v_a(t_{n-1}) - \frac{C_1}{C_2}v_i(t_{n-1}) \quad (9.52)$$

Equations (9.52) can be solved taking into account the initial values at the instants $t_{n-1}$ to give the following equation, where we have assumed $T = 1$ without any loss of generality:

$$v_o(t) = \hat{v}(n-1)(1 - e^{-KB(t-t_{n-1})}) + v_o(n-1)e^{-KB(t-t_{n-1})} \quad (9.53a)$$

From equations (9.48), (9.50) and (9.53a), we obtain

$$v_a(t) = \hat{v}(n-1)(1 - (1-K)e^{-KB(t-t_{n-1})}) + v_o(n-1)(1-K)e^{-KB(t-t_{n-1})} \quad (9.53b)$$

Substituting $t - t_{n-1} = \frac{1}{2}$ in equations (9.53), we obtain the values of $v_o(t)$ and $v_a(t)$ at the instant $(n - \frac{1}{2})$. It is evident that the time constant of the above transient is $1/KB$ or $(C_1 + C_2)/C_2B$.

During $\phi_1$, when the switches are thrown to the left, the circuit consists of just the OA and $C_2$. Using equations (9.48) and (9.50) and noting that the voltage across $C_2$ remains at $v_a(n - \frac{1}{2})$, we obtain the equation

$$\frac{1}{B}\frac{dv_o(t)}{dt} + v_o(t) = v_a(n - \frac{1}{2}) \quad (9.54)$$

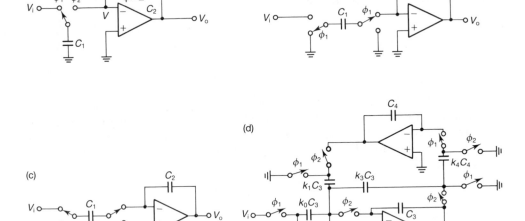

Fig. 9.50 Circuits used for analysis of gain-bandwidth effects: (a)–(c) SC integrators; (d) Martin–Sedra SC biquad.

Solving equation (9.54), we obtain

$$v_o(t) = v_a(n - \tfrac{1}{2})(1 - e^{-B(t - t_{n-1/2})} + v_o(n - \tfrac{1}{2})e^{-B(t - t_{n-1/2})} \qquad (9.55a)$$

$$v_a(t) = v_a(n - \tfrac{1}{2}) \qquad (9.55b)$$

The above equations are valid only during the period $t_{n-1/2} \leqslant t \leqslant t_n$. The time constant of this transient is $1/B$ and the values of $v_o(t)$ and $v_a(t)$ at the instant $nT$ can be obtained by substituting $(t - t_{n-1/2}) = \tfrac{1}{2}$ in equation (9.55a).

The voltages at the instants $t_{n-1}$, $t_{n-1/2}$ and $t_n$ available in equations (9.52), (9.53) and (9.55) can be used next to solve for the z-domain transfer functions of the SC integrator. Substituting for $\hat{v}(n - 1)$ from equation (9.52) in equation (9.53) with $t = t_{n-1/2}$ and defining

$$\frac{KBT}{2} = k_1$$

$$\frac{BT}{2} = k_2$$

we have

$$v_o(n - \tfrac{1}{2}) = v_a(n - 1)(1 - e^{-k_1}) + v_o(n - 1)e^{-k_1} - \frac{C_1}{C_2} v_i(n - 1)(1 - e^{-k_1}) \qquad (9.56a)$$

$$v_a(n - \tfrac{1}{2}) = v_a(n - 1)[1 - (1 - K)e^{-k_1}] - \frac{C_1}{C_2} v_i(n - 1)[1 - (1 - K)e^{-k_1}] + v_o(n - 1)(1 - K)e^{-k_1}$$

(9.56b)

From equations (9.55), we have at $t = t_n$,

$$v_o(n) = v_a(n - \tfrac{1}{2})(1 - e^{-k_2}) + v_o(n - \tfrac{1}{2})e^{-k_2}$$ (9.57a)

$$v_a(n) = v_a(n - \tfrac{1}{2})$$ (9.57b)

Using equations (9.56) in equations (9.57), we obtain

$$v_o(n) = \left[ v_a(n - 1) - \frac{C_1}{C_2} v_i(n - 1) \right][1 - (1 - K)e^{-k_1} - Ke^{-(k_1 + k_2)}]$$

$$+ v_o(n - 1)[(1 - K)e^{-k_1} + Ke^{-(k_1 + k_2)}]$$ (9.58a)

$$v_a(n) = \left[ v_a(n - 1) - \frac{C_1}{C_2} v_i(n - 1) \right](1 - (1 - K)e^{-k_1}) + v_o(n - 1)(1 - K)e^{-k_1}$$ (9.58b)

Taking the $z$-transforms of (9.58) and defining

$$\alpha = Ke^{-(k_1 + k_2)}$$

$$\delta_1 = e^{-k_1}(1 - K) + \alpha$$

(9.59)

we have the following equations:

$$V_o(z)(1 - \delta_1 z^{-1}) - V_a(z)z^{-1}(1 - \delta_1) = -\frac{C_1}{C_2} V_i(z)z^{-1}(1 - \delta_1)$$ (9.60a)

$$-V_o(z)z^{-1}(\delta_1 - \alpha) + V_a(z)[1 - z^{-1}(1 - \delta_1 + \alpha)] = -V_i(z)z^{-1}\frac{C_1}{C_2}(1 - \delta_1 + \alpha)$$ (9.60b)

Solving equations (9.60), we have

$$\frac{V_o(z)}{V_i(z)} = -\frac{C_1}{C_2} \frac{(1 - \delta_1)z}{(z - 1)(z - \alpha)}$$ (9.61)

It can be observed from equation (9.61) that a parasitic pole and a parasitic zero have been introduced due to the finite bandwidth of the OA.

The results are the same for the stray-insensitive non-inverting integrator of Figure 9.50(b) as well, except for the sign in equation (9.61). In the case of the inverting integrator of Figure 9.50(c) [9.49], the charge conservation equations derived for the case of Figure 9.50(a) remain the same except that equations (9.56) are true in the period $t_n \leqslant t \leqslant t_{n-1/2}$ and equations (9.57) are true in the period $t_{n-1} \leqslant t \leqslant t_{n-1/2}$. Accordingly, equations (9.57) have to be substituted in equations (9.56) to yield the transfer function

$$\frac{V_o(z)}{V_i(z)} = \frac{C_1/C_2}{1 - z^{-1}} \left\{ 1 - e^{-k_1} + Ke^{-k_1}z^{-1} \frac{[1 - e^{-(k_1 + k_2)}]}{[1 - z^{-1}Ke^{-(k_1 + k_2)}]} \right\}$$ (9.62)

(Note that, in deriving equation (9.62), it is assumed that the input is constant during $\phi_2$. This assumption may not be valid, in practice, because in high-order SC networks, the input comes from the output of another OA, which is also affected by the finite bandwidth.) It can be seen that by assuming $e^{-(k_1+k_2)} \ll 1$, approximate expressions for $V_o(z)/V_i(z)$ can be written from equations (9.61) and (9.62) as follows:

$$\frac{V_o(z)}{V_i(z)} = \frac{\mp (C_1/C_2)z^{-1}}{(1-z^{-1})}\left[1 - e^{-k_1}\frac{C_1}{C_1+C_2}\right] \qquad (9.63a)$$

and

$$\frac{V_o(z)}{V_i(z)} = \frac{-(C_1/C_2)}{(1-z^{-1})}\left[1 - e^{-k_1} + e^{-k_1}\left(\frac{C_2}{C_1+C_2}\right)z^{-1}\right] \qquad (9.63b)$$

for the integrators of Figures 9.50(a) or 9.50(b) and 9.50(c), respectively. These formulas indicate that for the integrators of Figure 9.50(a) or 9.50(b) there is only a magnitude error with no appreciable phase error, while the integrator of Figure 9.50(c) has both magnitude and phase errors. It is of interest to note the exponential relationship between the error terms and the amplifier bandwidth $B$ (since $k_1$ and $k_2$ are functions of the bandwidth of the OA).

It is easy to see that the results can be derived in a similar manner taking into account both finite d.c. gain and bandwidth of the OA for other first-order networks [9.46, 9.50]. The feedback switched capacitors, such as those that occur in a lossy SC integrator, can be considered to be additional input paths. The time constants of the transients are thus decided by the total capacitance connected to the virtual ground of the OA.

The results derived above for first-order networks can be used to evaluate the effect of finite bandwidth of the OAs on a two-integrator loop, as in the Martin–Sedra or Fleischer–Laker biquad. Note, however, that the results derived in the above analysis can be used only approximately. This arises because, for the two-integrator loop shown in Figure 9.50(d), the non-inverting integrator receives a step output during $\phi_2$ through $k_0 C_3$ and not through $k_1 C_3$ and $k_3 C_3$. With the simplifying assumption that the inputs through $k_1 C_3$ and $k_3 C_3$ are 'constants', the integrator's magnitude and phase errors evaluated individually can be used to predict the $\omega_p$ and $Q_p$ deviations. Representing the integrator non-ideal gains as

$$H_a(\omega) = \frac{H_i(\omega)_{\text{ideal}}}{[1 - m_i(\omega)]\,[e^{j\theta_i(\omega)}]} \qquad (9.64a)$$

where $m_i(\omega)$ is the magnitude error and $\theta_i(\omega)$ is the phase error, and approximating further, we obtain

$$H_a(\omega) = \frac{H_i(\omega)_{\text{ideal}}}{1 - m_i(\omega) - j\theta_i(\omega)} \qquad (9.64b)$$

Then, the errors in $\omega_p$ and $Q_p$ can be expressed as

$$\frac{\Delta\omega_p}{\omega_p} = \tfrac{1}{2}\left[m_1(\omega_p) + m_2(\omega_p)\right] \tag{9.65a}$$

and

$$Q_a = Q_p/\left[1 + Q_p\{\theta_1(\omega_p) + \theta_2(\omega_p)\}\right] \tag{9.65b}$$

Such an evaluation can be carried out for various OA bandwidths to realize the desired $\omega_p$ and $Q_p$. The results are not, however, obtainable analytically. Martin and Sedra have observed that the effects of finite OA bandwidth are negligible for $f_s/B < 1/5$. Thus, for given values of resonant frequency and $B$, one should choose as low a value of $f_s$ as possible. The effect of finite OA gain in SC biquad is similar to that in active RC filters, whereas the effect of finite bandwidth is much smaller. Martin and Sedra conclude that since the sampling frequency must be low to have reduced effects due to finite OA bandwidths, bilinear transformation type designs are preferable [9.49].

Geiger and Sanchez-Sinencio have considered the time-domain analysis for high-order networks, and the reader is referred to [9.51] for details.

Fischer and Moschytz have considered the effect of on-resistance of the switches as well as the finite bandwidth of the OA on SC network performance [9.46]. Their results also show that the integrator magnitude and phase errors can be kept to less than 1% when

$$A_o \geqslant 1000$$

$$B/f_s \geqslant 2$$

$$\frac{1}{4f_sK_fR_{on}C} \geqslant 10$$

where $K_fC$ is the feedforward capacitance of the integrator and $C$ is the feedback capacitance.

The basic results developed for SC integrators can be extended to the case of high-order leap-frog SC ladder filters. To this end, it is noted that the magnitude error of the integrators corresponds to uniform tolerance of the reactive elements of the LC ladder. It can be shown that the change in attenuation due to the errors in the reactive components can be written as follows [1.25]:

$$|\Delta\alpha|_{max} \leqslant \frac{8.68\,|m|\,|\rho|\,\omega\tau(\omega)}{1 - |\rho|^2}\ \text{dB} \tag{9.66}$$

where $m$ is the magnitude error of the integrator, $\rho$ is the reflection coefficient and $\tau(\omega)$ is the group delay. The phase errors of the integrators are equivalent to parasitic losses in the elements of the LC ladder. Thus the resulting attenuation deviation can be shown to be:

$$\Delta\alpha(\omega) \simeq 4.34\{[\theta_1(\omega) + \theta_2(\omega)]\,\omega\tau(\omega) + \tfrac{1}{2}[\theta_1(\omega) - \theta_2(\omega)]\,\text{Im}(\rho_1 + \rho_2)\}\ \text{dB} \tag{9.67a}$$

where $\theta_1$ and $\theta_2$ are respectively the phase errors of the integrators due to lossy inductor and lossy capacitor, $\rho_1$ and $\rho_2$ are the front-end and back-end reflection coefficients and Im represents the imaginary part. In the passband of the filter, the second term is negligible, leading to the approximate formula,

$$\Delta\alpha(\omega) \simeq 4.34\,[\theta_1(\omega) + \theta_2(\omega)]\,\omega\tau(\omega)\,\text{dB} \qquad (9.67b)$$

Thus inequality (9.66) and equations (9.67) can be used to evaluate the error in transfer function magnitude due to the finite bandwidth of the OA.

In the above sections, methods of taking into account several non-idealities of SC networks have been considered. We discuss in the next section a few applications of SC networks.

## 9.6   Applications of SC techniques

When the SC concept was introduced, it was intended to realize the function of a resistor; consequently, filter design using SCs has received considerable attention. However, even before the invention of the SC concept, several uses for switches and capacitors had been considered. For example, capacitor arrays and switches were found to be highly useful for realizing analog-to-digital (A/D) converters and digital-to-analog (D/A) converters. It is now common to employ switches and capacitors to realize various circuit functions other than filtering. Nevertheless, one important application of SC technology has been the realization of pulse-code modulation (PCM) filters. The need for these in substantial quantities in the telecommunications industry has resulted in numerous PCM filter designs, some of which even include PCM codec functions. Other applications of SC filters have been in dual tone multifrequency receivers (DTMF), four-quadrant multipliers, speech signal processing, etc. Some of these applications are briefly described in the following to highlight the potential utility of the design techniques presented so far.

### 9.6.1   PCM filters [9.4, 9.8, 9.15, 9.16, 9.18, 9.52–9.61]

In telephone switching and transmission systems, PCM techniques have become quite common in view of their various advantages. In these systems, the subscriber lines are connected to the digital switching network through subscriber-line units. These line units contain 2 wire/4 wire conversion circuits, two precision filters, and a codec circuit operating at a rate of 8 kHz.

The typical signal path in a PCM switching system is shown in Figure 9.51(a). The analog signal coming from the subscriber line interface circuit is first passed through a transmit filter where output is sampled at a rate of 8 kHz and coded to 8-bit PCM words in the encoder. The output of the encoder, after switching through the telephone exchange, reaches the decoder of the line unit of the receiving subscriber.

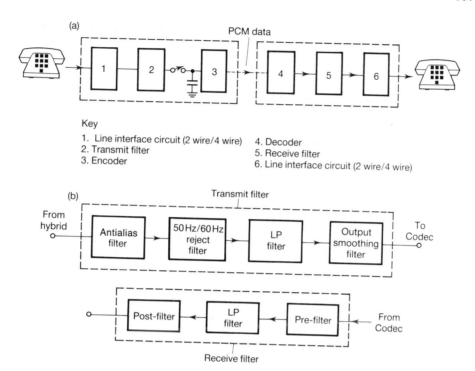

*Fig. 9.51* (a) Signal path in a typical PCM switching system; (b) PCM filter architecture.

For high-quality PCM systems, certain specifications have to be satisfied by Consultative Committee on International Telegraphy and Telephony (CCITT) or D3 channel requirements. A typical set of specifications satisfying both is shown in Table 9.3. The block diagram of the monolithic channel filter realizing these specifications is shown in Figure 9.51(b). SC techniques are useful for realizing the filtering functions in the transmit and receive paths. MOS D/A and A/D converters with relevant control logic circuits are used to perform the code functions.

**Table 9.3**  System specifications for transmit and receive PCM filters  (*adapted from [9.8]*, © 1980 IEEE).

| | | |
|---|---|---|
| Passband flatness | | +0.125 dB, −0.125 dB |
| | | 300 to 3200 Hz |
| Stopband rejection: | Transmit channel: | − 32 dB above 4600 Hz |
| | Receive channel: | − 28 dB above 4600 Hz |
| Transmit 50/60 Hz rejection | | > 20 dB |
| Absolute gain variation | | ± 0.1 dB |
| Dynamic range | | 85 dB |

The transmit filter is required to perform two functions. It must remove the line frequency noise (50/60 Hz) to at least $-20$ dB below the speech signal level. The transmit filter must pass 300–3400 Hz speech signals unattenuated and must provide a rejection of frequencies above 4.6 kHz (the aliasing frequency corresponding to maximum signal frequency of 3.4 kHz and a sampling frequency of 8 kHz) of at least 32 dB. It must also provide a smoothing section to remove the high-frequency noise (both clock and random noise). These signals will otherwise be aliased into the passband by the sampling process of the codec.

Each of the above filters can be realized in a suitable manner. A continuous-time antialiasing filter is required at the input which performs the bandlimiting for the SC filters that follow. It is customary to use active RC Sallen–Key filters with resistors realized by polysilicon regions. The accuracy of these resistors is important to maintain a flat passband response. Using polysilicon resistors, the ratio of the resistors can be accurately maintained even though their absolute values vary. Hence, $Q_p$ can be realized accurately, whereas $\omega_p$ deviates; however, this deviation has marginal effect on the system response, since the cutoff frequency is rather large (above 10 kHz).

For low-sensitivity designs, doubly terminated ladder filters are used as the basis for the design of SC ladder filters. These are used for both transmit and receive LP filters. The transmit LP filter is normally a fifth-order elliptic filter [9.4, 9.15, 9.16]. The 50/60 Hz rejection filter is usually a second-order or third-order Butterworth or Chebyshev HP filter [9.15]. In some designs, a second-order high-pass notch section has been employed [9.4]. The antialiasing and smoothing filters are second-order Sallen–Key active RC filters. The receive filter incorporates $(\sin x)/x$ correction, required to compensate the pass-band droop occurring because of the zero-order hold used at the output of the decoder. Thus, either a fifth-order elliptic filter suitably optimized to realize the required low-pass function as well as $(\sin x)/x$ correction or a sixth-order filter with one real zero, two real poles and two transmission zeros on the imaginary axis [9.62, 9.63] can be used. Alternatively, a sixth-order filter with only two transmission zeros could also be used [9.64].

The choice of clock frequencies for these SC filters is decided by several factors. Most of the early designs were based on LDI transformation, and used non-ideal termination resistances (see Chapters 6 and 7). Hence, certain errors in the frequency response were inevitable. In order to reduce these errors, a large sampling frequency was chosen. However, recent design techniques of SC ladder filters which facilitate *exact* design of bilinear transformation type responses, or LDI transformation type responses, can use even low sampling frequencies, thereby possibly reducing capacitor spread. Some designers have used cascade design of two third-order LP filters as well. These are based on bilinear transformation [9.55]. The choice of large sampling frequencies to implement non-ideal LDI SC ladder filters will result in large capacitor spread for the third-order 50/60 Hz rejection filters. Hence, in some designs, low sampling frequency for the third-order HP rejection filter and high sampling frequency for the fifth-order LP filter have been used. This configuration, however, uses the HP filter after the transmit LP filter and has the possible limitation that the clock feedthrough of the 50/60 Hz rejection SC filter at a relatively low frequency (8 kHz) may be aliased by the codec operating at a rate of 8 kHz. Also the amplitude of 50/60 Hz line frequency components can be large enough to degrade the dynamic range of the overall transmit section by possible overloading of the fifth-order LP filter. Further the broadband noise from the LP and HP filters is aliased into the passband even before sampling by codec. It is thus preferable to filter the low-frequency components as soon as they enter the filter [9.16]. Most designs use one of these two architectures; we will now discuss a few of these designs.

Consider first the architecture due to Black *et al.* [9.16], which has incorporated several techniques to improve noise performance as well as PSRR. This architecture, shown in Figure 9.51(b), employs a post-filter for the transmit section and pre-filter as well as post-filter for the receive filter. Further, the HP filter precedes the LP filter as in the Gray *et al.* design [9.4]. The HP filter is a fourth-order Chebyshev section clocked at 128 kHz. It provides good low-frequency rejection without introducing much noise. The large clock frequency helps to improve PSRR. The receive filter is a fifth-order elliptic LP filter using a clock frequency of 512 kHz. This large clock frequency reduces the foldover noise considerably.

The PSRR for an integrator is dependent on the ratio of total integrating capacitance to the total parasitic capacitance at the virtual ground of the OA, as already mentioned. Parasitic capacitances have been reduced in this design using minimum-size switches and shielding integrating node lines with ground planes whenever possible. Evidently, the size of the integrating capacitor $C_I$ has to be large for large PSRR. Note, however, that the $k\theta/C_I$ noise also increases for small capacitor values $C_I$. Hence, a compromise value of $C_I$ needs to be used.

One of the earliest designs identifying the need for codec and filter on one chip was due to Haque *et al.* [9.15], who have used the architecture where the HP filter follows the LP filter. This realizes transmit filter and codec on one chip and receive filter and decoder on another chip. Each chip contains its own phase-locked loop (PLL) to facilitate clock synchronization, thus permitting separate transmit and receive clocks to be employed.

Ohara *et al.* [9.57] have improved the PSRR by using an on-chip regulator for negative supply voltage. Also, the substrate voltage is decoupled from the a.c. voltages on the negative power supply using an active RC filter. These blocks are shown in Figure 9.52. The latter technique effectively swamps the low-frequency variations in the negative supply voltage. In this design also, the integrator capacitors were chosen large to improve the noise performance. Further, large gate areas were used for the input transistors.

The circuit shown in Figure 9.52(a) realizes a large capacitor while using only a 120 pF capacitor, due to the Miller effect. Thus, a single-section RC filter is realized. The regulator (Figure 9.52(b)) uses a potential divider formed by transistors $Q_1$ and $Q_2$. This reference potential ($-4.3$ V) is decoupled by an RC filter formed by $R$ and $C$. The next stage employs a differential amplifier with single-ended output and a buffer stage.

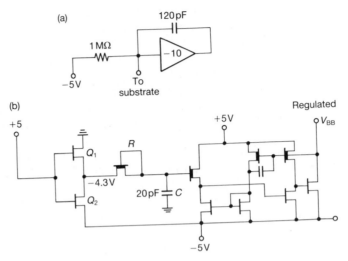

*Fig. 9.52* (a) A substrate decoupling filter; (b) negative supply regulator used for improving PSRR. *(Adapted from [9.57], © 1980 IEEE.)*

Iwata *et al.* [9.59] have described a two-chip system which employs an architecture similar to that in Figure 9.51, but using a second-order HP filter. The chip includes two PLLs as well, based on a 128 kHz voltage controlled multivibrator with an on-chip capacitor, edge-sensitive phase detector and 1/16 frequency divider. These internal clocks can synchronize to the 8 kHz sampling clocks for transmission and reception. Next-generation PCM combo systems have been realized in one chip itself.

Yamakido *et al.* [9.58] have developed a chip incorporating coder, decoder, filters, voltage references and signalling logic circuits. It has one cosine filter in the input before the fifth-order LP filter in the architecture of Figure 9.51. The speech output of the HP filter is fed to the encoder, whose PCM output word has been stored in an output register. This word can be shifted through an open drain output, using seven other chips operating at 2.048 MHz. The receive filter output is fed to an interpolating filter to reduce the sampling frequency noise. An autozero circuit is used to cancel the offset of the complete transmit section. This method is known as *sign bit integration* [9.64]. In order to reduce noise and crosstalk coupling, the transmit and receive sections are completely isolated.

In addition, in this chip, the output PCM bit rate can vary from 64 kHz to 2.048 MHz. The synchronization pulses (8 kHz) provide the codec, the time slot information and can have any width from 100 $\mu$s to 124.9 $\mu$s. The master clock (128 kHz) need not have any phase relationship with sync pulses and clock. A 10 dB improvement in PSRR is achieved by shielding both plates of the capacitors by p-wells. Also, the junction areas of switches were reduced as much as possible. For a frequency difference of 100 Hz, between transmit and receive clocks, the beat frequency and overtones could degrade the performance. Hence, only the first eight PCM pulses are cut out from the input PCM signal by gates close to the wire bounding pads for the transmit and receive clocks. This chip, however, does not employ PLLs. Iwata *et al.* have included these as well on their new-generation chip [9.59].

Gregorian *et al.* [9.60] have designed a chip which employs an architecture similar to that of their previous SC codec filter two-chip system, with slight modifications. This architecture is the same as that of the chip described in [9.58]. The analog LP filter performs bandlimiting to 512 kHz. The cosine filter performs 'decimation'. It is clocked at 512 kHz and allows the SC LP filter following it to be clocked at 256 kHz. More will be said about this decimation in the next section. The SC HP filter is clocked at 64 kHz to reduce the capacitor spread. Note that smoothing filters are not included in this design. The receive filter is clocked at 256 kHz. This chip has bandgap voltage reference for use in encoding/decoding functions. The clocks are derived from an external 1.544 MHz or 2.048 MHz system clock. In this chip, a supply-independent biasing scheme is used to improve the PSRR.

**Table 9.4** Performance details of a typical PCM filter for $\pm5v$ supplies at 25°C *(adapted from [9.8],* © 1980 IEEE).

| | Transmit filter | Receive filter (with $(\sin x)/x$ filter) |
|---|---|---|
| Total power consumption: | | |
| with power amplifier inactive | | 20 mW |
| with power amplifier driving $-20$ dBm into 600 Ω | | 25 mW |
| with power amplifier driving 0 dBm into 600 Ω | | 70 mW |
| Passband ripple (300–3200 Hz) | $\pm0.05$ dB | $\pm0.08$ dB |
| Stopband rejection ($>4600$ Hz) | $-34$ dB | $-42$ dB |
| Delay distortion (1000–2600 Hz) | 35 $\mu$s | 120 $\mu$s |
| 60 Hz rejection | 30 dB | — |
| 50 Hz rejection | 35 dB | — |
| 16 Hz rejection | 60 dB | — |
| Nominal 0 dBmO level at output | 2.26 V peak | 2.26 V peak |
| Total harmonic distortion at $+3$ dBmO, 1 kHz | $-50$ dB | $-50$ dB |
| Idle channel noise, filter only | 6 dBrnCO | 6 dBrnCO |
| Idle channel noise: | | |
| Transmit filter/codec/receive filter, end to end | | 12 dBrnCO |
| 1 kHz, absolute gain variation | $\pm0.1$ dB | $\pm0.1$ dB |
| Supply voltage coefficient of absolute 1 kHz gain | 0.04 dB V$^{-1}$ | 0.04 dB V$^{-1}$ |
| Power supply rejection of $V_{CC}$ at 1 kHz | 40 dB | 45 dB |
| Power supply rejection of $V_{BB}$ at 1 kHz | 30 dB | 35 dB |

Yet another architecture has been described by Senderowicz *et al.* [9.61]. This NMOS design employs fully differential SC circuits. In view of the symmetric nature of the circuits, all the power supply noise will be coupled to both outputs equally, making the theoretical PSRR infinite, if the output is sensed differently. Thus, at least 3 dB improvement is expected because of increases in signal level of 6 dB and in noise level of 3 dB only. The performance details of a typical PCM filter system are presented in Table 9.4 for completeness.

## 9.6.2 Decimation filters [9.65–9.68]

It has been mentioned before that continuous-time antialiasing filters are required at the input of the transmit filter. These are usually realized as active RC Sallen–Key

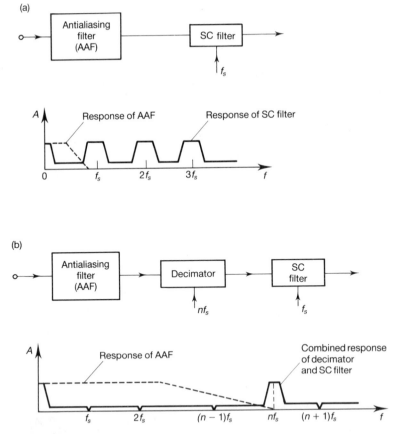

*Fig. 9.53* (a) SC filter antialiasing filter; (b) SC filter with combined antialiasing decimation filter.

unity-gain filters. Assuming a sampling frequency in the SC filter of 128 KHz to realize a fifth-order elliptic response, the antialiasing filter must have a rejection of at least 50 dB for the aliasing components of the speech band 300–3400 Hz. Hence, a second-order active RC Butterworth response for the above requirements requires an $\omega_p$ of 7.2 kHz and has an attenuation of 0.18 dB for 3.4 kHz. However, assume that the input is clocked at 512 kHz. Then the required $\omega_p$ of the antialiasing filter to realize an attenuation of 50 dB for the aliasing components of the speech band is 28.6 kHz. Further, the attenuation at the passband edge is 0.0068 dB, considerably less than that obtained using a sampling frequency of 128 kHz. Since the aliasing frequency in the modified case is very large, a simpler antialiasing filter can be used.

The proposed method can be appreciated by referring to Figure 9.53. The filter inserted between the antialiasing filter and the SC filter is called a *decimation filter* and introduces zeros at the clock rate. A circuit realization for this purpose is shown in Figure 9.54(a) [9.67–9.68]. The transfer function of this circuit is found to be (with the output measured just before short circuiting the feedback capacitor):

$$H(z) = \frac{V_o(z)}{V_i(z)} = \frac{C_1}{C_2} \left[ 1 + z^{-1/n} + z^{-2/n} + z^{-3/n} + \cdots + z^{-(n-1)/n} \right] z^{-1/(2n)}$$

$$= \frac{C_1}{C_2} \left[ \frac{1 - z^{-1}}{1 - z^{-1/n}} \right] \cdot z^{-1/(2n)} \qquad (9.68a)$$

Thus, the amplitude response is given by

$$|H(j\omega)| = \frac{C_1}{C_2} \frac{\sin\left(\dfrac{\omega T}{2}\right)}{\sin\left(\dfrac{\omega T}{2n}\right)} \qquad (9.68b)$$

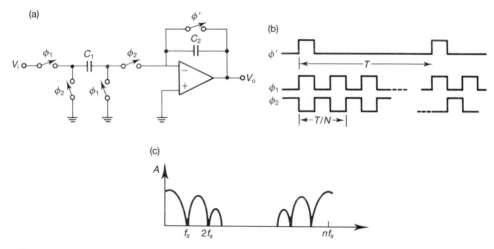

Fig. 9.54   (a) An SC decimator circuit;   (b) timing waveforms;   (c) frequency response. *(Adapted from [9.67], © 1981 IEE.)*

We first note that $|H(0)|$, the d.c. gain, is $(nC_1/C_2)$. Thus, to keep the d.c. gain unity, we set $nC_1 = C_2$. The frequency response of this decimation filter is shown in Figure 9.54(c). Due to the finite rolloff of the decimation filter, there is a slight error in the overall transmission at 3.4 kHz. Specifically, for a clock frequency of 512 kHz, it is 0.008 37 dB. Thus, the overall attenuation at 3.4 kHz amounts to 0.009 dB lower than the design with no decimation filter. The case for $n = 2$ can be simplified without necessitating additional clocking phases. This circuit, due to Gregorian [9.69], is shown in Figure 9.55(a). Note that the realized transfer function is

$$H(z) = -(1 + z^{-1/2}) \frac{C_1}{C_2} \qquad (9.69)$$

Eriksson [9.70] has suggested an improved decimator which realizes a 'squared cosine' response.

The circuit of Figure 9.55(b), which is only parasitic-compensated, realizes a transfer function given by

$$H(z) = \frac{Kz^{-1/2}(1 + az^{-1/2} + bz^{-1})}{(1 - z^{-1})} \qquad (9.70)$$

where

$$K = \frac{C_1}{C_5}$$

$$a = \frac{C_2}{C_1}$$

$$b = \frac{C_3}{C_1} = \frac{C_4}{2C_1}$$

The choice $a = 2$ and $b = 1$ gives a squared cosine filter in the sense that

$$H(j\omega) = 2K \frac{\cos^2(\omega T/8)}{\sin(\omega T/8)}$$

The decimation technique *decreases* the sampling rate. Another technique to *increase* the sampling rate exists and is known as *interpolation*. This is discussed in the next subsection.

## 9.6.3  Interpolation filters

Consider a signal sampled at a rate of $f_s/r$, which is intended to interpolate linearly between two sampling instants. Assuming that $r$ steps are required (see Figure 9.56(a))

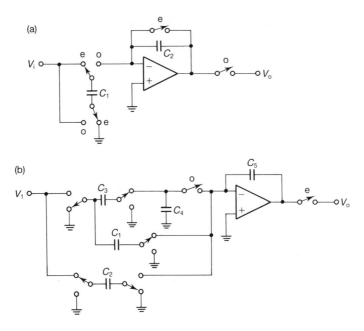

Fig. 9.55 (a) Decimation circuit ($n = 2$) due to Gregorian and Nicholson *(adapted from [9.69], © 1980 IEEE);* (b) Eriksson's decimator *(adapted from [9.70], © 1985 IEE).*

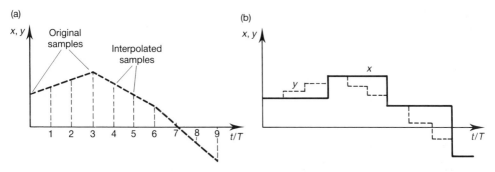

Fig. 9.56 Linear interpolation: (a) impulse-sampled signal; (b) sampled and held signal.

between the samples $x(nT) \triangleq x_n$ and $x((n-r)T)$, the transfer function of the interpolation filter is derived as follows.

The original samples are available at the time instants $x((n-r)T)$, $x(nT)$, $x((n+r)T)$, $x((n+2r)T)$, and so on. We can now define a sequence $\{u(nT)\} \triangleq \{u_n\}$ such that

$$\{u_n\} = \sum_{k=0}^{r-1} x_{(n-k)} \qquad (9.71)$$

This sequence can be seen to correspond to a sampled and held version of the original sampled signal $x(nT)$, when we note that the samples only exist at $x(nT)$ and $x((n \pm r)T)$, and that at all the remaining instants the samples are zero. It is intended to generate an output sequence $\{y(nT)\} \triangleq \{y_n\}$ such that

$$y_n - y_{n-1} = \frac{u_n - u_{n-r}}{r} = \frac{1}{r} \left\{ \sum_{k=0}^{r-1} x_{n-k} - \sum_{k=0}^{r-1} x_{n-k-r} \right\} \qquad (9.72)$$

(a)

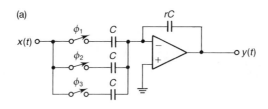

(b)

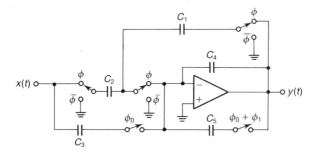

(c)

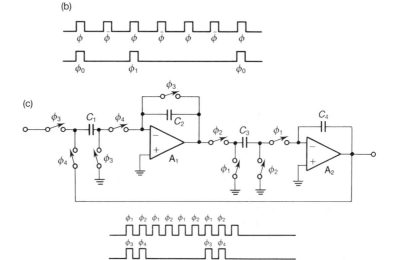

*Fig. 9.57* SC interpolators: (a) with no d.c. feedback; (b) with d.c. feedback; (c) two-OA structure with d.c. feedback. *(Adapted from [9.71], © 1981 IEE.)*

Taking $z$-transforms of equation (9.72), we obtain

$$H(z) = \frac{Y(z)}{X(z)} = \frac{1}{r} \left( \frac{1 - z^{-r}}{1 - z^{-1}} \right)^2 \tag{9.73}$$

We have thus derived the transfer function of the interpolation filter. It is possible to realize $H(z)$ as a cascade network of two FIR filters with all the tap weights chosen as equal. In the following, a few circuits will be described to realize the desired transfer function, where we assume that $x(t)$ is already a sampled and held signal, as shown in Figure 9.56(b). Hence, only $(1 - z^{-r})/(1 - z^{-1})$ needs to be realized [9.71].

A simple realization is shown in Figure 9.57(a). It can be seen that at each clock phase, a voltage step of $-\frac{1}{3}(x_n - x_{n-3})$ is generated when $r = 3$. It is possible to extend the above design for arbitrary $r$. One limitation of this circuit, however, is that no d.c. feedback is present. An alternative circuit without this drawback is shown in Figure 9.57(b). In this circuit requiring a complex clocking waveform, at each phase the input and feedback capacitors are adjusted such that

$$y_n - y_{n-1} = - \frac{(x_n - x_{n-r})}{3} \tag{9.74}$$

The reader is urged to verify this point.

An alternative two OA realization is shown in Figure 9.57(c). In this circuit, the voltage, $x_n - x_{n-r}$, is sampled and held on $C_2$ and equal increments of voltage $(x_n - x_{n-r})/r$ are obtained at the output of OA $A_2$ by choosing $C_2C_4 = C_1C_3r$. This realization requires a multiphase clock and uses two OAs. However, the input voltage need not be a sampled, or sampled and held signal, since $H(z)$ for this circuit is given by equation (9.73).

We conclude this section with a comment on the use of interpolation filters. They are especially useful for interconnecting SC networks operating at different clock frequencies. Interpolation circuits were first described by Gregorian and Nicholson [9.69], but they are not completely stray-insensitive and also are only described for $r = 2$.

## 9.6.4 Dual-tone multiple-frequency (DTMF) receivers [9.53, 9.56]

The basic function of this chip is to detect which button on the push-button telephone has been pressed. Each button, when depressed, generates a pair of frequencies, one each from two groups of frequencies called high-group and low-group. Hence, several filters are required to detect these frequencies. Two filters (known as *bandsplit filters*) are required for separating the low- and high-groups and eight filters are needed to detect the eight transmitted frequencies depending on the button pressed.

Two sixth-order Chebyshev band-elimination filters are required for bandsplit filters. Each could be realized by a cascade of three second-order notch filters. The eight BP filters required to select the different frequencies could be second-order stray-insensitive realizations. Thus, 28 poles are to be realized in addition to the 60 Hz reject and pre-emphasis functions usually required. The block diagram of a DTMF receiver is given in Figure 9.58. For details regarding a fully integrated DTMF receiver built on a single chip, the reader is referred to [9.56]. This application demonstrates the complexity of the functions that can be realized on a single IC using SC techniques.

## 9.6.5 MOS A/D and D/A converters

It is now common to include MOS A/D converters in the class of SC networks. Using exclusively switches, capacitors, comparators and OAs, we can realize successive-approximation type D/A and A/D converters and algorithmic A/D and D/A converters. First a typical A/D converter for five bits is described in brief [9.72–9.73].

The objective here is to obtain a digital word approximating an input voltage level. Three steps are required to achieve the desired function. Consider a binary weighted capacitor array, as shown in Figure 9.59. In the first step, the charging mode, all the capacitors are charged to the input voltage $V_{in}$ to be *encoded*, by connecting the top plates to ground and the bottom plates to $V_{in}$. In the second step, called the 'hold' mode, the top 'grounding' switch is opened and the bottom plates are grounded. Evidently, all the top plates are now at potential $-V_{in}$. The

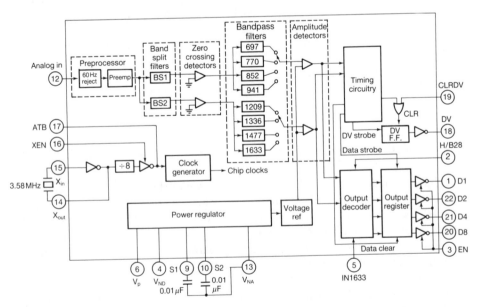

Fig. 9.58  Block diagram of a DTMF receiver *(adapted from [9.56],* © *1979 IEEE).*

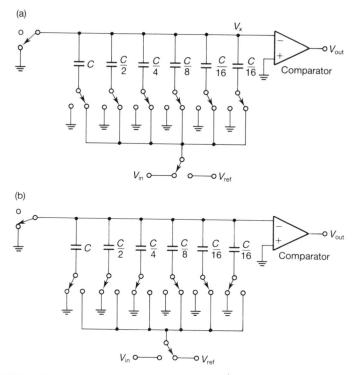

Fig. 9.59  MOS analog-to-digital conversion:  (a) a 5-bit A/D converter;  (b) final configuration for the output 00011. *(Adapted from [9.72], © 1975 IEEE.)*

comparator output indicates the sign bit. The third step is called the 'redistribution' mode, in which the bottom plate of the largest capacitor is raised to $V_{ref}$. Consequently, there is a voltage division between $C$ and all the other capacitors whose total capacitance is $C$. (It must be noted that the last $C/16$ capacitor is additional.) Thus, by this operation, the voltage at the top plate becomes

$$V_x = -V_{in} + \frac{V_{ref}}{2} \qquad\qquad (9.75)$$

If $V_x > 0$, then the comparator output is a logic 0 and if $V_x < 0$, the comparator output is a logic 1. Thus, the output of the comparator, the most significant bit (MSB) in this case is the value of the binary bit $b_0$ being tested. If MSB is zero, since $V_{in} < V_{ref}/2$, the switch associated with $C$ is returned to ground, while it will stay at $V_{ref}$ if MSB $= 1$. The procedure is next repeated for the other bits $b_1 b_2$, etc. (During these steps, the added voltages are $V_{ref}/4$, $V_{ref}/8$, etc.) Also, note that when a bit is zero, that particular capacitor is discharged. Figure 9.59(b) illustrates the final switch positions for the output 00011. It must be clear that $N$ redistributions are required to achieve an $N$-bit resolution.

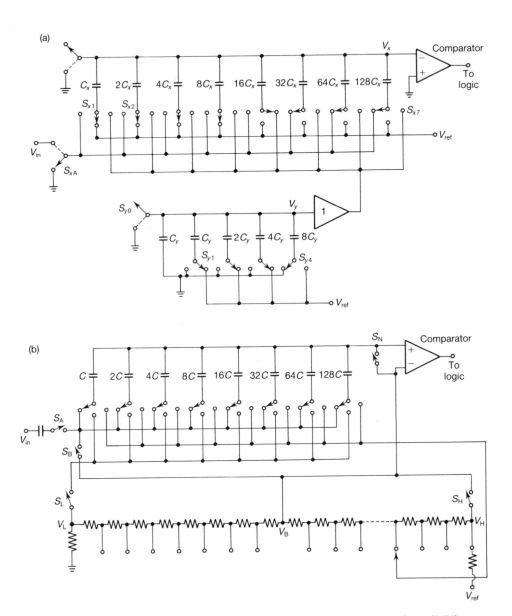

Fig. 9.60 Non-linear A/D converters using: (a) capacitor arrays *(adapted from [9.74],* © 1976 IEEE); (b) capacitor array and resistor network *(adapted from [9.52],* © 1979 IEEE).

The coding of a speech sample using non-linear coders such as PCM requires two operations: determination of the segment number and the step number in the segment. The coding process is controlled by a successive approximation register (SAR) and digital logic aided by a comparator to halt the conversion. The two operations mentioned above can be realized using two approaches. In one approach, shown in Figure 9.60(a), two SC arrays are utilized, while in another, a resistor string is employed together with one SC array (Figure 9.60(b)). For details, the reader is referred to [9.74] and [9.52] respectively.

In both these approaches, the weighted capacitor array requires capacitors ranging from $C$ to $2^N C$. Hence, for a 12-bit A/D converter, often required in digital filtering applications, the capacitor spread required is very large. The reduction of required capacitor spread can be accomplished in two ways. The method due to Ohri and Callahan [9.64] is presented in Figure 9.61. Note the manner in which two weighted capacitor arrays with a capacitor spread of 64 are used to realize a 13-bit D/A converter. The output of this converter can be expressed as

$$V_0 = \frac{V_{ref}}{128} \left[ \sum_{n=1}^{7} W_n C_n + \sum_{n=8}^{13} W_n \frac{C_n}{64} \right] \qquad (9.76)$$

In an alternative technique due to Singh et al. [9.75, 9.76], the classical $R-2R$ ladder technique is used as the starting point to synthesize a $C-2C$ ladder SC network. The advantage is the requirement of only two capacitor values with a spread of 2. This design is presented in Figure 9.62. The effect of parasitic capacitances can be reduced by using silicon gate technology. Singh et al. have extended this approach to realize a 13-bit non-linear A/D converter [9.76].

Algorithmic A/D converters, also known as *cycle* or *recirculating* A/D converters, operate on a different principle. The block diagram of a typical algorithmic converter is shown in Figure 9.63(a). It consists of an analog signal loop which contains a sample and hold amplifier, a multiply-by-two amplifier, a comparator and a reference subtraction circuit. The principle of coding is as follows. The input is first sampled and held and amplified by 2. Next, the amplified output is compared with a reference $V_{ref}$. Depending on the decision of the comparator, if $2V_{in} > V_{ref}$,

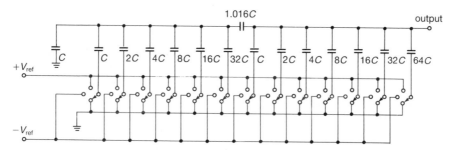

Fig. 9.61 Ohri and Callahan's 13-bit D/A converter *(adapted from [9.64], © 1979 IEEE)*.

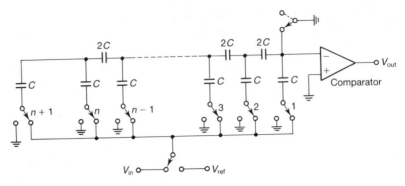

Fig. 9.62  Singh et al.'s linear A/D converter *(adapted from [9.75], © 1982 IEE).*

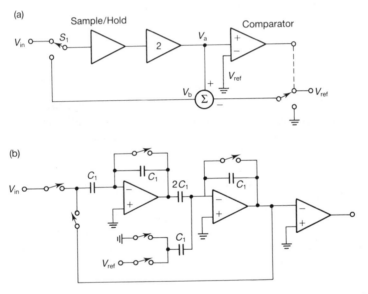

Fig. 9.63  (a) Block diagram of an algorithmic A/D converter;  (b) Li *et al.* implementation *(adapted from [9.77], © 1984 IEEE).*

the bit is set; otherwise it is made zero. The resultant signal (the difference between $2V_{in}$ and $V_{ref}$ in this case) is recirculated through the same process. Note that, this time, $V_{in}$ is disconnected (switch $S_1$ is in the lower position). Several realizations for these operations have been described.

The realization due to Li *et al.* [9.77] is shown in Figure 9.63(b), which uses matched capacitors. However, in this stray-insensitive realization, since the OA gains are realized by capacitor ratios, errors can be contributed by the inaccuracies in the matching of the capacitors. Additionally, offset voltages in the loop also contribute errors. Li *et al.* have shown that these gain errors can be eliminated at the expense of more clock cycles, by realizing gains independent of capacitor ratios. The clock-feedthrough effects can be eliminated by using fully differential topologies.

Several variations of the converter schemes are possible. A multiplying D/A converter (MDAC) [9.78] is shown in Figure 9.64. In this circuit, an analog input can be multiplied with a digital word. The digital word is used to control switches $S_1, S_2, \ldots, S_N$. The capacitors $C_1, C_2, \ldots, C_N$ are weighted in a binary fashion, i.e., $C_1 = C_0, C_2 = 2C_0, C_3 = 2^2 C_0, \ldots, C_N = 2^{N-1} C_0$. Thus a binary word applied to the gates of switches $S_1$–$S_N$ closes those switches having 1 as the control input. During $\phi_1$ the respective capacitors are charged to the value of the analog multiplicand and also the feedback capacitor $C_f$ is discharged. In $\phi_2$, the input is grounded so that the charges on all the capacitors $C_1$–$C_N$ are transferred to $C_f$. Thus the output of the OA is the multiplied value of the analog multiplicand and the digital multiplier. A four-quadrant multiplier can be built by adding another input capacitor $C_x = 2^{N-1} C_0$ and with $C_f = 2^{N-1} C_0$. This added capacitor $C_x$ always switches between ground on $\phi_1$ and input signal on $\phi_2$. Thus this works in anti-phase with the other switches. Any charge introduced by $C_N$ is cancelled by that due to $C_x$, when both are switched on. Thus, if MSB is 1, i.e., $S_N$ is on, the total contribution due to $C_N$ and $C_x$ is zero. If MSB is 0, i.e., $S_N$ is off, the total contribution due to $C_x$ and $C_N$ is $-1$. The weighted input due to the other switches and capacitors is summed to this value. This type of DAC is called an offset binary four-quadrant MDAC. When the analog input voltage in Figure 9.64 is a fixed voltage $V_{ref}$, the circuit realizes a D/A converter [9.79]. The circuit of Figure 9.64 can be used as a delay line, if all the capacitors are of equal value and the switches $S_1$ to $S_N$ are selected in succession, one at a time, and change state before the beginning of $\phi_2$.

## 9.6.6 Digital multipliers [9.80]

It is possible to multiply two digital words as well. Each word is fed to a capacitor array (binary weighted) similar to the previously described arrangements. The digital multiplication can be explained with reference to Figure 9.65. Let $x$ and $y$ be the two digital words to be multiplied. Each sets the capacitances $C_x$ and $C_y$. Two steps

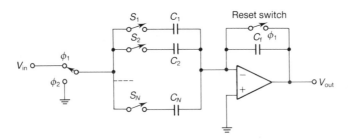

*Fig. 9.64* A multiplying D/A converter *(adapted from [9.78],* © 1983 IEEE).

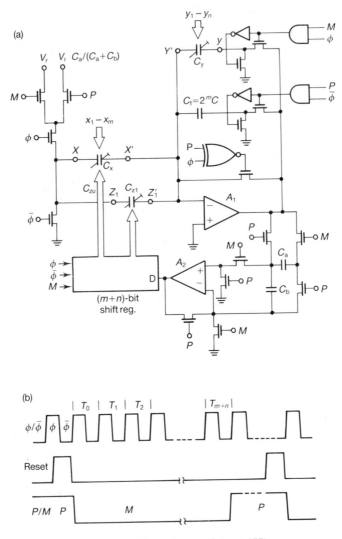

Fig. 9.65  A digital multiplier (adapted from [9.80], © 1983 IEE).

are required to perform the digital multiplication. In the first step, called the 'preset' step $(P)$, the array $C_x$, feedback capacitor $C_f$ and OA $A_1$ form a D/A converter, as discussed before. Thus, the multiplicand is converted to an analog voltage which is sampled at the output of $A_1$ and stored on $C_a$ in parallel with $C_b$. The OA $A_2$ acts as a buffer in this step, with its offset voltage stored on $C_a$ and $C_b$. In the next step, called the 'multiplication' step $(M)$, a D/A converter is realized using arrays $C_z$, $C_y$ and OA $A_1$. The capacitors $C_a$ and $C_b$ are connected now in series and hold the charge stored in the previous state. The OA $A_2$ is a comparator in this mode.

Note that $C_f$ does not come into the picture in this mode. Next, $m + n$ clock pulses are used to determine the product in the shift register $z$. Each $z_i$ bit is set to 1, storing the previous values of $z_i$. The resulting word programming the $C_z$ array, together with $C_y$, decides the output of $A_1$. The voltage across $C_b$ is now the weighted sum of the present output of $A_1$, $-V_r C_z/C_y$ and the output in the previous $P$ state. Evidently, when $V_b > 0$, $z_i$ is kept as 1, otherwise $z_i$ is zero. Thus, $m + n$ clock cycles result in making $V_b$ zero and then the $m + n$ shift register value is the product of $x$ and $y$ register values. Thus digital multiplication can be implemented using logic gates and SC networks.

## 9.6.7  SC FIR filters using delta modulation

FIR filters can be implemented using SC delays and summer blocks. Alternatively, a delta modulator can be utilized to realize finite impulse responses. This method, due to Peled and Liu [9.81], avoids the need for the multiplication required in computing the convolution sum:

$$y(nT) = \sum_{m=0}^{N-1} h_m x((n-m)T) \tag{9.77}$$

where $x(nT)$ and $y(nT)$ are the input and output sequences, and $\{h_m\}$ is the impulse response coefficients of the FIR filter. Consider next the block diagram of Figure 9.66(a). The delta modulator (DM) output can be written as

$$x(nT) = x((n-1)T) + \Delta x b(nT) \tag{9.78}$$

where $\Delta x$ is the DM step size. We also observe from Figure 9.66(a) that

$$v(nT) = v((n-1)T) + \sum_{m=0}^{N-1} h'_m b((n-m)T) \tag{9.79}$$

Substituting equation (9.78) in equation (9.79) and noting that $v(nT) = 0$ for negative values of $n$, we obtain

$$v(nT) = \sum_{m=0}^{N-1} \frac{h'_m}{\Delta x} x((n-m)T) \tag{9.80}$$

Hence, for $v(nT)$ to be equal to $y(nT)$, we need

$$h'_m = \Delta x h_m \tag{9.81}$$

Thus, for given impulse response coefficients, it is easy to compute $h'_m$. The SC realization of this block diagram is as shown in Figure 9.66(b). The switches $P_o$ and $Q_o$ are turned on and off so as to feed a positive or negative incremental charge depending on the bits $b_n$. The circuit around OA $A_1$ is a first-order offset-compensated LP network. The delta modulator also can be realized using SC techniques shown in Figure 9.66(c). A first-order LP SC network is used for the loop

(a)

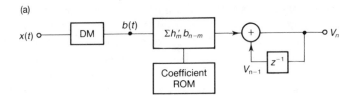

(b)

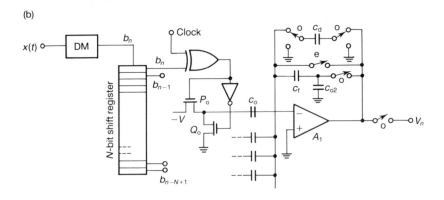

(c)

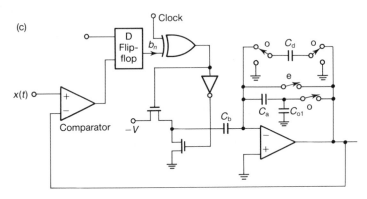

*Fig. 9.66* (a) FIR filter based on delta modulation; (b) implementation of (a); (c) implementation of DM. *(Adapted from [9.82],* © 1984 IEEE.)

filtering function in this delta modulator. For details, the reader is referred to Reddy and Swamy [9.82].

## 9.6.8  SC phase-locked loop [9.83]

A phase-locked loop is shown in Figure 9.67(a). It is clear that a voltage-controlled oscillator (VCO), an LP filter and comparator are required. The phase

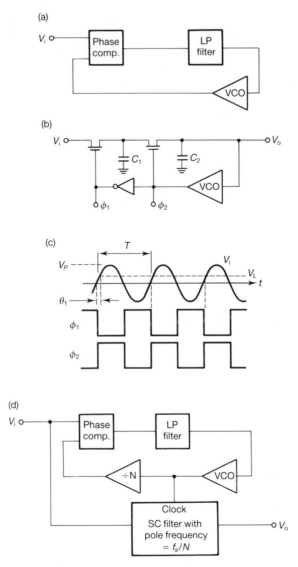

*Fig. 9.67* (a) Block diagram of a phase-locked loop (PLL); (b) SC implementation of (a); (c) waveforms when in lock; (d) modification to realize a tracking filter. (b) (c) (d). *(Adapted from [9.83], © 1981 IEEE.)*

detector detects the phase difference between the input signal and the VCO output, and the LP filtered output of the comparator (a d.c. signal) is used to adjust the VCO so that the VCO locks to the input frequency. Note that a small phase difference (called a static phase error) must exist between the input signal and the VCO output so that a finite d.c. voltage always exists to be able to change the VCO frequency from its free running value to the frequency of the input signal (see

Figure 9.67(c)). In SC techniques, the LP filter and the comparator can be combined together as illustrated in Figure 9.67(b). The input signal value $V_i$ is stored on $C_1$ in $\phi_1$. In $\phi_2$, the voltage on $C_2$ is made to approach that on $C_1$. Eventually, the filter output will settle to a d.c. value $V_o$. This voltage will be the value required to cause the VCO frequency to equal the input frequency.

Assuming that the input frequency is increased slowly, the amplitude of the samples across $C_1$ will increase thereby increasing the d.c. output voltage $V_o$. Thus the VCO frequency increases as desired. Note also that the static phase error increases in the process, the maximum limit being $\pi/2$. The phase-locked loop of Figure 9.67(b) can be used to design a tracking filter by employing the VCO output as the clock frequency of the tracking filter realized in the SC technique as shown in Figure 9.67(d).

## 9.6.9  A synchronous detector [9.83]

A synchronous detector or balanced modulator multiplies one input signal by the sign of another input signal. A modification of a first-order SC LP network can be used for this purpose (Figure 9.68). By choosing $\phi_a = \phi_1$, $\phi_b = \phi_2$, an inverting LP function is realizable, while with $\phi_a = \phi_2$, $\phi_b = \phi_1$, a non-inverting function is realized. Thus, alternating between these two modes results in an alternate multiplication of the signal by $+1$ or $-1$, before applying to the LP filter.

## 9.6.10  A discrete Fourier transformer [9.84]

Prior to the advent of SC filter technology, economical discrete Fourier transformers were realized using charge coupled devices and chirp $z$-transform algorithms [9.85]. However, the use of SC techniques to realize the same functions leads to certain advantages.

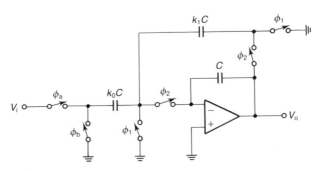

Fig. 9.68  An SC balanced modulator (adapted from [9.83], © 1981 IEEE).

The discrete Fourier transform (DFT) of a sequence $\{x(n)\}$ is defined as

$$X(k) = \sum_{n=0}^{N-1} x(n) e^{(-j2\pi/N)nk} \tag{9.82}$$

with $k = 0, 1, \ldots, N-1$. Each DFT point, i.e., the value of $X(k)$ for each $k$, can be seen to be the output of a transversal filter with tap weights $h_k(n)$, where

$$h_k(n) = e^{(-j2\pi/N)nk}, \qquad 0 \leqslant n \leqslant N-1 \tag{9.83}$$

The transfer function of the $k$th transversal filter is

$$H_k(z) = \sum_{n=0}^{N-1} h_k(n) z^{-n} = \frac{(1-z^{-N})}{(1-z^{-1}e^{(-j2\pi/N)k})} \tag{9.84}$$

The $H_k(z)$ in equation (9.84) can be alternatively expressed for various $k$s as follows:

$$H_0(z) = \sum_{n=0}^{N-1} z^{-n} = \sum_{n=0}^{N-1} M_{0,n} z^{-n} \tag{9.85a}$$

$$H_{N/2}(z) = \sum_{n=0}^{N-1} (-1)^n z^{-n} = \sum_{n=0}^{N-1} M_{N/2,n} z^{-n} \qquad \text{if } N \text{ is even} \tag{9.85b}$$

$$S_k(z) = \frac{H_k(z) + H_{N-k}(z)}{2} = \sum_{n=0}^{N-1} \cos\left(\frac{2\pi nk}{N}\right) z^{-n} = \sum_{n=0}^{N-1} M_{k,n} z^{-n} \tag{9.85c}$$

$$D_k(z) = \frac{H_k(z) - H_{N-k}(z)}{2} = \sum_{n=0}^{N-1} -\sin\left(\frac{2\pi nk}{N}\right) z^{-n} = \sum_{n=0}^{N-1} M_{N-k,n} z^{-n} \tag{9.85d}$$

where

$$k = \begin{cases} 1, 2, \ldots, (N/2) - 1, & \text{if } N \text{ is even} \\ 1, 2, \ldots, (N-1)/2, & \text{if } N \text{ is odd} \end{cases}$$

Note that $H_k(z)$ and $H_{N-k}(z)$ are combined together to make all the coefficients of the various transversal filters real. The real parts of the DFT points are given by $S_k(z)$, while the imaginary parts are given by $D_k(z)$. Thus, the DFT of the sequence $\{x(n)\}$ can be realized as a bank of transversal filters with tap weights as defined in equations (9.85). As an illustration, the block diagram of a four-point DFT and its SC version are shown in Figure 9.69. In this version, stray-insensitive delays due to Enomoto et al. [4.25] and three-phase clocking are utilized. Positive or negative tap weights can be realized by choosing the switching schemes appropriately for the input capacitors. The number of OAs required to implement an $N$-point DFT is $2N-1$. Note also that to avoid data truncation effects, windows can be used, and these multiplying coefficients can be included in the tap weights. For details, the reader is referred to Reddy and Swamy [9.84].

The above example demonstrates the capability of SC networks to perform several multiplications and additions in real time at a fast rate.

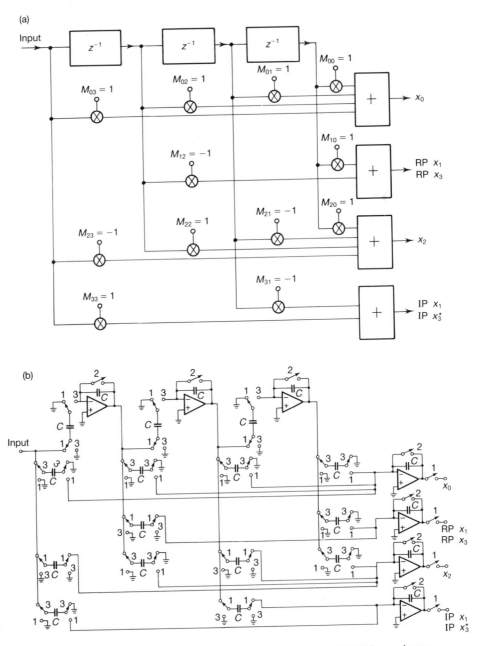

Fig. 9.69 (a) Frequency sampling approach to compute 4-point DFT (RP = real part, IP = imaginary part and $x^*$ = complex conjugate of $x$); (b) SC implementation of (a) (adapted from [9.84], © 1983 IEEE).

## 9.6.11  SC oscillators

An advantage of SC oscillators is that the frequency of oscillation can be varied by changing the clock frequency. A number of oscillator circuits have been discussed in Chapter 3. The two-integrator-loop biquads due to Martin and Sedra as well as Fleischer and Laker when $F$ and $E$ capacitors are deleted, are other possible oscillators.

Huertas *et al.* have proposed an SC oscillator circuit (Figure 9.70(a)) using unity-gain amplifiers [9.86]. Two non-inverting LDI integrators based on the trans-conductance concept are connected in a feedback loop. It can be shown that the circuit oscillates at a frequency given by

$$f_o = \frac{f_s}{2\pi} \cos^{-1} \left[ 1 - \frac{1}{2\left\{1 + \frac{C_1}{C_4}\left(1 + \frac{C_2}{C_3}\right)\right\}} \right] \tag{9.86}$$

and the condition for oscillation is not dependent on the capacitor ratios or values.

Colbeck has described a modification of the Fleischer–Laker SC oscillator (biquad with $E = 0$, $F = 0$) which does not stall at start [9.87]. This circuit, shown in Figure 9.70(b), uses positive feedback around OA $A_2$. The series capacitor $C_s$ is normally not required. However, its presence serves to attenuate the effect of positive feedback and reduce the capacitor ratios required. The transfer function due to the presence of $C_s$ can be derived to show that this circuit is a third-order network. The use of the on/off switch across the output and the input of OA $A_2$ makes the output of OA $A_2$ initially equal its offset voltage. Thus, the OA $A_1$ will saturate to one of the supply voltages. The circuit is thus in the maximum energy state and ready to oscillate when the on/off switch is open. The capacitors $C_{1B}$, $C_{2B}$ serve to control the frequency of oscillation.

We next discuss another modification of the Fleischer-Laker biquad with $F = 0$, which realizes an SC oscillator [9.88]. The starting point for this realization is the $E$-circuit (Figure 9.71(a)) designed for very large $Q_p$. The output $V_1$ can be fed to a comparator to yield a square wave, which will be in phase with an external square wave input fed at point $N_1$. Hence, the external input can be disconnected and the output of the comparator fed instead at $N_1$. Thus this circuit performs as an oscil-lator. The amplitude of oscillation, however, is not controllable since it is a function of the limits of saturation of the comparator. The modification shown in Figure 9.71(b) overcomes this limitation. In this circuit, the comparator is replaced by a hard limiter. An on-chip reference voltage $V_{ref}$ is employed which injects alter-nately a positive or negative incremental charge to sustain the oscillations. The sample and hold at the input of OA $A_3$ avoids a continuous feedback path when $\bar{X}$ is high. This circuit also requires 'start-up' circuitry as shown in Figure 9.71(b). For this purpose, a positive feedback loop around OA $A_1$ has been added through $C_2$ and $C_3$. Once the oscillations have built up, $M_1$ conducts, thus grounding the midpoint of $C_2$ and $C_3$ and removing them effectively from the circuit.

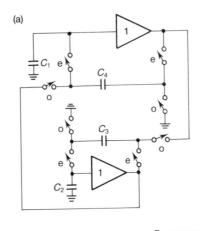

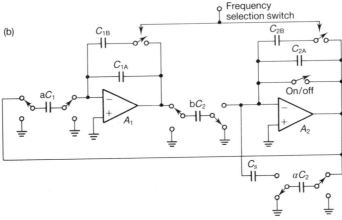

Fig. 9.70 (a) Huertas et al.'s SC oscillator (adapted from [9.86], © 1984 IEEE); (b) SC oscillator due to Colbeck (adapted from [9.87], © 1984 IEEE).

In the above paragraphs, SC sinusoidal oscillators have been discussed. A relaxation oscillator using SC techniques has been described by Martin [9.89] which we will consider next (Figure 9.72(a)). In this circuit, OA $A_1$ and $C$ act as an integrator, while OA $A_2$ acts as a comparator. Assume that at some instant, the output $V_1$ of the comparator is at the negative supply voltage $V_{ss}$; then $\alpha_2 C$ injects a charge periodically into $C$ so that $V_o$ decreases in steps of $\alpha_2 V_{ss}$. When $V_o$ becomes negative, the comparator changes state and switches to the supply voltage $V_{DD}$. Then, due to the coupling capacitor $\alpha_1 C$, a negative step is caused in $V_o$ of amplitude $\alpha_1(V_{DD} - V_{ss})$. Subsequently, $\alpha_2 C$ feeds negative charge packets in steps of $\alpha_2 V_{DD}$ until the comparator changes state once again. The cycle is thus repeated with an oscillation frequency given by $f_{osc} \simeq (\alpha_2/4\alpha_1)f_s$. The resulting waveforms are shown in Figure 9.72(b). The circuit can be modified to realize a VCO, as shown in Figure 9.72(c). The frequency increases with a positive control voltage and decreases

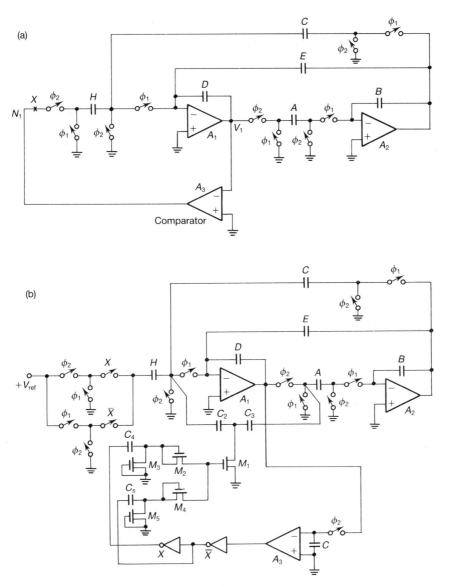

*Fig. 9.71* Fleischer–Ganesan–Laker SC oscillators: (a) basic circuit; (b) modifications to ensure start-up and to limit the amplitude of oscillations. *(Adapted from [9.88], © 1985 IEEE.)*

with a negative control voltage and is given by

$$f_{osc} \simeq \frac{\alpha_2}{4\alpha_1} f_s + \frac{\alpha_0}{4\alpha_1} \frac{V_{in}}{V_{DD}} f_s \qquad (9.87)$$

for $\alpha_0, \alpha_2 \ll \alpha_1$ and $V_{DD} = -V_{ss}$. Such VCOs are useful, e.g., in phase-locked loops.

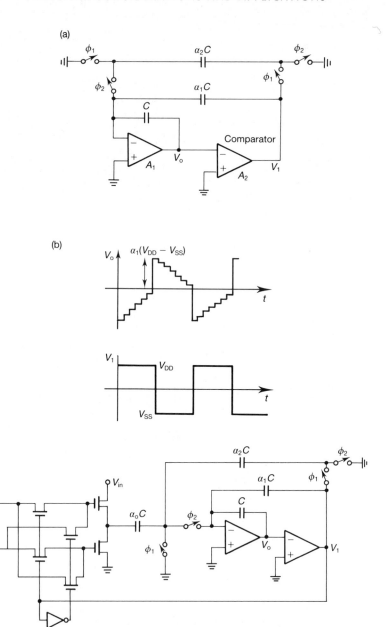

*Fig. 9.72* (a) SC relaxation oscillator due to Martin; (b) waveforms; (c) VCO based on (b). *(Adapted from [9.89], © 1981 IEEE.)*

## 9.6.12  SC modems

The modem is a key component in the world of data communications. Several low-bit-rate modems have been built using SC techniques. In this section, we briefly study the candidates among the sub-systems required to implement a modem system. Modems can be of frequency-shift keying (FSK) or quadrature phase-shift keying (QPSK) type. A typical FSK modem architecture is shown in Figure 9.73 [9.90]. Note that other architectures are also possible [9.91–9.93].

In the architecture proposed in [9.90], the transmit BP is a sixth-order SC filter while a second-order active RC Sallen–Key filter smoothes the output of this SC network. In the receive part, the BP filter is a fifteenth- or sixteenth-order SC network, incorporating amplitude and group delay equalization. This filter is made up of four blocks. The first block is a passive first-order RC section cascaded by a second-order SC elliptic filter. The second block is a third-order elliptic LP filter. The third block realizes a fourth-order delay equalizer, while the fourth block is a fifth-order elliptic HP filter. The demodulator also employs a fifth-order LP filter.

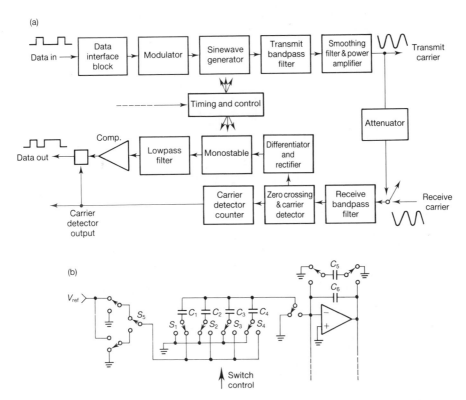

Fig. 9.73  (a) Typical FSK modem architecture;  (b) sine-wave generator using a sine-weighted capacitor array *(adapted from [9.90]*, © 1984 IEEE).

Further, the sine-wave generator uses a capacitor array controlled by a ROM, as shown in Figure 9.73(b). The four capacitors in the array are 'sine-weighted' and the reference is switched depending on the polarity of the sine wave form as required.

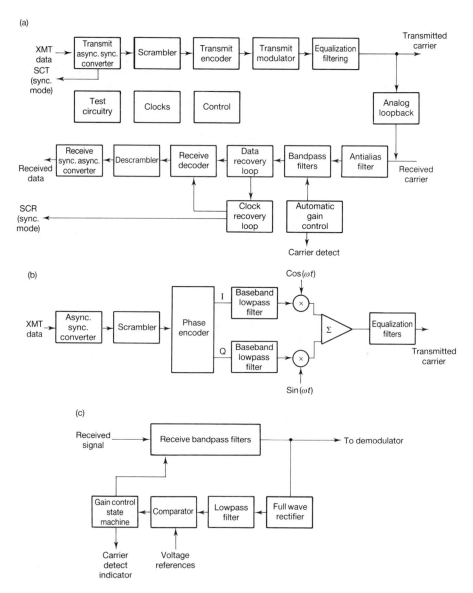

Fig. 9.74  (a) Block diagram of QPSK full duplex modem;  (b) differential QPSK transmitter;  (c) AGC circuit;  (d) coherent demodulator for QPSK;  (e) clock-recovery phase-locked loop  *(adapted from [9.93], © 1984 IEEE).*

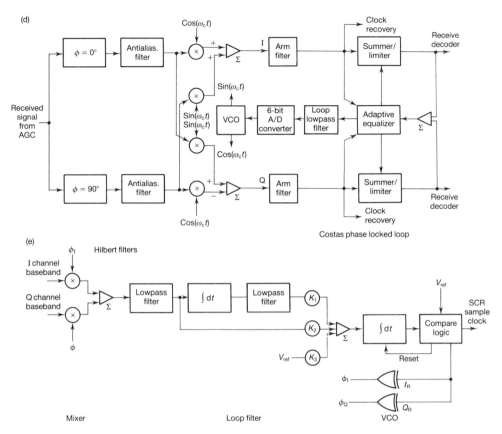

Fig. 9.74  (continued)

A QPSK modem architecture due to Hanson *et al.* [9.93] is presented in Figure 9.74(a). Note that this circuit has utilized only SC technology. Other sub-system architectures are shown in Figures 9.74(b)–9.74(e), to illustrate the complexity that can be realized in MOS technology.

Hosticka *et al.* [9.94] have described another FSK modem architecture using CMOS technology. The demodulator in their architecture employs PLLs, unlike the other modems where filters have been employed. A block diagram of the PLL used in this architecture is shown in Figure 9.75(a). In the SC phase comparator block of the PLL, depending on the reference square wave logic levels, the input is routed either directly or after inversion to the output. The VCO employed in the PLL is as shown in Figure 9.75(b). It employs a Schmitt trigger realized by OAs $A_2$ and $A_3$ together with switched capacitors, and an integrator realized using OA $A_1$. The control input to the VCO is added or subtracted from the Schmitt trigger output and integrated. When the Schmitt trigger output is low, the clock phases are A = $\bar{\phi}$ and B = $\phi$, while A = $\phi$ and B = $\bar{\phi}$ when the output is high. Depending on the output level of the Schmitt trigger, the integrator will be either inverting or non-inverting.

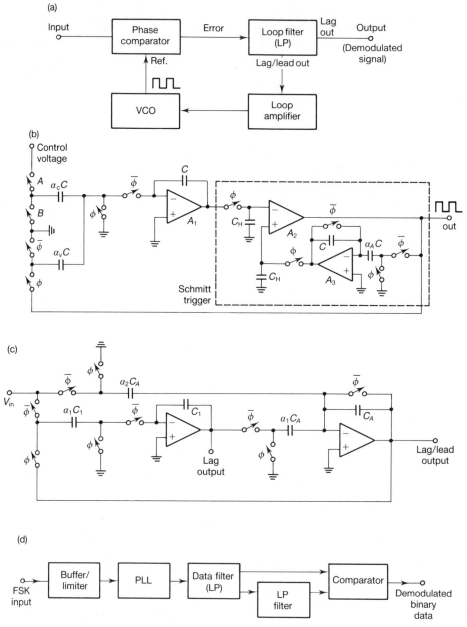

**Fig. 9.75**  (a) A block diagram of the PLL;  (b) an SC VCO;  (c) SC loop filter;  (d) FSK demodulator block diagram. *(Adapted from [9.94], © 1984 IEEE.)*

The loop filter of the PLL is as shown in Figure 9.75(c), which realizes a pole and a zero. The complete demodulator block diagram is presented in Figure 9.75(d). In order to reduce the power consumption, dynamic OAs have been employed throughout the realization of the FSK modem.

## 9.6.13  SC networks for speech analysis and synthesis

In the above sections, applications of SC networks for speech coding using non-linear/linear A/D converters and decoding have been considered. Other applications, such as audio spectrum analyzers, adaptive delta modulation (ADM) codecs, continuously variable slope delta modulation (CVSD) codecs, linear predictive coding (LPC) analyzers and synthesizers have also made extensive use of SC techniques. These will be briefly discussed next.

### SC speech spectrum analyzers

Lin *et al.* [9.95] have described an audio spectrum analyzer using exclusively SC techniques. This contains 16 second-order BP filters followed by rectifiers. The smoothed (third-order LP filtered) output of the rectifier is coded and the coded outputs of all the 16 paths are multiplexed to facilitate storage in a computer for speech recognition purposes. Bui *et al.* [9.96] have described a similar system having only seven channels, but employing fourth-order SC BP filters for speech filtering in each band. While these designs use separate SC filter for each channel, Kuraishi *et al.* [9.97] have described a 20-channel multiplexed spectrum analyzer. Each channel uses a sixth-order BP filter, rectifier and an eighth-order LP filter. The architecture of this chip is shown in Figure 9.76, in order to illustrate the complexity achievable in the technology.

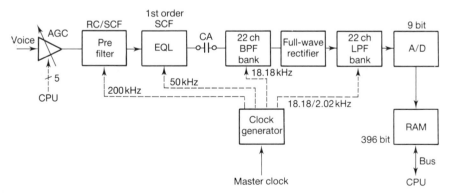

*Fig. 9.76*  Architecture of a 22-channel monolithic audio spectrum analyzer *(adapted from [9.97],* © *1984 IEEE).*

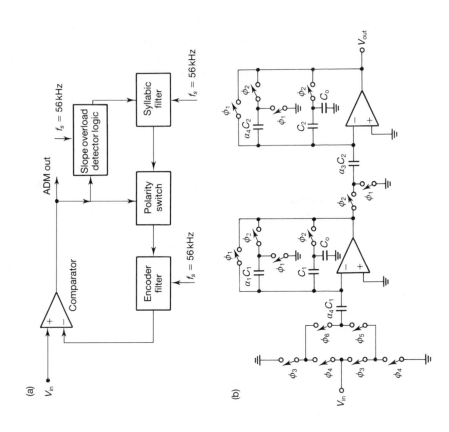

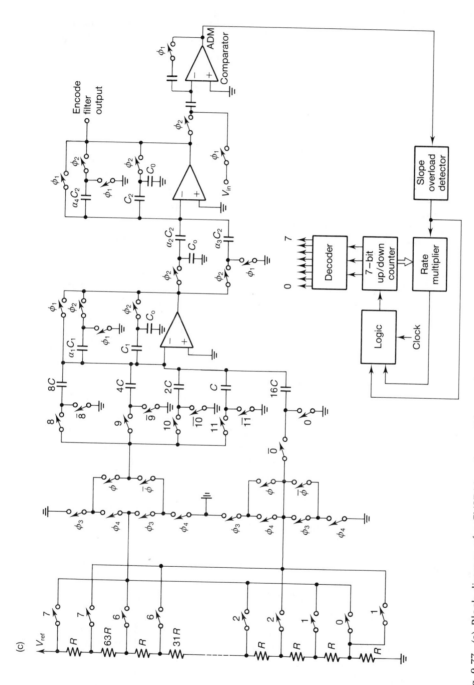

Fig. 9.77 (a) Block diagram of a CVSDM system; (b) encoder filter with polarity switch; (c) complete schematic representation of the CVSD modulator.

The prefilter uses a tenth-order SC LP filter preceded by a second-order active RC section. A first-order SC equalizer follows the prefilter, which provides pre-emphasis. The frequency range covered by the 20-channel BP filters is 350–7200 Hz. Full wave rectifiers and eighth-order LP filters have been used in each channel. A nine-bit A/D converter follows the LP filters. The LP filters have programmable cutoff frequencies too. The total filter order is 308 for the complete speech spectrum analyzer.

## SC adaptive delta modulators

Irie *et al.* [9.98] have described a single-chip ADM LSI codec, in which SC techniques have been used to realize the two-pole integrator for the coder and decoder sections, D/A converters and voltage shifters. The adaptation algorithm has been realized using digital logic circuits. Gregorian and Gord [9.99] have described a CVSDM chip. The block diagram of a CVSDM system is shown in Figure 9.77(a). The comparator output reflects the sign of the difference between the predicted and the input signals. The comparator output also actuates the polarity switch and applies a step with appropriate sign to the predictor. The output stream is examined in the slope overload detector to find three consecutive 1s or 0s and accordingly increases the step size. A first-order filter is sufficient in the encoder filter for simplicity, while a second-order predictor improves the SNR by about 4 dB. The syllabic filter is a two-pole filter realized as a cascade of two single-pole sections. The offset-compensated first-order SC networks due to Gregorian [9.33, 9.100] have been extensively used in these designs.

The encoder filter with polarity switch is shown in Figure 9.77(b). The switches clocked by $\phi_5$, $\phi_6$ perform the polarity selection; $\phi_5 = $ '0' and $\phi_6 = $ '1' make the first stage an inverting one, while $\phi_5 = $ '1' and $\phi_6 = $ '0' make it non-inverting. The clock waveforms are suitably chosen to eliminate clock-feedthrough effects. The syllabic filter used has digital implementation, using an MOS D/A converter, as shown in Figure 9.77(c). The up/down counter has been arranged such that it counts up one when the output of the slope overload detector is enabled and counts down at a rate proportional to its own current values. The resistor and capacitor arrays perform seven-bit non-linear conversion as already described.

## SC LPC analyzers

One method of speech coding [9.79] consists of modelling the vocal tract as a time-varying all-pole filter with an excitation which may be a periodic pulse train or a random noise source. The former source is called 'pitch generator'. This model is presented in Figure 9.78(a), in which the characteristics of the vocal tract are assumed to be stationary for a period of few milliseconds. The analysis of speech data in this period, called a 'frame', consists of determining the inverse of this all-pole filter. It is an all-zero filter and can be realized as a lattice filter structure, as

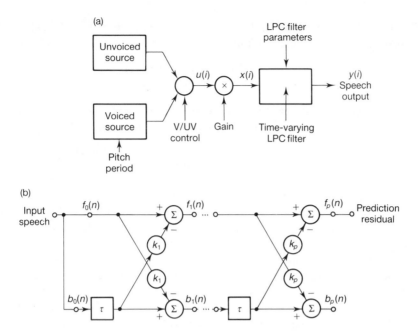

*Fig. 9.78*  (a) Speech synthesizer block diagram;  (b) LPC analysis filter.

shown in Figure 9.78(b). The lattice model fits the physical behaviour of the vocal tract as explained next.

The vocal tract can be assumed to be made up of about ten cylindrical tubes with varying cross-sections, through which air is pumped by the lungs. The varying cross-section of the adjacent tubes gives rise to forward and backward waves at every junction between adjacent tubes. Using the continuity of pressure and conservation of air mass, we can obtain [9.101]

$$f_m(n) = f_{m-1}(n) + k_m b_{m-1}(n-1)$$

$$b_m(n) = b_{m-1}(n) + k_m f_{m-1}(n)$$

(9.88)

These equations can be implemented by the lattice structure shown in Figure 9.78(b). In equations (9.88), $f$ and $b$ stand for forward and backward (reflected) waves, whereas $k_m$ denotes the reflection coefficient between tubes $m$ and $m-1$. The coefficients $k_m$ are determined by minimizing the error criterion,

$$Q = E[f_m^2(n) + b_m^2(n)]$$

where $E$ is the expectation operator. This analysis process, depicted in Figure 9.78(b), intends to 'whiten' the speech spectrum. The $f_m$ of the last stage is rich in pitch information and can be used to extract the pitch. The evaluation of $k_m$ mentioned above, obtained by minimizing $Q$, leads to the following equation:

$$k_m = \frac{-2E[f_{m-1}(n)b_{m-1}(n-1)]}{E[f^2_{m-1}(n)] + E[b^2_{m-1}(n-1)]} \tag{9.89}$$

The block which computes $k_m$ is called a 'correlator'. The expectation evaluation can be carried out using a first-order recursive filter with a transfer function given by

$$\frac{V_o(z)}{V_i(z)} = \frac{1/64}{1 - \frac{63}{64}z^{-1}} \tag{9.90}$$

This LP filter has a cutoff frequency of 20.2 Hz with a sampling frequency of 8 kHz. The division of the expectations in equation (9.89) can be carried out using look-up tables of logarithms and ROMs. Thus this portion of the correlator can be realized using digital logic. The implementation of equation (9.90) can be realized conveniently using SC techniques.

The operations of addition, subtraction, multiplication and delay are required in the lattice analysis filter. The reader will already have noted the various circuit arrangements possible to realize these functions. Specifically, the multiplication of an analog signal by a digital word can be realized using the multiplying D/A converter (MDAC). In their CMOS circuit, Fellman and Brodersen [9.78] have extensively used multiplexing techniques for reducing the number of OAs, while implementing this LPC analysis block.

## SC LPC synthesizer

Gregorian and Amir [9.79] have described a single-chip speech synthesizer which generates a speech signal from the LPC coefficients and pitch information. The block diagram of the chip is shown in Figure 9.79(a). A 10-pole lattice filter mode used for speech synthesis is shown in Figure 9.79(b). Note in particular the direction of data flow in the top and bottom branches in the block diagram. The SC implementation of the block diagram in Figure 9.79(b) is shown in Figure 9.79(c). The input to this block is a pulse train at the rate of pitch frequency or a pseudo-random bit generator. This input is multiplied by the gain in an MDAC whose output is fed to the synthesis filter.

The synthesis filter output samples every 125 $\mu$s. This period is divided into 20 cycles, where two cycles cater for all the computations required at each stage of the LPC filter. At the end of the twentieth cycle, the $b$-values are updated and stored in the '$b$-stack'. The various blocks in the SC implementation are, respectively, the arithmetic unit, a signal inverter controlled by the sign bit of $k_i$, the $b$-stack already mentioned, and forward prediction error sample and hold. The reader is urged to study the timing sequence required to implement the above algorithm. The speech output of the LPC synthesis filter is fed to a smoothing third-order elliptic SC LP filter, which is followed by a power amplifier.

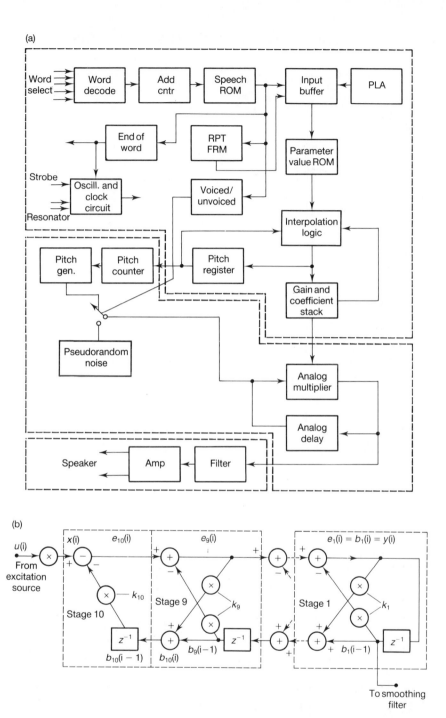

(a)

(b)

Fig. 9.79  (a) Block diagram of a single-chip LPC speech synthesizer;  (b) a 10-pole lattice filter model used for speech synthesis;  (c) SC implementation of (b).  *(Adapted from [9.79], © 1983 IEEE.)*

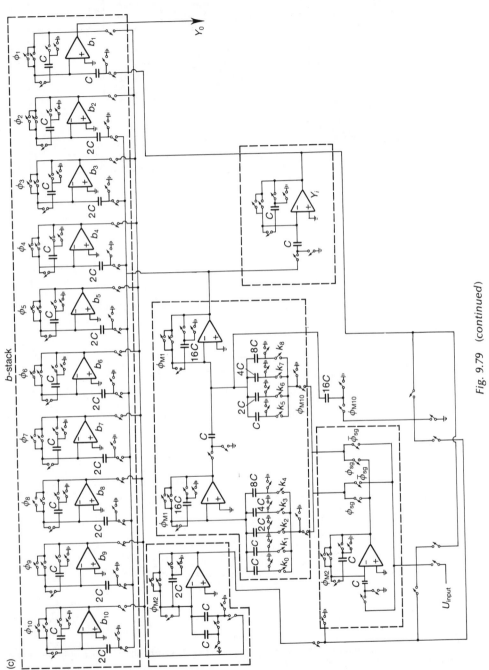

Fig. 9.79 (continued)

## 9.6.14 SC adaptive equalizers

Communications at high bit rates need correction for the imperfections introduced by the communication channel such as cables, and that introduced by bridge taps, which are spare T branches that may be left unconnected at the remote end thus generating reflection (i.e., echo). This correction is accomplished by using $\sqrt{f}$ equalizers and bridge-tap equalizers (known conventionally as BT equalizers). The former arises because of skin effects of the conductors carrying data signals. SC techniques have attracted considerable attention for these applications.

A typical adaptive equalizer for use in digital communications networks can be realized as shown in Figure 9.80(a). Several architectures described in the literature have realized either $\sqrt{f}$ equalizers or BT equalizers or both using SC techniques, a few of which we will briefly consider in what follows.

The $\sqrt{f}$ equalizer compensates for a line loss of typically up to 42 dB at 100 kHz (Nyquist frequency for 200 Kbit s$^{-1}$ data). This equalization can be done in coarse and fine steps. In the design due to Ishikawa *et al.* [9.102], the coarse as well as the fine automatic gain control (AGC) circuits are each realized by two first-order SC networks (Figure 9.80(b)). The coarse AGC circuit adjusts the gain by 3.0–4.0 dB per step, while the fine AGC circuit adjusts the gain at Nyquist frequency by 0.3–0.4 dB per step. The filter coefficients are adjusted by using programmable SCs for all the eight capacitors. The peak voltage of the equalized pulses is used to adapt the coefficients by comparing it with a reference voltage. The circuit is designed to

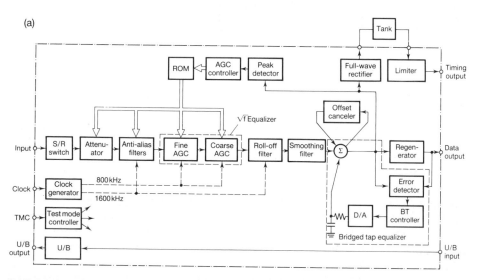

*Fig. 9.80* (a) Block diagram of an adaptive line equalizer; (b) SC $\sqrt{f}$ equalizer implementation; (c) block diagram of a bridged-tap equalizer (heavy lines represent the SC portion) (a) and (b) *(adapted from [9.102], © 1984 IEEE)* (c) *(adapted from [9.103], © 1984 IEEE).*

(b)

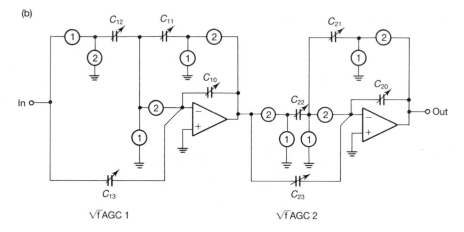

√f̄ AGC 1          √f̄ AGC 2

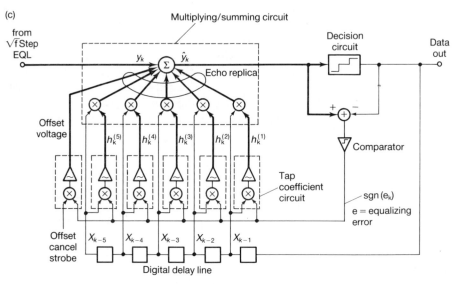

*Fig. 9.80  (continued)*

work at a clock frequency of 800 kHz. A rolloff filter follows the fine AGC circuit, which is a fourth-order LP filter. A prefilter is used to antialias the subsequent $\sqrt{f}$ equalizer sections. This prefilter employs an active RC filter, a decimator (cosine filter) and a second-order LP filter. The BT equalizer, however, uses digital logic exclusively in this design.

Takatori *et al.* [9.103] have also used SC techniques to realize a $\sqrt{f}$ equalizer and a part of the BT equalizer. A time-domain decision feedback equalizer has been used to implement the latter. It has an all-pole transfer function, the realization of which is shown in Figure 9.80(c). Only the convolution summing part is realized using a simple single-amplifier SC network. The $\sqrt{f}$ equalizer is a third-order SC network which consists of a rolloff function as well. For details regarding the realization of the BT and $\sqrt{f}$ equalizer design, the reader is referred to [9.103, 9.104].

Nakayama *et al.* [9.105] have studied in detail the adaptation algorithm for an SC line equalizer. Their equalizer uses a coarse $\sqrt{f}$ equalizer employing an *E*-type Fleischer–Laker biquad with all capacitors replaced by programmable arrays. The fine $\sqrt{f}$ equalizer is a first-order SC LP network. The rolloff filter is a fourth-order SC LP network. The prefiltering and smoothing filters are active RC filters.

## 9.7 Conclusions

We conclude this chapter on the practical considerations and applications of SC filters with a few remarks. The application of switches and capacitors to realize analog or digital or both analog and digital functions can be judiciously used according to the ingenuity of the designer. Several such attempts have been described in this chapter. It is hoped that in future, jobs other than filtering may continue to receive close practical attention. Some technological challenges still remain to be pursued, for example, the realization of frequency division multiplexing (FDM) filters, which have to meet stringent requirements regarding sensitivity to temperature, power supply voltages, noise, etc. Using improved design techniques and eliminating problems associated with OAs, such as limited slew rate, it is hoped that most analog processing functions will in future be easily implemented in CMOS technology. We can expect to see hybrid processors which can deal with analog or digital signals and do any operation on either digital or analog data. SC networks fabricated in GaAs technology have been reported [9.106]. As such, it is hoped that complex systems not yet realizable in monolithic form may in future be implemented. The progress made in sub-micron technology in the digital circuit area may well be used to advantage in the design of these complex hybrid processors.

## 9.8 Exercises

9.1 Using the small signal equivalent circuit, derive the expressions for differential voltage gain and common-mode gain for an NMOS differential amplifier using (a) enhancement loads and (b) depletion loads. Derive the conditions for maximum gain.

9.2 Using the small signal equivalent circuit, determine the output impedance of the output stage comprising transistors $T_{23}-T_{26}$ in Figure 9.4.

9.3 Evaluate the gain enhancement resulting in an inverter with transconductance boosted by additional transistors as in Figure 9.5 (see the circuit comprising $T_{10}-T_{14}$).

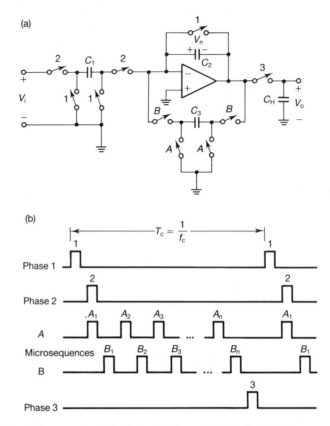

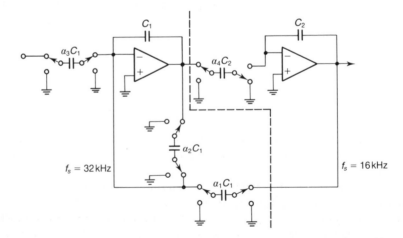

Fig. 9.81 SC circuit for exercise 9.5 *(adapted from [9.107], © 1980 IEE).*

Fig. 9.82 Schematic of a multisampling SC LP filter *(adapted from [9.95], © 1983 IEEE).*

9.4 Derive the expressions for amplitude and phase errors of stray-insensitive SC integrators of non-inverting and inverting types, taking into account the switch resistances and OA finite bandwidth.

9.5 In some applications, using a small capacitor ratio and large sampling frequency, a desired time constant has to be realized. In such applications, the time

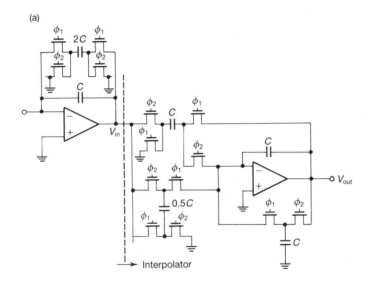

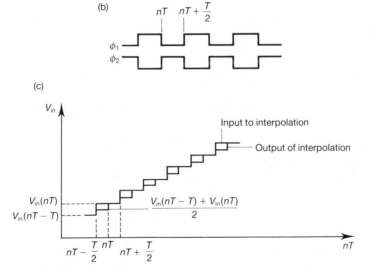

*Fig. 9.83* A circuit for interpolation by 2 *(adapted from [9.69],* © 1980 IEEE).

constant $\tau = KC/f_cC_R$ can be realized instead of $Cf_c/C_R$, where $K \geqslant 1$. A circuit for realizing such a function is shown in Figure 9.81. Derive the transfer function of this circuit and compare it with a conventional integrator [9.107].

9.6 The use of different sampling rates in the first and second integrator circuits in a two-integrator loop results in a third-order function. Thus one OA can be saved by this technique, as shown in Figure 9.82. Derive the transfer function of this circuit [9.95].

9.7 Gregorian and Nicholson [9.69] have described a circuit for interpolation by 2. This circuit is presented in Figure 9.83. Analyze the circuit and compare it with the realization based on the Ghaderi *et al.* technique [9.71].

## 9.9  References

[9.1] P.W. Fry, A MOST integrated differential amplifier, *IEEE Journal of Solid-State Circuits*, SC-4, 3, 166–8, June 1969.

[9.2] Y.P. Tsividis and P.R. Gray, An integrated NMOS operational amplifier with internal compensation, *IEEE Journal of Solid-State Circuits*, SC-11, 6, 748–754, December 1976.

[9.3] D. Senderowicz, D.A. Hodges and P.R. Gray, High-performance NMOS operational amplifier, *IEEE Journal of Solid-State Circuits*, SC-13, 6, 760–6, December 1978.

[9.4] P.R. Gray, D. Senderowicz, H. Ohara and B.M. Warren, A single chip NMOS dual channel filter for PCM telephony applications, *IEEE Journal of Solid-State Circuits*, SC-14, 6, 981–91, December 1979.

[9.5] I.A. Young, A high-performance all-enhancement NMOS operational amplifier, *IEEE Journal of Solid-State Circuits*, SC-14, 6, 1070–7, December 1979.

[9.6] E. Toy, An NMOS operational amplifier, *IEEE International Solid-State Circuits Conference Digest of Technical Papers, Philadelphia*, 134–5, 1979.

[9.7] Y.P. Tsividis, D.L. Fraser and J.E. Dziak, A process insensitive high-performance NMOS operational amplifier, *IEEE Journal of Solid-State Circuits*, SC-15, 6, 921–8, December 1980.

[9.8] I.A. Young, A low-power NMOS transmit/receive IC filter for PCM telephony, *IEEE Journal of Solid-State Circuits*, SC-15, 6, 997–1005, December 1980.

[9.9] K. Hsieh, P.R. Gray, D. Senderowicz and D. Messerschmitt, A low-noise chopper-stabilized differential switched-capacitor filtering technique, *IEEE Journal of Solid-State Circuits*, SC-16, 6, 708–15, December 1981.

[9.10] D. Senderowicz and J.M. Huggins, A low-noise NMOS operational amplifier, *IEEE Journal of Solid-State Circuits*, SC-17, 6, 999–1008, December 1982.

[9.11] G. Goto, S. Fujii, T. Nakamura, T. Tsuda and S. Baba, An NMOS operational amplifier for an output buffer of analog LSI, *IEEE Journal of Solid-State Circuits*, SC-17, 5, 948–51, October 1982.

[9.12] P.V. Calzalori, G. Masetti, C. Turchetti and M. Severi, Integrated NMOS stage with low output impedance, *Electronics Letters*, 17, 6, 218–19, 19 March 1981.

[9.13] G. Smarandoiu, D.A. Hodges, P.R. Gray and G.F. Landsburg, CMOS pulse-code-modulation voice codec, *IEEE Journal of Solid-State Circuits*, SC-13, 4, 504–10, August 1978.

[9.14] W. Steinhagen and W.L. Engl, Design of integrated analog CMOS circuits – Multi-channel telemetry transmitter, *IEEE Journal of Solid-State Circuits*, SC-13, 6, 799–805, December 1978.

[9.15] Y.A. Haque, R. Gregorian, R.W. Blasco, R.A. Mao and W.E. Nicholson, Jr, A two chip PCM voice codec with filters, *IEEE Journal of Solid-State Circuits*, SC-14, 6, 961–9, December 1979.

[9.16] W.C. Black, Jr, D.J. Allstot and R.A. Reed, A high performance low-power CMOS channel filter, *IEEE Journal of Solid-State Circuits*, SC-15, 6, 929–38, December 1980.

[9.17] V.R. Saari, Low-power high-drive CMOS operational amplifiers, *IEEE Journal of Solid-State Circuits*, SC-18, 1, 121–7, February 1983.

[9.18] D.G. Marsh, B.K. Ahuja, T. Misawa, M.R. Dwarakanath, P.E. Fleischer and V.R. Saari, A single-chip CMOS PCM codec with filters, *IEEE Journal of Solid-State Circuits*, SC-16, 4, 308–15, August 1981.

[9.19] R.D. Jolly and R.H. McCharles, A low-noise amplifier for switched-capacitor filters, *IEEE Journal of Solid-State Circuits*, SC-17, 6, 1192–4, December 1982.

[9.20] J. Bertails, Low frequency noise considerations for MOS amplifier design, *IEEE Journal of Solid-State Circuits*, SC-14, 4, 773–6, August 1979.

[9.21] T.C. Choi, R.T. Kaneshiro, R.W. Brodersen, P.R. Gray, W.B. Jett and M.W. Wilcox, High-frequency CMOS switched-capacitor filters for communications applications, *IEEE Journal of Solid-State Circuits*, SC-18, 6, 652–64, December 1983.

[9.22] F. Krummenacher and J.L. Zufferey, High gain CMOS cascode operational amplifier, *Electronics Letters*, 16, 6, 232–3, 13 March 1980.

[9.23] F. Krummenacher, High gain CMOS OTA for micropower SC filters, *Electronics Letters*, 17, 4, 160–2, 19 February 1981.

[9.24] F. Krummenacher, Micropower SC biquadratic cell, *IEEE Journal of Solid-State Circuits*, SC-17, 3, 507–12, June 1982.

[9.25] A. Budak, *Passive and Active Network Analysis and Synthesis*, Houghton Mifflin, 1974.

[9.26] M.A. Copeland and J.M. Rabaey, Dynamic amplifier for MOS technology, *Electronics Letters*, 15, 10, 301–2, 10 May 1979.

[9.27] B.J. Hosticka, Novel dynamic CMOS amplifier for switched-capacitor integrators, *Electronics Letters*, 15, 17, 532–3, 16 August 1979.

[9.28] B.J. Hosticka, Dynamic amplifiers in MOS technology, *Electronics Letters*, 15, 25, 819–20, 6 December 1979.

[9.29] M.G. Degrauwe, J. Rijmenants, E.A. Vittoz and H.J. Deman, Adaptive biasing CMOS amplifiers, *IEEE Journal of Solid-State Circuits*, SC-17, 3, 522–8, June 1982.

[9.30] Y.P. Tsividis, Design considerations in single-channel MOS analog integrated circuits – a tutorial, *IEEE Journal of Solid-State Circuits*, SC-13, 3, 383–91, June 1978.

[9.31] P.R. Gray and R.G. Meyer, MOS operational amplifier design – a tutorial review, *IEEE Journal of Solid State Circuits*, SC-17, 6, 969–82, December 1982.

[9.32] P.R. Gray, Basic MOS operational amplifier design – an overview, *Analog MOS Integrated Circuits*, IEEE Press, 1980.

[9.33] R. Gregorian and S.C. Fan, Offset free high-resolution D/A converter, *Proc. 14th Asilomar Conference on Circuits, Systems and Computers*, 316–19, November 1980.

[9.34] M.H. White, D.R. Lampe, F.C. Blaha and I.M. Mack, Characterization of surface channel CCD image arrays at low light levels, *IEEE Journal of Solid-State Circuits*, SC-9, 1, 1–13, February 1974.

[9.35] K.K.K. Lam and M.A. Copeland, Noise cancelling switched-capacitor (SC) filtering technique, *Electronics Letters*, 19, 20, 810–11, 29 September 1983.

[9.36] G.C. Temes and K. Haug, Improved offset-compensation schemes for switched-capacitor circuits, *Electronics Letters*, 20, 12, 508–9, 7 June 1984.

[9.37] K. Watanabe and K. Fujiwara, Offset-compensated switched-capacitor circuits, *Electronics Letters*, 20, 19, 780–1, 13 September 1984.

[9.38] S. Erikson and K. Chen, Offset-compensated switched-capacitor leapfrog filters, *Electronics Letters*, 20, 18, 731–3, 30 August 1984.

[9.39] R.C. Yen and P.R. Gray, A MOS switched-capacitor instrumentation amplifier, *IEEE Journal of Solid-State Circuits*, SC-17, 6, 1008–13, December 1982.

[9.40] K. Martin, New clock feedthrough cancellation technique for analogue MOS switched-capacitor circuits, *Electronics Letters*, 18, 1, 39–40, 7 January 1982.

[9.41] J.H. Fischer, Noise sources and calculation techniques for switched-capacitor filters, *IEEE Journal of Solid-State Circuits*, SC-17, 4, 742–52, August 1982.

[9.42] B. Furrer and W. Guggenbuhl, Noise analysis of a switched-capacitor biquad, *AEU*, 37, 35–40, January–February 1983.

[9.43] H.W. Bode, *Network Analysis and Feedback Amplifier Design*, Van Nostrand, 1945.

[9.44] R. Plodeck, U.W. Brugger, D.C. von Grunigen and G.S. Moschytz, SCANAL – a program for the computer aided analysis of switched-capacitor networks, *Proc. IEE, Electronics Circuits and Systems*, Part G, 128, 277–85, December 1981.

[9.45] J. Lau and J.I. Sewell, Inclusion of amplifier finite gain and bandwidth in analysis of switched-capacitor filters, *Electronics Letters*, 16, 12, 462–3, 5 June 1980.

[9.46] G. Fischer and G.S. Moschytz, On the frequency limitations of SC filters, *IEEE Journal of Solid-State Circuits*, SC-19, 4, 510–18, August 1984.

[9.47] K. Nagaraj, K. Singhal, T.R. Viswanathan and J. Vlach, Reduction of finite gain effects in switched-capacitor filters, *Electronics Letters*, 21, 5, 644–6, 18 July 1985.

[9.48] G.C. Temes, Finite amplifier gain and bandwidth effects in switched-capacitor filters, *IEEE Journal of Solid-State Circuits*, SC-15, 3, 358–61, June 1980.

[9.49] K. Martin and A.S. Sedra, Effect of the opamp finite gain and bandwidth on the performance of switched-capacitor filters, *IEEE Transactions on Circuits and Systems*, CAS-28, 8, 822–9, August 1981.

[9.50] E. Sanchez-Sinencio, J. Silva-Martinez and R. Alba-Flores, Effects of finite operational amplifier gain-bandwidth product on a switched-capacitor filter, *Electronics Letters*, 17, 14, 509–10, 9 July 1981.

[9.51] R.L. Geiger and E. Sanchez-Sinencio, Operational amplifier gain-bandwidth product effects on the performance of switched-capacitor networks, *IEEE Transactions on Circuits and Systems*, CAS-29, 2, 96–106, February 1982.

[9.52] J.T. Caves, C.H. Chan, S.D. Rosenbaum, L.P. Sellars and J.B. Terry, A PCM voice codec with on-chip filters, *IEEE Journal of Solid-State Circuits*, SC-14, 1, 65–73, February 1979.

[9.53] M.J. Callahan, Jr, Integrated DTMF receiver, *IEEE Journal of Solid-State Circuits*, SC-14, 1, 85–90, February 1979.

[9.54] P.J. Schwarz, V. Blatt and J. Fox, An integrated multichannel approach to PCM, *IEEE Journal of Solid-State Circuits*, SC-14, 6, 953–60, December 1979.

[9.55] R. Gregorian and W.E. Nicholson, Jr, CMOS switched-capacitor filters for a PCM voice codec, *IEEE Journal of Solid-State Circuits*, SC-14, 6, 970–80, December 1979.

[9.56] B.J. White, G.M. Jacobs and G.F. Landsburg, A monolithic dual tone multifrequency receiver, *IEEE Journal of Solid-State Circuits*, SC-14, 6, 991–7, December 1979.

[9.57] H. Ohara, P.R. Gray, W.M. Baxter, C.F. Rahim and J.L. McCreary, A precision low power PCM channel filter with on-chip power supply regulation, *IEEE Journal of Solid-State Circuits*, SC-15, 6, 1005–13, December 1980.

[9.58] K. Yamakido, T. Suzuki, H. Shirasu, M. Tanaka, K. Yasumari, J. Sakaguchi and S. Hagiwara, A single chip CMOS filter/codec, *IEEE Journal of Solid-State Circuits*, SC-16, 4, 302–7, August 1981.

[9.59] A. Iwata, H. Kikuchi, K. Uchimura, A. Morino and M. Nakajima, A single chip CODEC with switched-capacitor filters, *IEEE Journal of Solid-State Circuits*, SC-16, 4, 315–21, August 1981.

[9.60] R. Gregorian, G.A. Wegner and W.E. Nicholson, Jr, An integrated single-chip PCM voice codec with filters, *IEEE Journal of Solid-State Circuits*, SC-16, 4, 322–32, August 1981.

[9.61] D. Senderowicz, S.F. Dryer, J.H. Huggins, C.F. Rahim and C.A. Laber, A family of differential NMOS analog circuit for a PCM codec filter chip, *IEEE Journal of Solid-State Circuits*, SC-17, 6, 1014–23, December 1982.

[9.62] M.S. Nakhla, Approximation of low-pass filters with frequency dependent input gain characteristic, *IEEE Transactions on Circuits and Systems*, CAS-26, 3, 198–202, March 1979.

[9.63] R. Friedman, R.W. Daniels, R.J. Dow and O.H. McDonald, RC active filters for the D3 channel bank, *Bell System Technical Journal*, 54, 507–29, March 1975.

[9.64] K.B. Ohri and M.J. Callahan, Integrated PCM codec, *IEEE Journal of Solid-State Circuits*, SC-14, 1, 38–46, February 1979.

[9.65] C.A.T. Salama, VLSI technology for telecommunications, *IEEE Journal of Solid-State Circuits*, SC-16, 4, 253–60, August 1981.

[9.66] K. Martin and A.S. Sedra, Easing prefiltering requirements of SC filters, *Electronics Letters*, 16, 16, 613–14, 31 July 1980.

[9.67] D.C. von Grunigen, U.W. Brugger and G.S. Moschytz, Simple switched-capacitor decimation circuit, *Electronics Letters*, 17, 1, 30–1, 8 January 1981.

[9.68] D.C. von Grunigen, R. Sigg, M. Ludwig, U.W. Brugger, G.S. Moschytz and H. Melchoir, Integrated switched-capacitor low-pass filter with combined anti-aliasing decimation filter for low frequencies, *IEEE Journal of Solid-State Circuits*, SC-17, 6, 1024–9, December 1982.

[9.69] R. Gregorian and W.E. Nicholson, Jr, Switched-capacitor decimation and interpolation circuits, *IEEE Transactions on Circuits and Systems*, CAS-27, 6, 509–14, June 1980.

[9.70] S. Eriksson, SC filter circuit with decimation of sampling frequency, *Electronics Letters*, 21, 11, 484–5, 23 May 1985.

[9.71] M.B. Ghaderi, G.C. Temes and S. Law, Linear interpolation using CCDs or switched-capacitor filters, *Proc. IEE*, Part G, 128, 213–15, August 1981.

[9.72] J.L. McCreary and P.R. Gray, All-MOS charge redistribution analog to digital conversion techniques – Part I, *IEEE Journal of Solid-State Circuits*, SC-10, 6, 371–9, December 1975.

[9.73] R.E. Suarez, P.R. Gray and D.A. Hodges, All-MOS charge redistribution analog to digital conversion techniques – Part II, *IEEE Journal of Solid-State Circuits*, SC-10, 6, 379–85, December 1975.

[9.74] Y.P. Tsividis, P.R. Gray, D.A. Hodges and J. Chack, Jr, A segmented $\mu$-255 law PCM voice encoder using NMOS technology, *IEEE Journal of Solid-State Circuits*, SC-11, 6, 740–7, December 1976.

[9.75] S.P. Singh, A. Prabhakar and A.B. Bhattacharyya, C-2C ladder voltage dividers for application in all-MOS A/D converters, *Electronics Letters*, 18, 12, 537–8, 10 June 1982.

[9.76] S.P. Singh, A. Prabhakar and A.B. Bhattacharyya, Modified C-2C ladder voltage divider for application in PCM A/D converters, *Electronics Letters*, 19, 19, 788–90, 15 September 1983.

[9.77] P.M. Li, M.J. Chin, P.R. Gray and R. Castello, A ratio-independent algorithmic analog-to-digital conversion technique, *IEEE Journal of Solid-State Circuits*, SC-19, 6, 828–36, December 1984.

[9.78] R.D. Fellman and R.W. Brodersen, A switched-capacitor adaptive lattice filter, *IEEE Journal of Solid-State Circuits*, SC-18, 1, 46–56, February 1983.

[9.79] R. Gregorian and G. Amir, A single chip speech synthesizer using a switched-capacitor multiplier, *IEEE Journal of Solid-State Circuits*, SC-18, 1, 65–75, February 1983.

[9.80] K. Watanabe and G.C. Temes, Switched-capacitor digital multiplier, *Electronics Letters*, 19, 7, 33–4, 20 January 1983.

[9.81] A. Peled and B. Liu, A new approach to the realization of non-recursive digital filters, *IEEE Transactions on Audio and Electroacoustics*, AV-21, 6, 477–84, December 1973.

[9.82] N. Sridhar Reddy and M.N.S. Swamy, Switched-capacitor realization of FIR filters, *IEEE Transactions on Circuits and Systems*, CAS-31, 4, 417–19, April 1984.

[9.83] K. Martin and A.S. Sedra, Switched-capacitor building blocks for adaptive systems, *IEEE Transactions on Circuits and Systems*, CAS-28, 6, 576–84, June 1981.

[9.84] N. Sridhar Reddy and M.N.S. Swamy, Switched-capacitor realization of a discrete Fourier transformer, *IEEE Transactions on Circuits and Systems*, CAS-30, 4, 254–5, April 1983.

[9.85] W.L. Eversole, D.J. Mayer, P.W. Bosshart, M. de Wit, C.R. Hewes and D.D. Buss, A completely integrated 2-point chirp z-transform, *IEEE Journal of Solid State Circuits*, SC-13, 6, 822–31, December 1978.

[9.86] J.L. Huertas, A.R. Vazquez and B.P. Verdu, A novel SC oscillator, *IEEE Transactions on Circuits and Systems*, CAS-31, 3, 310–12, March 1984.

[9.87] R.P. Colbeck, A CMOS low-distortion switched-capacitor oscillator with instantaneous start up, *IEEE Journal of Solid-State Circuits*, SC-19, 6, 996–8, December 1984.

[9.88] P.E. Fleischer, A. Ganesan and K.R. Laker, A switched-capacitor oscillator with precision amplitude control and guaranteed start-up, *IEEE Journal of Solid-State Circuits*, SC-20, 2, 641–7, April 1985.

[9.89]  K. Martin, A voltage-controlled switched-capacitor relaxation oscillator, *IEEE Journal of Solid-State Circuits*, SC-16, 4, 412–14, August 1981.

[9.90]  C.A. Laber and P.P. Lemaitre, A monolithic 1200 baud FSK CMOS modem, *IEEE Journal of Solid-State Circuits*, SC-19, 6, 861–9, December 1984.

[9.91]  A.K. Takla and Y.A. Haque, A single chip 300 baud FSK modem, *IEEE Journal of Solid-State Circuits*, SC-19, 6, 846–54, December 1984.

[9.92]  K. Yamamoto, Shoji Fujii and K. Matsuoka, A single chip FSK modem, *IEEE Journal of Solid-State Circuits*, SC-19, 6, 855–61, December 1984.

[9.93]  K. Hanson, W.A. Severin, E.R. Klinkovsky, D.C. Richardson and J.R. Hoschchild, A 1200 Bit/s QPSK full duplex modem, *IEEE Journal of Solid-State Circuits*, SC-19, 6, 878–87, December 1984.

[9.94]  B.J. Hosticka, W. Brockherde, U. Kleine and G. Zimmer, Switched capacitor FSK modulator and demodulator in CMOS technology, *IEEE Journal of Solid-State Circuits*, SC-19, 3, 389–96, June 1984.

[9.95]  L.T. Lin, H.F. Tseng, D.B. Cox, S.S. Viglione, D.P. Conrad and R.G. Runge, A monolithic audio spectrum analyzer, *IEEE Journal of Solid-State Circuits*, SC-18, 1, 40–5, February 1983.

[9.96]  N.C. Bui, J.J. Monbaron and J.G. Michel, An integrated voice recognition system, *IEEE Journal of Solid-State Circuits*, SC-18, 1, 75–81, February 1983.

[9.97]  Y. Kuraishi, K. Nakayama, K. Miyordera and T. Okamura, A single chip 20-channel speech spectrum analyzer using a multiplexed switched-capacitor filter bank, *IEEE Journal of Solid-State Circuits*, SC-19, 6, 964–70, December 1984.

[9.98]  K. Irie, T. Uno, K. Uchimura and A. Iwata, A single chip ADM LSI codec, *IEEE Journal of Solid-State Circuits*, SC-18, 1, 33–9, February 1983.

[9.99]  R. Gregorian and J.G. Gord, A continuously variable slope adaptive delta modulation codec system, *IEEE Journal of Solid-State Circuits*, SC-18, 6, 692–700, December 1983.

[9.100] R. Gregorian, An offset-free switched-capacitor biquad, *Microelectronics Journal*, 13, 37–40, August–September 1982.

[9.101] C.R. Hewes, R.W. Brodersen and D.D. Buss, Applications of CCD and switched-capacitor filter technology, *Proc. IEEE*, 67, 10, 1403–15, October 1979.

[9.102] M. Ishikawa, T. Kumura and N. Tamaki, A CMOS adaptive line equalizer, *IEEE Journal of Solid-State Circuits*, SC-19, 5, 788–93, October 1984.

[9.103] H. Takatori, T. Suzuki, F. Fujii and M. Ogawa, A CMOS line equalizer for a digital subscriber loop, *IEEE Journal of Solid-State Circuits*, SC-19, 6, 906–12, December 1984.

[9.104] T. Suzuki, H. Takatori, H. Shirasu, M. Ogawa and N. Kunimi, A CMOS switched capacitor variable line equalizer, *IEEE Journal of Solid-State Circuits*, SC-18, 6, 700–6, December 1983.

[9.105] K. Nakayama, Y. Sato and Y. Kuraishi, Design techniques for switched-capacitor adaptive line equalizers, *IEEE Transactions on Circuits and Systems*, CAS-32, 8, 759–66, August 1985.

[9.106] S.A. Harrold, I.A.W. Vance and D.G. Haigh, Second-order switched-capacitor band-pass filter implemented in GaAs, *Electronics Letters*, 21, 11, 494–6, 23 May 1985.

[9.107] T.R. Viswanathan and T.L. Viswanathan, Increasing the clock frequency of switched-capacitor filters, *Electronics Letters*, 16, 9, 316–17, 24 April 1980.

## 9.9.1  Further reading

[9.108] R. Castello and P.R. Gray, Performance limitations in switched-capacitor filters, *IEEE Transactions on Circuits and Systems*, CAS-32, 9, 865–76, September 1985.

[9.109] U. Georgiev and K. Stantchev, Adequate SC model of Opamp with finite dc gain and bandwidth suitable for two opamp SC biquad analysis, *Electronics Letters*, 22, 19, 996–7, 11 September 1986.

[9.110] K. Martin, L. Ozcolak, Y.S. Lee and G.C. Temes, Differential switched-capacitor amplifier, *IEEE Journal of Solid-State Circuits*, SC-22, 1, 104–6, February 1987.

[9.111] P.M. Van Peteghem, Improved clock buffer with high PSRR for SC circuit applications, *Electronics Letters*, 25, 1, 15–16, 5 January 1989.

[9.112] J. Robert, G.C. Temes, F. Krummenacher, V. Valencic and P. Deval, Offset and clock feedthrough compensated switched-capacitor integrators, *Electronics Letters*, 21, 20, 941–3, 26 September 1985.

[9.113] G.C. Temes, Simple formula for estimation of minimum clock-feedthrough error voltage, *Electronics Letters*, 22, 20, 1069–70, 25 September 1986.

[9.114] J. Robert and P. Deval, A second-order high-resolution incremental A/D converter with offset and charge injection compensation, *IEEE Journal of Solid-State Circuits*, 23, 3, 736–41, June 1988.

[9.115] K. Watanabe and S. Ogawa, Clock-feedthrough compensated sample/hold circuits, *Electronics Letters*, 24, 19, 1226–8, 15 September 1988.

[9.116] B.J. Sheu and C. Hu, Switch-induced error voltage on a switched capacitor, *IEEE Journal of Solid-State Circuits*, SC-19, 4, 519–25, August 1984.

[9.117] A.G. Hall, Correlation and aliasing in the noise-cancelling switched-capacitor filtering technique, *Electronics Letters*, 21, 20, 932–3, 26 September 1985.

[9.118] W.M.C. Sansen, H. Qiuting and K.A. Halonen, Transient analysis of charge transfer in SC filters – gain error and distortion, *IEEE Journal of Solid-State Circuits*, SC-22, 2, 268–76, April 1987.

[9.119] J. Goette and W. Guggenbuhl, Noise performance of SC integrators assuming different operational transconductance amplifier (OTA) models, *IEEE Transactions on Circuits and Systems*, 35, 8, 1042–8, August 1988.

[9.120] H. Walscharts, L. Kustermans and W.M.C. Sansen, Noise optimization of switched-capacitor biquads, *IEEE Journal of Solid-State Circuits*, SC-22, 3, 445–7, June 1987.

[9.121] F. Maloberti and F. Montecchi, Low-frequency noise reduction in SC ladder filters, *Electronics Letters*, 18, 15, 674–5, 22 July 1982.

[9.122] F. Maloberti and F. Montecchi, Low-frequency noise reduction in time shared switched-capacitor ladder filters, *Proc. IEE*, Part *G*, 132, 2, 39–45, April 1985.

[9.123] G. Fischer and G.S. Moschytz, SC filters for high frequencies with compensation for finite-gain amplifiers, *IEEE Transactions on Circuits and Systems*, CAS-32, 10, 1050–6, October 1985.

[9.124] K. Nagaraj, T.R. Viswanathan, K. Singhal and J. Vlach, Switched-capacitor circuits with reduced sensitivity to amplifier gain, *IEEE Transactions on Circuits and Systems*, CAS-34, 5, 571–4, May 1987.

[9.125] J.B. Hughes, N.C. Bird and R.S. Soin, A receiver IC for a 1 + 1 digital subscriber loop, *IEEE Journal of Solid-State Circuits*, SC-20, 3, 671–9, June 1985.

[9.126] J.A. Guinea and D. Senderowicz, A differential narrow-band switched capacitor filtering technique, *IEEE Journal of Solid-State Circuits*, SC-17, 6, 1029–38, December 1982.

[9.127] B.B. Bhattacharyya and R. Raut, Analysis of switched-capacitor networks containing operational amplifiers with finite DC gain and gain-bandwidth product values, *Proc. IEE*, Part *G*, 130, 4, 114–24, August 1983.

[9.128] J.S. Martinez, E. Sanchez-Sinencio and A.S. Sedra, Effects on the performance of a pair of SC biquads due to the opamp gain-bandwidth product, *ISCAS, Rome*, 375–6, 1982.

[9.129] E.P. Rudd and R. Schaumann, An analysis program for switched-capacitor filters including the effects of opamp bandwidth and switch resistance, *ISCAS, Rome*, 13–16, 1982.

[9.130] H. Kuneida, Effects of finite gain-bandwidth products on switched-capacitor networks approached via equivalent representations, *ISCAS, Rome*, 464–7, 1982.

[9.131] E. Sanchez-Sinencio, R.L. Geiger, J. Silva-Martinez and A.S. Sedra, Minimization of gain-bandwidth product effects in switched-capacitor filters, *ISCAS, Rome*, 468–71, 1982.

[9.132] E. Sanchez-Sinencio, J. Silva-Martinez and R.L. Geiger, Biquadratic SC filters with small GB effects, *IEEE Transactions on Circuits and Systems*, CAS-31, 10, 876–84, October 1984.

[9.133] M. Lamkemeyer, W. Brockherde, B.J. Hosticka and P. Richert, Switched-capacitor noise-shaping coders of high-order for A/D conversion, *Electronics Letters*, 21, 22, 1039–40, 24 October 1985.

[9.134] H. Matsumoto and K. Watanabe, Switched-capacitor algorithmic digital-to-analog converters, *IEEE Transactions on Circuits and Systems*, CAS-33, 7, 720–4, July 1986.

[9.135] S.P. Singh, A. Prabhakar and A.B. Bhattacharyya, C-2C ladder based D/A converters for PCM codecs, *IEEE Journal of Solid-State Circuits*, 22, 6, 1197–1200, December 1987.

[9.136] H. Onodera, T. Tateishi and K. Tamaru, A cyclic A/D converter that does not require ratio-matched components, *IEEE Journal of Solid-State Circuits*, 23, 1, 152–8, February 1988.

[9.137] G. Troster and D. Herbst, Error cancellation technique for capacitor arrays in A/D and D/A converters, *IEEE Transactions on Circuits and Systems*, 35, 6, 749–51, June 1988.

[9.138] S.P. Singh, A. Prabhakar and A.B. Bhattacharyya, Design methodologies for C-2C ladder based D/A converters for PCM codecs, *Proc. IEE*, Part *G*, *ECS*, 135, 4, 133–40, August 1988.

[9.139] H. Matsumoto and K. Watanabe, Improved switched-capacitor algorithmic analogue to digital converter, *Electronics Letters*, 21, 10, 430–1, 9 May 1985.

[9.140] V.F. Dias, J.E. Franca and J.C. Vital, High speed digital-to-analogue converter using passive switched-capacitor algorithmic conversion, *Electronics Letters*, 24, 17, 1063, 18 August 1988.

[9.141] K. Watanabe and G.C. Temes, A switched-capacitor multiplier/divider with digital and analog outputs, *IEEE Transactions on Circuits and Systems*, CAS-31, 9, 796–800, September 1984.

[9.142] F. Maloberti, Switched-capacitor building blocks for analogue signal processing, *Electronics Letters*, 19, 7, 263–5, 31 March 1983.

[9.143] M.G.R. Degrauwe and F.H. Salchli, A multipurpose micropower SC filter, *IEEE Journal of Solid-State Circuits*, SC-19, 3, 343–9, June 1984.

[9.144] T. Enonoto and M.A. Yasumoto, Integrated MOS four-quadrant analog multiplier using switched-capacitor technology for analog signal processor IC's, *IEEE Journal of Solid-State Circuits*, SC-20, 4, 852–9, August 1985.

[9.145] A. Cichocki and R. Unbehauen, MOS SC microsystem for generating trigonometrical functions and their inverses, *Electronics Letters*, 22, 20, 1056–7, 25 September 1986.

[9.146] A.A. Abidi, Linearization of voltage controlled oscillators using switched-capacitor feedback, *IEEE Journal of Solid-State Circuits*, SC-22, 3, 494–6, June 1987.

[9.147] W.B. Mikhael and S. Tu, Continuous and switched-capacitor multiphase oscillators, *IEEE Transactions on Circuits and Systems*, CAS-31, 3, 280–93, March 1984.

[9.148] J.H. Fischer, J.L. Sonntag, J.S. Lavranchuk, D.P. Ciolini, A. Ganesan, D.G. Marsh, W.E. Keasler, J. Plany and L.H. Young, Line and receiver interface circuit for high-speed voice band modems, *IEEE Journal of Solid-State Circuits*, SC-22, 6, 982–9, December 1987.

[9.149] C.C. Shih, K.K. Lam, K.L. Lee and R.W. Schalk, A CMOS 5-v analog front-end for 9600-bit/s facsimile modems, *IEEE Journal of Solid-State Circuits*, SC-22, 6, 990–5, December 1987.

[9.150] W. Oswald and J. Mulder, Dual tone and multifrequency generator with on chip filters and voltage reference, *IEEE Journal of Solid-State Circuits*, SC-19, 3, 379–88, June 1984.

[9.151] B.K. Ahuja, E.C. Samson, N. Attaie and G. Nair, An analog front-end for a two-chip 2400 bits/s voice-band modem, *IEEE Journal of Solid-State Circuits*, SC-22, 6, 996–1003, December 1987.

[9.152] W.B. Wilson, H.Z. Massoud, E.J. Swanson, R.T. George and R.B. Fair, Measurement and modelling of charge feedthrough in *n*-channel MOS analog switches, *IEEE Journal of Solid-State Circuits*, SC-20, 6, 1206–13, December 1985.

[9.153] B.J. Sheu, J.H. Shieh and M. Patil, Modelling charge injection in MOS analog switches, *IEEE Transactions on Circuits and Systems*, CAS-34, 2, 214–16, February 1987.

[9.154] J.H. Shieh, M. Patil and B.J. Sheu, Measurement and analysis of charge injection in MOS analog switches, *IEEE Journal of Solid-State Circuits*, SC-22, 2, 271–81, April 1987.

[9.155] G. Wegmann, E.A. Vittoz and F. Rahali, Charge injection in analog MOS switches, *IEEE Journal of Solid-State Circuits*, SC-22, 6, 1091–7, December 1987.

[9.156] D.B. Ribner and M.A. Copeland, Biquad alternatives for high-frequency switched-capacitor filters, *IEEE Journal of Solid-State Circuits*, SC-20, 6, 1085–95, December 1985.

[9.157] R. Castello and P.R. Gray, A high-performance micropower switched-capacitor filter, *IEEE Journal of Solid-State Circuits*, SC-20, 6, 1122–32, December 1985.

[9.158] K.A.I. Halonen, W.M.C. Sansen and M. Steyaert, A micropower fourth order elliptical switched-capacitor filter, *IEEE Journal of Solid-State Circuits*, SC-22, 2, 164–73, April 1987.

[9.159] K. Matsui, T. Matsuura, S. Fukasawa, Y. Izawa, Y. Toba, N. Miyake and K. Nagasawa, CMOS video filters having switched-capacitor 14 MHz circuits, *IEEE Journal of Solid-State Circuits*, SC-20, 6, 1096–1102, December 1985.

[9.160] K.L. Lee and R.G. Meyer, Low-distortion switched-capacitor filter design techniques, *IEEE Journal of Solid-State Circuits*, SC-20, 6, 1103–13, December 1985.

[9.161] M.S. Tawfik and P. Senn, A 3.6 MHz cut-off frequency CMOS elliptic low-pass switched-capacitor ladder filter for video communication, *IEEE Journal of Solid-State Circuits*, SC-22, 3, 378–84, June 1987.

[9.162] C.A. Lish, A Z-plane Lerner switched-capacitor filter, *IEEE Journal of Solid-State Circuits*, SC-19, 6, 888–92, December 1984.

[9.163] C.K. Wang, R.P. Castello and P.R. Gray, A scalable high-performance switched-capacitor filter, *IEEE Transactions on Circuits and Systems*, CAS-33, 2, 167–74, February 1986.

[9.164] Y.S. Lee and K.W. Martin, A switched-capacitor realization of multiple FIR filters on a single chip, *IEEE Journal of Solid-State Circuits*, 23, 2, 536–42, April 1988.

[9.165] P.M. von Peteghem, I. Verbauwhede and W. Sansen, A micropower high-performance SC building block for integrated low-level signal processing, *IEEE Journal of Solid-State Circuits*, SC-20, 4, 837–44, August 1985.

[9.166] L.E. Larson, K.W. Martin and G.C. Temes, GaAs switched-capacitor circuits for high-speed signal processing, *IEEE Journal of Solid-State Circuits*, SC-22, 6, 971–81, December 1987.

[9.167] K.R. Laker, A. Ganesan and P.E. Fleischer, Design and implementation of switched-capacitor delay equalizers, *IEEE Transactions on Circuits and Systems*, CAS-32, 7, 700–11, July 1985.

[9.168] C.F. Rahim, C.A. Laber, B.L. Pickett and F.J. Baechtold, A high-performance custom standard cell CMOS equalizer for telecommunications applications, *IEEE Journal of Solid-State Circuits*, SC-22, 2, 174–80, April 1987.

[9.169] P.E. Allen, H.A. Rafat and S.A. Bily, A switched-capacitor waveform generator, *IEEE Transactions on Circuits and Systems*, CAS-32, 1, 103–5, January 1985.

[9.170] K. Nagaraj and R.E. Turner, Precision switched-capacitor attenuator, *IEEE Transactions on Circuits and Systems*, CAS-34, 4, 446–7, April 1987.

[9.171] M.J. Hasler, M. Saghafi and A. Kaelin, Elimination of parasitic capacitances in switched-capacitor circuits by circuit transformations, *IEEE Transactions on Circuits and Systems*, CAS-32, 5, 467–75, May 1985.

[9.172] H. Qiuting and W. Sansen, Design techniques for switched-capacitor broad-band phase-splitting networks, *IEEE Transactions on Circuits and Systems*, CAS-34, 9, 1096–1102, September 1987.

[9.173] P.M. von Peteghem, On the relationship between PSRR and clock feedthrough in SC filters, *IEEE Journal of Solid-State Circuits*, 23, 4, 997–1004, August 1988.

[9.174] P.M. von Peteghem and W.M.C. Sansen, Power consumption versus filter topology in SC filters, *IEEE Transactions on Circuits and Systems*, CAS-33, 2, 150–7, February 1986.

[9.175] R.K. Hester, K.S. Tan and C.R. Hewes, A monolithic data acquisition channel, *IEEE Journal of Solid State Circuits*, SC-18, 1, 57–65, February 1983.

[9.176] G.J. Smolka, L.J. Rademacher, G. Weinberger, J. Bareither, U. Grehl and G. Geiger, A 384 kbits/s ISDN burst transreceiver, *IEEE Journal of Solid-State Circuits*, SC-22, 6, 1004–10, December 1987.

[9.177] J. Assael, P. Senn and M.S. Tawfik, A switched-capacitor filter silicon compiler, *IEEE Journal of Solid-State Circuits*, 23, 1, 166–74, February 1988.

[9.178] F. Krummenacher, A high resolution capacitance-to-frequency converter, *IEEE Journal of Solid-State Circuits*, SC-20, 3, 666–70, June 1985.

[9.179] K. Watanabe, H. Matsumoto and K. Fujiwara, Switched-capacitor frequency-to-voltage and voltage-to-frequency converters, *IEEE Transactions on Circuits and Systems*, CAS-33, 8, 836–8, August 1986.

[9.180] W.C. Black, Jr, Floating integrator techniques for switched-capacitor circuits employing reset cycles, *Electronics Letters*, 21, 24, 1126–7, 21 November 1985.

[9.181] K. Haug, F. Maloberti and G.C. Temes, Switched-capacitor integrators with low finite-gain sensitivity, *Electronics Letters*, 21, 24, 1156–7, 21 November 1985.

[9.182] H. Matsumoto and K. Watanabe, Spike-free switched-capacitor circuits, *Electronics Letters*, 23, 8, 428–9, 9 April 1987.

[9.183] S.J. Harrold, Switch driver circuit suitable for high-order switched-capacitor filters implemented in GaAs, *Electronics Letters*, 24, 15, 982–4, 21 July 1988.

[9.184] B.J. Hosticka, W. Brockherde, U. Kleine and R. Schweer, Design of nonlinear analog switched capacitor circuits using building blocks, *IEEE Transactions on Circuits and Systems*, CAS-31, 4, 354–68, April 1984.

[9.185] J.L. Huertas, L.O. Chua, A.B. Rodriguez-Vasquez and A. Reuda, Nonlinear switched-capacitor networks: basic principles and piecewise linear design, *IEEE Transactions on Circuits and Systems*, CAS-32, 4, 305–19, April 1985.

[9.186] A.B. Rodriguez-Vasquez, J.L. Huertas and L.O. Chua, Chaos in a switched-capacitor circuit, *IEEE Transactions on Circuits and Systems*, CAS-32, 10, 1083–5, October 1985.

[9.187] D.B. Ribner, M.A. Copeland and M. Milkovic, 80 MHz low-offset fully differential and single-ended Opamps, *Proc. IEEE Custom Integrated Conference*, 74–5, 1983.

[9.188] K. Nagaraj, Slew-rate enhancement technique for CMOS output buffers, *Electronics Letters*, 25, 1304–5, 14 September 1989.

[9.189] G. Fischer, Analog FIR filters by switched-capacitor techniques, *IEEE Transactions on Circuits and Systems*, CAS-37, 6, 808–14, June 1990.

[9.190] J.A. Fischer, A high-performance CMOS power amplifier, *IEEE Journal of Solid-State Circuits*, SC-20, 6, 1200–5, December 1985.

[9.191] G. Lainey, R. Saintlaurens and O. Senn, Switched-capacitor second-order noise shaping coder, *Electronics Letters*, 19, 149–50, 17 February 1983.

[9.192] M. Mallya and J.H. Nevin, Design procedures for a fully differential folded-cascode CMOS operational amplifier, *IEEE Journal of Solid-State Circuits*, SC-25, 6, 1737–40, December 1989.

[9.193] J.A. Fischer and R. Koch, A highly linear CMOS buffer amplifier, *IEEE Journal of Solid-State Circuits*, SC-22, 3, 330–4, June 1987.

[9.194] T.S. Adez, H.C. Yang, J.J. Young, C. Yu and D.J. Allstot, A family of high-swing CMOS operational amplifiers, *IEEE Journal of Solid-State Circuits*, SC-25, 6, 1683–7, December 1989.

[9.195] S.L. Wong and C.A.T. Salama, An efficient CMOS buffer for driving large capacitive loads, *IEEE Journal of Solid-State Circuits*, SC-21, 3, 464–9, June 1986.

[9.196] R. Klinke, B.J. Hosticka and H.L. Pfleiderer, A very high sensitive CMOS operational amplifier, *IEEE Journal of Solid-State Circuits*, SC-24, 3, 744–7, June 1989.

[9.197] J.N. Babanezhad and R. Gregorian, A programmable gain/loss circuit, *IEEE Journal of Solid-State Circuits*, SC-22, 6, 1082–90, December 1987.

[9.198] M. Steyaert and W. Sansen, A high dynamic-range CMOS opamp with low distortion output structure, *IEEE Journal of Solid-State Circuits*, SC-22, 6, 1204–7, December 1987.

[9.199] D.A. Ribner and M.A. Copeland, Design techniques for cascoded CMOS opamps with improved PSRR and common-mode input range, *IEEE Journal of Solid-State Circuits*, SC-19, 6, 919–25, December 1984.

[9.200] B.K. Ahuja, An improved frequency compensation technique for CMOS operational amplifiers, *IEEE Journal of Solid-State Circuits*, SC-18, 6, 629–33, December 1983.

[9.201] M. Milkovic, Current-gain high-frequency CMOS operational amplifiers, *IEEE Journal of Solid-State Circuits*, SC-29, 4, 845–51, August 1985.

[9.202] K. Nagaraj, Large swing CMOS buffer amplifier, *IEEE Journal of Solid-State Circuits*, SC-24, 1, 181–3, February 1989.

[9.203] B.S. Song, A 10.7 MHz switched-capacitor bandpass filter, *IEEE Journal of Solid-State Circuits*, SC-24, 2, 320–4, April 1989.

[9.204] H.C. Yang and D.J. Allstot, Considerations for fast settling operational amplifiers, *IEEE Transactions on Circuits and Systems*, CAS-37, 3, 326–34, March 1990.

[9.205] C. Toumazon and D.G. Haigh, Design of GaAs operational amplifiers for analog sampled data applications, *IEEE Transactions on Circuits and Systems*, CAS-37, 7, 922–35, July 1990.

[9.206] A.R. Barlow, K. Takasuka, Y. Nambu, T. Adachi, J. Konno, M. Nishimoto, S. Suzuki, K. Nemoto and K. Takashima, An integrated switched-capacitor signal processing design system, *IEEE Journal of Solid State Circuits*, SC-25, 2, 346–52, April 1990.

[9.207] S.D. Levy, P.J. Hurst, P. Ju, J.M. Huggins and C.R. Cole, A single chip 5-V 2400 b/s modem, *IEEE Journal of Solid State Circuits*, SC-25, 3, 632–43, June 1990.

[9.208] K. Suyama, S.C. Fang and Y.P. Tsividis, Simulation of mixed switched-capacitor/digital networks with signal-driven switches, *IEEE Journal of Solid State Circuits*, SC-25, 6, 1403–13, December 1990.

[9.209] R. Castello and L. Tomasini, 1.5-V high-performance SC filters in BiCMOS technology, *IEEE Journal of Solid State Circuits*, SC-26, 7, 930–6, July 1991.

[9.210] R. Becker and J. Mulder, SIGFRED: a low-powered DTMF and signal frequency detector, *IEEE Journal of Solid State Circuits*, SC-26, 7, 1027–37, July 1991.

[9.211] C.Y. Wu, P.H. Lu and M.K. Tsai, Design techniques for high frequency CMOS switched-capacitor filters using non-op-amp-based unity gain amplifiers, *IEEE Journal of Solid State Circuits*, SC-26, 10, 1460–6, October 1991.

[9.212] A. Kaelin, J. Goette, W. Guggenbuhl and G. Moschytz, A novel capacitance assignment procedure for the design of sensitivity-and-noise optimized SC filters, *IEEE Transactions on Circuits and Systems*, CAS-38, 11, 1255–69, November 1991.

[9.213] P. Erratico, Silicon technologies and design methodologies for telecom applications: today's status and future perspectives, *Proceedings of the European Solid-State Circuits Conference – ESSCIRC*, 47–58, November 1991.

[9.214] L. Ping, R.C.J. Taylor, R.K. Henderson and J.I. Sewell, Design of a switched-capacitor filter for a mobile telephone receiver, *IEEE Journal of Solid State Circuits*, SC-27, 9, 1294–8, September 1992.

[9.215] P. Hurst, R.A. Levinson and D.J. Block, A switched-capacitor delta-sigma modulator with reduced sensitivity to op-amp gain, *IEEE Journal of Solid State Circuits*, SC-28, 6, 691–6, June 1993.

[9.216] J.L. Huertas, A. Rueda and D. Vazquez, Testable switched-capacitor filters, *IEEE Journal of Solid State Circuits*, SC-28, 7, 710–24, July 1993.

[9.217] R.P. Martins, J.E. Franca and F. Maloberti, An optimum CMOS switched-capacitor antialiasing decimating filter, *IEEE Journal of Solid State Circuits*, SC-28, 9, 962–9, September 1993.

APPENDIX A

# Analysis of SC networks using SPICE

In this appendix, the analysis of SC networks using a general-purpose circuit analysis program, SPICE, is considered. Recall that Laker's equivalent circuit method can be used to derive the $z$-domain equivalent circuit from a given SC circuit. It is also known that this $z$-domain equivalent circuit can be converted to an $s$-domain equivalent circuit using a bilinear $s \to z$ transformation [2.14]. However, in this method, the frequency scale is warped and analysis in the time domain is not possible. Nelin [A.1] has suggested an alternative, elegant method which handles the frequency-domain as well as time-domain analysis using SPICE. This is briefly discussed in what follows.

Consider the SC HP/LP network shown in Figure A.1(a), and its $z$-domain equivalent circuit shown in Figure A.1(b). This equivalent circuit employs resistors and storistors ($z^{-1/2}$ elements). A $z^{-1/2}$ branch, shown in Figure A.2(a), can be redrawn using a voltage dependent current source and a $T/2$ delay element, as shown in Figure A.2(b). The resulting complete $z$-domain simulation network of Figure A.1(a) is shown in Figure A.2(c). The input file of the SPICE program implementing the circuit of Figure A.2(c) is shown in Figure A.3(a).

In this program, the title of the program is given in line 1, while lines 3–5 describe the resistors in the $z$-domain equivalent circuit of Figure A.2(c) with all values normalized to 1 k$\Omega$. In line 2, the input voltage source is described to be in between nodes 1 and 0 with an amplitude of 1 V. The voltage-dependent current sources in Figure A.2(c) are described in lines 6, 8 and 10, together with the time-delay block description in lines 7, 9 and 11. In the voltage-dependent current source description, e.g., line 6, rewritten here for convenience,

$$\text{G1} \quad 3 \ 1 \ 4 \ 0 \ 1\text{M}$$

G1 stands for a current source whose output current flows between terminals 3 and 1 for an applied voltage between nodes 4 and 0 with the output resistance of the source being 1 M$\Omega$. Note that the current source direction has to be taken into account appropriately, since $-z^{-1/2}$ elements have to be realized as well as $z^{-1/2}$ elements, as is seen in Figure A.2(c). The time-delay block description – for

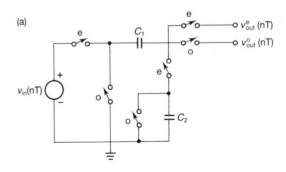

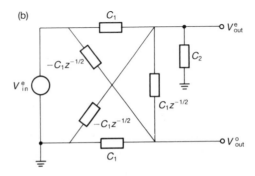

*Fig. A.1* (a) First-order SC LP/HP network; (b) Laker's equivalent circuit of (a). *(Adapted from [A.1], © 1983 IEEE.)*

example, in line 7 – indicates that the input is between terminals 1 and 3, and the output is at terminal 4. Next, the frequency range of a.c. response computation is a linear scale with 99 steps ranging from 1 kHz to 99 kHz, as shown in line 12, with print and plot statements available in the next two lines.

The sub-circuit description is based on Figure A.2(d), where the input is seen to be applied between terminals 1 and 2 and output at terminal 4 as indicated in line 15. The input voltage applied between terminals 1 and 2 is used to realize a dependent voltage source of gain unity between terminals 3 and 0 (see line 16). In the next line, the delay element is described which provides the output between terminals 4 and 0 corresponding to an input between terminals 3 and 0. The characteristic impedance $Z_0$ of the transmission line is 1 kΩ and the time delay realized is half the clock period (5 μs in the case of a sampling frequency of 100 kHz). The next line describes the termination resistance of the transmission line model. The resulting responses are plotted in Figure A.4, where the LP and HP outputs are clearly seen.

The above program only computes the frequency response. If, however, the time-domain response is desired for a given input, SPICE can be utilized. The second line

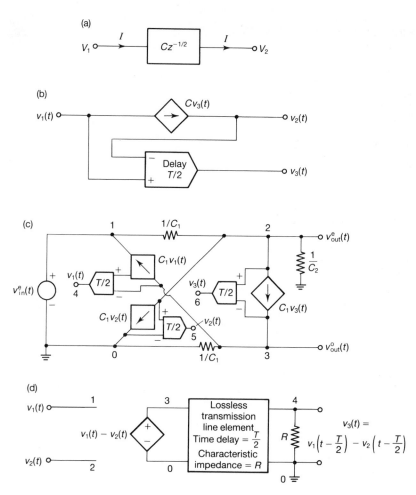

Fig. A.2  (a) Storistor of half-period delay;  (b) continuous-time model of (a); (c) simulation of Figure A.1(b);  (d) implementation of (b). *(Adapted from [A.1], © 1983 IEEE.)*

in the input file of Figure A.3(a) needs to be modified as

$$\text{VIN 1 0 \quad PULSE(V1 V2 TD TR TF PW PER)}$$

where

$V1, V2$ = pulse lower and upper voltage levels
$\quad$ $TD$ = time after which pulse has to be applied after reference time instant
$\quad$ $TR$ = rise time of the pulse
$\quad$ $TF$ = fall time of the pulse
$\quad$ $PW$ = pulse width
$\quad$ $PER$ = period of the pulse train

(a)

```
HIGH-PASS/LOW-PASS SC NETWORK
VIN 1 0 AC 1
R1 1 2 1K
R2 3 0 1K
R3 2 0 1K
G1 3 1 4 0 1M
X1 1 3 4 TD
G2 2 0 5 0 1M
X2 0 2 5 TD
G3 2 3 6 0 1M
X3 2 3 6 TD
.AC LIN 99 1K 99K
.PRINT AC VDB(2) VDB(3)
.PLOT AC VDB(2) VDB(3)
.SUBCKT TD 1 2 4
E1 3 0 1 2 1
T1 3 0 4 0 Z0=1K TD=5U
RO 4 0 1K
.ENDS TD
.END
```

(b)

```
*VIN 1 0 AC 1V
VIN 1 0 PULSE( 0 1 0 0 0 5U 1K )
R1 1 2 1K
R2 3 0 1K
R3 2 0 1K
G1 3 1 4 0 1M
X1 1 3 4 TD
G2 2 0 5 0 1M
X2 0 2 5 TD
G3 2 3 6 0 1M
X3 2 3 6 TD
.WIDTH OUT=80
*.AC LIN 99 1K 99K
.TRAN 1U 60US
.PRINT TRAN V(2) V(3) V(1)
.PLOT TRAN V(2) V(1)
.PLOT TRAN V(3) V(1)
.SUBCKT TD 1 2 4
E1 3 0 1 2 1
T1 3 0 4 0 Z0=1K TD=5U
RO 4 0 1K
.ENDS TD
.END
```

*Fig. A.3* (a) SPICE input file of simulation of Figure A.1(a) for frequency response evaluation *(adapted from [A.1]*, © 1983 IEEE); (b) SPICE input file of simulation of Figure A.1(a) for time-domain response evaluation; (c) time-domain response computed with (b).

(c)

| Time | $v(2)$ | $v(3)$ | $v(1)$ |
|---|---|---|---|
| 0.000E+00 | 0.000E+00 | 0.000E+00 | 0.000E+00 |
| 1.000E-06 | 5.000E-01 | 0.000E+00 | 1.000E+00 |
| 2.000E-06 | 5.000E-01 | 0.000E+00 | 1.000E+00 |
| 3.000E-06 | 5.000E-01 | 0.000E+00 | 1.000E+00 |
| 4.000E-06 | 5.000E-01 | 0.000E+00 | 1.000E+00 |
| 5.000E-06 | 5.000E-01 | 0.000E+00 | 1.000E+00 |
| 6.000E-06 | 5.000E-01 | -5.000E-01 | 1.000E+00 |
| 7.000E-06 | 4.235E-16 | -5.000E-01 | 8.470E-16 |
| 8.000E-06 | 0.000E+00 | -5.000E-01 | 0.000E+00 |
| 9.000E-06 | 0.000E+00 | -5.000E-01 | 0.000E+00 |
| 1.000E-05 | 0.000E+00 | -5.000E-01 | 0.000E+00 |
| 1.100E-05 | -2.500E-01 | -5.000E-01 | 0.000E+00 |
| 1.200E-05 | -2.500E-01 | -8.470E-16 | 0.000E+00 |
| 1.300E-05 | -2.500E-01 | 0.000E+00 | 0.000E+00 |
| 1.400E-05 | -2.500E-01 | 2.115E-18 | 0.000E+00 |
| 1.500E-05 | -2.500E-01 | -3.330E-16 | 0.000E+00 |
| 1.600E-05 | -2.500E-01 | -2.500E-01 | 0.000E+00 |
| 1.700E-05 | -7.638E-16 | -2.500E-01 | 0.000E+00 |
| 1.800E-05 | 0.000E+00 | -2.500E-01 | 0.000E+00 |
| 1.900E-05 | 0.000E+00 | -2.500E-01 | 0.000E+00 |
| 2.000E-05 | -1.804E-16 | -2.500E-01 | 0.000E+00 |
| 2.100E-05 | -1.250E-01 | -2.500E-01 | 0.000E+00 |
| 2.200E-05 | -1.250E-01 | 5.421E-17 | 0.000E+00 |
| 2.300E-05 | -1.250E-01 | 1.300E-04 | 0.000E+00 |
| 2.400E-05 | -1.250E-01 | -4.441E-18 | 0.000E+00 |
| 2.500E-05 | -1.250E-01 | -1.388E-16 | 0.000E+00 |
| 2.600E-05 | -1.250E-01 | -1.250E-01 | 0.000E+00 |
| 2.700E-05 | 1.355E-17 | -1.250E-01 | 0.000E+00 |
| 2.800E-05 | -9.683E-05 | -1.250E-01 | 0.000E+00 |
| 2.900E-05 | 3.507E-05 | -1.250E-01 | 0.000E+00 |
| 3.000E-05 | -6.180E-17 | -1.250E-01 | 0.000E+00 |
| 3.100E-05 | -6.250E-02 | -1.250E-01 | 0.000E+00 |
| 3.200E-05 | -6.250E-02 | 2.646E-17 | 0.000E+00 |
| 3.300E-05 | -6.250E-02 | -6.212E-04 | 0.000E+00 |
| 3.400E-05 | -6.250E-02 | 2.219E-04 | 0.000E+00 |
| 3.500E-05 | -6.250E-02 | -2.056E-16 | 0.000E+00 |
| 3.600E-05 | -6.250E-02 | -6.250E-02 | 0.000E+00 |
| 3.700E-05 | 2.065E-17 | -6.250E-02 | 0.000E+00 |
| 3.800E-05 | -3.808E-04 | -6.250E-02 | 0.000E+00 |
| 3.900E-05 | 5.699E-05 | -6.250E-02 | 0.000E+00 |
| 4.000E-05 | -1.031E-16 | -6.250E-02 | 0.000E+00 |
| 4.100E-05 | -3.125E-02 | -6.250E-02 | 0.000E+00 |

Fig. A.3   (continued)

| Time | $v(2)$ | $v(3)$ | $v(1)$ |
|---|---|---|---|
| 4.200E-05 | -3.125E-02 | 1.420E-17 | 0.000E+00 |
| 4.300E-05 | -3.125E-02 | -8.272E-04 | 0.000E+00 |
| 4.400E-05 | -3.125E-02 | 5.540E-05 | 0.000E+00 |
| 4.500E-05 | -3.125E-02 | -8.826E-17 | 0.000E+00 |
| 4.600E-05 | -3.125E-02 | -3.125E-02 | 0.000E+00 |
| 4.700E-05 | 6.938E-18 | -3.125E-02 | 0.000E+00 |
| 4.800E-05 | -4.122E-04 | -3.125E-02 | 0.000E+00 |
| 4.900E-05 | -3.701E-05 | -3.125E-02 | 0.000E+00 |
| 5.000E-05 | -4.106E-17 | -3.125E-02 | 0.000E+00 |
| 5.100E-05 | -1.563E-02 | -3.125E-02 | 0.000E+00 |
| 5.200E-05 | -1.563E-02 | 6.938E-15 | 0.000E+00 |
| 5.300E-05 | -1.563E-02 | -6.536E-04 | 0.000E+00 |
| 5.400E-05 | -1.562E-02 | -1.338E-04 | 0.000E+00 |
| 5.500E-05 | -1.563E-02 | -2.645E-17 | 0.000E+00 |
| 5.600E-05 | -1.562E-02 | -1.563E-02 | 0.000E+00 |
| 5.700E-05 | 3.469E-18 | -1.563E-02 | 0.000E+00 |
| 5.800E-05 | -3.029E-04 | -1.562E-02 | 0.000E+00 |
| 5.900E-05 | -1.062E-04 | -1.563E-02 | 0.000E+00 |
| 6.000E-05 | -1.174E-17 | -1.563E-02 | 0.000E+00 |

Fig. A.3  (continued)

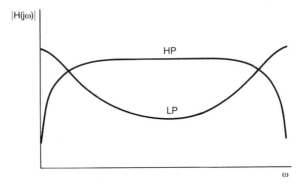

Fig. A.4  LP and HP responses computed by SPICE for the circuit of Figure A.1(a) *(adapted from [A.1],* © 1983 IEEE*)*.

The listing of the program is presented in Figure A.3(b), and the printout of the response is shown in Figure A.3(c) for the case where a 5 μs pulse of 1 V amplitude is applied during the first even slot. The outputs of the LP and HP responses can be seen to correspond to the impulse response sequence, viz.,

$$-\tfrac{1}{2}, \ -\tfrac{1}{4}, \ -\tfrac{1}{8}, \ -\tfrac{1}{16}, \ \dots$$

$$\tfrac{1}{2}, \ -\tfrac{1}{4}, \ -\tfrac{1}{8}, \ -\tfrac{1}{16}, \ \dots$$

respectively.

We next consider a circuit employing an OA as shown in Figure A.5(a), whose $z$-domain and continuous-time equivalent circuits are presented in Figures A.5(b) and A.5(c). Note that in deriving the circuit in Figure A.5(c), the simplified storistor model is employed. This simplification is possible only in situations where the storistor feeds the virtual ground of the OA, as shown in Figure A.6. The corresponding input file is presented in Figure A.7, where the last few lines describe the sub-circuit to realize the simplified storistor model as well as the sub-circuit of Figure A.8. This latter circuit models the $(\sin \omega T/2)/(\omega T/2)$ response of the sample and hold circuit, which is invariably present at the output of the final SC network. The network of Figure A.8 realizes the transfer function of the zero-order hold

$$H(s) = \frac{1 - e^{-sT}}{s}$$

The results of the simulation of the program shown in Figure A.7 with and without the zero-order hold are presented in Figure A.9.

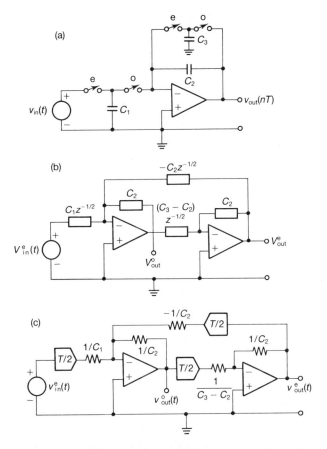

Fig. A.5 (a) Lossy integrator; (b) z-domain equivalent circuit; (c) continuous-time equivalent circuit. (Adapted from [A.1], © 1983 IEEE.)

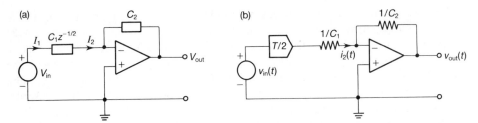

*Fig. A.6* Simplification of storistor model feeding the virtual ground of an OA *(adapted from [A.1],* © *1983 IEEE).*

```
LOSSY SC INTEGRATOR
VIN 1 0 AC 1
R1 2 3 1K
R2 3 4 1K
R3 5 6 -2K
R4 6 7 1K
R5 8 3 -1K
E1 4 0 3 0 -10E10
E2 7 0 6 0 -10E10
X1 1 2 UTD
X2 4 5 UTD
X3 7 8 UTD
X4 4 9 SNX
X5 7 10 SNX
.AC LIN 99 1K 99K
.PRINT AC VD8(4) VD8(7)
.PLOT AC VD8(4) VD8(7)
.PRINT AC VD8(9) VD8(10)
.PLOT AC VD8(9) VD8(10)
.SUBCKT UTD 1 3
TD 1 0 2 0 ZO=1K TD=5U
RO 2 0 1K
EO 3 0 2 0 1
.ENDS UTD
.SUBEKT SNX 1 5
R1 1 4 10K
R2 3 4 -10K
X1 1 2 UTD
X2 2 3 UTD
E1 5 0 4 0 -10E10
C1 4 5 IN
.ENDS SNX
.END
```

*Fig. A.7* SPICE input file of the lossy integrator *(adapted from [A.1],* © *1983 IEEE).*

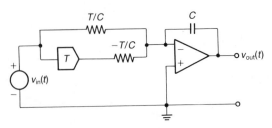

*Fig. A.8* Continuous-time implementation for the simulation of a sample-and-hold of $T_s$ *(adapted from [A.1],* © *1983 IEEE).*

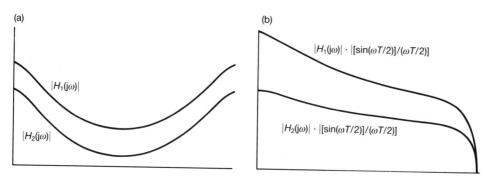

*Fig. A.9* Frequency responses computed by SPICE for the circuit of Figure A.5: $H_1(z) = V_{out}^e(z)/V_{in}^e(z)$, $H_2(z) = V_{out}^o(z)/V_{in}^e(z)$ (a) without zero-order hold; (b) with zero-order hold. *(Adapted from [A.1], © 1983 IEEE.)*

In a similar manner, SPICE can be utilized to analyze even $n$-phase SC networks. It is useful to recall that Fischer's noise models can be employed in the above circuits in order to model noise as well. However, the non-ideal effects of the OA, especially bandwidth, cannot be simulated using the above procedure because $z$-domain equivalents of SC circuits are derived assuming ideal OAs.

# Reference

[A.1] B.D. Nelin, Analysis of switched-capacitor networks using general-purpose circuit simulation programs, *IEEE Transactions on Circuits and Systems*, CAS-30, 43–8, January 1983.

# Computer-aided design of Fleischer–Laker biquads

The versatility and practical utility of the Fleischer–Laker stray-insensitive biquad has already been demonstrated. It may be recalled that this biquad is available in two topologies ($E$ and $F$), each having its own advantage depending on the specifications to be realized. The design procedure is available in the form of general design tables, indicating the simple choice of capacitor values (e.g., that do not involve matching, that reduce the number of capacitors and number of switches). These simple choices by themselves lead to several possible types of circuit in $E$ and $F$ topologies (see Section 5.1.1). Considerable effort is required to select the best topology, in the sense of minimum chip area (i.e., total capacitance), if the calculations involving the various steps – namely, (a) preliminary evaluation of capacitor values, (b) scaling for optimal dynamic range, (c) scaling for minimum total capacitance and (d) reduction of capacitors by transformations – are to be carried out manually. Hence, a computer program in Fortran (developed by the first author) [B.1] is given for obtaining the optimal design. This appendix briefly describes this computer program.

The second-order sampled data transfer function to be realized, or the $s$-domain transfer function together with the sampling frequency (for a bilinear transformation type design), could be interactively given as input data to the program. The $z$-domain transfer function obtained by bilinear transformation or directly given as input data can be realized as the $T$ or $T'$ output of the $E$- or $F$-type biquads. While choosing the capacitor values for the four feedforward capacitors – $G$, $H$, $I$ and $J$ – a search is conducted for all possible choices with the aim that at most three capacitors are used. To realize a given $H_d(z)$, it can be shown, in general, that seven overall choices are possible:

1. $E$ circuit with output at $T$;
2. $F$ circuit with output at $T$;
3. $E$ circuit with output at $T'$ ($J = 0$);
4. $E$ circuit with output at $T'$ ($H = 0$);
5. $E$ circuit with output at $T'$ ($I = 0$);

6. $E$ circuit with output at $T'$ $(G = 0)$;
7. $F$ circuit with output at $T'$.

Once the capacitor values for each of these seven options are evaluated, the 'other' output (i.e., $T'$ for $T$ design, and $T$ for $T'$ design) is evaluated to obtain its maximum value using the formulas presented in Table 1.5. The next step is to scale the capacitor values to equalize the maxima of $T$ and $T'$. The resulting capacitor values are checked at this point for their positiveness for all the cases, so as to exclude impossible designs. The remaining designs are scaled for minimum total capacitance. The total capacitance is used as a 'figure of merit' for choosing the design that requires minimum chip area. Before the last step, care is taken to ensure the elimination of one capacitor for choices such as $G = H$ or $I = J$.

The complete design procedure described above is repeated for the case with the signs of all terms in the numerator opposite to those in the previous case, in which case additional possibilities may exist. The designer can choose then the best of these two final choices.

This computer program has been successfully used for designing SC filters. As an illustration, a design example is considered. The BP transfer function to be realized is chosen as in Example 5.1.

*Example B.1*
The total capacitances of the seven designs possible are, respectively, $46.66C_u$, $32.62C_u$, $-$ , $39.62C_u$, $39.62C_u$, $222.98C_u$ and $32.197C_u$, and thus the design with output at $T'$ and $F$-type circuit is chosen with the following capacitor values:
$A = 14.914C_u$,     $B = 11.97C_u$,     $C = 1.1978C_u$,     $D = F = 1$,     $E = G = H = 0$,
$I = J = K = 2.114C_u$.

Note that, instead of two capacitors $I$ and $J$, a single unswitched capacitor $K$ can be used. Designs with the signs of all terms in the numerator changed also yield the same design having minimum total capacitance.

# Reference

[B.1] P.V. Ananda Mohan, Computer-aided design of stray-insensitive second-order switched-capacitor filters, *Electronics Letters*, 21, 15, 635–6, 18 July 1985.

```
C      BILINEAR/DIRECT DIGITAL FLEISCHER-LAKER BIQUAD DESIGN
       DIMENSION TP(15),TQ(15),TR(15),TU(15),TV(15),TT(15),TTT(15),
     9 TTB(15),TTC(15),TTD(15),TTE1(15),TB(15),TC(15),TD(15),TE(15),
     5 TTE2(15),TMAX(15),R(15)
       DIMENSION AAA(15),AAB(15),AAC(15),AAD(15),AAE(15),AAF(15),
     1 AAG(15),AAH(15),AAI(15),AAJ(15),TMU(15)
     2 SCA(15),SCB(15),SCC(15),SCD(15),SCE(15),SCF(15),SCG(15),SCH(15),
     3 SCI(15),SCJ(15),TOT(15),QM(15),AQM(15),QMAX(15),AQMAX(15),
     6 AAAP(15),AABP(15),AACP(15),AADP(15),AAEP(15),AAFP(15),AAGP(15),
     7 AAHP(15),AAIP(15),AAJP(15),NMIN(15),
     9 SCX(15),SCL(15)
       DOUBLE PRECISION TP,TQ,TR,TU,TV,TT,R
       DOUBLE PRECISION NEW1,NEW2,NEW3,NEW4,NEW5,NEW6,
     2 TMAX,TTT,TTB,TTC,TTD,TTE1,TTE2,TB,
     7 TC,TD,TE,AAA,AAB,AAC,AAD,AAE,
     8 AAF,AAG,AAH,AAI,AAJ,SCK,SCL,
     9 SCA,SCB,SCC,SCD,SCE,SCF,SCG,SCH,
     6 SCI,SCJ,NMIN
       WRITE(6,211)
211    FORMAT(2X, 'DO YOU WISH TO GIVE ANALOG DOMAIN SPEC OR
     2 DIGITAL TRANSFER FUNCTION COEFFICIENTS? IF ANALOG WRITE
     3 1,ELSE 2')
       READ(6,*) TYPE
       IF (TYPE.EQ.1.0) GO TO 724
       IF (TYPE.EQ.2.0) GO TO 725
724    WRITE(6,698)
698    FORMAT(1X, 'WRITE AA,AB,AC,AD,AE')
       READ (6,*) AA,AB,AC,AD,AE
       AF=1700.0
       AFS=128000.0
       AK=(1/TAN(22.0*AF/(7.0*AFS)))*44*1700.0/7
       AG=AK*AK+AD*AK+AE
       AH=(AA*AK*AK+AB*AK+AC)/AG
       AI=(2*AA*AK*AK-2*AC)/AG
       AJ=(AA*AK*AK-AB*AK+AC)/AG
       AL=(2*AK*AK-2*AE)/AG
       AM=(AK*AK-AD*AK+AE)/AG
       GO TO 726
725    WRITE(6,699)
699    FORMAT(1X, 'WRITE AH,AI,AJ,AL,AM')
       READ (6,*) AH,AI,AJ,AL,AM
726    AAE(1)=1.0-AM
       AAF(1)=0.0
       AAC(1)=1.0+AM-AL
       AAE(2)=0.0
       AAF(2)=(1.0/AM)-1.0
       AAC(2)=1.0+(1-AL)/AM
       M=0.0
8011   IF (AH) 9001,9000,9002
9000   AH1=AH
       AI1=AI
       AJ1=AJ
       GO TO 9002
9001   AH1=-AH
       AI1=-AI
       AJ1=-AJ
9002   CONTINUE
C      E CIRCUIT T OUTPUT MAIN
       AAI(1)=AH1
       IF (AJ1) 30,10,20
10     AAJ(1)=0
       AAH(1)=0
       GO TO 70
```

```
20        AAJ(1)=AJ1
          AAH(1)=0.0
          GO TO 70
30        AAJ(1)=0.0
          AAH(1)=-AJ1
70        CONTINUE
          AAG(1)=AAI(1)+AAJ(1)-AI1
C         F CIRCUIT T OUTPUT MAIN
          AAT(2)=AH1/AF
          IF (AJ1) 60,40,50
40        AAJ(2)=0
          AAH(2)=0
          GO TO 80
50        AAJ(2)=AJ1/AM
          AAH(2)=0
          GO TO 80
60        AAJ(2)=0
          AAH(2)=-AJ1/AM
80        CONTINUE
          AAG(2)=AAI(2)+AAJ(2)-AI1/AM
C         F CIRCUIT TDASH OUTPUT MAIN
          AAE(3)=AAE(1)
          AAC(3)=AAC(1)
          AAF(3)=AAF(1)
C         E CIRCUIT WITH TDASH OUTPUT WITH J=0
          AAJ(3)=0.0
          IF (AJ.LE.0.0) GO TO 4001
          IF (AJ.GT.0.0) GO TO 4002
4001      AAH(3)=AJ
          AAI(3)=(AH-AI-AAH(3))/AAC(3)
          AAG(3)=AAI(3)*AAE(3)-AAH(3)-AI
          GO TO 7009
4002      AAH(3)=AJ
          AAI(3)=(-AH+AI-AAH(3))/AAC(3)
          AAG(3)=AAI(3)*AAE(3)-AAH(3)+AI
7009      CONTINUE
C         F CIRCUIT TDASH OUTPUT MAIN WITH H=0
          AAE(4)=AAE(1)
          AAF(4)=AAF(1)
          AAC(4)=AAC(1)
          AAH(4)=0.0
          IF (AJ.LE.0.0) GO TO 4005
          IF (AJ.GT.0.0) GO TO 4006
4005      AAJ(4)=-AJ/AAE(4)
          AAI(4)=(AAJ(4)*AAC(4)+AAJ(4)*AAE(4)+AI-AH)/AAC(4)
          AAG(4)=AAI(4)*AAC(4)+AAE(4)+AH
          GO TO 7008
4006      AAJ(4)=AJ/AAE(4)
          AAI(4)=(AAJ(4)*AAC(4)+AAJ(4)*AAE(4)-AI+AH)/AAC(4)
          AAG(4)=AAI(4)*(AAC(4)+AAE(4))-AH
7008      CONTINUE
C         E CIRCUIT WITH I=0 TDASH OUTPUT MAIN
          AAE(5)=AAE(1)
          AAF(5)=AAF(1)
          AAC(5)=AAC(1)
          AAT(5)=0.0
          IF (AH.LE.0.0) GO TO 7015
          IF (AH.GT.0.0) GO TO 7016
7015      AAG(5)=-AH
          AAJ(5)=(AAG(5)+AI-AJ)/AAC(5)
          AAH(5)=AAE(5)*AAJ(5)+AJ
          GO TO 7017
7016      AAG(5)=AH
          AAJ(5)=(AAG(5)-AI+AJ)/AAC(5)
          AAH(5)=AAE(5)*AAJ(5)+AJ
```

```
7017    CONTINUE
C       E CIRCUIT WITH TDASH OUTPUT MAIN G=0
        AAE(6)=AAE(1)
        AAF(6)=AAF(1)
        AAC(6)=AAC(1)
        IF (AH.GT.0.0) GO TO 7018
        IF (AH.LE.0.0) GO TO 7019
7018    AAI(6)=AH/(AAC(6)+AAE(6))
        AAJ(6)=(AI-AJ-AAI(6)*AAE(6))/AAC(6)
        AAH(6)=AAE(6)*AAJ(6)-AJ
        GO TO 7020
7019    AAI(6)=-(AAC(6)+AAE(6))/AH
        AAJ(6)=(AJ-AI-AAI(6)*AAE(6))/AAC(6)
        AAH(6)=AAE(6)*AAJ(6)+AJ
7020    CONTINUE
C       F CIRCUIT TDASH OUTPUT MAIN
        AAE(7)=AAE(2)
        AAF(7)=AAF(2)
        AAC(7)=AAC(2)
        IF (AJ.LE.0.0) GO TO 1201
        GO TO 1202
1201    AAH(7)=-(AAF(7)+1.0)*AJ
        IF(AH) 1300,1301,1302
1300    AAI(7)=0.0
        AAG(7)=-AH
        GO TO 1304
1301    AAG(7)=0.0
        AAI(7)=0.0
        GO TO 1304
1302    AAG(7)=0.0
        AAI(7)=(AAF(7)+1.0))*AH/AAC(7)
1304    AAJ(7)=((AI+AAH(7))*(AAF(7)+1.0)+AAG(7))/AAC(7)
        GO TO 1309
1202    AAH(7)=AJ*(AAF(7)+1.0)
        IF (AH) 1305,1306,1307
1305    AAI(7)=0.0
        AAG(7)=AH
        GO TO 1308
1306    AAI(7)=0.0
        AAG(7)=0.0
        GO TO 1308
1307    AAG(7)=0.0
        AAI(7)=-AH*(AAF(7)+1.0)/AAC(7)
1308    AAJ(7)=((AAF(7)+1.0)*(AAH(7)-AI)+AAG(7))/AAC(7)
1309    CONTINUE
C       TDASH OUTPUT
        DO 191 I=1,2
        TP(I)=(AAI(I)*AAC(I)+AAI(I)+AAE(I)-AAG(I)-AAF(I)*AAG(I))/(1.0+AAF(I))
        TQ(I)=-(AAF(I)*AAH(I)+AAH(I)+AAG(I)-AAJ(I)*AAC(I)-AAJ(I)*AAE(I)
      6 -AAI(I)*AAE(I))/(1.0+AAF(I))
        TR(I)=(AAE(I)*AAJ(I)-AAH(I))/(1.0+AAF(I))
        TU(I)=(2.0+AAF(I)-AAE(I)-AAC(I))/(1.0+AAF(I))
        TV(I)=(1.0-AAE(I))/(1.0+AAF(I))
191     CONTINUE
C       T OUTPUT
        DO 192 I=3,7
        TP(I)=AAI(I)/(AAF(I)+1.0)
        TQ(I)=(AAI(I)+AAJ(I)-AAG(I))/(AAF(I)+1.0)
        TR(I)=(AAJ(I)-AAH(I))/(AAF(I)+1.0)
        TU(I)=(2.0+AAF(I)-AAE(I)-AAC(I))/(1.0+AAF(I))
        TV(I)=(1.0-AAE(I))/(1.0+AAF(I))
192     CONTINUE
C       DESIRED TRANSFER FUNCTION MAXIMUM VALUE EVALUATION
        TP(8)=AH
        TQ(8)=AI
```

```
            TR(8)=AJ
            TU(8)=AL
            TV(8)=AK
            DO 100 I=1,8
            TT(I)=(TP(I)+TQ(I)+TR(I))/(1+TU(I)+TV(I))
            TB(I)=2*(TP(I)-TR(I))/(1+TU(I)+TV(I))
            TC(I)=(TP(I)-TQ(I)+TR(I))/(1+TU(I)+TV(I))
            TD(I)=2*(1-TV(I))/(1+TU(I)+TV(I))
            TE(I)=(1-TU(I)+TV(I))/(1+TU(I)+TV(I))
            NEW1 = 4.0 * (TD(I) * TB(I) * TE(I))
            NEW2 = 8.0 * TT(I) * TC(I) * TE(I)
            NEW3 = 2.0 * (TB(I) * TB(I)) * (TD(I) * TD(I))
            NEW4 = 4.0 * TT(I) * TC(I) * (TD(I) * TD(I))
            NEW5 = 4.0 * (TC(I) * TC(I))
            NEW6 = 4.0 * (TT(I) * TT(I)) * (TE(I) * TE(I))
            TTB(I) = NEW1-NEW2-NEW3+NEW4+NEW5+NEW6
            TTT(I) = (TD(I)**4)-(4*TE(I)*(TD(I)**2))
            TTC(I) = (TB(I)**4)-(4*(TB(I)**2)*TT(I)*TC(I))
            TTD(I)=SQRT((TTB(I)**2)-(4*TTT(I)*TTC(I)))
            TTE1(I)=(-TTB(I)+TTD(I))/(2*TTT(I))
            TTE2(I)=(-TTB(I)-TTD(I))/(2*TTT(I))
            R(I)=AMAX1(TTE1(I),TTE2(I))
            TMAX(I)=SQRT(R(I))
            IF (TMAX(I).EQ.0.0) TMAX(I)=0.00000001
100         CONTINUE
            DO 295 I=1,7
            TMU(I)=TMAX(8)/TMAX(I)
295         CONTINUE
            DO 111 I=1,2
            AAA(I)=1.0/TMU(I)
            AAD(I)=1.0/TMU(I)
            AAB(I)=1.0
111         CONTINUE
            DO 113 I=3,7
            AAA(I)=1.0
            AAD(I)=1.0
            AAB(I)=1.0/TMU(I)
            AAF(I)=AAF(I)/TMU(I)
            AAC(I)=AAC(I)/TMU(I)
            AAE(I)=AAE(I)/TMU(I)
113         CONTINUE
            DO 114 I=1,7
            QMAX(I)=AMAX1(AAC(I),AAD(I),AAE(I),AAG(I),AAH(I))
            AQMAX(I)=AMAX1(AAA(I),AAB(I),AAF(I),AAI(I),AAJ(I))
            IF (AAC(I).GT.0.0) AACP(I)=AAC(I)
            IF (AAC(I).LE.0.0) AACP(I)=QMAX(I)
            IF (AAD(I).GT.0.0) AADP(I)=AAD(I)
            IF (AAD(I).LE.0.0) AADP(I)=QMAX(I)
            IF (AAE(I).GT.0.0) AAEP(I)=AAE(I)
            IF (AAE(I).LE.0.0) AAEP(I)=QMAX(I)
            IF (AAG(I).GT.0.0) AAGP(I)=AAG(I)
            IF (AAG(I).LE.0.0) AAGP(I)=QMAX(I)
            IF (AAH(I).GT.0.0) AAHP(I)=AAH(I)
            IF (AAH(I).LE.0.0) AAHP(I)=QMAX(I)
            IF (AAA(I).GT.0.0) AAAP(I)=AAA(I)
            IF (AAA(I).LE.0.0) AAAP(I)=AQMAX(I)
            IF (AAB(I).GT.0.0) AABP(I)=AAB(I)
            IF (AAB(I).LE.0.0) AABP(I)=AQMAX(I)
            IF (AAF(I).GT.0.0) AAFP(I)=AAF(I)
            IF (AAF(I).LE.0.0) AAFP(I)=AQMAX(I)
            IF (AAI(I).GT.0.0) AAIP(I)=AAI(I)
            IF (AAI(I).LE.0.0) AAIP(I)=AQMAX(I)
            IF (AAJ(I).GT.0.0) AAJP(I)=AAJ(I)
            IF (AAJ(I).LE.0.0) AAJP(I)=AQMAX(I)
            QM(I)=AMIN((AACP(I),AADP(I),AAEP(I),AAGP(I),AAHP(I))
```

```
          AQM(I)=AMIN1(AAAP(I),AABP(I),AAFP(I),AAIP(I),AAJP(I))
          SCC(I)=AAC(I)/QM(I)
          SCD(I)=AAD(I)/QM(I)
          SCE(I)=AAE(I)/QM(I)
          SCG(I)=AAG(I)/QM(I)
          SCH(I)=AAH(I)/QM(I)
          SCA(I)=AAA(I)/AQM(I)
          SCB(I)=AAB(I)/AQM(I)
          SCF(I)=AAF(I)/AQM(I)
          SCI(I)=AAI(I)/AQM(I)
          SCJ(I)=AAJ(I)/AQM(I)
          IF ((SCG(I)-SCH(I)).EQ.0.0) GO TO 3001
3004      IF ((SCI(I)-SCJ(I)).EQ.0.0) GO TO 3002
          GO TO 3003
3001      SCH(I)=0.0
          SCK(I)=SCG(I)
          GO TO 3004
3002      SCJ(I)=0.0
          SCL(I)=SCI(I)
3003      CONTINUE
          NMIN(I)=AMIN1(SCA(I),SCB(I),SCC(I),SCD(I),SCE(I),SCF(I),
     6    SCG(I),SCH(I),SCI(I),SCJ(I))
          IF (NMIN(I).LT.0.0) GO TO 351
          GO TO 354
351       SCA(I)=10000.0
          SCB(I)=10000.0
          SCC(I)=10000.0
          SCD(I)=10000.0
          SCE(I)=10000.0
          SCF(I)=10000.0
          SCG(I)=10000.0
          SCH(I)=10000.0
          SCI(I)=10000.0
          SCJ(I)=10000.0
354       CONTINUE
          TOT(I)=SCA(I)+SCB(I)+SCC(I)+SCD(I)+SCE(I)+SCF(I)+SCG(I)+
     6    SCH(I)+SCI(I)+SCJ(I)
114       CONTINUE
          TCAP=AMIN1(TOT(1),TOT(2),TOT(3),TOT(4),TOT(5),TOT(6),TOT(7))
          WRITE(6,320) TOT(1),TOT(2),TOT(3),TOT(4),TOT(5),TOT(6),TOT(7)
320       FORMAT(7(2X,F15.6))
          IF(TCAP.EQ.TOT(1)) GO TO 2000
          IF(TCAP.EQ.TOT(2)) GO TO 2001
          IF(TCAP.EQ.TOT(3)) GO TO 2002
          IF(TCAP.EQ.TOT(4)) GO TO 2003
          IF(TCAP.EQ.TOT(5)) GO TO 2004
          IF(TCAP.EQ.TOT(6)) GO TO 2005
          IF(TCAP.EQ.TOT(7)) GO TO 2006
2000      TCAP=TOT(1)
          WRITE(6,90)
90        FORMAT(2X, 'E CIRCUIT T OUTPUT MAIN NEEDS MINIMUM TOTAL CAPACITANCE')
          WRITE(6,110) SCA(1),SCB(1),SCC(1),SCD(1),SCE(1),SCF(1),SCG(1),
     4    SCH(1),SCI(1),SCJ(1),TOT(1),SCK(1),SCL(1)
110       FORMAT(13(2X,F15.6))
          GO TO 900
2001      TCAP=TOT(2)
          WRITE(6,120)
120       FORMAT(2X, 'F CIRCUIT WITH T OUTPUT MAIN NEEDS MINIMUM
     1    TOTAL CAPACITANCE')
          WRITE(6,921) SCA(2),SCB(2),SCC(2),SCD(2),SCE(2),SCF(2),
     5    SCG(2),SCH(2),SCI(2),SCJ(2),TOT(2),SCK(2),SCL(2)
921       FORMAT(13(2X,F15.6))
          GO TO 900
2002      TCAP=TOT(3)
          WRITE(6,121)
```

```
121      FORMAT(2X, 'E CIRCUIT WITH TDASH OUTPUT WITH J=0 NEEDS
      6 MINIMUM TOTAL CAPACITANCE')
         WRITE(6,131) SCA(3),SCB(3),SCC(3),SCD(3),SCE(3),SCF(3),SCG(3),
      5 SCH(3),SCI(3),SCJ(3),TOT(3),SCK(3),SCL(3)
131      FORMAT(13(2X,F15.6))
         GO TO 900
2003     TCAP=TOT(4)
         WRITE(6,132)
132      FORMAT(2X, 'E CIRCUIT WITH TDASH OUTPUT MAIN WITH H=0 NEEDS
      7 MINIMUM TOTAL CAPACITANCE')
         WRITE(6,133) SCA(4),SCB(4),SCC(4),SCD(4),SCE(4),SCF(4),SCG(4),
      4 SCH(4),SCI(4),SCJ(4),TOT(4),SCK(4),SCL(4)
133      FORMAT(13(2X,F15.6))
         GO TO 900
2004     TCAP=TOT(5)
         WRITE(6,134)
134      FORMAT(2X, 'E CIRCUIT WITH I=0 TDASH OUTPUT MAIN NEEDS
      5 MINIMUM TOTAL CAPACITANCE')
         WRITE(6,139)SCA(5),SCB(5),SCC(5),SCD(5),SCE(5),SCF(5),

      7 SCG(5),SCH(5),SCI(5),SCJ(5),TOT(5),SCK(5),SCL(5)
139      FORMAT(13(2X,F15.6))
         GO TO 900
2005     TCAP=TOT(6)
         WRITE(6,137)
137      FORMAT(2X, 'E CIRCUIT WITH G=0 T OUTPUT MAIN NEEDS
      6 MINIMUM TOTAL CAPACITANCE')
         WRITE(6,136) SCA(6),SCB(6),SCC(6),SCD(6),SCE(6),SCF(6),
      5 SCG(6),SCH(6),SCI(6),SCJ(6),TOT(6),SCK(6),SCL(6)
136      FORMAT(13(2X,F15.6))
         GO TO 900
2006     TCAP=TOT(7)
         WRITE(6,239)
239      FORMAT(2X, 'F CIRCUIT WITH TDASH OUTPUT MAIN NEEDS
      6 MINIMUM TOTAL CAPACITANCE')
         WRITE(9,939)SCA(7),SCB(7),SCC(7),SCD(7),SCE(7),SCF(7),

      5 SCG(7),SCH(7),SCI(7),SCJ(7),TOT(7),SCK(7),SCL(7)
939      FORMAT(13(2X,F15.6))
         GO TO 900
900      CONTINUE
         AH=-AH
         AI=-AI
         AJ=-AJ
         M=M+1.0
         IF (M.LT.2.0) GO TO 8011
         WRITE(6,711)
711      FORMAT(2X, 'IF LAST TWO PRINTED VALUES ARE NON-ZERO,
      7 NOTE THAT G,H AND I,J ARE TO BE IGNORED AND K,L HAVE TO BE USED')
         STOP
         END
```

# Other books on switched capacitor circuits and systems

G.S. Moschytz (ed.), *MOS Switched-Capacitor Filters: Analysis and Design*, IEEE Press, 1984.

P.E. Allen and E. Sanchez-Sinencio, *Switched Capacitor Circuits*, Van Nostrand Reinhold, 1984.

R. Gregorian and G.C. Temes, *Analog MOS Integrated Circuits for Signal Processing*, Wiley, 1986.

R. Unbehauen and A. Cichocki, *MOS Switched-Capacitor and Continuous-Time Integrated Circuits and Systems*, Springer-Verlag, 1989.

# Subject index

# Author index

(Only those authors whose work has been dealt with in the book are included)